Integrated Management of Fruit Crops and Forest Nematodes

Integrated Management of Plant Pests and Diseases

Published:

Volume 1
General Concepts in Integrated Pest and Disease Management
edited by A. Ciancio and K.G. Mukerji
ISBN 978-1-4020-6060-1

Volume 2
Integrated Management and Biocontrol of Vegetable and Grain Crops Nematodes
edited by A. Ciancio and K.G. Mukerji
ISBN 978-1-4020-6062-5

Volume 3
Integrated Management of Diseases Caused by Fungi, Phytoplasma and Bacteria
edited by A. Ciancio and K.G. Mukerji
ISBN 978-1-4020-8570-3

Forthcoming:

Volume 5
Integrated Management of Arthropod Pests and Insect Borne Diseases
edited by A. Ciancio and K.G. Mukerji

Integrated Management of Fruit Crops and Forest Nematodes

Edited by

A. Ciancio
C.N.R., Bari, Italy

and

K.G. Mukerji
University of Delhi, India

Springer

Editors
Aurelio Ciancio
Consiglio Nazionale delle Ricerche
Istituto per la Protezione delle Piante
Sezione di Bari
Via G. Amendola, 122/D
70126 Bari
Italy
ciancio@area.ba.cnr.it

K.G. Mukerji
University of Delhi
Dept. Botany
New Delhi-110007
India
kgmukerji@rediffmail.com

ISBN: 978-1-4020-9857-4 e-ISBN: 978-1-4020-9858-1

Library of Congress Control Number: 2009920105

Printed on acid-free paper

9 8 7 6 5 4 3 2 1

springer.com

CONTENTS

Preface xv

SECTION 1 - Tropical Fruit Crops and Commodities

1 Integrated Management of Banana Nematodes 3

Patrick Quénéhervé

1. Introduction 3
   1.1. Botanical and Economical Backgrounds on Bananas 4
   1.2. Integrated Nematode Management: Concept Definition and Applications 7
2. Dessert Bananas for Export 8
   2.1. Geographic Distribution of Associated Nematode Species 8
   2.2. Basic Studies on Nematode Biology 10
   2.3. Damage and Economic Importance 13
   2.4. Nematode Control: The Golden Age of Nematicides (1960-1990) 15
   2.5. Research of an Alternative to Chemical Control 16
   2.6. Future Prospects 17
3. Nematodes on Bananas in Asia and Oceania 17
   3.1. Introduction 17
   3.2. Nematode Species 18
   3.3. Importance and Damage Potential 21
   3.4. Nematode Management 21
   3.5. Future Prospects 22
4. Nematodes on Bananas in Africa 22
   4.1. The Nematode Problem 22
   4.2. Nematode Species Occurrence 23
   4.3. Importance and Potential Damage 26
   4.4. Current Nematode Management 27
      4.4.1. Clean Planting Material 27
      4.4.2. Cultivated Fallow and Alternate Cropping 28
      4.4.3. Mulching and Fertilisers 28
   4.5. Future Prospect 29
5. Nematodes on Banana in America 30
   5.1. The Nematode Problem 30
   5.2. The Nematode Species Occurrence 31
   5.3. Importance and Damage Potential 32
   5.4. Nematode Management 33
   5.5. Future Prospects 33

6. Future and Common Strategies 34
6.1. Plant Health Measures 34
6.2. The Search for Sources of Resistance to Nematodes 36
6.2.1. Resistance to the Burrowing Nematode *R. similis* 37
6.2.2. Resistance to the Lesion Nematode *Pratylenchus* spp. 37
6.2.3. Resistance to the Root-Knot Nematode *Meloidogyne* spp. 38
6.2.4. Resistance to the Spiral Nematode *Helicotylenchus multicinctus* 38
6.3. Tolerance to Nematodes 38
6.4. New Synthetic Banana Hybrids and Their Response to Nematodes 39
6.5. Resistance Mechanisms and Plant Defence 40
6.6. Transgenic Resistance 41
6.7. Biological Control 41
6.7.1. Soil Treatment with Antagonistic Microorganisms 42
6.7.2. Induction of In Planta Suppressiveness 42
6.8. Cultural Practice Improvements 44
7. Conclusion 44
References 45

2 Nematodes of Tropical Fruit crops in Venezuela 63

Renato Crozzoli

1. Introduction 63
2. Nematodes by Crop 64
2.1. Acerola (*Malpighia glabra*) 64
2.2. Avocado (*Persea americana*) 65
2.3. Banana and Plantain (*Musa* AAA and AAB) 65
2.4. Breadfruit (*Artocarpus altilis*) 66
2.5. Cashew (*Anacardium occidentale*) 66
2.6. Citrus (*Citrus* spp.) 66
2.7. Coconut (*Cocos nucifera*) 71
2.8. Date Palm (*Phoenix dactylifera*) 73
2.9. Fig (*Ficus carica*) 73
2.10. Grapevine (*Vitis* sp.) 73
2.11. Guava (*Psidium guajava*) 74
2.12. Mango (*Mangifera indica*) 75
2.13. Papaya (*Carica papaya*) 76
2.14. Passionfruit (*Passiflora edulis* f. sp. *flavicarpa*) 77
2.15. Peach (*Prunus persica*) 78
2.16. Pineapple (*Ananas comosus*) 79
2.17. Sapodilla (*Manilkara zapota*) 80
2.18. Tamarind (*Tamarindus indica*) 80
References 80

3 **Concepts in Management of Tree Crops Nematodes in Fruit Production Systems** **85**

Juan Carlos Magunacelaya

1. Introduction 85
2. Good Quality of Root, and Balanced Plants 88
3. Concepts of Resistance and Tolerance 88
4. Rootstocks and Nematodes 89
5. Nematodes Management Strategies 90
   5.1. Virgin Soils 90
   5.2. Fumigations 91
   5.3. Replants 91
6. Treatments 92
   6.1. Fumigation 92
      6.1.1. Doses and Dates 92
      6.1.2. Application 93
      6.1.3. Concentration 93
      6.1.4. Volume of Treated Soil 94
      6.1.5. Elicitors of Resistance 94
7. Soil Conditions 95
8. Diagnosis 95
   8.1. Plant Vigor 95
   8.2. Test Pits Analysis 95
   8.3. Sampling Strategies 96
   8.4. Treatments 96
   8.5. Minirhizotrons 97
9. Other Management Strategies 98
References 98

4 **Nematodes Management in Coffee Production Systems** **101**

Nahum Marbán-Mendoza

1. Introduction 101
2. Cropping Systems 105
   2.1. Natural or Mountain System 105
   2.2. Traditional Polyculture System 106
   2.3. Specialized System 108
   2.4. Commercial Polyculture System 109
   2.5. Sunlight System 109
3. Phytosanitary Aspects of the Coffee Crop 110
   3.1. Nematodes Affecting Coffee 110
   3.2. Coffee Nematodes Investigations 114
References 117

5 Nematodes of Cacao and Their Integrated Management.......................... 119

Samuel B. Orisajo

1. Introduction.. 119
1.1. Production... 119
1.2. Climatic Requirements... 120
1.3. Cultivation Techniques and Practices.. 121
2. Nematodes of Cacao... 122
2.1. Root-Knot Nematodes.. 122
2.1.1. Geographic Distribution.. 122
2.1.2. Species.. 122
2.1.3. Damage Symptoms... 123
2.1.4. "Morte Subite" or Sudden Death.. 123
2.2. Other Nematodes Parasitic on Cacao... 124
2.3. Disease Complexes.. 124
3. Nematode Control in Cacao... 124
3.1. Integrated Management Approach to Nematode Control.................. 127
3.1.1. Resistant Planting Materials... 127
3.1.2. Production of Nematode-Free Seedlings in the Nursery.............. 128
3.1.3. Use of Nematicides in the Field... 128
3.1.4. Organic Amendments.. 128
3.1.5. Biological Control.. 130
4. Conclusions and Outlook for Future Containment............................. 130
References.. 131

6 Managing Nematodes in Citrus Orchards 135

Larry W. Duncan

1. Introduction.. 135
2. *Tylenchulus semipenetrans*... 136
2.1. Slow Decline Symptoms.. 136
2.2. Biology and Ecology... 137
2.3. Interactions with Soil Organisms... 141
2.4. Biotypes and Rootstock Resistance... 142
2.5. Economic Importance and Crop Loss Prediction............................... 143
2.6. Management... 145
2.6.1. Sampling and Extraction.. 145
2.6.2. Sanitation and Exclusion.. 146
2.6.3. Cultural Practices... 147
2.6.4. Fumigants and Nematicides... 148
3. *Radopholus similis* and *R. citri*... 149
3.1. Spreading Decline Symptoms.. 150
3.2. Biology and Ecology... 151
3.3. Interactions with Soil Organisms... 152

3.4. Biotypes and Resistance 152
3.5. Economic Importance 153
3.6. Management 153
4. *Pratylenchus* spp. 154
5. *Belonolaimus longicaudatus* 157
6. *Meloidogyne* spp. 158
7. *Hemicycliophora* spp. 159
8. Dorilaimid Species 160
References 160

SECTION 2 - Temperate Fruit Crops and Forest Nematodes

7 Integrated Management of Nematodes Parasitic on *Prunus* spp. 177

Soledad Verdejo-Lucas and M. Talavera

1. Introduction 177
2. Parasitic nematodes affecting *Prunus* spp. 178
2.1. Root-Knot Nematodes 178
2.2. Root Lesion Nematodes 179
2.3. Ring Nematodes 180
2.4. Dagger Nematodes 180
3. Management of Nematodes and Control Methods 181
3.1. Prevention and Quarantine 181
3.2. Pre-planting Measures 182
3.2.1. Fallow 182
3.2.2. Crop Rotation 182
3.2.3. Site Preparation 182
3.2.4. Soil Solarization 183
3.2.5. Biofumigation 183
3.2.6. Steam 183
3.2.7. Soil Fumigation 184
3.2.8. Chemical Control with Non-fumigants Nematicides at Pre-planting 185
3.2.9. Seedling Treatments 185
3.2.10. Resistance 185
3.3. Post Planting Measures 187
3.3.1. Chemical Control 187
3.3.2. Biological Control 188
3.3.3. Cultural Methods 189
4. Integrated Management 189
5. Future Prospects 190
References 190

8 Selection and Application of Resistant Germplasm for Grapevine Nematodes Management 195

Daniel Esmenjaud and A. Bouquet

1. Introduction........ 195
2. The Root-Knot Nematodes *Meloidogyne* spp. 196
   2.1. Biology, Ecology, Symptoms and Control........ 196
   2.2. Selection and Breeding of Resistant Rootstocks........ 197
      2.2.1. Evidence of Resistant *Vitis* and *Muscadinia* Material........ 197
      2.2.2. Breeding for Resistance........ 197
      2.2.3. Genetics of Resistance........ 198
      2.2.4. Resistance Mechanisms and Durability........ 199
3. GFLV Vector Nematode *Xiphinema index*........ 200
   3.1. *Xiphinema index* and the Other Virus Vector Nematodes on Grape... 200
   3.2. Biology, Vection and Classical Control of *X. index*........ 201
      3.2.1. Biology and Vection........ 201
      3.2.2. Classical Control........ 201
   3.3. Selection and Breeding of Resistant Rootstocks........ 202
      3.3.1. *Vitis* Breeding........ 202
      3.3.2. *Vitis* × *Muscadinia* Breeding........ 204
         3.3.2.1. Resistance Features of *Muscadinia rotundifolia*........ 204
         3.3.2.2. *V. vinifera* × *M. rotundifolia* F_1 Hybrids Obtained in California........ 204
         3.2.2.3. *V. vinifera* × *M. rotundifolia* F_1 and Backcrosses Obtained in France........ 205
   3.3. Future Prospects........ 206
4. Resistance to Other Nematodes and Rootstock Control of Multiple Nematode Pests........ 206

References........ 208

9 Management of Virus-Transmitting Nematodes with Special Emphasis on South-East Europe........ 215

Tatyana Bileva, Boryana Choleva, Sue Hockland, Aurelio Ciancio

1. Introduction........ 215
   1.1. Geographic Distribution and Spread........ 217
2. Vector and Virus Diagnosis 221
   2.1. Morphological Identification of Virus Vectors........ 221
   2.2. Transmission Assays........ 222
   2.3. Molecular Detection........ 222
3. Integrated Management........ 225
   3.1. Prevention and Quarantine Procedures........ 225
   3.2. Agronomic Practices........ 226
   3.3. Chemical Control........ 226
   3.4. Nematode Resistance in Plants........ 227

4. Organic Management 228
4.1. Organic and Natural Products........ 228
4.2. Biofumigation and Nematicidal Plants........ 232
4.3. Biological Control Agents........ 234
Conclusions 237
References........ 237

10 Integrated Management of Pistachio Nematodes........ 243

Saddigheh Fatemy

1. Introduction........ 243
2. Pistachio Production and Properties........ 243
2.1. Pistachio Production 243
2.2. Nutritional Value........ 244
3. Nematodes Management........ 246
3.1. Pistachio Nematodes........ 246
3.2. Management........ 248
References........ 250

11 Pine Wilt Disease and the Pinewood Nematode, *Bursaphelenchus xylophilus*........ 253

Manuel M. Mota, Kazuyoshi Futai and Paulo Vieira

1. Introduction........ 253
2. PWN Distribution and Disease Dissemination........ 256
3. Pinewood Nematode Taxonomy........ 259
3.1. Morphological Approaches........ 259
3.2. Molecular Approaches........ 260
4. PWN Introduction in Portugal and the EU........ 262
4.1. Dispersal of the PWN Within the Affected Area in Portugal........ 263
5. Control Measures for PWD........ 263
5.1. Control Measures Before the Discovery of PWN as the Causal Agent........ 264
5.2. Control Measures After the Discovery of the PWN and its Vector, *Monochamus alternatus*........ 266
5.2.1. Physical Control........ 266
5.2.2. Chemical Control........ 266
5.2.3. Biological Control........ 267
5.2.4. Breeding of Resistant Hosts........ 268
6. A Blind Spot in PWD Control Strategy: The Asymptomatic Carrier and its Sophisticated Detection Method........ 269
7. Conclusions........ 270
References........ 270

12 Olive Nematodes and Their Control.. 275

Nicola Sasanelli

1. Introduction... 275
2. Symptoms and Pathogenicity of Nematodes on Olive............................. 280
3. Interactions Between Nematodes and Soil Fungi 288
4. Management and Control 293
   4.1. Prevention and Use of Nematodes-Free Propagation Material......... 293
   4.2. Use of Resistant Cultivars or Rootstocks... 294
   4.3. Grafting.. 296
   4.4. Biological Control... 297
   4.5. Soil Solarization.. 298
   4.6. Biofumigation... 300
   4.7. Soil Amendments.. 301
   4.8. Nematicidal Plants.. 302
   4.9. Soil Steam Sterilization.. 304
   4.10. Chemical Control.. 304
      4.10.1. Fumigants.. 305
      4.10.2. Non Fumigant Nematicides.. 307
5. Conclusions... 307

References.. 308

Index... 317

CONTRIBUTORS

Tatyana Bileva
Agricultural University,
4000 Plovdiv, 12 Mendeleev str.,
1164 Sofia, Bulgaria

A. Bouquet
INRA, UMR-DIAPC, Equipe Vigne,
34060 Montpellier Cedex 1,
France

Boryana Choleva
Sofia University, Faculty of Biology,
8 Dr.Tzankov Blv.,
1164 Sofia, Bulgaria

Aurelio Ciancio
Istituto per la Protezione delle Piante,
CNR, Via Amendola 122/D
70126 Bari, Italy

Renato Crozzoli
Universidad Central de Venezuela,
Facultad de Agronomía,
Instituto de Zoología Agrícola,
Laboratorio de Nematología Agrícola,
Apdo. 4579, Maracay, Venezuela

Larry W. Duncan
University of Florida, IFAS,
Citrus Research and Education Center,
700 Experiment Station Road,
Lake Alfred, FL 33850 USA

Daniel Esmenjaud
INRA, UMR IPMSV,
Equipe de nématologie,
06903 Sophia-Antipolis Cedex,
France

Saddigheh Fatemy
Nematology Department
Plant Protection Diseases Research
Institute, P.O. Box 1454
Tehran 19395 Theran, Iran

Kazuyoshi Futai
Laboratory of Environmental
Mycoscience,
Graduate School of Agriculture,
Kyoto University, Sakyo-ku, Kyoto,
606-8502, Japan

Sue Hockland
Plant Health Group,
Central Science Laboratory,
Sand Hutton,
York YO41 1LZ, UK

Juan Carlos Magunacelaya
Universidad Católica de Valparaíso,
Facultad de Ciencias,
Avda. Brasil 2950,
Valparaíso, Chile

Nahum Marbán-Mendoza
Universidad Autónoma Chapingo,
Posgrado en Protección Vegetal,
Chapingo, Edo. de México,
CP 56230, México

Manuel M. Mota
NemaLab-ICAM, Departamento de
Biologia,
Universidade de Évora,
7002-554 Évora,
Portugal

Patrick Quénéhervé
IRD, UMR RPB
Pôle de Recherche
Agroenvironnementale
de la Martinique, BP 214,
97285 Le Lamentin, Martinique, France

Samuel B. Orisajo
Plant Pathology Division
Cocoa Research Institute of Nigeria
PMB 5244, Ibadan,
Oyo State, Nigeria

Nicola Sasanelli
Istituto per la Protezione delle Piante,
CNR, Via G. Amendola 122/D,
70126 Bari, Italy

Miguel Talavera
IFAPA, Centro Camino de Purchil,
Apdo. 2027,
18080 Granada, Spain

Soledad Verdejo
IRTA, Protecció Vegetal
Ctra de Cabrils Km 2
E- 08348 Cabrils,
Barcelona, Spain

Paulo Vieira
NemaLab-ICAM, Departamento de
Biologia,
Universidade de Évora,
7002-554 Évora,
Portugal

PREFACE

This series originated during a visit of prof. K. G. Mukerji to the CNR Plant Protection Institute at Bari, Italy, in November 2005. Both editors convened to produce a series of five volumes focusing, in a multi-disciplinary approach, on recent advances and achievements in the practice of crop protection and integrated pest and disease management.

This fourth Volume deals with management of nematodes parasitic of tree crops, and includes a section on tropical fruit crops and commodities, as well as a second section on tree crops from more temperate areas. The latter also includes a chapter updating the current knowledge about the pine wood nematode, *Bursaphelenchus xylophilus*. Volume 4 flanks Volume 2 of this IMPD series, which focused on management of vegetable and grain crops nematodes.

Nematodes are a very successful, diversified and specialised animal group, present in nature in any ecological niche. Among nematode species, only a reduced number feeds on plants, of which a few species cause severe economic impacts on crop productions. Plant parasitic nematodes represent an important concern for a broad range of agricultural productions and systems, worldwide. This statement explains the attention devoted in last decades to nematodes, and the research and technical efforts invested for their control. As for other disciplines included in plant protection, nematology is now in a mature stage in which the initial optimism deriving from the widespread use of chemicals and fumigants lent space to a more pragmatic, comprehensive and integrated vision of control and management, including traditional approaches like resistance-based management or innovative approaches like biocontrol or use of natural compounds.

A wide literature already covers chemical or biological control of nematodes, but there is still a need for a more holistic vision of management, accounting for different experiences and solutions, developed worldwide. In this series we attempted to fill this gap aiming at providing an informative coverage for a broad range of agricultural systems which coexist in the world today, focusing on solutions fitting the corresponding background economies. Chapters are mainly organized and centered on crops and/or regional problems, ranging from nematodes of tropical crops like banana, cocoa and coffee, to species affecting more widespread crops like citrus. Regional aspects are described in chapters dealing with tropical fruit or commodity productions (Venezuela, Mexico, Nigeria) or with export-oriented cropping systems (Chile). Chapters in Section 2 review nematodes and related management options for more temperate crops, i.e. *Prunus* spp., grapevine, pistachio and olive nematodes, with a chapter on the management options for virus-transmitting nematode species. Emphasis was also given to long-term technological solutions, updating the actual knowledge on the application of resistant germplasm in several of the cited crops.

In the first chapter, the integrated management of banana nematodes is reviewed, starting from the botanical and economical backgrounds of this crop. Dessert banana crops for export and the geographic distribution of associated nematode species are revised. Concepts definition and applications are then discussed, in the light of integrated nematode management. Damage and economic importance of main nematode species and control strategies are reviewed, with reference to nematicide use. The search for alternatives to chemical control are then illustrated, with basic studies on nematode biology for the different species. Nematode problems of banana crops in Africa, Asia, Oceania and America are reviewed, focusing on the occurrence, importance and potential damage caused by main species. Current management options, including the production and dissemination of clean planting material, the application of cultivated fallow and alternate cropping systems, as the use of mulching and fertilisers, are then reviewed. Future and common strategies and plant health measures applied are revised, with emphasis on the search for sources of resistance to the burrowing nematode *R. similis*, the lesion nematode *Pratylenchus* spp., root-knot *Meloidogyne* spp. and the spiral nematode *Helicotylenchus multicinctus*. The nematode tolerance, the production of new synthetic banana hybrids and their response to nematodes are also examined. Finally, resistance and plant defence mechanisms are reviewed, followed by transgenic resistance, biological control and antagonistic microorganisms, induction of suppressiveness and available improvements for cultural practices.

In the following chapters, problems and solution applied on a regional scale for management of nematodes of fruit and commodity crops follow, illustrating some case-studies ranging from South to Central America and West Africa. A compehensive review of the main nematode species of tropical fruit crops is given in Chapter 2, describing the case-study of Venezuela. In the following chapter, concepts in management in export-oriented, fruit production systems in Chile are reviewed, focusing on the conservation of soil fertility by means of irrigation and fertilization, placing nematodes management options in a more general view of root and plant protection, as well as soil fertility conservation. In Chapter 4, the broad diversity of coffee cropping systems of Mexico is analysed, with a review of the main properties and problems of natural or mountain systems, including traditional polyculture and specialized systems, commercial polyculture and sunlight system. The main phytosanitary aspects of these coffee cropping systems are reviewed, focusing on nematodes and related investigations.

Nematode problems of cocoa production systems in West Africa are revised in Chapter 5. After discussing the production, climatic requirements, cultivation techniques and practices, the main nematode species attacking cacao are reviewed, with data on their geographic distribution, damage and symptoms. Apart of root-knot nematodes, other nematode parasites and related disease complexes are examined. The options for management and control in cacao are then reviewed, focusing on the integrated approach to nematode control, the use of resistant planting material, the production of nematode-free seedlings in nurseries, the use of nematicides in the field, and of organic amendments and biological control.

In the following chapter the status of nematodes management in citrus orchards is reviewed. This chapter deals with the citrus nematode, *Tylenchulus semipenetrans* and the related slow decline symptoms. Other nematode species of citrus are also examined, including *Radopholus similis* and *R. citri*, *Pratylenchus* spp., *Belonolaimus longicaudatus*, *Meloidogyne* spp., *Hemicycliophora* spp. and dorilaimid species. Data are provided on their biology and ecology, on the interactions with other soil organisms, biotypes, rootstock resistance, economic importance and crop loss prediction. Management, sampling and extraction techniques are also reviewed, together with sanitation practices and exclusion, as well as cultural practices, use of fumigants and nematicides.

In Section 2, six further chapters deal with nematodes of temperate fruit crops, with a revision of forest nematodes management, mainly updating the situation for *B. xylophilus*, a major emerging problem in Europe. The integrated management of nematodes parasitic on *Prunus* spp. is reviewed in Chapter 7, dealing with root-knot, lesion, ring and dagger nematodes. The chapter focus is on management and control methods based on prevention and quarantine. Pre-planting measures are reviewed, including fallow, crop rotation, site preparation, soil solarization, biofumigation, steam application, soil fumigation, chemical control with non-fumigants nematicides, seedling treatments and resistance. Post-planting measures examined include chemical and biological control, cultural methods and integrated management.

In the following chapter, the selection and application of resistant germplasm for management of nematodes attacking grapevine is reviewed. The chapter focus is on root-knot nematodes and the GFLV vector *Xiphinema index*. For root-knot nematodes, data on their biology, ecology, symptoms and control are provided. The selection and breeding of resistant rootstocks is then reviewed, with data on resistant *Vitis* and *Muscadinia* material. The advances in breeding for resistance, as well as the genetics of resistance mechanisms and its durability are also discussed. The chapter then reviews the biology, vection and classical control methods of *X. index* and other virus vector nematodes of grapevine, focusing on the selection and breeding for *Vitis* and *Vitis* × *Muscadinia* resistant rootstocks. The resistance features of *Muscadinia rotundifolia* are then discussed, together with the properties of the *V. vinifera* × *M. rotundifolia* F_1 hybrids obtained in California and France. Data on resistance to other nematodes and rootstock control of multiple nematode pests are also provided.

Given the importance of virus-vector nematodes, the following chapter reviews the management of virus-transmitting species with special emphasis on South-East Europe. The geographic distribution and spread of main species is reviewed, focusing on vectors and virus diagnostic techniques, including vectors identification, transmission assays, molecular detection and integrated management. Concepts in prevention and quarantine are then discussed, together with the main practices available for management, like agronomic and chemical control, exploitation of nematode resistance sources available in plants, and organic management. Data on assays with organic and natural products are then discussed, together with biofumigation, use of nematicidal plants and potentials of biological control agents.

In Chapter 10 a further regional and specific agricultural issue concerning pistchio production is reviewed, in reference to nematodes management in the Middle East. Pistachio crops are important sources of nutrients and income for local producers. The distribution of pistachio nematodes and the management options available are listed, including agronomic management, use of resistant rootstocks, biological control, as well as soil solarization.

In the following chapter the situation for the pine wood nematode, *B. xylophilus*, is reviewed. Pine wilt disease (PWD), caused by *B. xylophilus*, is one of the most severe disease affecting *Pinus* spp. in the Far East, North America and now the European Union (Portugal). In some countries, such as Japan, PWD was catastrophic, destroying native pine species at such an extent that some areas had to be totally replaced by other tree species. *Bursaphelenchus xylophilus*, endemic to North America where it causes minor damage, was introduced in Japan in the early XXth century and then spread to mainland Asia. Since its first arrival in the EU this nematode has been monitored and efforts are continuously provided to halt its spreading in the european continent. Experience from Japanese control actions include aerial spraying of insecticides to control the insect vector (the Cerambycid beetle *Monochamus alternatus*), direct injection of nematicides to the trunk of infected trees (mainly for added-value trees), slashing and burning of areas out of control, beetle traps, biological control and tree breeding programs. In Portugal, the damage, although lower than in Asia, is still significant and PWD has caused severe losses to the forestry industry. In this chapter, a brief history of PWD is provided, mapping its spread in Japan and to other East Asian nations, as well as updating the situation for Portugal. The economic impact of PWD is reviewed, in relation to the world importance of forestry and conifer production and trade. Inspection and quarantine issues are then discussed. The PWN biology and life cycle are reviewed, together with its relationship with the insect vector. Data on the taxonomy and progress using molecular biology techniques are also provided. The pine resistance and susceptibility to the nematode are also reviewed, including pathogenicity and the potentials of breeding programs. The authors also provide a comprehensive review about the control of PWN and its insect vector, with methods like insecticide spraying, nematicide injection, biological control and breeding for resistance. A discussion on the results achieved by means of management actions worldwide is also provided.

Finally, in the last chapter, the pathogenicity, geographic distribution and damage of nematodes associated with olive are revised, for species within the genera *Gracilacus*, *Helicotylenchus*, *Heterodera,* *Meloidogyne*, *Ogma*, *Pratylenchus*, *Rotylenchulus*, *Tylenchulus*, and *Xiphinema*. Research data on olive nematodes are reviewed, focusing on the effects of parasitism by root-knot nematodes, plant growth, cultivars and rootstocks susceptibility, nematodes interactions with the soil-borne pathogen *Verticillium dahliae*, replant problems and control strategies. These include chemical and biological control, solarization, use of soil amendments and organic management, biofumigation and application of nematicidal plants.

In conclusion, we acknowledge the authors for providing a broad range of data on nematode management solutions available worldwide in different agricultural systems. Thanks to the efforts and will of many nematologists studying and applying advanced solutions in their long term research efforts and field practices, we hope we were able to provide a tool useful in the deployment of environment friendly and sustainable menagement practices for the main crops and parasites listed. Our hope is that this volume will result useful and helpful for interested readers and students, inspiring and supporting research efforts invested in their field and laboratory work.

A. Ciancio
K. G. Mukerji

Section 1

Tropical Fruit Crops and Commodities

PATRICK QUÉNÉHERVÉ

INTEGRATED MANAGEMENT OF BANANA NEMATODES

IRD, UMR RPB
"Résistance des Plantes aux Bioagresseurs"
Pôle de Recherche Agroenvironnementale
de la Martinique, BP 214,
97285 Le Lamentin, Martinique, France

Abstract. Botanical and economical backgrounds on dessert and non-dessert bananas, together with basic concepts for nematode management, are provided, including the geographic distribution of main banana nematode species in Asia, Oceania, Africa and Americas. Basic studies on the biology, damage, economic importance and control of nematodes are then discussed, with reference to the burrowing nematode *Radopholus similis*, the lesion nematodes *Pratylenchus* spp., root-knot nematodes *Meloidogyne* spp., and the spiral nematode *Helicotylenchus multicinctus*. The use of nematicides is reviewed and the research on alternatives to chemical control is discussed. Current nematode management strategies focus on the use of clean planting material, fallow and alternate croppings, application of mulching and fertilisers. Future and common strategies include best plant health measures, the identification of sources of resistance and plant defence mechanisms, including transgenic resistance. Other management strategies concern biological control through soil treatment with microbial antagonists, induction of in-planta suppressiveness and improvements in cultural practices. Tolerance to nematodes, use of new synthetic banana hybrids and their response to parasitism are also reviewed.

1. INTRODUCTION

Plant-parasitic nematodes are widespread and are among the most damaging pests of all banana varieties, causing not only severe crop losses in commercial banana plantations for export but also seriously limiting the production and viability of other banana types. Numerous reviews have already been written on the nematode problems in bananas (Wardlaw, 1961; Champion, 1963; Blake, 1969; Stover, 1972; Roman, 1978; Jones, 2000; Gowen & Quénéhervé, 1990; Gowen et al., 2005) and most of the knowledge of banana nematodes arose quite exclusively from their management on dessert bananas (*Musa* AAA Cavendish group) cultivated in large plantations for export.

A. Ciancio & K. G. Mukerji (eds.), *Integrated Management of Fruit Crops and Forest Nematodes*, 3-61.

In this chapter, we will try to widen these views by considering the different aspects of nematode management in respect both to the type of cultivated bananas and the geographic situation.

1.1. Botanical and Economical Backgrounds on Bananas

After rice, wheat and corn, bananas are the fourth most widely consumed food for humans and the majority of cultivated bananas are grown for local consumption in private gardens and smallholdings in mixed cropping systems. Bananas are cultivated in more than 130 countries and provide staple food and steady cash income to million people. Bananas, monocotyledons belonging to the *Musa* genus, are large herbaceous perennials with underground rhizomes (or corms) from which abundant roots and vegetative buds grow. The aerial part consists of leafy 'trunks' (or pseudostem), which eventually bear bunches.

Bananas can be divided into two main categories, the dessert bananas, mostly eaten fresh, and the non-dessert bananas, including cooking and brewing bananas. In general, pure stands of cooking and dessert types only occur where there is access to export or local markets or where bananas make a major contribution to the diet. From a pest management point of view, the division is even more precise and clearly opposes dessert bananas grown for export to all other banana types.

Most cultivated bananas within the genus *Musa* arose from the Eumusa section. The Eumusa group of species is the largest and most wide-ranging section of the genus and comprises some eleven species being found throughout South East Asia, from India to the Pacific Islands (Horry et al., 1997). Some other edible *Musa* varieties, including the Fe'i banana cultivars, are derived from wild species within the Australimusa section. However, most edible cultivars are derived from two ancestor species, *Musa acuminata* (A genome) and *Musa balbisiana* (B genome) (Simmonds & Shepherd, 1955).

Edible diploid and triploid *M. acuminata* cultivars were largely disseminated by humans (Simmonds, 1960) to native areas of *M. balbisiana*, resulting in natural hybridization and in the formation of hybrid progeny with the genome AB, AAB, and ABB. Consequently, a very diverse selection of *Musa* cultivars is thought to have arisen in South East Asia along with the earliest developments of agriculture many thousand years ago (Price, 1995). The number of different clones has been estimated to be 400-500 (Perrier & Tezenas du Montcel, 1990).

The main genomic groups and sub-groups with some important cultivars are summarized in Table 1, with their uses and geographical distribution (adapted from Simmonds, 1966). This wide genomic diversity, combined with a wide and worldwide human dispersal, have led to very different broad systems of banana cultivation and pest management, depending on local conditions (tropical or subtropical regions; native or introduced crops; productions for export, local market or subsistence; cultivated varieties for dessert, cooking or even brewing).

In 2003, the total world production was estimated at over 100 million metric tons, of which dessert bananas represented 56 %. Only 14 % of this world production is grown for commercial export, so the rest, over 86 %, comprises a wide

Table 1. Main genomic groups of banana with some important cultivars, their uses and geographical distribution (adapted from Simmonds, 1966).

Genome	Sub-group	Cultivar	type	Distribution
AA	sucrier	Pisang mas, Figue sucrée	sweet dessert	Worldwide
		Pisang lilin	dessert	Indonesia, Malaysia
		Pisang berangan, Lakatan	dessert	Indonesia, Malaysia, Philippines
AAA	Gros-Michel	Gros-Michel, Highgate	dessert	Wordlwide
	Cavendish	Dwarf Cavendish, Giant Cavendish, Robusta, Pisang masak hijau	dessert	Exporting countries wordlwide
	Red	Red, Green-red	dessert	Wordlwide
	Mutika Lujugira	Intundu, Mujuba	brewing-cooking	Central & East Africa, Colombia
	Matooke	Nshakara, Nyoya	cooking	East Africa
	Mbire	Mbire, Kisubi	brewing	East Africa
	Ibota	Yangambi Km5	dessert	Indonesia, Africa
AB	Ney poovan	Ney poovan, Safet velchi, Kunnan, Sukari, Lady's finger	dessert	India, East Africa
AAB	Plantain	French plantain, Horn plantain, False Horn	cooking	Central & West Africa, India, Latin America, Carribean
	Pisang kelat	Pisang kelat, Thiruvanthapuram	dessert	India, Malaysia
	Pisang raja	Pisang raja	cooking	Malaysia, Indonesia
	Mysore	Poovan, Mysore	dessert	India
	Chuoi Xiem	?	dessert	Asia
	Silk	Figue pomme, Maça, Silk	dessert	Wordlwide
	Pome	Prata	dessert	Asia, Australia, West Africa, Brasil
	Popoulou	Popoulou	cooking	Pacific
	Laknao	Laknao	cooking	Philippines
	Pisang nangka	Pisang nangka	cooking	Malaysia
ABB	Bluggoe	Bluggoe, Matavia, Poteau, Cacambou	cooking	Wordlwide
	Pelipita	Pelipita	cooking	Philippines, Latin America
	Pisang awak	Fougamou	dessert	India, Thailand, Philippines, East Africa
	Peyan	Peyan	cooking	Philippines, Thailand
	Saba	Saba	cooking	Philippines, Indonesia, Malaysia

Table 2. Estimates of banana production and Cavendish exportation (adapted from Lescot, 2004).

	Cooking bananas		*Dessert bananas*		*Total*	*Bananas for export*	
	AAB	*ABB*	*Cavendish*	*others*		*Cavendish*	*cooking*
North America	0	9 000	400	100	9 500	428 449	1
Central America	932 000	108 000	5 860 162	167 000	7 067 162	3 903 124	109 202
South America	5 424 570	299 400	10 729 070	5 010 060	21 463 100	6 346 533	207 195
Caribbean	873 096	563 152	1 310 097	222 024	2 968 369	549 667	17 801
East Africa	1 287 451	13 995 956	2 023 593	734 960	18 041 960	15 089	13
West - Central Africa	7 991 102	963 963	1 970 757	410 788	11 210 610	559 451	463
North Africa - Middle East	2	3 022	1 471 568	1 071	1 475 663	189 259	0
Asia	1 067 020	9 795 840	20 728 071	5 338 285	36 929 216	1 826 981	9
Oceania	1 130	688 200	269 705	85 350	1 044 385	1 282	0
Europe	1	5	440 191	5	440 202	393 878	73 868
Total	17 576 372	26 426 538	44 803 614	11 969 643	100 650 167	13 785 264	408 551

range of banana varieties and crop systems (Lescot, 2004). Table 2 (adapted from Lescot, 2004) illustrates both estimates of banana production and Cavendish export. It shows the importance of banana cultivation in the different parts of the world, from the most intensive production systems for export of Cavendish bananas to the subsistence production of brewing bananas for local consumption.

As a consequence, it is obvious that banana diseases and pest management are also very diverse and depend primarily on the local conditions of cultivation.

1.2. Integrated Nematode Management: Concept Definition and Applications

All definitions agree that Integrated Pest Management (IPM) is a general approach which first assesses the pest situation, evaluates the advantages and disadvantages of pest management options and then implements a system of complementary management actions used in combination to control pests, with an emphasis on methods that are least injurious to the environment and most specific to the particular pest. For example, nematode-resistant plant varieties, regular monitoring for nematodes, judicious use of pesticides, biological control, and good stand management practices may be used alone or in combination to control or prevent particular forms of nematode damage. IPM is a dynamic system that is adaptable to diverse management approaches. In these approaches, the pest management decisions are taken by the individual producer, business entity or government agency but are influenced by the diversity of public and private values.

Historically, some of the most important nematode management practices were scientifically sound very early for commercial bananas, but their practical application was difficult, due to the absence of certain techniques (e.g. in vitro culture) or basic biological knowledge (e.g. nematode survival and dispersal, transitional host plants). For example, early as the sixties, Loos and contemporaries laid the basis of nematode management measures for controlling the burrowing nematode on dessert bananas and already recommended planting clean seed material on uninfested land (Loos & Loos, 1960a).

Bananas are attacked by many species of plant parasitic nematodes but only a few cause damage of economic importance. Worldwide, the nematode species known to cause, in the broad sense, the most serious damage to bananas are the migratory endoparasites, *Radopholus similis*, the lesion nematodes *Pratylenchus coffeae* and *P. goodeyi*, the endoparasite *Helicotylenchus multicinctus* and the sedentary parasite *Meloidogyne* spp. In addition to these five major species, some other species have been reported to be associated with *Musa* spp. throughout the world. Depending on local conditions, the associated damage of any of these nematode species may be locally important where their densities are high.

As for any other pest or parasite, nematode relationships with bananas, including damage, depend on environmental conditions, susceptibility of the host and pathogenicity of the nematode considered. In the last 50 years, many efforts have been made in nematology to collect these basic biological data and to test new nematode management practices on bananas. These efforts were particularly important on dessert bananas for export but, thanks to some national and international research institutes and to the banana and plantain section (formerly INIBAP, International Network for the Improvement of Banana and Plantain) of Biodiversity International (formerly IPGRI, International Plant Genetic Resources Institute), these efforts are now very considerable on all the other banana types.

In this chapter, the different nematode management approaches will be reviewed as specific procedures on commercial dessert bananas, as regional options due to the specificity of the different cropping systems (e.g. Asia and Oceania, Africa, America and the Caribbean) and as shared strategies and future approaches common to these different banana cropping systems.

2. DESSERT BANANAS FOR EXPORT

The first exported bananas from Central America arrived on the west coast of the United States before 1870 and by 1905 almost 1 M tons had already been imported (USA: 740000 tons; Great Britain: 115000 tons) from Central America but also from Jamaica and the Canary Islands (Simmonds, 1960; Champion, 1963). At this time, the variety 'Gros-Michel', a triploid *Musa* AAA originating from Malaysia, was the favourite variety in all commercial banana plantations.

Following the spread of Panama disease (*Fusarium oxysporum* f. sp. *cubense*) in the seventies, all the commercial plantations changed over from the susceptible cultivar 'Gros Michel' to the resistant cultivars from the Cavendish subgroup, which are still cultivated (Jones, 2000). However, different authors in Central America (Leach, 1958; Whehunt et al., 1978), India (Rajendran et al., 1979) and West Africa (Mateille, 1992; 1993) had already noticed that the variety 'Gros-Michel' was less sensitive to *R. similis* than the newly introduced Cavendish varieties.

At present, the main producing countries of export bananas are localized in Central and South America (Guatemala, Costa Rica, Ecuador, Colombia) and in Southeast Asia (Philippines), where these Cavendish varieties are grown in intensive monoculture mostly for export (14.2 M tons in 2003). Ecuador alone accounts for more than one third of the international banana exports. However, the tonnages of these Cavendish bananas (a world production of more than 44.8 M tons in 2003) are even greater when grown for the local market in countries such as India, China, Brazil, Indonesia, Mexico and Egypt (Lescot, 2004).

Most of these bananas grown for export belong to the Cavendish subgroup and are cultivated in the humid tropics, with a uniform warm climate on flat lowlands with deep and well-drained soils.

2.1. Geographic Distribution of Associated Nematode Species

The nematode problem on commercial bananas was observed very early and soon received much attention from researchers in Latin America and the Caribbean, as dessert bananas were cultivated for export to North America and Europe from 1870 (Champion, 1963). Ashby (1915) in Jamaica was the first author to describe appropriately nematode symptoms in banana rhizomes as a 'Black head disease of bananas'. The same year, Cobb completed the nematode description using soil samples taken earlier from around banana roots from Fiji, described as *Tylenchulus similis* (Cobb, 1893) and additional specimens from Hawaii and Jamaica. Following this early discovery, the burrowing nematode *Radopholus similis* was progressively observed in almost all dessert banana producing areas of the world: in the French West Indies, Jamaica and Trinidad (Mallamaire, 1939; Leach, 1958; Scotto la Massèse, 1968); in the large plantations of the United Fruit Company of Central America (Stover & Fielding, 1958; Holdeman, 1960); in Brazil (Carvalho, 1959); in Belize (Pinochet & Ventura, 1977); in West Africa (Mallamaire, 1939; Luc & Vilardebo, 1961), the Caribbean (Ayala & Roman, 1963; Decker & Casamayor, 1966; Stoyanov, 1967; Edmunds, 1968), Surinam (Maas, 1969), India (Nair et al., 1966) and Asia (Timm, 1965; O'Bannon, 1977).

Blake (1961) suggested that the burrowing nematode was first introduced into Australia in infested banana plants imported from Fiji between 1860 and 1910. In 1972, Stover advanced the explanation that the recent and widespread dissemination of *R. similis* began soon after the progressive replacement of the variety 'Gros Michel' by the Cavendish varieties. As an example, while already present in the Philippines, the occurrence of *R. similis* increased dramatically when large amounts of infested planting materials of giant Cavendish were imported from Central America in the early seventies (Boncato & Davide, 1980; Davide, 1992).

Recently, Marin et al. (1998a) reviewed the spread of bananas in Latin America and the Caribbean and its relationship to the occurrence of *R. similis*. Diseases caused by *R. similis* were also known as "spreading decline of citrus" in Florida, USA (Suit & DuCharme, 1953) and "yellows disease of pepper" in Bangka, Indonesia (van der Vetch, 1950). Throughout the world, *R. similis* has also been recovered from the roots of many other hosts, including important cultivated crops (tea, coffee, pepper), ornamentals and weeds (Gowen et al., 2005).

Besides the widespread occurrence of the burrowing nematode *R. similis*, some other nematode species are also able to cause economic damage on dessert bananas.

After *R. similis*, the spiral nematode *Helicotylenchus multicinctus* is probably the most damaging nematode on bananas. This species, originally described by Cobb in 1893 as *Tylenchus multicinctus*, has been frequently found in mixed populations with *R. similis* throughout the tropics and the subtropics on all varieties of bananas. Its geographical distribution follows almost exactly that of *R. similis* (McSorley & Parrado, 1986; Bridge, 1993) while its abundance depends both on the presence or absence of the burrowing nematode *R. similis* and on the soil organic matter content (Vilardebo & Guérout, 1976; Quénéhervé, 1988). Its economic importance has been acknowledged mostly in bananas growing in subtropical conditions, such as in Israel (Minz et al., 1960), South Africa (Jones, 1979) and Florida (McSorley & Parrado, 1986). *Helicotylenchus multicinctus* should be regarded as the main parasitic nematode on bananas in the absence of lesion nematodes (*Radopholus* and *Pratylenchus*) and where environmental conditions are suboptimal for the crop in relation to latitude, temperature and rainfall.

Among the lesion nematodes from the genus *Pratylenchus*, only *P. coffeae* and *P. goodeyi* are recognized as damaging species, and cause similar symptoms on bananas as the burrowing nematode. Zimmerman (1898) was the first to describe as *Tylenchus coffeae* the species infesting coffee plants in Java, whereas Cobb observed and described the species as *Tylenchus musicola* in roots of plantains in Grenada in 1919. Since then, *P. coffeae* has been recorded worldwide on bananas (Bridge, 1993). This nematode is a pan-tropical species and a major pest of economic crops such as coffee, banana and fruit trees, tuber crops and ornamentals (Luc et al., 2005). While the distribution of the burrowing nematodes was mostly associated with commercial plantations of Cavendish varieties, the distribution of the lesion nematode *P. coffeae* seems mostly associated with plantains, rather than Cavendish varieties.

Pratylenchus goodeyi was first observed in roots of dessert bananas in the Canary Islands by de Guiran & Vilardebo (1962) and later in Crete (Vovlas et al., 1994). Since then, this species has been observed on highland bananas in East Africa (Gichure & Ondieki, 1977; Walker et al., 1984; Bridge, 1988a) and Cameroon (Price & Bridge, 1995) in addition to its presence on *Ensete* in Ethiopia (Peregrine & Bridge, 1992). More recently, the species was also reported from subtropical areas of Australia (Stanton et al., 2001). The presence of *P. goodeyi* on bananas seems conditioned by the altitude and the latitude, presumably in relation to soil temperature (Price & Bridge, 1995).

All banana varieties are hosts of the root-knot nematodes belonging to the *Meloidogyne* genus, which attack many economically important crops and cause deformations and stunting of the roots. They were first reported to occur on bananas, in Egypt and Southeast Asia, by Delacroix (1901). In general, the root-knot nematodes are more likely to cause damage in subtropical conditions such as in Crete (Vovlas et al., 1994), Lebanon (Sikora & Schlosser, 1973), North Yemen (Sikora, 1979), South Africa (Jones & Milne, 1982) and Taiwan (Lin & Tsay, 1985) and in greenhouse production systems of Morocco (Janick & Ait-Oubalou, 1989) and the Canary Islands (Pinochet et al., 1998).

In tropical conditions, root-knot nematodes are more likely to be found in great numbers on Cavendish varieties in absence or near-absence of burrowing or lesion nematodes such as on sandy loam soils in the Philippines (Davide, 1980) or sandy soils of West Africa (Quénéhervé, 1988). Currently, in the French West Indies, they are reported in large numbers only on new Cavendish plantations established from tissue culture plants, after a fallow or a rotation period. In Asia, Boncato and Davide (1980) in the Philippines and Razak (1994) in Malaysia also reported large populations of root-knot nematodes on commercial Cavendish plantations.

Other species of minor incidence on dessert bananas include *Rotylenchulus reniformis*, *Hoplolaimus pararobustus*, *H. seinhorsti* and *Heterodera oryzicola*. In the islands of Madagascar and La Réunion, a nematode species, *Zygotylenchus taomasinae* has been found in association with *R. similis* in banana plantations (Vilardebo & Guérout, 1976).

2.2. Basic Studies on Nematode Biology

Outstanding studies on biology and life-cycle of the burrowing nematode *R. similis* and histological observations were first conducted by Blake in Australia (1961; 1966) and Loos while working at the United Fruit Co., in Honduras (Anonymous,

1957; Loos & Loos, 1960b). In these studies, the authors described how nematodes could invade, feed and reproduce in the cells of the cortex along the entire length of the roots and in the rhizome. Nematodes, while migrating in the cortical parenchyma but not in the stele, cause cavities which then coalesce to appear as necrotic tunnels. The migration and egg laying seem governed by nutritional and biochemical factors, as nematodes move in the parenchyma in search of healthy tissue, away from the necrosis (Blake, 1961). Loos (1962) was the first to describe the complete life-cycle from eggs to eggs in 20-25 days at a temperature range of 24-32°C, with the eggs hatching after 8-10 days and the completion of the juvenile stages in 10-13 days.

Increases of nematode populations in banana roots are thought to be the result of several factors (see: Gowen et al., 2005, for a review) but clearly the renewal of the root system following bursts of root growth is the main factor in the population build-up of *R. similis*. Any factor, endogenous or exogenous, which favours root emergence on banana plants, contributes to this increase (Quénéhervé, 1993a).

The existence of different biotypes of *R. similis* was first illustrated by the physiological differences in reproductive capabilities and morphological variations among *R. similis* populations. This hypothesis was extensively studied in Central America and the Caribbean (Edwards & Wehunt, 1971; Pinochet, 1979; Tarté et al., 1981; Kaplan & O'Bannon, 1985; Pinochet, 1987; Sarah et al., 1993; Fallas et al., 1995; Hahn et al., 1996; Marin et al., 1999). Different biotypes of *R. similis* are now widely recognised and certainly could explain the discrepancies observed worldwide in damage levels, in terms of yield losses, plantation longevity, transitional hosts and nematode management efficacy. Until recently, it was recognized that *R. similis* had two races, one non-pathogenic to citrus and another pathogenic either on citrus and banana, the former *R. citrophilus* (DuCharme & Birchfield, 1956). Recent research does not support the existence of a sibling species (Kaplan & Opperman, 1997; Valette et al., 1998). Nevertheless, these different biotypes of *R. similis* were also observed on other plants than bananas and led to inconsistent results in terms of the host status of some weeds (Edwards & Whehunt, 1971; Keetch, 1972; O'Bannon, 1977; Inomoto, 1994), and of very important rotation crops too (sugarcane, pineapple, forage crops e.g. *Bracharia* sp).

The interaction with other pathogens was studied since the increase in Panama disease (caused by *F. oxysporum* f. sp. *cubense*), in presence of the burrowing nematode was observed early on the cultivar 'Gros Michel' (Newhall, 1958). Soon after that, and beginning in the sixties, the cultivar 'Gros Michel' was completely replaced by banana varieties from the Cavendish subgroup following the spread of Panama Disease into every commercial plantation in Latin America and the Caribbean. Since then, many studies have described and assessed the pathogenic effects of fungi alone or in combination with nematodes on the *Musa* AAA, from the subgroup Cavendish (Brun & Laville, 1965; Stover, 1966; Booth & Stover, 1974; Pinochet & Stover, 1980; Loridat, 1989).

The presence of *R. similis* on hosts other than *Musa* was also investigated and Christie, in 1958, published the first list of putative hosts of *R. similis*, including cultivated crops and weeds. While this topic was extensively studied in Florida from a quarantine point of view (O'Bannon, 1977; Lehman, 1980; Esser et al., 1984), similar studies were gradually carried out in every banana producing country as a prerequisite for nematode management (Ayala & Roman, 1963; Keetch, 1972; Rivas & Roman, 1985; Zem, 1983). More recently, a study conducted in Martinique clearly shows how weeds could be significant reservoirs of plant parasitic nematodes, including *R. similis* and *P. coffeae* in banana fields (Quénéhervé et al., 2006).

The survival of *R. similis* in soils was studied in citrus soils in Florida (Tarjan, 1961) and banana soils in Honduras (Loos, 1961). These authors demonstrated that *R. similis*, which does not have a specialized survival strategy (e.g. quiescence, cryptobiosis), was not able to survive more than 6 months in the soil, in absence of host roots or pieces of live corms. The corms, used as seeds or planting materials, have been known to be a major means of dissemination of banana nematodes for many years in Latin America (Loos & Loos, 1960a), Australia (Blake, 1961) and Africa (Quénéhervé & Cadet, 1985b). In a study conducted in the Ivory Coast on the cultivar 'Poyo', most of the nematodes were localized in the outer part of the corm but a significant proportion (11 %) was found at depths ranging from 3 to 7 cm, well protected against any physical nematode control method (e.g. paring, heat-treatment) (Quénéhervé & Cadet, 1985a).

During the last decade, most of the studies on the biology of nematodes found on export bananas were mainly conducted in Costa Rica with the Corporación Bananera Nacional (CORBANA) (Araya et al., 1999; Araya & De Waele, 2004; Moens et al., 2006).

The biology of *R. similis* was extensively studied as the major nematode problem on export bananas, and relatively little information exists on the biology of the other nematode species. The biology of the lesion nematodes *Pratylenchus* spp. was mostly studied on non-export bananas and will be considered later. The spiral nematode, often encountered together with *R. similis* in dessert bananas, feeds on the outer cells of the root cortex and produces small, characteristic discoloured necrotic lesions (Luc & Vilardebo, 1961), but it is also able to complete its life-cycle within the cortical part of the root (Zuckerman & Strich-Harari, 1963). In contrast to *R. similis*, histological changes seem to be confined to parenchyma cells close to the epidermis (Orion et al., 1999).

The biology and life-cycle of root-knot nematodes are not documented on bananas but should not differ from those described for other hosts. In thick and fleshy primary roots, roots deformations and stunting can be very important, with many females and egg masses occurring within the same gall. In general, root-knot nematodes occur in banana roots in mixed populations of nematode genera and species (Pinochet, 1977; Cofcewicz et al., 2004a; 2004b; 2005) and their populations are greater on the distal part of the banana root system, as a reflection of the competition occurring with the other nematode species (Santor & Davide, 1982; Quénéhervé, 1990). Pinochet (1977) suggested that extensive colonization by *R. similis* might contribute to the inhibition of the *Meloidogyne* spp. development, by reducing the feeding sites and interrupting their life cycle in roots, near the rhizome.

For all these species, as with *R. similis*, survival occurs on infected corms or on tissue remaining from the previous crop, and infected planting material is also the primary means of dissemination (Quénéhervé & Cadet, 1985a, 1985b).

2.3. Damage and Economic Importance

The importance of nematodes as a widespread cause of banana losses was first reported in Jamaica by Leach (1958), who emphasized how destructive the burrowing nematode *R. similis* was for banana production, attributing to this pest the widely distributed disease know as "Black head toppling disease". Loos (Anonymous, 1957) was the first to describe root symptoms and associated damage with the presence of *R. similis* in banana roots, since *"the lesioning of the primary roots together with the girdling and death of those roots which anchor the plant to the ground make the plant prone to 'tip over' under wind pressure"*.

Nematodes affect banana plant growth and yield by damaging the root system, and increases in population densities of some nematode species (e.g. burrowing and lesion nematodes) are most often associated with increased root necrosis, reduced root biomass and toppling of the plants. Bananas infected with plant-parasitic nematodes are therefore less able to take up water and nutrients, resulting in stunting, delayed maturation time and reduced bunch size. Depending on the nematode species mixture and on environmental factors, the damage can vary from a slight and hidden lengthening of the vegetative period to the most obvious symptom of attack by lesion nematodes, which is the toppling over of banana plants.

From a mechanistic approach, it is possible to define three successive levels of nematode damage (Quénéhervé, 1993a, 1993b):

1. A lengthening of the vegetative phase: the different phenological intervals (lag between planting and flowering, harvest and flowering of ratoons, harvest to harvest etc.) are lengthened without significant reduction in plant size, bunch weight, number of harvested bunches and total harvest. This minor damage is mostly ignored, except in commercial plantations, where the number of boxes is monitored such as in Central and Latin America.

2. A lengthening of the vegetative phase with a reduction in the total harvest: in this case there are two sub-levels according to the reduction in the number of harvestable bunches (bunches that are non-exportable because of poor quality or immature delayed fruits), in addition to the reduction in the average plant size and bunch weight. This type of damage is often observed in commercial plantations in West Africa.

3. A lengthening of the vegetative phase, with a reduction either in the total harvest and in the longevity of the plantation: this third level is the same as above but now it is irreversible, due to the destruction of plants which are uprooted or whose growth is too severely delayed. When infested with the highly pathogenic strain of the burrowing nematode and in absence of any nematode control, this third level of damage is observed almost worldwide on dessert bananas.

However, in some regions, irreversible damage due to uprooted plants bearing fruits can occur very early with gusty winds. The probability of observing this type of damage with *R. similis* is highest in the Caribbean and Central America, as compared to other continental banana producing areas of the world.

After the replacement of cv. 'Gros Michel' by more nematode-susceptible Cavendish cultivars, crop losses were estimated on the basis of the yield improvement after nematicide treatments in the different producing countries. These reported yield responses varied greatly from 15 to 275% (Gowen & Quénéhervé, 1990). These differences may be due to the several biotic and abiotic factors affecting the nematode population dynamics that were extensively studied such as the soil type (Stover & Fielding, 1958; Ayala & Roman, 1963; Guérout, 1975; Davide, 1980; McSorley & Parrado, 1981; Quénéhervé, 1988), the nematode species and biotype (cf. above), the host plant physiology (Guérout, 1972; Jaramillo & Figueroa, 1974; Hugon et al., 1984; Mateille et al., 1984; Quénéhervé, 1993a) and the climate (Jimenez, 1972; Jaramillo & Figueroa, 1974, 1976 ; Vilardebo, 1976; Marcelino et al., 1978; Hutton, 1978; McSorley & Parrado, 1981 ; Badra & Caveness, 1983; Davide & Marasigan, 1985; Hugon et al., 1984; Quénéhervé, 1989a; 1989b).

Besides *R. similis*, some other species have been shown to cause damage to dessert bananas for export. In the Canary Islands on sandy and loamy soils, root-knot nematodes can cause yield reductions of over 20 %, while the lesion nematode *P. goodeyi*, widespread at altitudes above 300-500 m, causes serious root damage in the three major banana producing Canary Islands, with 16 % yield reduction (Rodriguez, in: Pinochet et al., 1998). In the Philippines, yield reductions based on bunch weights ranging from 26.4 % to 57.1 % were observed after inoculation with the root-knot nematode *M. incognita* (Davide & Marasigan, 1985). In greenhouse experiments, significant reductions in plant growth (Jonathan et al., 2000) and alteration of the concentration of macro- and micronutrients in leaves (Cofcewicz et al., 2004) were observed after inoculation with root-knot nematodes. In Israel and Cyprus, yield reductions ranging from 18 % (Minz et al., 1960) to 30 % (Phillis: in Gowen & Quénéhervé, 1990) have been observed with *H. multicinctus*.

Due to the almost constant superiority in numbers of the burrowing nematodes on Cavendish bananas, the assessment of yield losses due to other nematode species has always been neglected and certainly underestimated, such as for *H. multicinctus* (Moens et al., 2006). In recent experiments conducted in Costa Rica on cv. 'Grande Naine' (*Musa* AAA), *H. multicinctus* reduced the mean root weight by 13 % compared to uninoculated plants, *M. incognita* increased the mean root weight by 6.7 % and *P. coffeae* did not significantly decrease the mean root weight. In a microplot experiment, only plants infected with *R. similis* showed a significant root weight reduction of 66 %, after 12-15 months.

Damage is assessed by choosing an appropriate nematode extraction technique (Gowen & Edmunds, 1973; Whyte & Gowen, 1974; Vilardebo, 1974; Quimi & Villacis, 1977; Escobar & Rodriguez-Kabana, 1980; Araya, 2002) and type and place for root sampling (Quénéhervé & Cadet, 1986; Araya et al., 1999; Moens et al., 2001) or a standardized sampling method for pesticide or resistance screening trials (Vilardebo, 1974; Carlier et al., 2002). Obviously, the choice of any assessment method depends both on objectives (research or laboratory routine diagnosis) and laboratory facilities. In the absence of laboratory facilities, the visual assessment of damage is also possible by recording the incidence of banana plant uprooting per hectare and per month (Tarté & Pinochet, 1981). This technique is still currently used in large plantations where uprooted banana plants are monitored weekly. As an example, in Costa Rica the incidence of uprooted plants can reach 5.5 $\% \cdot ha^{-1} \cdot week^{-1}$ without treatments, while this percentage is lowered to 0.3-0.5 % with nematicide applications (G. Fallas, pers. comm.). Techniques of visual assessment of necrosis on roots and rhizomes were also developed in America (Stover, 1972; Tarté & Pinochet, 1981), Africa (Bridge, 1988a; Bridge & Gowen, 1993; Speijer & Gold, 1996) and Australia (Broadley, 1979). These methods (percentage of necrotic roots) combined with nematode countings are applied in most of the banana nematode monitoring programmes in Latin America (Araya, 2002) and Australia (Stanton et al., 2001).

2.4. Nematode Control: The Golden Age of Nematicides (1960-1990)

The early investigations on the control of nematodes in banana soils were conducted in Africa and Australia (Blake, 1961) with two fumigant nematicides (D-D, dichloropropane-dichloropropene; EDB, ethylene dibromide), which gave a 30-40 % yield increase (Vilardebo, 1959; Champion, 1963). Very soon, these fumigants were replaced by DBCP (1,2-dibromo-3-chloropropane) because it was the only fumigant nematicide which was not phytotoxic and could therefore be applied prior to planting, or onto established crops, to control *R. similis* (Luc & Vilardebo, 1961).

In Central America and as early as the sixties, Loos and contemporaries laid the basis of the nematode management measures for controlling the burrowing nematode on bananas, and already recommended planting clean seed plants on uninfested land (Anonymous, 1957; Loos & Loos, 1960a). This objective was tentatively first achieved using physical (paring, heat treatment) and chemical (dipping in a nematicide) methods, in order to clean the planting material. This use of DBCP was recommended in the Windward Islands (Edmunds, 1968), while some phytotoxicity after dipping with DBCP was observed in Honduras, leading to its replacement by heat treatment (Hildreth, 1962).

Between 1960 and 1978, the fumigant nematicide DBCP was used extensively on commercial bananas in Africa, Latin America and the Caribbean, and treatments were normally applied twice a year using hand-held injectors in which the fumigant was injected around individual plants. Wehunt and Edwards (1968) reported yield increases in Central America varying from 14 to 86 %. In parallel, research efforts were concentrated on the evaluation of the newly released non-fumigant nematicides (organophosphates and carbamates), mostly systemic, used as seed treatment (Vilardebo & Robin, 1969; Coates, 1971; Guérout, 1975) or as soil treatment after planting (Vilardebo, 1970; Guérout, 1970; Gowen, 1975; 1979; Figueroa & Mora, 1977).

At present, the application of non-fumigant nematicides still remains the most used nematode control worldwide on dessert bananas for export, with granular or liquid nematicides applied through the sure-fill system and hand-held applicators to ensure safe application. In most countries, governments require all nematicides to be used only where banana plantation companies exercise close supervision of workers handling and applying the chemicals.

In the past these treatments were mostly applied at fixed times of the year, but now they are applied on the basis of nematode incidence, of banana plant uprooting and/or nematode counts in the roots, in an effort aiming at minimizing nematicide applications. It is interesting to note that no universal threshold level in terms of nematodes per unit of roots has been suggested. For *R. similis*, this threshold level can vary from place to place: from 1000 per 100 g of roots in the Ivory Coast (Guérout, 1972) and Martinique; from 4000 to 6000 in plantations of Costa Rica (Fallas pers. com); from 10000 in the Philippines (Davide, 1992) and the Windward Islands (Gowen, pers. com.) and from 20000 in Honduras and Panama (Pinochet, 1987) as a response to regional differences in *R. similis* pathogenicity.

2.5. Research of an Alternative to Chemical Control

During recent decades there have been many changes in the management of banana nematodes in large commercial plantations (e.g., loss of important non-fumigant nematicides and homologation of one new nematicide only; absence of a still effective nematicide alternative, e.g. biological control; increased concerns related to nematicide applications for environmental quality (product, soil, water) and human health). These problems were very important in the frequently replanted crop systems of the Caribbean, compared to the large plantations of Latin America or Asia (Philippines) where bananas are grown continuously without replanting. As a consequence, the search for an alternative to chemical treatments has been quite intense in the Caribbean.

Efforts were concentrated on replanting practices, using tissue culture plants on cleaned soils. The concept was proposed very early (Loos & Loos, 1960a) but its application only became feasible since the availability of disease-free tissue culture plants, through the meristem culture technique. Since that period, hot-water treatment, following peeling away of all lesions from the corms, became a standard practice in many parts of Central America and the Caribbean (Stover, 1972) with inconsistent results. As an example, four years after the establishment of new plantations in Belize with heat-treated seeds imported from Honduras in areas where bananas had never been grown before, the infestation rate by *R. similis* was already 43.1 % (Pinochet & Ventura, 1977).

In the meantime, many cultural practices were tried in Latin America and Caribbean regions in order to free the soil from *R. similis*. These practices included:

- flood fallowing in Surinam and the Ivory Coast, prior to replanting (Maas, 1969; Sarah et al., 1983)
- dry- or bare-fallow (Loos, 1961; Edwards, 1963; Salas et al., 1976)
- weed fallow (Chabrier & Quénéhervé, 2003)
- cultivated fallow with Pangola grass (Stoyanov, 1967; Roman et al., 1978) and Sudan grass (Ternisien & Melin, 1989)
- rotation with other crops such as sugarcane (Loos, 1960; Stoyanov, 1967), cassava (Zem & Alves, 1983) or pineapple (Sarah, 1989)

Nevertheless, these efficient rotation crops are still rarely practised because of the high cost of planting and maintaining the rotation crop along with the inability to develop marketable rotation crops. Following these studies, some improvements in the fallow setting-up were made to ensure the elimination of the burrowing nematode from the soil. This was achieved by a previous individual chemical destruction of each plant before the mechanical destruction of the banana plantation (Chabrier & Quénéhervé, 2003). In the French West Indies, the use of these practices not only extended the longevity of the plantations, but also drastically reduced (by 63 % from 1996 to 2004) the application of nematicides (Chabrier et al., 2005).

2.6. Future Prospects

For more than fifty years, many (and probably the most important) advances in the knowledge and management of banana nematodes were obtained in Latin America, the Caribbean and in West Africa, beginning in the labs of the United Fruit Co. in La Lima, Honduras, in the field of the Banana Board of Jamaica at Bodles, as well as in Guinea and the Ivory Coast. Currently, most of outstanding research on nematodes of banana for export is now conducted in Costa Rica and in the French Antilles, with new challenges. The golden age of nematode control with nematicides is definitely behind us. There is a global tendency to replace the former nematode control by a wider view of 'sustainable nematode management' (Sikora et al., 2005). This trend will certainly increase under the pressure of consumers and commercial firms in order to improve quality and diversity of dessert bananas for export. The breeding of new dessert banana varieties, not only resistant to Black Sigatoka but also to burrowing and lesion nematodes, is certainly the next step.

3. NEMATODES ON BANANAS IN ASIA AND OCEANIA

3.1. Introduction

Given their size and diversity, Asia and Oceania are more a cultural concept incorporating a number of regions and peoples than homogeneous, physical entities. Asia can be divided into different sub-regions in which some of the major banana producing countries are found, such as South Asia (India subcontinent), Southeast Asia (mainland and archipelago) and Eastern Asia with China. Oceania is a geographical region consisting of numerous islands including Australia. It is traditionally divided into the Australasian, Melanesian, Micronesian and Polynesian archipelagos. Southeast Asia is considered to be the centre of origin of *Musa* species and of domestication of the edible banana (Jones, 2000) and Melanesia is the centre of origin of the burrowing nematode *R. similis*, the most detrimental plant-parasitic nematode associated with bananas worldwide. Paradoxically, Asia was also, until recently, the world region where the least number of studies had been made on banana nematodes. This was mainly because very few countries grew bananas for export until recently. In 2004, banana production in Asia and Oceania was estimated at 38 M tons (95 % of non-export bananas) produced in more than 35 countries with some countries such as India (the largest banana producer in the world, at 16.4 M tons), China (5.8 M tons), the Philippines (ranked fifth among the world's major export banana countries, with 1.7 M tons out of 5.5 M tons) and Indonesia (3.8 M tons) being the most important producing countries for dessert bananas, Cavendish and others, in the world (Lescot, 2004).

In Asia, banana is an indigenous crop to many countries, especially from Southeast Asia, planted everywhere for thousands of years by smallholder farmers while wild species are commonly found in the primary and secondary forests. As a centre of origin of *Musa*, the genomic diversity of cultivated bananas is very wide (*Musa* genome AA, AAA, AB, AAB, ABB). There is also a wide diversity of banana lines in Oceania, especially in Papua New Guinea, the Solomon Islands and Vanuatu (*Musa* genome AA, AAA, AAB and Fe'i group of the Australimusa section). These bananas are very important in terms of nutrition, cultural significance and traditional use in medicine. Until recently, the edible cultivars were mostly grown as a subsistence crop to provide small incomes and to contribute to the nutrition of the population. However, in recent decades, four distinct production systems can be roughly distinguished (Valmayor, 1990):

- a backyard production system characterized by a wide diversity of banana cultivars and very minimal inputs.
- a mixed-cropping production system in which bananas are intercropped with annual crops (taro, ginger, sweet potato, bean, corn, etc.) or perennials plants (rubber, Durian trees , coconut, arecanut, etc.).
- a commercial smallholder monoculture production system with some minimal management practices (fertilizing, weeding, etc.).
- a corporate farm production system strictly intended for the export market of dessert Cavendish bananas.

Since the development of the market economy in Asia, banana production for domestic consumption and export is also considered as a new opportunity in terms of economic value and often ranks now in the top ten of the total fruit production of these countries.

3.2. Nematode Species

Until recently, there was a general lack of information on the nematode species associated with local banana cultivars in Asia as there was no public or private priority in terms of funding for research and development in comparison with export crops (e.g. rubber, oil palm, cocoa, coffee).

Paradoxically, it was very early that Cobb (1893) completed the first description of the burrowing nematode described as *Tylenchulus similis*, from specimens found in soil around banana roots from Fiji in Melanesia. Following this early discovery in Oceania, banana nematodes including *R. similis* have only been reported lately from bananas in Australia (Blake, 1972), Samoa (Orton Williams, 1980), Tonga (Kirby et al., 1980), Papua New Guinea (Bridge & Page, 1984) and the Solomon Islands (Bridge, 1988b).

In South Asia, the occurrence of *R. similis* on bananas was first reported from the Kerala district in India (Nair et al., 1966) and from Sri Lanka (Gnanapragasam et al., 1991). In fact, extensive surveys in India revealed that the root lesion nematode *P. coffeae* was the predominant species and ranked first in prominence and importance. This species was followed by the root knot nematode, the spiral nematode and the burrowing nematode. Subsequently, the burrowing nematode was reported from almost all banana-growing states including isolated pockets like the Andaman Island (Khan, 1999; Sundararaju et al., 2005). In Bangladesh, the main nematode species reported on banana is *R. similis* (Mian, 1986).

In Southeast Asia, the detection of *R. similis* on bananas also occurred lately after its previous detection on other crops (Table 3): in the Philippines (Timm, 1965; Boncato & Davide, 1980), Malaysia (Larter & Allen, 1953; O'Bannon, 1977; Winoto & Sauer, 1982), Thailand (Timm, 1965; Prachasaisoradej et al., 1994) and Indonesia (O'Bannon, 1977; Hadisoeganda, 1994). In the Philippines, all species except *R. similis* were generally associated with native banana cultivars (Davide & Gargantiel, 1974) while *R. similis* was found widely associated with the Cavendish bananas. Often crops, which are good hosts of *R. similis* but also of *P. coffeae* and *Meloidogyne* spp., including banana, ginger, turmeric, betel vine, coconut and arecanut were intercropped with pepper, as in India (Koshy et al., 2005).

In Eastern Asia, *R. similis* has not yet been detected until now on bananas from Vietnam and China. In Vietnam, all the common species associated with banana were identified on both wild and cultivated bananas. The most frequently species found were *Helicotylenchus dihystera*, *Meloidogyne incognita* and *Rotylenchulus reniformis*, while the lesion nematode *P. coffeae* and the spiral nematode *H. multicinctus* were also found rather infrequently (Chau et al., 1997). However, indigenous populations of *R. similis* were recently reported from coffee in two Vietnam provinces (Nguyet et al., 2003). In China, with the notable exception of *R. similis*, the banana root-knot nematodes *M. javanica* and *M. arenaria* occur in sandy fields in Hainan and Fujian provinces, as well as *Rotylenchulus reniformis* and *Helicotylenchus* spp. (Linbing et al., 2004).

The lesion nematode *P. coffeae*, was first reported on abaca in the Philippines (Taylor & Loegering, 1953). It was reported to cause damage to young bananas in Malaysia (Winoto, 1976) in combined infestation with *Meloidogyne incognita*.

Beside these common species, Charles and Venkitesan (1984) first reported the occurrence of a cyst nematode, *Heterodera oryzicola*, on banana (Musa AAB) in the state of Kerala, India, where this nematode is also one of the major pests on rice.

Among the *Meloidogyne* species, *Meloidogyne graminicola*, one of the major pests of rice in the Philippines and other Asian countries (Bridge et al., 1990) can also be found associated in large numbers with some common banana cultivars in the Philippines like Saba, Latundan and Lakatan (Reversat & Soriano, 2002).

Table 3. Early detections of **Radopholus similis** *on banana and on some other crops.*

	America	*Caribbean*	*Africa*	*South Asia*	*Southern Asia*	*Pacific*
Anthurium						Hawaii (Sher, 1954)
Arecanut				India (Kumar et al., 1971)		
Avocado	Florida (DuCharme & Suit, 1953)			India (Jasy & Koshy, 1992)		
Banana	Panama (Newhall, 1958)	Jamaica (Ashby, 1915)	Guinea (Mallamaire, 1939)	India (Nair et al., 1966)	Malaysia (Larter & Allen, 1953)	Fiji (Cobb, 1893)
Betel Vine				India (Koshy & Sosamma, 1975)		
Black Pepper					Indonesia (Van der Vetch, 1950)	
Citrus	Florida (Suit & DuCharme, 1953)					
Coconut				Sri Lanka (Ekanayake, 1964)		
Coffee					Indonesia (Zimmermann, 1898)	
Date Palm				India (Lal & Mathur, 1986)		
Faba Bean				India (Sosamma & Koshy, 1977)		
Food Legumes				India (Sikora et al., 2005)		
Ginger	Florida (Hart, 1956)					Fiji (Butler & Vilsoni, 1975
Kava						Fiji (Kirby et al., 1980)
Persimmon	Florida (McSorley, 1981)					
Sugarcane						Hawaii (Cobb, 1909)
Swam Taro						Guam (Jackson, 1987)
Sweet Potato				India (Koshy & Jasy, 1991)		
Tamarind				India (Sosamma & Koshy, 1977)		
Taro						Fiji (Kirby et al., 1980)
Tea					Indonesia (Zimmermann, 1898)	
Turmeric				India (Sosamma et al., 1979)		
Yam						Fiji (Butler & Vilsoni, 1975)

3.3. Importance and Damage Potential

Limited information is available on the nematode damage to native bananas from Asia and Oceania since most studies were carried out on the Cavendish banana. However, despite favourable environmental conditions for banana production, the average yield is often very low.

In South Asia, nematodes constitute one of the major limiting factors to banana production in India, with reported yield reductions up to 41 % for *R. similis* (Nair, 1979), 44 % for *P. coffeae* (Sundararaju & Cannayane, 2003), 34 % for *H. multicinctus* (Rajendran & Sivakumar, 1996), 20-56 % for the cyst nematode *H. oryzicola* (Charles & Venkitesan, 1993) and 31 % for *M. incognita* (Jonathan & Rajendran, 2000).

In the Philippines, most of the studies to evaluate the pathogenic capabilities of nematodes commonly associated with banana were conducted on dessert Cavendish banana (Davide & Marasigan, 1985). In Vietnam, *Meloidogyne* spp. seems to have an adverse effect on the growth of native banana cultivars, while the effect of *P. coffeae* on Musa plant growth is unclear (Van den Berg et al., 2002).

3.4. Nematode Management

In Asia, and particularly Oceania, the banana cultivation is basically a smallholder enterprise of small size and, except in home gardens where bananas benefit from the regular application of animal manure and household refuse, these banana-cropping systems receive little or no inputs. In general, management practices that include nematode control are used less extensively in commercial smallholder plantations than on corporate farms, which rely almost exclusively on the use of chemicals to control nematodes on export Cavendish bananas.

In the Philippines, the government has decreed that all nematicides in the country should be for institutional use only, where plantation companies exercise close supervision of labourers handling and applying the chemicals. Alternative control measures were also conducted to explore the potential of botanical nematicides and of biological control agents against the nematodes (Villanueva, 2004).

The use of suckers or rhizomes as seed stock is the main practice among smallholders in Asia and Oceania. Due to this practice, the spread of pests such as nematodes is difficult to control and/or eliminate and often production becomes poorer from one cycle to the next, while nematode populations build up over the years.

However, in China, more than 100 commercial laboratories produce millions of tissue-cultured plants for most banana plantations. Eighty to ninety percent of tissue culture plants are used for new plantings. Some of these tissue-cultured plants, issued with a certification ISO-9001, are even exported to other countries (Linbing et al., 2004).

The burrowing nematode is the most important nematode on bananas in Australia. Current management options mostly include a rotation, application of the registered nematicides, fallow and the use of clean planting material. The prospect of a financial return from fallows has raised enormous interest in the use of fallows (Rhodes grass, Digitgrass) for management of the burrowing nematode.

3.5. Future Prospects

One important fact to consider is that the centre of origin of the burrowing nematode is undoubtedly located in the Pacific Rim islands and that the nematode has already been reported from many primary crops other than bananas. This information is crucial in terms of nematode management and future prospects.

In most countries of Asia and Oceania, growers are not aware of the prime importance of the quality of planting material. Suckers are mostly collected from old banana fields without knowing their disease status. Some nematode species can cause extensive root damage on native banana. In general, infested plants exhibit stunted growth, premature defoliation and carry small bunches and fruits. In addition, nematodes can cause decay and death of the proximal parts of the roots and the plants are prone to toppling over, specially when bearing bunches or during windy weather, because of inadequate anchorage. There is definitely a need for the provision of pest-free banana seeds from local extension or research services to ensure that all material used for planting by the farmer is free of nematodes. Everywhere there is also an increase in growers wanting to evaluate new varieties to explore potential new markets. In Australia, the banana industry faces a changing consumer focus with more emphasis on environmental protection and sustainability while pressure from pests and diseases still increases. Current and future research into pests and diseases, as well as industry development, all rely on the use of disease-free banana varieties.

4. NEMATODES ON BANANAS IN AFRICA

4.1. The Nematode Problem

The first evidence of *Musa* on the African continent comes from the discovery of ancient banana phytoliths, distinctive microscopic silica bodies that accumulate in plant cells. According to new phytolith evidence from Uganda, it appears that humans may have brought bananas to eastern Africa during the fourth millennium BC (Lejju et al., 2006). Now, it is commonly assumed that not only Arab traders but also traders from India and from the Indonesian peninsula brought diverse banana clones to the east coast of Africa and Madagascar and then across the continent to the west coast (Simmonds, 1966).

Nowadays, and after this early introduction on the African continent, approximately one-third of the total world production of bananas (98 % of non-export bananas, 29.3 M tons in 2004) is produced in sub-Saharan Africa (Lescot, 2004). These bananas, particularly important in the humid forest and mid-altitude regions, are produced mostly for subsistence purposes by smallholder farmers *i)* under systems of shifting cultivation in West and Central Africa, *ii)* in permanent farming systems in East Africa where they are often grown in association with coffee or cocoa tree crops or *iii)* everywhere as backyard/garden crops. Generally, these production systems are characterized by no or very low inputs.

While less diverse than in Asia, a relative range of genetic diversity of bananas is observed in Africa, with different types specifically adapted to different sub-regions. In West and Central Africa, cultivars of the plantain subgroup *Musa* AAB (False Horn, French Horn) predominate in the humid lowlands while in East Africa, endemic highland bananas (*Musa* AAA) and diverse brewing cultivars (*Musa* ABB) predominate (Table 4).

Over recent years, banana yield and plantation longevity have been gradually declining in sub-Saharan Africa. Many pests, diseases and abiotic constraints (declining soil fertility, high soil acidity) were observed on bananas in Africa and not only affected production but also led to an increased frequency of land clearing. Currently, among the diseases, one of the major constraints to banana production is black leaf streak (or black Sigatoka) caused by the fungus *Mycosphaerella fijiensis*. All traditional banana cultivars of West and Central Africa are very susceptible and this particular disease causes severe leaf necrosis, increasing gradually with the age of the plantation, leading to 33-76% yield losses (Carlier et al., 2000). Major pests include the banana weevil *Cosmopolites sordidus* and nematodes: *H. multicinctus*, *Meloidogyne* spp., *R. similis*, *P. goodeyi* and *P. coffeae*. These species affect the root system functionality at two levels: anchorage and ability to take up and transport water and nutrients.

The first studies dealing with the nematode associated with bananas in sub-Saharan Africa were very scarce and preliminary (Luc & de Guiran, 1960; Luc & Vilardebo, 1961). Since then, several extensive surveys provided reference data on species occurrence and densities for the different countries (Speijer & Fogain, 1999).

4.2. Nematode Species Occurrence

In sub-Saharan Africa, the most commonly occurring nematode species on bananas is *H. multicinctus*, which is found in 70-100% of samples (Table 4), while declining at altitudes above 1500 meters above sea level in East Africa (Speijer & Fogain, 1999). As already mentioned for Cavendish bananas, this nematode species is always found in mixed populations, often with root-knot nematodes, and its abundance depends primarily on the presence and abundance of other nematode species, particularly the burrowing and lesion nematode species.

Whereas the geographical distribution of *R. similis* follows closely the distribution of dessert bananas cultivated for export (e.g. Cavendish), the distribution of this species on other banana types in Africa differs widely from place to place. In West Africa, the occurrence of *R. similis* has increased on plantain types in recent decades from nil (Caveness, 1967; Fademi & Bayero, 1993) to 46 % in Nigeria (Speijer et al., 2001) whereas it remains absent in Gambia (Merny et al., 1974); from 2-9 % in the mid-west to 43-52 % in the south east of Ivory Coast (Adiko, 1988; Adiko & N'Guessan, 2001) and at least by 39 % in Cameroon (Bridge et al., 1995).

Table 4. Occurrence (%) of the different banana nematode species in Africa

Region	*Musa group*	*Hm*	*Rs*	*Pc*	*Pg*	*Msp*	*Hp*	*Reference*
Ivory Coast (mid-west)	AAB	99	9	2	0	98	1	Adiko, 1988
Ivory Coast (mid-west)	AAB	100	2	2	0	93	0	Adiko & N'Guessan, 2001
Ivory Coast (southeast)	AAB	99	52	7	0	93	3	Adiko, 1988
Ivory Coast (southeast)	AAB	96	43	35	0	90	19	Adiko & N'Guessan, 2001
Ghana	AAB	98	32	66	0	56	0	Speijer & Fogain, 1999
Cameroon	AAB	38	39	5	33	36	44	Bridge et al, 1995
Nigeria	AAB	100	0	68	0	+	+	Fademi & Bayero, 1993
Nigeria (southern)	AAB	100	46	49	0	68	64	Speijer et al., 2001
Kenya	EAHB	75	42	8	62	17	0	Seshu Reddy et al.,1997
Kenya	ABB	62	20	25	55	50	0	Seshu Reddy et al.,1997
Rwanda	EAHB	60	40	0	80	0	0	Bagabe et al., 1997
Rwanda	ABB	40	40	0	60	20	0	Bagabe et al., 1997
South Africa & Swaziland	AAA & ABB	93	9	3	0	94	0	Daneel et al., 2003
Uganda (central)	AAA & ABB	88	71	0	47	35	0	Speijer & de Waele, 2001
Uganda (central)	ABB	88	76	0	52	82	0	Speijer & de Waele, 2001
Uganda (southeast)	EAHB	83	54	8	96	54	12	Kashaija et al., 1994
Tanzania	EAHB	100	50	0	100	80	0	Speijer & Bosch, 1996
Zanzibar	mixed	100	75	68	0	42	0	Rajab et al., 1999

Hm: *Helicotylenchus multicinctus*; Rs: *Radopholus similis*; Pc: *Pratylenchus coffeae*; Pg: *Pratylenchus goodeyi*; Msp: *Meloidogyne* sp.; Hp: *Hoplolaimus pararobustus*; + = presence.

Radopholus similis is also present in Ghana, but its occurrence is mainly localised in the western region (Brentu et al., 2004). In East Africa, the occurrence of *R. similis*, absent from the region prior to the 1960s (Price, 2006), seems now to be greater, ranging from 42 to 76 % (Table 4) while declining rapidly at altitudes above 1400 meters above sea level. In South Africa, its occurrence is still fairly limited (9 %) in home garden bananas (Daneel et al., 2003).

The lesion nematode *P. coffeae* occurs widely throughout the tropics and is a significant pest of some primary crops (e.g. yams and tubers). The species is only found in pockets on bananas in Ivory Coast, Ghana, Nigeria, Kenya and Zanzibar (Speijer & Fogain, 1999). It was first reported on bananas from Ghana (Addoh, 1971) while being absent from plantains in Nigeria (Caveness, 1967). In 1988, it was reported in the Ivory Coast only near the Ghana border both on bananas (Fargette & Quénéhervé, 1988) and plantains (Adiko, 1988). There is no doubt that the occurrence of this species on bananas and plantains is now increasing as illustrated by the 35 % occurrence in the south-east of the Ivory Coast in 2001 (Table 4).

Pratylenchus goodeyi is regarded as a species indigenous to Africa (Table 4), where it is recognized as an important pest of highland bananas in East Africa (Gichure & Ondieki, 1977; Bridge, 1988a; Speijer & Fogain, 1999) and in Cameroon (Bridge et al., 1995). This nematode species is also a major pest of bananas in the Canary Islands (de Guiran & Vilardebo, 1962) and has been found in Egypt (Oteifa, 1962) and in Crete (Machon & Hunt, 1985). The distribution of *P. goodeyi* is closely linked to altitude and temperature, since *P. goodeyi* is rarely observed below 800 meters above sea level and its occurrence in western Africa is restricted to the highlands of Cameroon (Price & Bridge, 1995).

Meloidogyne spp. occur widely throughout the tropics on bananas and also are significant pests of numerous crops (Luc et al., 2005). In Africa, they mostly occur on banana roots together with other nematode species and are likely to be found in great numbers in absence (or limited density) of the burrowing or lesion nematodes (Table 3) due to competition phenomena (Quénéhervé, 1990).

Hoplolaimus pararobustus also shows a distribution in pockets (Table 4) with an considerable occurrence in Nigeria and Cameroon and an increasing occurrence in the south-eastern Ivory Coast, from 3 to 19 % (Adiko & N'Guessan, 2001). While scarcely present in 1961 in the Ivory Coast, the occurrence of this species was already over 80 % on dessert bananas in 1988, presumably after the introduction of infested Cavendish material from Cameroon (Fargette & Quénéhervé, 1988).

All these studies show that the nematode problem is changing rapidly, mainly with the increasing occurrence of the burrowing and lesion nematodes, in areas and on banana varieties formerly free of these pests (Price, 2006). During the last fifty years, the increasing occurrence of *R. similis* was mainly due to the dissemination and exchange of infested planting materials (e.g. dessert Cavendish bananas inter-planted with other banana varieties), locally facilitated by the improved means of communication (roads and trucks) among the different banana production areas and between countries, during the establishment of new commercial plantations from place to place with infested planting materials (Sarah, 1989; Marin et al., 1998).

The situation is presumably similar with the increasing distribution of *P. coffeae* and *H. pararobustus* through infested planting materials. In fact, all these nematode species can infest banana corms deeply and in abundance, reaching a depth of more than 7 cm (Quénéhervé & Cadet, 1985a). Therefore they are totally protected during transport and against some of the primary nematode management procedures, such as root removal and surface paring of corms.

4.3. Importance and Potential Damage

It is always difficult to partition the damage according to species or species mixtures. In the 1980s, only *H. multicinctus* and *Meloidogyne* spp. were considered important pests of plantains in Nigeria (Caveness & Badra, 1980) and Ivory Coast (Adiko, 1988), and yield increases ranging from 61 to 98 % were observed after nematicide treatments of established plantains infested with these nematode species (Caveness & Badra, 1980; Badra & Caveness, 1983). In East Africa, production losses ranging from 15 to 50 % have been associated with *R. similis* and *H. multicinctus* attack on East African Highland bananas (EAHB)(Speijer et al., 1999; Speijer & De Waele, 2001). Results of path analysis showed that *H. multicinctus* was also a severe constraint, second in importance to *R. similis* in terms of root death and necrosis (Ssango et al., 2004). Recently, its own importance has been assessed in micro-plot evaluations and greenhouse experiments and indicates low (26 %) to zero effect on vegetative growth and yield loss (Brentu et al., 2004; Adiko, 2005). Nevertheless, these experimental results, although consistent with some field observations and trials (Barekye et al., 2000), need to be confirmed with different *Musa* cultivars and in different experimental conditions.

The pathogenicity of *H. pararobustus*, often present in low densities in the roots, to either plantain or banana has not yet been observed (Price, 1994b). Plant toppling can be considered as the major loss factor for banana production, and is mostly associated with the presence of the burrowing nematode *R. similis* or the lesion nematodes *P. coffeae* and *P. goodeyi*.

In Ghana, a total production loss of 70 % (associated toppling incidence 60 %) was observed after inoculation of plantains with the lesion nematode *P. coffeae* (Brentu et al., 2004). Plantain yield losses ranging from 25-64 % for the first crop to 50-90 % for the successive crop cycles were reported from Ghana (Udzu, in: Coyne et al., 2005). In a field experiment in Cameroon, the total production losses in the first and second cycles were 60 and 51 % respectively (associated toppling incidence of 18 and 53 %) (Fogain, 2000).

In Tanzania, *P. goodeyi* has been associated with plant toppling of highland bananas (Bridge, 1988a) and has been implicated as a cause of the cultivar shifts from indigenous highland bananas to newly introduced 'Pisang awak' and 'Gros Michel' cultivars (Speijer & Bosch, 1996). As mentioned by Speijer et al. (1999), when plant toppling occurred on a mat, the chance for this mat to produce a harvestable bunch in the following cycle is highly reduced, thus diminishing the plantation longevity.

4.4. Current Nematode Management

Integrated pest management (IPM) strategies offer the most suitable and efficient means by which small-scale farmers can control pest and disease attack. IPM strategies are also environment friendly, and should provide a highly desirable alternative to pesticide application in highly populated areas. In general, three main types of nematode management are envisaged. These include prevention with the use of clean planting material, cultural control with a particular focus on soil fertility treatments, and host plant resistance. During recent years in Africa, the combined efforts of regional research networks such as IITA (International Institute for Tropical Agriculture) and CARBAP (Centre Africain de Recherche sur le Bananier et le Plantain) has led to the development and adoption of user-friendly techniques in terms of nematode management to the benefit of banana and plantain growers.

4.4.1. Clean Planting Material

Farmers depend on natural regeneration of plants for the supply of planting materials. However, poor soil fertility, combined with high nematode and weevil infestation, not only slow down this natural regeneration in numbers but also lead to the production of suckers of poor health and quality. The most sophisticated way to obtain nematode-free planting materials is by using plants micropagated in vitro. However this method will certainly be restricted, for a long time yet, to only certain banana clones and to high value crops, such as commercial bananas. Nevertheless, other methods of propagating banana plants have been improved during the last decade.

The use of in vivo seedbed techniques increases the rate of banana multiplication in the field, but it carries the risk of multiplying contaminated materials. In Cameroon, CARBAP has developed a new detached corm technique for in vivo mass multiplication easily usable by growers. This technique allows the activation of latent buds and the quick production of large quantities of healthy planting material, at least free of nematodes and black weevils, in soil-less culture conditions (Kwa, 2003). Thanks to CARBAP and IITA, this detached corm technique has been instrumental in the recent increase of banana production and hybrid dissemination process both in Cameroon and Nigeria (Tenkouano et al., 2006).

In the absence of nematode-free planting material, paring is certainly the first and easiest prophylactic measure to apply. Complete root removal followed by a severe paring to discard all the necrotic and discoloured areas of the corms should be done before any use of planted materials infested with either nematodes or black weevils. This sanitation method can be combined with sun exposure: the storage of peeled rhizome for 2 weeks prior to planting (Quénéhervé & Cadet, 1985b) can complete this elimination of surface-living nematodes. However, neither paring nor sun exposure will completely eliminate nematodes from the deepest infested layers of the corms, and these physical methods cannot be applied to small suckers in order to avoid loss of regrowth and vigor.

Other physical methods include the hot water treatment of planting materials. Mallamaire (1939) was the first to suggest immersing banana suckers in water at

65°C for 5 minutes to eliminate *R. similis*. The hot water treatment technique was then improved (Blake, 1961) and widely recommended (55° C for 25 minutes) to farmers (Colbran, 1967) in Australia and Central America with minor adjustments. Nevertheless, its application to commercial bananas in Africa was not considered to be as feasible and successful as treatment with nematicides (Melin & Vilardebo, 1973). If its application on commercial bananas seemed difficult and uneconomic, its application to other banana types was absolutely unrealistic and scarcely applied. However the technique has been drastically simplified recently in East and West Africa (immersion in boiling water, 30 seconds) using local materials to treat infested suckers and has led to significant improvements in yield (Tenkouano et al., 2006).

4.4.2. Cultivated Fallow and Alternate Cropping

Unlike the situation in Asia, the fact that *R. similis* was rarely found on other primary crops in large numbers outside banana roots suggests that some management strategies (e.g. crop rotation) should be tried for better control. However, these management strategies are still rarely adopted since available land is scarce and farmers are usually reluctant to grow other crops than banana. Many studies were conducted in Cameroon and West Africa: natural fallow followed by a 3-4 month groundnut crop was recommended (Sarah, 1989) but only if the natural fallow lasted for a long time. As a substitute for natural fallow the spontaneous weed *Chromolaena odorata* was also used as a cover crop to eliminate *R. similis* from the soil before the replanting of dessert bananas (Sarah, 1989). In Cameroon, alternate cropping with maize and groundnut showed heavy infestation with *R. similis* in the following plantain crop (Price, 1994). Further studies with alternating crops demonstrated that maize and okra maintained a high level of nematode infestation and that groundnut and soya beans were similar to natural fallow, while only sweet potato and amaranth crops were able to suppress *R. similis* for almost 18 months. In terms of plantain yields over two experiments during two cycles and compared to a permanent plantain crop, this strategy of alternate cropping allowed significant yield increases of 57-96 % with sweet potato or amaranths, of 33-47 % with maize or okra, while increases were 38-42 % under natural fallow (Achard, personal communication). Similarly, sweet potato and Irish potato were also found to be non-hosts of *P. goodeyi* while intercropped with highland bananas (Price, 1994). A study conducted in Uganda with some plants reported as antagonistic or suppressive to nematodes (*Canavalia ensiformis*, *Mucuna pruriens*, *Tephrosia vogelli*) and cultivated as legume intercrops do not show significant advantages in banana production and no benefit in terms of nematode control or spatial distribution of banana roots and nematodes (Kashaija et al., 2004).

4.4.3. Mulching and Fertilisers

As mulching improves soil physical structure and therefore soil fertility, nematode damage to roots appears to restrict the growth potential of bananas. A study carried out in Nigeria suggested that mulching might mitigate the impact of nematodes on bananas only when applied to low fertility systems (McIntyre et al., 2000). In Uganda on highland bananas, the presence of nematodes reduced the average production by 32 % without mulch and by 30 % with mulch, but the average yield increase with mulch was over 65 % (Speijer et al., 1999). In a recent experiment in Nigeria, only the mulched plants, with a low level of infestation, reached harvest (71 % of dead plants in the highly infested non-mulched plants compared to only 1 % in the lightly infested mulched plant in the first cycle) (Coyne et al., 2005b).

Promising results were obtained with the use of *Tithonia diversifolia*, a shrub of the family Asteraceae, easily recognisable and widely distributed along farm boundaries in the humid and subhumid tropics of Central America and Africa. Its use as mulch led to a significant decrease in nematode damage and improved yield (Coyne et al., 2005a). All these studies confirmed the highly damaging nature of nematodes to banana production in Africa and the importance of the systematic evaluation of different organic mulches to improve banana plant vigor and longevity.

Recent studies indicate that nematode infestations need to be controlled before fertilizer use becomes profitable in terms of banana fruit yields (Smithson et al., 2001).

4.5. Future Prospect

Almost everywhere in Africa, except in permanent highland banana production systems, bananas are still established after a slash and burn preparation of the land and are seldom maintained for more than one cycle of production. Bananas are shifting from the status of a perennial crop to that of an annual crop. The reasons for abandoning the crop before it ratoons are numerous and comprise biotic (pests and diseases) and abiotic constraints (declining soil fertility, high soil acidity). During recent decades, population pressure in Africa has also led to a shortening of the fallow periods and increased the need for banana planting material, which is often the vector of pests and diseases. With this social and environmental situation, is prevention a lost cause?

In theory, IPM strategies offer the most suitable and efficient means by which small-scale farmers can control pest and disease attack. IPM strategies should also be environment friendly, and should provide a highly desirable alternative to pesticide application in heavily populated areas. Fortunately, the use of pesticides has never been a realistic nematode control method for smallholders in Africa. At present, nematode management includes the use of clean planting material, the establishment of nematode-free nurseries, crop rotation with a particular focus on soil fertility treatments and the development of host plant resistance. In recent years, the efforts of international research networks such as IITA and CARBAP has led to the development and adoption of user-friendly techniques to mitigate nematode damage and other problems, for the benefit of banana and plantain growers in Africa.

As we have seen, the recent spread of banana nematodes such as *R. similis* still increases and the efficiency of intra-continent domestic quarantine seems totally inadequate. Only the massive distribution of pest-free tissue culture plants can prevent the further spread of nematode species and allow the distribution of new dessert banana and plantain hybrids resistant to Black Sigatoka but also resistant or tolerant to nematodes and other pests and diseases. All these improvements will only be possible through the coordination of strong regional and international research networks.

5. NEMATODES ON BANANA IN AMERICA

5.1. The Nematode Problem

Marin et al. (1998) wrote an in-depth review of the different hypotheses for the dissemination of bananas in Latin America and the Caribbean. According to their findings and although no exact dates can be assigned to their introduction, it is likely that bananas were introduced early in the 1500s to the New World in Hispaniola island (now the Dominican Republic) by the Portuguese settlers via the Cap Verde and Canary Islands. From the sixteenth to the nineteenth centuries, European traders carried bananas all over tropical America. According to Simmonds (1960), the first bananas identified in the New World were the 'Silk Fig' (Figue Pomme, *Musa* AAB) and the 'French plantain' (*Musa* AAB), which were present in the West Indies in the seventeenth century. Some very important dessert banana clones such as 'Gros Michel' and cultivars of Cavendish (*Musa* AAA) were introduced directly from Asia into Martinique island in the nineteenth century and then distributed widely in Central America and the Caribbean islands, before being adopted by the banana trade (Simmonds, 1960).

After this late introduction into Latin America and the Caribbean, approximately one-third of the total world production of bananas (63.3 % of non-export bananas, 31.5 M tons in 2004) is now produced in the Americas in more than 33 countries (Lescot, 2004). The leading banana-producing countries are Brazil, Ecuador and Colombia with 6.5, 5.9 and 5.2 M tons, respectively, being produced in 2003.

Depending on the country, banana production is dominated by different banana types (Table 1). In Ecuador, 77 % is dessert bananas for export from the Cavendish subgroup. In Brazil, bananas are mostly cultivated for the local market (96.3%) and comprise different types such as the 'Silk Fig' (Figue Pomme, *Musa* AAB), the 'Figue sucrée' (*Musa* AA), and the 'Prata' (Pome, AAB). In Colombia, besides the Cavendish bananas for export, other bananas such as the cultivar "Gros Michel" and cultivars of the plantain subgroup *Musa* AAB (French Horn, False Horn) are particularly important in mid-altitude regions, where they are often grown in association with other crops such as coffee. Cooking bananas such as 'Bluggoe' in Cuba and 'Pelipita' (*Musa* ABB) are also very important in Latin America and the Caribbean.

As in other banana producing areas, many pests, diseases and abiotic constraints (declining soil fertility, high soil acidity) are observed on bananas in Latin America and the Caribbean. At present, besides the major world constraint to banana production, the 'black leaf streak' or 'black Sigatoka', other pests include the banana weevil *C. sordidus*, and the nematodes *H. multicinctus*, *Meloidogyne* spp., *R. similis* and *P. coffeae*.

5.2. The Nematode Species Occurrence

In Latin America and the Caribbean, nematode surveys on non-export bananas were very scarce and detailed studies on their relative abundance are lacking (Table 5). In areas free of *R. similis*, the main nematode species reported belong to the *Pratylenchus*, *Meloidogyne* and *Helicotylenchus* genera (Stover, 1972) beside other minor species (Roman, 1978).

As reported previously, the rapid spread of the burrowing nematode *R. similis* is closely linked to the dissemination of dessert bananas cultivated for export. Its introduction into Latin America is believed to have occurred with infested plants of 'Gros Michel', originally introduced into Martinique from Southeast Asia early in the 1800s and then transferred to Jamaica in about 1835. From Jamaica, this cultivar 'Gros Michel' and associated nematodes were exported to Cuba, Colombia (1892) and Surinam (1904) and then widely distributed in Central America and the Caribbean for the banana trade (Marin et al., 1998). Although infestations were present, the symptoms associated with *R. similis* on the banana roots and corms were not described until 1957 (Anonymous, 1957; Loos & Loos, 1960b).

Table 5. Occurrence (%) of banana nematode species on plantains in America.

Region	*Musa group*	*Hm*[*]	*Rs*	*Pc*	*Msp*	*Reference*
Venezuela	mixed	75	0	19	56	Torrealba, 1969
Venezuela	mixed	31	14	20	12	Yepez et al., 1972
Costa Rica	AAB	58	71	65	50	Araya & Cheves, 1997
French Guiana	AAB	29	14	50	64	Quénéhervé (not publ.)

[*]Hm: *Helicotylenchus multicinctus*; Rs: *Radopholus similis*; Pc: *Pratylenchus coffeae*; Msp: *Meloidogyne* sp.

As soon as banana and plantain production became business-related, the crops were mostly cultivated intensively in lowland areas and the presence of *R. similis* on plantains usually arose through the proximity of dessert bananas or through infested soil or planting materials. In Puerto Rico, Ayala and Roman (1963) found *R. similis* widely distributed on bananas and plantains. Loof (1964) first recorded the presence of *Radopholus* sp. in Venezuela on *Musa* sp. and Yepez et al. (1972) suggested its introduction into Venezuela occurred circa 1966, with infested planting material from Honduras.

In Honduras, *R. similis*, while very frequent on dessert banana, was reported to occur less frequently on plantains, unlike the lesion nematode *P. coffeae* which was the most important nematode species found associated with root and rhizome injury on plantains but also on coffee and citrus (Wehunt & Edwards, 1968; Pinochet & Ventura, 1980).

In southern Florida, the most prevalent species on bananas is the spiral nematode *H. multicinctus* while *R. similis* is very infrequent (McSorley, 1979).

The root lesion nematode *P. coffeae* was first observed on roots of plantains in Grenada and described by Cobb in 1919. As reported by Stover (1972), this species was frequently found associated with root injury in plantains (*Musa* AAB, ABB) in Central America. Histopathological studies by Pinochet (1978) showed that the destruction of the cortical parenchyma of plantain roots by *P. coffeae*, leading to large cavities eroded and detached from the vascular tissues, was similar to the effects described by Blake (1961; 1966) for *R. similis*, with typical cell discoloration followed by the dark necrotic lesions on the roots that appeared 6 days after nematode inoculations.

Besides this lesion nematode, root-knot nematodes are also encountered on bananas and plantains in Central America (Pinochet, 1977) and Brazil (Zem & Alves, 1978) in mixed populations. Cofcewicz et al. (2004b) in a study of different banana producing areas of Brazil (*Musa* AAA, AAB) provided an outline of the diversity of root-knot nematodes parasitizing *Musa*, showing the prevalence of *M. javanica* (61.7 %), *M. incognita* (32.2 %) and *M. arenaria* (4.3 %). A similar study conducted in the Caribbean indicated the prevalence of *M. arenaria* (61.9 %) followed by *M. incognita* (34.3 %) (Cofcewicz et al., 2005).

The spiral nematode, *H. multicinctus*, was first recorded as damaging to plantains in Cuba (Stoyanov, 1967). The nematode attacks and feeds on the outer cells of the root cortex and produces small necrotic lesions (Luc & Vilardebo, 1961).

The reniform nematode *R. reniformis* has also been reported to be pathogenic to plantains in Puerto Rico (Roman, 1978).

5.3. Importance and Damage Potential

Yield decline of plantains caused by the lesion nematode *P. coffeae* was first described from Cuba (Stoyanov, 1967) and Trinidad (Ogier & Merry, 1970). In Honduras, Stover (1972) observed a 455 % increase in uprooted plants of 'Horn plantain' (Musa AAB) in *R. similis*-infested plots and a 62 % increase in uprooted plants in *P. coffeae*-infested plots compared to nematode-free plots, with no effect on fruit weight in a three-year experiment. Depending on the presence of *R. similis* and on the soil fertility, the plantation longevity varied from more than 10 years to only 2-3 years in the Dominican Republic.

In the same conditions of poor soil fertility and with *P. coffeae*, plantation longevity of plantains rarely exceeds 2-3 years in French Guiana (Queneherve, unpublished).

The fungi associated with nematode lesions on plantains are the same as those found on dessert bananas (Pinochet & Stover, 1980). Conversely, bananas such as the cultivar 'Gros Michel' and cultivars of the plantain subgroup *Musa* AAB (French Horn, False Horn) growing in association with other crops such as coffee in the mid-altitude regions of Colombia, and highland bananas called 'Guineo' do not suffer from nematode problems (Grisales & Lescot, 1993; Price, 1999).

The pathogenicity of different *Meloidogyne* species was studied on different banana cultivars (triploids AAA-group, triploids AAB-group and tetraploid AAAB-group) in Brazil and it was found that all species partially affected plant growth and altered the concentration of macro- and micronutrients in leaves (Cofcewicz et al., 2004).

5.4. Nematode Management

In Latin America and the Caribbean, except on dessert bananas for export, very little research has been done on banana nematode management. When nematode control was practised, usual recommendations followed those already made for dessert bananas. Roman (1978) reviewed the different experiments with nematicides in Latin America and the Caribbean.

After chemical treatment, large yield improvements were observed in Jamaica with 119 % over one cycle (Hutton & Chung, 1973) and in Puerto Rico, with yield increases of 207-275 % over three years on plantains cv 'Maricongo' (*Musa* AAB) (Roman et al., 1977). Since that time, when chemicals were applied in commercial plantains, most changes simply concerned new chemicals, following those used on dessert bananas.

In the Dominican Republic, some field experiments were done on possible crops to rotate with plantains to control banana nematodes. These studies showed that *i)* the burrowing nematode *R. similis* was recovered from continuous plantings of beans and corn after 6 months, but not from sorghum, tobacco, cassava, Pangola grass, sugarcane or grapefruit, *ii)* the lesion nematode *Pratylenchus* sp. was suppressed under cassava and *iii) Meloidogyne* sp. was suppressed by Pangola grass (Smith & Thames, 1969).

In Brazil, Bringel and Silva (2000) showed the antagonistic properties of some rotation crops (*Crotalaria juncea*, *C. spectabilis*, *Mucuna nivea*, *M. atterima*) towards the spiral nematode *H. multicinctus*.

5.5. Future Prospects

America and the Caribbean, while now producing almost one third of the total world production of bananas, were the latest continent and islands where bananas and their associated nematodes were introduced. This could explain the relatively narrow host range of the burrowing nematode *R. similis* and spiral nematode *H. multicinctus* on primary crops other than bananas (Table 3). As a result, IPM strategies including the use of clean planting material, the establishment of nematode-free nurseries and appropriate rotation crops should be successful in the eradication of *R. similis*, as already observed in some former contaminated areas. On the other hand, the research on nematode resistance will have to focus on the lesion nematode

P. coffeae, as this nematode is already replacing *R. similis* in terms of damage and occurrence on non-export bananas in Latin America and the Caribbean.

6. FUTURE AND COMMON STRATEGIES

As illustrated above, except for some geographical areas, the current options for nematode control on dessert bananas for export are still quite limited to a better use of pesticides through practical improvements (e.g., chemical formulation and dosage, application procedure, decision of nematicide application after nematode and/or damage monitoring). On non-export bananas, the range of options for nematode management is more directed towards prophylactic methods and regional improvements in cultural practices (e.g. crop rotation, fallowing) than on chemical treatments. Nevertheless in the future, nematode management for bananas should converge towards similar plant health measures and IPM options, such as the use of resistant or tolerant varieties, the distribution of clean plants obtained by tissue culture as well as the development of biological control methods in order to limit the use of pesticides.

6.1. Plant Health Measures

According to the International Plant Protection Convention (IPPC), phytosanitary measures include any legislation, regulation or official procedure whose purpose is to prevent the introduction and/or spread of plant pests and to be applied to regulated pests. Among the different nematode species encountered on banana, the burrowing nematode, *R. similis*, is qualified as a 'quarantine pest' in more than 55 countries, mostly because the occurrence of a physiological race of *R. similis* able to infest and damage citrus in Florida has prompted a worldwide ban of this nematode especially in citrus-growing countries (Hockland et al., 2006). In some countries, specific restrictions are imposed against other endoparasitic root lesion nematodes (*P. coffeae*, *P. goodeyi*). However, the dissemination of *R. similis* and other banana nematodes first occurred very early in Asia and has continued since. Beginning in the sixteenth century, early travellers, traders and more recently, research scientists, disseminated these nematodes with infested plant materials all over the world, such as in Asia (Khan, 1999), in Africa (Price, 2006) and America (Marin et al., 1998).

Radopholus similis is a polyphagous species that will feed and reproduce in the roots of more than 400 plant species in most of the tropical and subtropical areas of the world. As illustrated in table 2, this species has been found associated with many primary crops mostly in Asia and the Pacific. After its early discovery on banana in Fiji by Nathan A. Cobb (1893), *R. similis* was found associated with coffee and tea plants in Indonesia (Zimmerman, 1898). Its presence as a potential pest of tea has been confirmed since then in Sri Lanka, India, China, Zimbabwe and South Africa (Gnanapragasam & Mohotti, 2005). In the Pacific, *R. similis* was also observed in Fiji on sugarcane (Cobb, 1915), on yam and ginger (Butler & Vilsoni, 1975), on taro (Kirby et al., 1980) and on swamp taro in Guam (Jackson, 1987). Currently the presence of *R. similis*, causing dry rot of yam tubers, seems only restricted to Papua New Guinea, New Caledonia, Fiji and Solomon Islands (Bridge et al., 2005). According to Williams (1969), *R. similis* was recorded from sugarcane from Hawaii, Louisiana and Florida (USA), Cuba, India, the Philippines and Australia. However, this species is no longer considered as a pest of sugarcane (Cadet & Spaull, 2005) although some records of *R. similis* on sugarcane could suggest the existence of a biotype or 'sugarcane race'. Similar observations were made by Godfrey (1931) with records of a 'citrus race' of *R. similis* able to attack pineapple in Florida, while this species is also not considered as a pest of pineapple worldwide (Sipes et al., 2005).

The first and major evidence of plant damage was observed when *R. similis* was responsible of the loss of 22 million pepper vines within 20 years in Bangka Island, Indonesia, due to the 'pepper yellows disease' (Van der Vecht, 1950), a severe disease of pepper (*Piper nigrum*) subsequently reported from Malaysia, Thailand, India and Sri Lanka (Koshy et al., 2005). In India but also in some other countries of Asia, many plant species, used as live standards for pepper vines (coconut, arecanut) or intercropped with pepper (banana, ginger, turmeric, betel vine, food legume) were also recognised as primary hosts for *R. similis* (Table 2). This fact, in addition to its dissemination through infested banana plants (Khan, 1999), is certainly of major importance in the widespread dissemination of *R. similis* in India and Southeast Asia.

At the same time in Florida, Suit and DuCharme (1953) identified *R. similis* as the causal agent of the very severe "spreading decline of citrus" and differentiated this 'citrus race', able to parasitize banana from the distinct but more widespread 'banana race' for which citrus is not a host (DuCharme & Birchfield, 1956). On ornamentals, *R. similis* was first reported to occur on anthurium by Sher (1954) in Hawaii and is one of the major pests of *Anthurium andreanum*, characterised by root necrosis, stunting of plants and chlorosis. This important disease known as "anthurium decline" was mostly reported from Hawaii (Aragaki et al., 1984) and from the Caribbean (Bala & Hosein, 1996; Quénéhervé et al., 1997). The nematode is well known as a pest of foliage ornamentals belonging to the Araceae, Marantaceae and Zingiberaceae. Sixteen palms including coconuts are already reported as hosts of the burrowing nematode. Among them arecanut or betel nut (*Areca catechu*) growing in India and southeast Asia is highly infested with *R. similis*, particularly when intercropped with banana, black pepper, cardamon, coconut and cocoa (Griffith et al., 2005). Other hosts include weeds, acting either as transitional or primary hosts. All these records illustrate the importance of the quarantine regulations concerning not only *R. similis* but also other banana nematodes liable to become major pests on some other important crops.

In accordance with the principles of the IPPC, most of the countries around the world have developed their own plant health and quarantine regulations and now the international movement of soil and infested plants (e.g. banana planting materials, black pepper cuttings, anthurium cuttings) should be totally banned. Therefore, these basic principles of exclusion still seem always difficult to apply at the borders of many countries from Asia, Africa and America and there is no domestic quarantine to limit the dissemination of infested banana planting materials within some large countries (e.g. Brazil, Colombia, Ivory Coast, Nigeria, Uganda, India) or archipelagos (e.g. Indonesia, Polynesia). In order to avoid the new introduction or dissemination of banana nematodes or other pests, only pest-free tissue culture should be now authorized for transfer among and within countries. In parallel, prophylactic measures should be taken in the research stations to ensure the establishment of nematode-free nurseries, before any further distribution to farmers within the country.

6.2. The Search for Sources of Resistance to Nematodes

Due to increasing concern about environment contamination by pesticides, the search for both plant resistance and/or tolerance to plant-parasitic nematodes of bananas is now a major challenge, with many research teams involved. Currently, screening for nematode resistance is an ongoing process, particularly as newly-developed banana hybrids become available. The *Musa* germplasm screening, while formerly restricted to searching for resistance against *R. similis*, is also developed for some other nematode species (*P. coffeae*, *P. goodeyi*, *Meloidogyne* spp., *H. multicinctus*).

Historically, the first search for possible sources of resistance was conducted in the 1960s: the 'Banana Breeding Scheme' in Jamaica at Bodles produced a series of tetraploid banana hybrids bred specifically for desirable factors such as disease resistance or fruit characteristics (dessert banana). Among these, the cultivar 'Bodles Altafort' (Osborne, 1962) that was obtained from a cross between cultivars 'Gros Michel' and 'Pisang lilin' was promising against some diseases, but further results indicated different degrees of susceptibility to nematodes rather than true resistance (Gowen, 1976). Following this early work, the most significant contribution in this field was made in Honduras at the FHIA on the field screening of numerous cultivars and the first discovery of nematode resistance in the diploids *Musa* AA from the 'Pisang jary buaya' group (Wehunt et al., 1978).

In recent decades, different procedures and guidelines for the screening of *Musa* germplasm have been set up (Pinochet, 1988b; Sarah et al., 1992; Speijer & De Waele, 1997; Marin et al., 2000; Elsen et al., 2002; Quénéhervé et al., 2006). In parallel, several successive results of resistance screenings were published: in Asia (Davide & Marasigan, 1985; Van den Bergh et al., 2002; Elsen et al., 2002; Nguyet et al., 2002; Krishnamoorthy & Kumar, 2005), Europe (Pinochet et al., 1998), Latin America and the Caribbean (Binks & Gowen, 1996; Costa et al., 1998; Marin et al., 2000; Moens et *al.*, 2003; Viaene et al., 2003; Moens et al., 2005), Africa (Price, 1994b; Fogain & Gowen, 1997; Fogain, 1996; Stoffelen et al., 2000) and Australia (Stanton, 1999), in search for different sources of resistance to nematodes. As mentioned by Gowen et al. (2005), inconsistencies in the results may be due to the highly variable environmental conditions and biological materials (plants and nematodes).

Some authors (Mateille, 1990; Stanton, 1999) also indicated that results of screening studies done on young tissue culture plants might not be consistent with studies with older plants. It is reasonable to think that results of early resistance screenings can only be indicative of a tendency that should be confirmed through multi-site field experiments.

6.2.1. Resistance to the Burrowing Nematode R. similis

The first resistance source to *R. similis* was found in the diploid 'Pisang jari buaya', accession III-116 (Wehunt et al., 1978). In spite of many breeding difficulties (e.g. male sterility and low female fertility, difference between accessions from different geographical origins), this source of resistance was used to create the resistant diploid 'SH-3142' (Pinochet & Rowe, 1979), that led by successive crossing to the cultivar 'Goldfinger' (tetraploid SH-3481 or FHIA-01) but also to other interesting tetraploid cultivars (Pinochet, 1988; Rowe & Rosales, 1994).

From a practical and breeding standpoint, Pinochet and Rowe (1979) already mentioned that the synthetic diploid 'SH-3142' was not only more resistant than its parents but was also pollen fertile and produced several seeds per bunch. Following this work, some discrepancies were observed in the field on the level of resistance to *R. similis* of the tetraploid cultivars (Stanton, 1994; Binks & Gowen, 1996, Marin et al., 1998b). This fact, among other undesirable traits (e.g. consumer acceptance, susceptibility to *P. coffeae*), limited the commercial development of these cultivars and confirmed that the resistance, if any, will be certainly difficult to handle directly in a breeding programme (Pinochet, 1988a). Beside this first source of resistance, the cultivar 'Yangambi Km5' (*Musa* AAA group Ibota) was reported to be partially resistant to *R. similis* (Sarah et al., 1992; Fallas & Marban-Mendoza, 1994; Price, 1994b; Fogain & Gowen, 1998).

Hahn et al. (1996), indicated that cultivar 'Yangambi Km5', although not totally resistant to *R. similis*, was able to tolerate nematode parasitism. In fact, the damage caused by *R. similis* on the banana root system (% of root necrosis) was always lower on this cultivar than on susceptible cultivars, by 5 % to 85 % (Fogain & Gowen, 1997), or 10.5-19 % to 48-56 % (Dochez et al., 2006). Other sources of potential resistance to *R. similis* were found in two other diploids, the cultivars 'Paka' (*Musa* AA) and 'Kunnan' (*Musa* AB) (Collingborn & Gowen, 1997). In a recent study, Dochez et al. (2006) found ten new potential sources of resistance to *R. similis* within *Musa* diploids (AA) and triploids (AAA, ABB) from Papua New Guinea, Malaysia and the Philippines.

6.2.2. Resistance to the Lesion Nematode Pratylenchus *spp.*

Besides its resistance to *R. similis*, the cultivar 'Yangambi Km5' (*Musa* AAA group Ibota) was also reported to be partially resistant to *P. goodeyi* (Fogain & Gowen, 1998; Pinochet et al., 1998) and to *P. coffeae* (Collingborn & Gowen, 1998). This is a remarkable feature since most frequently, resistance is found to be effective to a single nematode species. Unfortunately, due to some breeding incompatibilities, this cultivar is not really used in banana breeding programs. Similarly, cultivars 'Paka' and 'Kunnan' were also found resistant to *P. coffeae* (Collingborn & Gowen, 1997). In field trials conducted in Cameroon, Price (1994b) reported some triploid cultivars

'Banane Cochon' (AAA), 'Gros Michel' (AAA) and 'Big Ebanga' (AAB) to be partially resistant while most of the plantain cultivars (AAB and ABB) and cultivar 'Pisang jari buaya' were equally susceptible to *P. goodeyi.*

6.2.3. Resistance to the Root-Knot Nematode Meloidogyne *spp.*

Very little information is available on the existence of sources of resistance or tolerance to root-knot nematodes in *Musa*, although some screening studies were carried out in Indonesia (Hadisoeganda, 1994), Brazil, (Costa et al., 1998), the Canary Islands (Pinochet et al., 1998) and Vietnam (Stoffelen et al., 2000a; 2000b; Van den Bergh et al., 2002).

In the Philippines, Davide and Marasigan (1985) found nine cultivars assigned as 'resistant' to *M. incognita*. However, although these cultivars showed gall indices and root nematode densities lower than the control, their real host status needs to be confirmed using standardized procedures (Speijer & De Waele, 1997; Quénéhervé et al., 2006).

6.2.4. Resistance to the Spiral Nematode Helicotylenchus multicinctus

In Costa Rica, Moens et al., (2005) were the first to assess the host response of *Musa* cultivars to *H. multicinctus* and found a resistance response in the cultivar 'Tjau lagada'.

6.3. Tolerance to Nematodes

The existence of 'tolerance' or varietal susceptibility of cultivated bananas to nematodes was first observed by the response of the cv. 'Gros Michel' which apparently was less susceptible to nematode damage than Cavendish cultivars (Leach, 1958; Stover, 1972). In 1978, Wehunt et al., confirmed these observations and also showed that moderate susceptibility to high level of resistance to *R. similis* might be found in wild diploids and diploid cultivars. Gowen (1976) was the first to mention that tetraploid cultivars bred in Jamaica exhibited better vigor and were less susceptible to nematodes than others.

Many workers (Price, 1994; Fogain, 1996) observed a higher susceptibility of plantains to nematodes than Cavendish cultivars. Swennen et al., (1986) related this higher susceptibility to the quality of their root systems, which are less vigorous than those of *Musa* AAA. Similar observations were made with FHIA tetraploids (Rowe & Rosales, 1994). The results of the numerous nematode screenings (see above) among *Musa* germplasm definitely confirmed this huge varietal susceptibility.

The rapid development of the meristem culture technique has revolutionized banana propagation (Israeli et al., 1995) and commercial tissue culture laboratories (France, Israel, Republic of South Africa, Taiwan, Costa Rica, etc...) produce millions of banana plantlets throughout the world. Since the work of Champion (1963) and Stover (1972) it is widely accepted that varieties from the Cavendish subgroup were highly and equally susceptible to nematodes.

At present several different clones of Cavendish are widely distributed, sometimes under the same name ('Grande Naine', 'William', 'Poyo', 'Americani', 'Dwarf Cavendish'), while exhibiting slight phenotypic differences depending on their geographic origin but without any data on pest susceptibility.

In 1990, scientists from CIRAD, while working in collaboration with a tissue culture laboratory, selected within the 'Grande Naine' bananas from Martinique, Guadeloupe, but also from Africa (Ivory Coast, Cameroon) some peculiar plants based on several interesting criteria locally defined (dwarfism, hardiness, vigor, drought or cold tolerance, productivity, bunch conformation, finger size etc.). As a result, several clones of 'Grande Naine' were selected and evaluated against nematodes in greenhouse and field experiments.

A natural mutant of 'Grande Naine' cv. 'MA13' demonstrated significantly lower susceptibilities to *R. similis* and *P. coffeae* in addition to its good horticultural characteristics (Quénéhervé, unpublished).

As most of the banana-producing countries are now trying to reduce their use of pesticides for the sake of environmental and human safety, it is important to select the best clones to cultivate in terms of resistance to pests and parasites. As illustrated in nematode population dynamics studies (Quénéhervé, 1993a) and in modelling studies (Tixier et al., 2005), any plant or environmental characteristic which reduces the multiplication rate of nematodes, is a step forwards a global reduction use of nematicides.

6.4. New Synthetic Banana Hybrids and Their Response to Nematodes

In most banana growing parts of the world, different *Musa* breeding programs are developed to create new synthetic hybrids primarily resistant to Sigatoka leaf spot diseases, such as in Africa (IITA; CARBAP), Latin America (FHIA; Embrapa), the Caribbean (CIRAD) and Asia (Tamil Nadu Agricuture University). As soon as these new hybrids are released, they are evaluated for their reaction to the burrowing nematode *R. similis* and other nematode species.

In Honduras, different bred genotypes were evaluated in pot tests for resistance and tolerance to *R. similis* (Viaene et al., 2003). These tests confirmed once again the resistance status of the synthetic hybrid FHIA-01 to *R. similis* and the resistance of the male parents (diploids 'SH-3142', SH-3362, SH-3648, SH-3723) and female parents (Calcutta 4, Prata Enana) used in the *Musa* improvement programme of FHIA. The same synthetic hybrid FHIA-01 was already reported as tolerant to *P. goodeyi* (Pinochet et al., 1998). This hybrid has been already distributed for experiments in many countries (Honduras, Costa Rica, Cuba, Ecuador, Brasil, Nigeria, Australia, South Africa, Taiwan, Canary Islands).

In Uganda, IITA is developing a breeding program for the production of new hybrids of highland bananas (EAHB), and those which have the resistant 'Pisang jari buya' cultivar in their pedigree are very promising in terms of resistance to *R. similis* (Dochez et al., 2000).

In Nigeria, the hybrid 'Pita-14' is currently distributed to farmers (Coyne et al., 2005a).

In India, at the Tamil Nadu Agricultural University, recent results (Krishnamoorthy & Kumar, 2005) indicate the breeding of some resistant and tolerant diploids to *R. similis* that could be used in their future breeding programs.

In the Caribbean, CIRAD is currently releasing new synthetic hybrids of dessert bananas (*Musa* AAA), resistant to Sigatoka leaf spot diseases and highly tolerant to nematodes (*R. similis* and *P. coffeae*). All these synthetic hybrids, originally bred for the resistance to Sigatoka disease from a common pool of resistant parents, often share also a better tolerance to nematode than current cultivars.

Unfortunately, banana streak disease, caused by several distinct badnavirus species, has severely hindered international *Musa* breeding programmes, as new hybrids were frequently infected with this virus, curtailing any further exploitation. This infection is thought to arise from viral DNA integrated into the nuclear genome of *Musa balbisiana* (B genome) of the wild species, contributing to many of the cultivars currently grown (Geering et al., 2005).

6.5. Resistance Mechanisms and Plant Defence

It is more and more recognized that plant defence responses to plant-parasitic nematodes have the potential to become part of the management strategies to increase plant productivity and that both constitutive and induced defence mechanisms can be observed in plants (Giebel, 1982; Veech, 1982). Within the plant metabolism, the phenylpropanoid pathway that produces different phenolic compounds (e.g. tannins, anthocyans) is involved in the plant's defence against abiotic and biotic factors (Treutter, 2006).

On bananas, Mateille (1994) first suggested that the compatibility to *R. similis* of a susceptible cultivar 'Poyo' was due to a high polyphenol oxidase (PPO) activity, while the relative incompatibility of a less susceptible cultivar 'Gros Michel' was due to a higher peroxidase activity. He also found higher numbers of cells with phenolic contents in the cv 'Gros Michel' compared to the susceptible cv. 'Poyo' using histochemical studies (Mateille, 1994b). In subsequent studies, these results (e.g. callose and phenol accumulation) were confirmed on susceptible 'Poyo' and partially resistant 'Yangambi km5' cultivars (Valette et al., 1997).

On the other hand, the resistant cv 'Pisang jari buaya', in which fewer preformed phenolic cells were found but larger numbers of cells with lignified walls, suggested a different resistance mechanism (Fogain & Gowen, 1996). The production of phenyphenalenone phytoalexins (Binks et al., 1997) and of proanthocyanidins (Collingborn et al., 2000) after infection with nematodes or fungi (Luis, 1998) were also reported. Wuyts et al., (2003) tried to elucidate the biochemical basis for nematode resistance in bananas and concluded that after nematode infection, *i)* constitutive lignification and induced cell wall strengthening were similar in susceptible and resistant cultivars, *ii)* cells containing flavonols increased in the central cylinders of resistant cultivars.

In a recent study, Wuyts et al., (2006) confirmed through in vitro bioassays the effect of several phenylpropanoid compounds on chemotaxis, motility and hatching of migratory and sedentary nematode species. In this study, several flavonols and lignin-related compounds were found repellent to *R. similis*.

At present the mechanisms by which constitutive or induced root cell compounds are active against plant-parasitic nematodes are still largely unknown. However this research should benefit from the discovery of new sources of resistance in the *Musa* germplasm, in order to find biochemical links among resistance mechanisms against nematodes.

The use of elicitors of plant defence leading to systemic acquired resistance (SAR) is in its infancy (Sticher et al., 1997) but promising results should also arise in the coming years, as already shown in pineapple (Chinnasri et al., 2006).

6.6. Transgenic Resistance

Until recently, the only way to obtain nematode-tolerant or resistant cultivars was through conventional plant breeding, while the prospects for genetically engineered nematode-resistant banana cultivars were already understood (De Waele et al., 1994). A decade later, Atkinson et al. (2004) successfully transformed Cavendish bananas using *Agrobacterium tumefaciens* in order to express a protein engineered rice cystatin (OcI deltaD86) of value for control of plant parasitic nematodes. When ingested by nematodes, this protein, a cystein proteinase inhibitor, impairs digestion of dietary protein and then reduces the multiplication of nematodes. That was already demonstrated on sedentary endoparasites such as *M. incognita*, *Globodera pallida*, *Heterodera schachtii* and *R. reniformis* (Urwin et al., 1997; 2000; 2001).

This first work on transformed Cavendish bananas showed that eight of 115 lines were able to reduce *R. similis* multiplication and expressed detectable levels of cystatin in their roots, with one of these promising lines providing a resistance level of 70 ± 10 % (Atkinson et al., 2004).

While still controversial among banana consumers, this type of partial resistance, induced through transgenic transformation, will certainly be deployed in the future alongside conventional banana breeding (Tripathi, 2003), due to its enormous potential. It is also noteworthy that the cystatin used in this work has already been donated on a royalty-free basis to resource poor small banana farmers in Africa (Atkinson et al., 2001).

As mentioned by the FAO (Anonymous, 2001), the most compelling reason for adopting genetic transformation in bananas is to reduce the use of fungicides and insecticides. It is for these constraints that genetic constructs are already recognized and attempts at their incorporation in Cavendish (and other bananas) are advanced but protected under commercial secrecy agreements.

6.7. Biological Control

Biological control of plant-parasitic nematodes has long been considered as an alternative to chemicals, especially because of the environmental and health concerns associated with these chemicals. Plant-parasitic nematodes have many natural enemies in the soil and early research on biological control focused mainly on microorganisms which are predacious (e.g. trapping fungi) and parasitic (e.g. *Pasteuria penetrans*) towards sedentary endoparasites (e.g., *Meloidogyne* spp.). Among all groups of plant-parasitic nematodes, migratory endoparasites such as *R. similis* and *Pratylenchus* spp. are the most difficult to control with natural enemies (Stirling, 1991).

As an alternative to chemicals, these biocontrol agents were first applied as soil treatments but the industrial attempts were all unsuccessful. Ongoing research is now directed to biocontrol agents able to induce *in planta* suppressiveness (Sikora & Pocasangre, 2005). These biocontrol agents should be able to colonize permanently either the rhizosphere or the roots and to induce direct or indirect nematode control or to promote the natural plant defence against plant-parasitic nematodes. The currently potential biocontrol agents include parasitic fungi, rhizobacteria, mycorrhizae and endophytic fungi.

6.7.1. Soil Treatment with Antagonistic Microorganisms

An isolate of a parasitic fungus *Paecilomyces lilacinus* (Pl251) originating from the Philippines was the first to be developed commercially (Tandigan & Davide, 1986; Davide, 1988) and used against banana nematodes. This same strain of parasitic fungi (Pl251) is now sold in many countries under several trade names as water dispersible granules made up of 10^{10} viable spores of *P. lilacinus* per gram, but published data on its long-term efficacy on banana nematodes under field conditions are still lacking.

Recent experiments conducted in Martinique, in fields heavily infested with *R. similis*, *P. coffeae* and *M. arenaria*, failed to show any effect either on nematode populations or banana yields (Chabrier, personal communication). On the other hand in Cuba, in a recently established banana plantation with low initial nematode populations, the preventative use of *P. lilacinus* on tissue culture led to good nematode control and increased the yield by 25 % (Fernandez et al., 2005).

In Cuba, the application on banana fields on a large scale of a particular strain of *Bacillus thuringiensis* (Bt var. Kurstaki, strain LBT-3) gave an average nematode reduction of 87 % two months after treatments (Fernandez et al., 2005). The trapping fungi *Arthrobotrys* sp. have also been found promising on plantain in a laboratory experiment (Lopez et al., 2000). Under controlled conditions the use of the strain of *Corynebacterium paurometabolum* (C-924) led to 85 % *R. similis* reduction and in the field, yields of treated plants were significantly higher than those of the control plants, with increases of 106 % for the bacteria and 66 % for the nematicide treatment (Fernandez et *al*, 2005).

6.7.2. Induction of In Planta *Suppressiveness*

Among the rhizobacteria, the fluorescent *Pseudomonas* spp. constitute a major group, certain strains of which have been demonstrated to act positively on plants either by promoting their growth or by inhibiting root pathogens (Kloepper et al.,

1980). An experimental study on bananas showed promising results in terms of reduction of root invasion and repulsion of *R. similis* (Aalten et al., 1998).

Arbuscular mycorrhizal fungi (AMF), which are obligate symbionts, increase the plant's capacity to take up water and mineral nutrients (e.g. soluble phosphates) from the soil, especially under poor fertility conditions (Jaizme-Vega, 1999). The inoculation of banana plants with AMF has shown positive growth responses in the early vegetative stage (Declerck et al., 1994; Rodriguez-Romero et al., 2005). The studies on the interaction with plant-parasitic nematodes showed both a suppressive effect in the nematode population build-up and nematode damage in the presence of AMF on bananas (Umesh et al., 1988; Jaizme-Vega & Pinochet, 1997; Jaizme-Vega et al., 1997; Pinochet et al., 1997, Fogain & Njifenjou, 2003). From the different studies, it is clear that if the migratory nematodes can be harmed by the presence of AMF, the development of AMF can also be harmed by migratory nematodes (Elsen et al., 2003).

Since both plant-parasitic nematodes and AMF colonize the root tissues, the competition for food resources should be considered, either directly due to root necrosis or indirectly due to structural and physiological root alteration. The possibility of in-vitro mass production of AMF (Declerck et al., 1996) may allow massive inoculation of young plantlets in nurseries.

Most plants harbour endophytic fungi (e.g. *Fusarium* and *Trichoderma* spp.) that live part of their life cycle inside the plant, without producing disease symptoms, and can even develop mutualistic relationships with the plant acting as antagonists to various pests and diseases (Sikora, 1992).

Among the naturally occurring avirulent endophytic fungi on bananas, avirulent strains of *F. oxysporum* are the most promising and many studies have shown nematode control through induced systemic resistance (Vu et al., 2006) in greenhouse trials in Africa (Dubois et al., 2004; Paparu et al., 2006) and Latin America (Pocasangre et al., 2000; Zum Felde et al., 2004). Secondary metabolites produced by these *F. oxysporum* strains were strongly inhibitory to the movement and hatching of *R. similis* in a recent study (Athman et al., 2006). The use of these avirulent fungal endophytes is very promising, especially if these endophytes are able to persist over cropping cycles after inoculation.

More data on parameters associated with the use of these new biocontrol agents and their mode of action are necessary to understand the mechanisms underlying the incidence of these different microorganisms on the different nematode species on bananas. Nevertheless from a practical standpoint, due to the promotion of tissue culture derived plants, not only in the commercial banana industry but also for smallholders through regional banana networks, this new approach in nematode management should be easily applicable to any banana production systems.

At least, endophytes and/or AMF and/or rhizobacteria should be artificially inoculated into tissue culture plants to give a better start to the banana plantation and increase host tolerance to pests and diseases. However, these biological products will certainly have to follow the same biosafety and homologation procedures as chemical products.

6.8. Cultural Practice Improvements

Most of the listed cultural practices were already described above as regional strategies and they will only be briefly summarized below.

The priority is in the use of clean banana seeds. The revolution observed with the distribution of millions of dessert banana plantlets following the rapid development of the meristem culture technique (Israeli et al., 1995) is still limited to commercial dessert bananas. In parallel, new hybrids of non-export bananas are introduced into farmers' fields in on-farm demonstration plots by the different research institutes in Africa (FHIA, IITA, CARBAP) with funding from several development investors. The challenge is now to ensure permanent access to clean banana plantlets and new hybrids to farmers worldwide via public or non-governmental public extension service.

The use of rotation crops and fallowing should be encouraged whenever possible. Replanting on highly infested soils is worthless. The use of plants that are antagonistic or detrimental to the development of plant parasitic nematodes is currently gaining most interest from research institutes, especially in areas where these plants are readily available and accessible. However, even if these plants are inexpensive and provide a valuable nematode management option, their adoption and usefulness will mainly depend on their economic or agronomic value (e.g. vs soil fertility) and if the farmers can derive some benefit from their presence beyond nematode management.

Treatments with nematicides, as part of nematode management strategies in some cropping systems (e.g. commercial dessert bananas), needs to be applied more rationally. Their intensive use in the past led to different drawbacks, e.g. soil and water contamination, loss of efficacy through microbial biodegradation. They should only be applied as control means of last resort on the basis of nematode incidence (percentage of uprooted plants) and/or numbers of nematodes in roots, in an effort to minimize nematicide applications. In older banana fields with plants at various developmental stages, the treatment could be applied individually after harvest *i)* to improve efficacy by application at the ideal time and *ii)* to minimize the risk of leaching and acute pollution (Quénéhervé et al., 1991; Quénéhervé, 1993a). However, the adoption of this practice will mainly depend of the willingness of the banana companies, often reluctant to modify any cultural procedures involving workers.

Recently, the model SIMBA-NEM (Tixier et al., 2006) has been designed to simulate the population dynamics of *R. similis* and *P. coffeae* on *Musa* spp. This model, able to predict long-term nematode population size for a range of conditions, is already a very helpful tool for designing sustainable and more environment-friendly banana cropping systems (e. g. optimization of the effect of nematicide applications on commercial bananas).

7. CONCLUSION

Banana farmers, from subsistence farming to commercial production, are typically faced with a multitude of problems. Nematode problems on bananas are widespread and can severely affect crop productivity and longevity. The different approaches to nematode management (e.g. cultural practices, use of nematicide, plant resistance and biological control) have all their interests, depending on the banana cropping systems. Until recently, commercial banana growers producing fruits for the international banana trade relied almost exclusively on the regular application of nematicides as pre- or post-plant treatments in the planting holes or around the established plants. However, the golden age of chemical control with nematicides is definitely behind us for many well-understood reasons in terms of environmental security and human health. The hierarchy and range of management tactics are now widened and differ greatly between export and non-export banana and in the different parts of the world.

In Asia and Oceania, centres of origin of both *Musa* spp. and burrowing nematode *R. similis*, the huge potential of diversity among wild and cultivated bananas has yet to be explored in order to select cultivars that can be grown without nematode control and still yield enough to be economic despite nematode damage.

In Africa, America and the Caribbean, cultural practices that include pest avoidance through international and domestic quarantine should slow down the dissemination, not only of the burrowing nematode but also of the lesion nematodes.

In export banana, very soon the application of diverse cultural practices including systematic use of pest-free vitroplants, fallows, rotation crops and biological control (e.g. in planta suppressivness) should totally replace the chemical control of nematodes, to respond to the new requirements in terms of quality and safety of the international banana trade.

Definitely host resistance, which is an environment friendly management tactic that has much potential, needs to be more effectively used and the development of disease-resistant and high yielding banana hybrids should constitute the most significant scientific achievement of the near future. This is particularly true in Africa in terms of food security impact .

REFERENCES

Aalten, P. M., Vitour, D., Blanvillain, D., Gowen, S. R., & Sutra, L. (1998). Effect of rhizosphere fluorescent *Pseudomonas* strains on plant-parasitic nematodes *Radopholus similis* and *Meloidogyne* spp. *Letters in Applied Microbiology*, 27, 357-361.

Abdul Karim, S. (1994). Status of nematode problems affecting banana in Malaysia. In: Valmayor, R.V., Davide, R.G., Stanton, J.M., Treverrow, N.L., Rao, V.N. (Eds) *Banana Nematodes and Weevil Borers in Asia and the Pacific*. INIBAP/ASPNET, Los Baños, Philippines, pp. 74-78.

Addoh, P. G. (1971). The distribution and economic importance of plant parasitic nematodes in Ghana. *Ghana Journal of Agriculture Science*, 4, 21-32.

Adiko, A. (1988). Plant-parasitic nematodes associated with plantain, *Musa paradisiaca* (AAB) in the Ivory Coast. *Revue de Nématologie*, 11, 109-113.

Adiko, A. (2005). Pathogenicity of *Helicotylenchus multicinctus* on plantain, Musa (AAB). *Russian Journal of Nematology*, 13, 55-59.

Adiko, A., & N'Guessan, A.B. (2001). Evolution de la nématofaune du bananier plantain (Musa AAB) en Côte d'Ivoire. *InfoMusa*, 10(2), 26-27.

Anonymous (1957). Plant parasitic nematodes and their association with bananas. *United Fruit Company, Research Department Extension News Letter*, 4, 8-16.

Anonymous (2001). *Biotechnology and banana production*. FAO downlodable word document available: at http://www.fao.org/docrep/meeting/004/y1896e.htm#P37_6810.

Aragaki, M., Apt, W. J., Kunimoto, R. K., Ko, W. H., & Uchida, J. Y. (1984). Nature and control of *Anthurium* decline. *Plant Disease*, 68, 509-511.

Araya, M. (2002). Metodología utilizada en el laboratorio de nematología de CORBANA S.A. para la extracción de nematodos de las raíces de banano (*Musa* AAA) y plátano (*Musa* AAB). *CORBANA*, 28, 97-110.

Araya, M., & Cheves, A. (1997). Determinacion de los nematodos fitoparasitos del platano (*Musa* AAB, clon Falso Cuerno) en la zona atlantica de Costa Rica. *CORBANA*, 22(47), 27-33.

Araya, M., & De Waele, D. (2004). Spatial distribution of nematodes in three banana (*Musa* AAA) root parts considering two root thickness in three farm management systems. *Acta Oecologica*, 26, 137-148.

Araya, M., Vargas, A., & Cheves, A. (1999). Nematode distribution in roots of banana (*Musa* AAA cv Valery) in relation to plant height, distance from the pseudostem and soil depth. *Nematology*, 1, 711-716.

Ashby, S. F. (1915). Notes on diseases of cultivated crops observed 1913-1914 – Banana Diseases. *Bulletin Department of Agriculture, Jamaica*, 2, 316-317.

Athman, S., Dubois, T., Viljoen, A., Labuschagne, N., Coyne, D., Ragama, P., et al. (2006). *In vitro* antagonism of endophytic *Fusarium oxysporum* isolates against the burrowing nematode *Radopholus similis*. *Nematology*, 8, 627-636.

Atkinson, H. J., Green, J., Cowgill, S & Levesley, A. (2001). The case of genetically modified crops with a poverty focus. *Trends Biotechnol.*, 19, 91-96.

Atkinson, H. J., Grimwood, S., Johnston, K. & Green, J. (2004). Prototype demonstration of transgenic resistance to the nematode *Radopholus similis* conferred on banana by a cystatin. *Transgenic Research*, 13, 135-142.

Ayala, A., & Roman, J. (1963). Distribution and host range of the burrowing nematode in Puerto Rican soils. *J. Agric. Univ. P. Rico*, 47, 28-37.

Badra, T., & Caveness, F.E. (1983). Effects of dosage sequence on the efficacy of nonfumigant nematicides on plantain yields, and nematode seasonal fluctuations as influenced by rainfall. *Journal of Nematology*, 15, 496-502.

Bagabe, M., Speijer, P. R., Uwimpuhwe, B. & Gold, C. S. (1997). Potential nematode effect on planned *Musa* cultivar changes in Rwanda (Abstr.). *African Plant Protection*, 3(2), 107.

Bala, G., & Hosein, F. (1996). Plant parasitic nematodes associated with anthuriums and others tropical ornamentals. *Nematropica*, 26, 9-14.

Barekye, A., Kashaija, I. N., Tushemereirwe, W. K., & Adipala, E. (2000). Comparison of damage levels caused by *Radopholus similis* and *Helicotylenchus multicinctus* on bananas in Uganda. *Annals of Applied Biology*, 137, 273-278.

Beugnon, M., & Vilardebo, A. (1973). Les nématodes du bananier à Madagascar. *Fruits*, 28, 607-612.

Binks, R. H., & Gowen, S. R. (1996). Field evaluation of nematode infestation in *Musa* germplasm at FHIA (La Lima, Honduras). *InfoMusa*, 5, 15-17.

Binks, R. H., Greenham, J. R., Luis, J. G. & Gowen, S. R. (1997). A phytoalexin from roots of *Musa acuminata* var Pisang sipulu. *Phytochemistry*, 45, 47-49.

Blake, C. D. (1961). Root rot of bananas caused by *Radopholus similis* (Cobb) and its control in New South Wales. *Nematologica*, 6, 295-310.

Blake, C. D. (1966). The histological changes in banana roots caused by *Radopholus similis* and *Helicotylenchus multicinctus*. *Nematologica*, 12, 129-137.

Blake, C. D. (1969). Nematode parasites of banana and their control. In: Peachey, J. E. (Ed), *Nematodes of Tropical Crops*. Commonwealth Agricultural Bureaux, Slough, UK: 109-132

Blake, C. D. (1972). Nematode diseases of banana plantations. In: Webster, J.M. (Ed), *Economic Nematology*. Academic Press, New York, USA: 245-267

Boncato, A. A., & Davide, R. G. (1980). *Radopholus similis* on Cavendish banana in Davao del Norte: Host range and relative distribution and density. *Philippine Agriculturist*, 63,111-119.

Booth, C., & Stover, R. H. (1974). *Cylindrocarpon musae* sp. nov., commonly associated with burrowing nematode (*Radopholus similis*) lesions on banana. *Trans. Br. Mycol. Soc.* , 63, 503-507.

Brentu, C. F., Speijer, P. R., Green, K. R., Hemeng, B. M. S., De Waele, D., & Coyne, D. L. (2004). Micro-plot evaluation of the yield reduction potential of *Pratylenchus coffeae*, *Helicotylenchus multicinctus* and *Meloidogyne javanica* on plantain cv. Apantu-pa (*Musa* spp., AAB-group) in Ghana. *Nematology*, 6, 455-462.

Bridge, J. (1988a). Plant nematode pests of banana in East Africa with particular reference to Tanzania. In: INIBAP (Ed.), *Nematodes and the borer weevil in bananas: present status of research outlook.* Workshop Proceedings, Bujumbura, Burundi, December 7-11, 1987. INIBAP Montpellier, France: 35-39.

Bridge, J. (1988b). Plant parasitic nematode problems in the Pacific Islands. *Journal of Nematology*, 20, 173-183.

Bridge, J. (1993). Worldwide distribution of the major nematode parasites of bananas and plantains. In : Gold, C.S. and Gemmil, B. (Eds), *Biological and Integrated Control of Higland Banana and Plantain Pests and Diseases.* IITA, Cotonou, Benin: 85-198.

Bridge, J., & Gowen, S.R. (1993). Visual assessment of plant parasitic nematode and weevil damage on bananas and plantains. In : Gold, C.S. and Gemmil, B. (Eds), *Biological and Integrated Control of Higland Banana and Plantain Pests and Diseases.* IITA, Cotonou, Benin:147-154

Bridge, J., & Page, S. (1984). Plant nematode pests of crops in Papua New Guinea. *Journal of Plant Protection in the Tropics*, 1, 99-109.

Bridge, J., Coyne, D. L., & Kwoseh, C. H. (2005). Nematodes parasites of tropical root and tuber crops (Excluding Potatoes). In M. Luc, R. A. Sikora, and J. Bridge (Eds), *Plant Parasitic Nematodes in Subtropical and Tropical Agriculture*, 2nd Edition. CAB International, Wallingford, UK: 221-258

Bridge, J., Price, N. S., & Kofi, P. (1995). Plant parasitic nematodes of plantain and other crops in Cameroon, West Africa. *Fundam. Appl. Nematol.*, 18, 251-260.

Bringel, J. M. M., & Silva, G.S. (2000). Efeito antagonico de algumas especies de plantas a *Helicotylenchus multicinctus. Nematologia Brasileira*, 24, 179-181.

Broadley, R. A. (1979). A simple method for estimating banana root rot. *Australasian Plant Pathology*, 8, 24-25.

Brooks, A. N. (1954). Host range of the burrowing nematode internationally and in Florida. *Citrus Ind.*, 35(12), 7-8,14.

Brun, J., & Laville, E. (1965). Etude de la mycoflore du bananier Poyo.II, Côte d'Ivoire, Guadeloupe, Mali. *Fruits d'Outre Mer*, 20, 123-128.

Butler, L., & Vilsoni, F. (1975). Potential hosts of the burrowing nematode in Fiji. *Fiji Agricultural Journal*, 37, 38-39.

Cadet, P. & Spaull, V. W. (2005). Nematode parasites of sugarcane. In : Luc, M., Sikora, R.A., and Bridge, J. (Eds), *Plant Parasitic Nematodes in Subtropical and Tropical Agriculture*, 2nd Edition. CAB International, Wallingford, UK: 645-674

Carlier, J., De Waele, D., & Escalant, J. V. (2002). Global evaluation of *Musa* germplasm for resistance to Fusarium wilt, *Mycosphaerella* leaf spot diseases and nematodes. In : Vezina, A. & Picq, C. (Eds) *INIBAP Technical Guidelines 6*, Montpellier, France.

Carlier, J., Fouré, E., Gauhl, F., Jones, D. R., Lepoivre, P., Mourichon, X., et al. (2000). Fungal diseases of the foliage. In: Jones, D. R. (Ed.), *Diseases of Banana, Abaca and Ens*et. CAB International, Wallingford, UK: 37-141

Carvalho, J. C. (1959). O nematoide cavernicola e o seu aparecimiento em Sao Paulo. *O Biologico*, 26, 226-228.

Caveness, F. E. (1967). End of tour progress report on the nematology project, Ministry of Agriculture and National Resources, Western Region, Nigeria.

Caveness, F. E., & Badra, T. (1980). Control of *Helicotylenchus multicinctus* and *Meloidogyne javanica* in established plantain and nematode survival as influenced by rainfall. *Nematropica*, 10, 10-15.

Chabrier, C., & Quénéhervé, P. (2003). Control of the burrowing nematode (*Radopholus similis* Cobb) on banana: impact of the banana field destruction method on the efficiency of the following fallow. *Crop Protection*, 22, 121-127.

Chabrier, C., Hubervic, J., & Quénéhervé, P. (2002). Evaluation of fosthiazate (Nemathorin® 10G) for the control of nematodes in banana fields in Martinique. *Nematropica*, 32, 137-147.

Champion, J. (1963). Le bananier. Maisonneuve et Larose (Ed.), Paris, France.

Charles, J. S. K., & Venkitesan, T. (1984). New hosts of *Heterodera oryzicola* Rao & Jayaprakash, 1978, in Kerala. *Indian Journal of Nematology*, 14, 181-182.

Charles, J. S. K., & Venkitesan, T. (1993). Pathogenicity of *Heterodera oryzicola* (Nemata: Tylenchina) towards banana (*Musa* AAB cv. Nendran). *Fundamental and Applied Nematology*, 16, 359-365.

Chau, N. N., Thanh, N. V., De Waele, D., & Geraert, E. (1997). Plant-parasitic nematodes associated with banana in Vietnam. *International Journal of Nematology*, 7, 122-126.

Chinnasri, B., Sipes, B. S., & Schmitt, D. P. (2006). Effects of inducers of systemic acquired resistance on reproduction of *Meloidogyne javanica* and *Rotylenchulus reniformis* in pineapple. *Journal of Nematology*, 38, 319-325.

Christie, J. R. (1959). Plant nematodes : their bionomics and control. Agricultural Experiment Stations, University of Florida, Gainesville, 256 pp.

Coates, P. L. (1971). Effects of treatments of banana corms with a systemic nematicide. *PANS*, 18, 165-170.

Cobb, N. A. (1893). Nematodes, mostly Australian and Fijian. Linnean Society of New South Wales. *MacLeay Memorial Volume*, 252-308.

Cobb, N. A. (1915). *Tylenchulus similis*, the cause of a root disease of sugarcane and banana. *Jour. Agric. Res.*, 4, 561-568.

Cofcewicz, E. T, Carneiro, R. M. D. G., Cordero, C. M. T., Quénéhervé, P., & Faria, J. L. C. (2004a). Reaçao de cultivares de bananeira a diferentes especies de nematoides das galhas. *Nematologia Brasileira*, 28, 11-22.

Cofcewicz, E. T., Carneiro, R. M. D. G., Castagnone-Sereno, P., & Quénéhervé, P . (2004b). Enzyme phenotype and genetic diversity of root-knot nematodes parasitizing *Musa* in Brazil. *Nematology*, 6, 85-95.

Cofcewicz, E. T., Carneiro, R. M. D. G., Randig, O, Chabrier, C., & Quénéhervé, P . (2005). Diversity of *Meloidogyne* spp. on *Musa* in Martinique, Guadeloupe and French Guiana. *Journal of Nematology*, 37, 313-322.

Colbran, R. C. (1967). Hot water tank for treatment of banana planting material. Queensland Department of Primary Industry, Division of Plant Industry, Brisbane, Australia, Advisory Leaflet 924.

Collingborn, F. M. B. & Gowen, S. R. (1997). Screening Indian cultivars for resistance to *Radopholus similis* and *Pratylenchus coffeae*. *InfoMusa*, 6, 3.

Collingborn, F. M. B., Gowen, S. R., & Mueller-Harvey, I. (2000). Investigation into the biochemical basis for nematode resistance in roots of three *Musa* cultivars in response to *Radopholus similis* infection. *Journal of Agricultural and Food Chemistry*, 48, 5297-5301.

Costa, D. D. C., Oliveira, S., & Alves, F. R. (1998). Reaçao de genotipos de bananeira (*Musa* spp.) a *Radopholus similis* e *Meloidogyne incognita*. *Nematologia Brasileira*, 22(2), 49-57.

Coyne, D., Kajumba, C., & Kagoda, F. (2005a). Nematode management at the International Institute of Tropical Agriculture in East Africa. In: Blomme, G., Gold, C. & Karamura, E. (Eds) *Farmer-participatory testing of banana integrated pest management options for sustainable banana production in Eastern Africa*. INIBAP, Montpellier, France:141-148

Coyne, D. L., Rotimi, O., Speijer, P., De Schutter, B., Dubois, T., Auwerkerken, A., et al. (2005b). Effects of nematode infection and mulching on the yield of plantain (*Musa* spp., AAB-group) ratoon crops and plantation longevity in southeastern Nigeria. *Nematology*, 7, 531-541.

Daneel, M., Dillen, N., Husselman, J., De Jager, K., & De Waele, D. (2003). Recensement des populations de nématodes dans les jardins de bananiers en Afrique du Sud et au Swaziland. *InfoMusa*, 12, 8-11.

Davide, R. G., & Gargantiel, F. T. (1974). Survey of nematodes associated with banana in the Philippines. *Phil. Phytopath.* 10, 1-2.

Davide, R. G., & Marasigan, L. Q. (1985). Yield loss assessment and evaluation of resistance of banana cultivars to the nematode *Radopholus similis* Thorne and *Meloidogyne incognita* Chitwood. *Phil. Agr.*, 63, 335-349.

Davide, R. G. (1980). Influence of cultivars, age, soil texture and pH on *Meloidogyne incognita* and *Radopholus similis* on banana. *Plant Dis.*, 64, 571-573.

Davide, R. G. (1988). Nematode problems affecting agriculture in the Philippines. *Journal of Nematology*, 20, 214-218.

Davide, R. G. (1992). Studies on nematodes affecting bananas in the Philippines. Philippine Agriculture and Resources Research Foundation, Inc. Los Baños, Laguna, Philippines: 175 pp.

Davide, R. G. (1994). Status of nematode and weevil borer problems affecting banana in the Philippines. In: Valmayor, R.V., Davide, R.G., Stanton, J.M., Treverrow, N.L., Rao, V.N. (Eds) *Banana Nematodes and Weevil Bores in Asia and the Pacific*. INIBAP/ASPNET, Los Baños, Philippines: 79-88

De Guiran, G., & Vilardebo, A. (1962). Le bananier aux Iles Canaries. IV. Les nématodes parasites. *Fruits*, 17, 263-277.

De Waele, D., Sagi, L., & Swennen, R. (1994). Prospects to engineer nematode resistance in banana. In: R. V. Valmayor, R. G. Davide, J. M. Stanton, N. L. Treverrow & V. N. Rao (Eds), *Banana Nematodes and Weevil Borers in Asia and the Pacific: Proceedings of a Conference workshop on Nematode and Weevil Borers Affecting Bananas in Asia and the Pacific*. INIBAP/ASPNET, Selangor, Malaysia: 204-216.

Decker, H., & Casamayor, R. (1966). Observaciones sobre incidiencia de nematodos parasitos del platano en Cuba. *Centro Boletin, UCLV, Villa Clara, Cuba*, 1, 7-32.

Declerck, S., Strullu, D. G., & Plenchette, C. (1996). *In vitro* mass production of the arbuscular mycorrhizal fungus *Glomus versiforme* associated with Ri T-DNA transformed carrot roots. *Mycological Research*, 100, 1237-1242.

Declerck, S., Devos, B., Delvaux, B., & Plenchette, C. (1994). Growth response of micropropagated plants to VAM inoculation. *Fruits*, 49, 103-109.

Delacroix, G. (1901). Pathologie végétale, maladies vermiculaires. I. Sur une maladie vermiculaire des bananiers en Egypte. II. Sur une maladie du poivrier (*Piper nigrum*) en Cochinchine. *Agriculture prat. Pays chauds*, 1, 672-680.

Dochez, C., Speijer, P., Hartman, J., Vuylsteke, D., & De Waele, D. (2000). Criblage d'hybrides de bananiers resistants à *Radopholus similis*. *InfoMusa*, 9(2), 3-4.

Dochez, C., Dusabe, J., Whyte, J., Tenkouano, A., Ortiz, R., & De Waele, D. (2006). New sources of resistance to *Radopholus similis* in *Musa* germplasm from Asia. *Australasian Plant Pathology*, 35, 481-485.

Dubois, T., Gold, C. S., Coyne, D., Paparu, P., Mukwaba, E., Athman, S., et al. (2004). Merging biotechnology with biological control: Banana Musa tissue culture plants enhanced by endophytic fungi. *Uganda Journal of Agricultural Sciences*, 9, 445-451.

DuCharme, E. P., & Birchfield, W. (1956). Physiologic races of the burrowing nematode. *Phytopathology*, 46, 615-616.

Edmunds, J. E. (1968). Nematodes associated with bananas in the Windward Islands. *Tropical Agriculture, Trinidad*, 45, 119-124.

Edwards, D. L. (1963). Dry fallow experiments. United Fruit, Department of Research, Annual Report: 75-77.

Edwards, D. L., & Wehunt, E. J. (1971). Host range of *Radopholus similis* from banana areas of Central America with indications of additional races. *Plant Disease Reporter*, 55, 415-418.

Elsen, A., Baimey, H., Swennen, R., & De Waele, D. (2003). Relative mycorrhizal dependency and mycorrhiza-nematode interaction in banana cultivars (*Musa* spp.) differing in nematode susceptibility. *Plant and Soil*, 256, 303-313.

Elsen, A., Stoffelen, R., Thi Tuyet, N., Baimey, H., Dupre de Boulois, H., & De Waele, D. (2002). In vitro screening for resistance to *Radopholus similis* in *Musa* spp. *Plant Science*, 163, 407-416.

Escobar, J., & Rodriguez-Kabana, R. (1980). Comparacion de un metodo de flotacion con uno de tamizado para la determinacion de *Radopholus similis* en raices de banano. *Nematropica*, 10, 86-88.

Esser, R. P., Taylor, A. L., & Holdeman, Q. L. (1984). Characterization of burrowing nematodes *Radopholus similis* for regulatory purposes. *Nematology circular of the Florida Department of Agriculture and Consumer Services*, n° 113.

Fademi, O. A., & Bayero, A. (1993). Nématodes parasites des plantains et des bananiers dans le sud-ouest et le centre du Nigeria. *InfoMusa*, 2(2), 13-14.

Fallas, G., & Marban-Mendoza, N. (1994). Response of three cultivars and one hybrid of Musa to *Radopholus similis* in Costa Rica. *Nematropica*, 24, 161-164.

Fallas, G., Sarah, J-L., & Fargette, M. (1995). Reproductive fitness and pathogenicity of eight *Radopholus similis* isolates on banana plants (Musa AAA cv Poyo). *Nematropica*, 25, 135-141.

Fargette, M., & Quénéhervé, P. (1988). Populations of nematodes in soils under banana, cv Poyo, in the Ivory Coast. 1. The nematofauna occurring in the banana producing areas. *Revue de Nématologie*, 11, 239-244.

Fernandez, E., Mena, J., Gonzalez, J., & Marquez, M. E. (2005). Biological control of nematodes in banana. In: D. W. Turner and F. E. Rosales (Eds), *Banana root system: towards a better understanding for its productive management*. INIBAP, Montpellier, France: 193-200.

Figueroa, A., & Mora, R. A. (1977). Efectos de nematicidas en las poblaciones de nematodos y en la produccion de banano. *Nematropica*, 7, 26-31.

Fogain, R. (1996). Screenhouse evaluation of *Musa* for susceptibility to *Radopholus similis*. In: E. A. Frison, J. P. Horry & D. De Waele (Eds), *Evaluation of Plantains AAB and Diploid AA, AB and BB.*

Proceeding of the Workshop on New Frontiers in Resistance Breeding for Nematode, Fusarium and Sigatoka. INIBAP, Montpellier, France: 79-86.

Fogain, R. (2000). Effect of *Radopholus similis* on plant growth and yield of plantains (Musa AAB). *Nematology*, 2, 129-133.

Fogain, R., & Gowen, S. R. (1996). Investigations on possible mechanisms of resistance to nematodes in Musa. *Euphytica*, 92, 375-381.

Fogain, R., & Gowen, S. R. (1997). Damage to roots of *Musa* cultivars by *Radopholus similis* with and without protection of nematicides. *Nematropica*, 27, 27-32.

Fogain, R., & Gowen, S. R. (1998). Yangambi Km5 (*Musa* AAA, Ibota subgroup), a possible source of resistance to *Radopholus similis* and *Pratylenchus goodeyi. Fundamental and Applied Nematology*, 21, 75-80.

Fogain, R., & Njifenjou, S. (2003). Effect of a mycorrhizal *Glomus* sp. on growth of plantain and on the development of *Radopholus similis* under controlled conditions. *African Plant Protection*, 9, 27-30.

Geering, A. D., Olszewski, N. E., Harper, G., Lockhart, B. E., Hull, R., & Thomas, J. E. (2005). Banana contains a diverse array of endogenous badnaviruses. *Journal of General Virology*, 86, 511-520.

Gichure, E., & Ondieki, J. J. (1977). A survey of banana nematode in Kenya. *Zeitschfrift fur Pflanzenkrankheiten und Pflanzenschutz*, 84, 724-728.

Giebel, J. (1982). Mechanisms of resistance to plant nematodes. *Annual Review of Phytopathology*, 20, 257-279.

Gnanapragasam, N. C., & Mohotti, K. M. (2006). Nematode parasites of tea. In: M. Luc, R. A. Sikora and J. Bridge (Eds). *Plant Parasitic Nematodes in Subtropical and Tropical Agriculture*, 2nd Edition. CAB International, Wallingford, UK: 581-609.

Gnanapragasam, N. C., Prematunga, A. K., & Herath, U. B. (1991). Preliminary survey for alternative hosts of the burrowing nematode, *Radopholus similis* in the tea areas of Sri Lanka. *Afro-Asian Journal of Nematology*, 1, 114-115.

Godfrey, G. H. (1931). The host plants of the burrowing nematode *Tylenchus similis*. *Phytopathology*, 21, 315-322.

Gowen, S. R. (1975). Improvement of banana yield with nematicides. Proceedings 8th British Insecticide and Fungicide Conference, 1975, Brighton, UK: 121-125.

Gowen, S. R. (1976). Varietal responses and prospect for breeding nematode resistant banana varieties. *Nematropica*, 6 , 45-49.

Gowen, S. R. (1979). Some considerations of problems associated with the nematode pests of bananas. *Nematropica*, 9, 45-49.

Gowen, S. R., & Edmunds, J. E. (1973). An evaluation of some simple extraction techniques and the use of hydrogen peroxide for estimating nematode populations in banana roots. *Plant Disease Reporter*, 57, 678-681.

Gowen, S., & Quénéhervé, P. (1990). Nematode parasites of bananas, plantains and abaca. In: M.Luc, R. A. Sikora, and J. Bridge (Eds), *Plant Parasitic Nematodes in Subtropical and Tropical Agriculture.* CAB International, Wallingford, UK: 431-460.

Gowen, S., Quénéhervé, P., & Fogain, R. (2005). Nematode parasites of bananas and plantains. In: M. Luc, R. A. Sikora and J. Bridge (Eds), *Plant Parasitic Nematodes in Subtropical and Tropical Agriculture*, 2nd Edition. CAB International, Wallingford, UK: 611-643.

Griffith, R., Giblin-Davis, R. M., Koshy, P. K., & Sosamma, V. K. (2005). Nematode parasites of Coconut and other Palms. In: M. Luc, R.A. Sikora and J. Bridge (Eds), *Plant Parasitic Nematodes in Subtropical and Tropical Agriculture*, 2nd Edition. CAB International, Wallingford, UK: 493-527.

Grisales, F., & Lescot, T. (1993). Le bananier plantain dans la zone centrale de Colombie. *InfoMusa*, 2(2), 11-12.

Guérout, R. (1970). Etude de trois nouveaux nématicides en bananeraie. *Fruits*, 25, 767-779.

Guérout, R. (1972). Relations entre les populations de *Radopholus similis* Cobb et la croissance du bananier. *Nematropica*, 27, 331-337.

Guérout, R. (1975). The soil and its importance on the effect of nematicidal treatments on banana yields. *Nematropica*, 5, 22-23.

Hadisoeganda, W. W. (1994). Status of nematode and weevil problems affecting banana in Indonesia. In: R. V. Valmayor, R. G. Davide, J. M. Stanton, N. L. Treverrow and V. N. Rao (Eds), *Banana Nematodes and Weevil Bores in Asia and the Pacific.* INIBAP/ASPNET, Los Baños, Philippines: 63-67.

Hahn, M. L., Sarah, J.-L., Boisseau, M., Vines, N. J., Wright, D. J., & Burrows, P. R. (1996). Reproductive fitness and pathogenicity of selected *Radopholus* populations on two banana cultivars. *Plant Pathology*, 45, 223-231.

Hildreth, R. C. (1962). Certified banana "seed". *Tropical Agriculture Trinidad*, 39, 103-107.

Hockland, S., Inserra, R. N., Millar, L., & Lehman, P. S. (2006). International Plant Health: Putting Legislation into Practice. In: R. N. Perry and M. Moens (Eds), *Plant Nematology*. CAB International, Wallingford, UK: 327-345.

Holdeman, Q. L. (1960). Nematodes that attack bananas in United Fruit Company divisions. *United Fruit Company, Research Dept. Extension News Letter*, v°7, n°1.

Horry, J. P., Ortiz, R., Arnaud, E., Crouch, J. H., Ferris, R. S. B., Jones, D. R., et al. (1997). Banana and Plantain. In D. Fucillo, L. Sears and P. Stapleton (Eds), *Biodiversity in Trust. Conservation and use of plant genetic resources in CGIAR centres*. Cambridge University Press, UK: 67-81.

Hugon, R., Ganry, J., & Berthe, G. (1984). Dynamique de population du nématode *Radopholus similis* en fonction du stade de développement du bananier et du climat. *Fruits*, 39, 251-253.

Hutton, D. G., & Chung, D. C. (1973). Effects of post-planting application of the nematicide DBCP to plantains. *Nematropica*, 3, 46-50.

Hutton, D. G. (1978). Influence of rainfall on some plantain nematodes in Jamaica. *Nematropica*, 8, 34-39.

Inomoto, M. M. (1994). Reacoes de algumas plantas ao nematoide cavernicola. *Nematologia Brasilieira*, 18, 21-27.

Israeli, Y., Lahav, E., & Reuveni, O. (1995). In vitro culture of bananas. In: Gowen, S. (Ed.) *Bananas and Plantains*. Chapman and Hall, London, UK: 147-178,

Jackson, V. G. H. (1987). Corm rot of *Cyrtosperma* in Guam. Report to the Federative States of Micronesia and Palau.

Jaizme-Vega, M. C. (1999). Application of arbuscular Mycorrhizal Fungi in micropropagated banana. In: F. E.Rosales, S. C.Tripon and J. Cerna (Eds), *Organic/environmentally friendly banana production* . INIBAP, Montpellier, France: 103-117.

Jaizme-Vega, M. C. & Pinochet, J. (1997). Growth response of banana to three mycorrhizal fungi in *Pratylenchus goodeyi* infested soil. *Nematropica*, 27, 69-76.

Jaizme-Vega, M. C., Tenoury, P., Pinochet, J., & Jaumont, M. (1997). Interactions between the root knot nematode *Meloidogyne incognita* and *Glomus mosseae* in banana. *Plant Soil*, 196, 27-35.

Janick, J., & Ait-Oubahou, A. (1989). Greenhouse production of banana in Morocco. *HortScience*, 24, 22-27.

Jaramillo, R. (1987). Comments on nematological research on Musa spp in Latin America and the Caribbean. In : *Nematodes and the borer weevil in bananas : present status of research and outlook* (pp 41-46). INIBAP, Montpellier, France.

Jaramillo, R., & Figueroa, A. (1974). Analisis armonico de la densidad de poblacion de *Radopholus similis* (Cobb) Thorne en la zona bananera de Guapiles, Costa Rica. *Turrialba*, 24, 402-407.

Jaramillo, R., & Figueroa, A. (1976). Relacion entre el balance hydrico y la poblacion de *Radopholus similis* (Cobb) Thorne en la zona bananera de Guapiles, Costa Rica. *Turrialba*, 26, 187-192.

Jimenez, M. F. (1972). Fluctuaciones anuales de la poblacion de *Radopholus similis* (Cobb) Thorne en la zona bananera de Potoci, Costa Rica. *Nematropica*, 2, 33-40.

Jonathan, E. I., & Rajendran, G. (2000). Assesment of avoidable yield loss in banana due to root-knot nematode *Meloidogyne incognita*. *Indian J. Nematol.*, 30,162-164.

Jonathan, E. I., Barker, K. R., & Abd-El-Aleem, F. F. (1999). Host status of banana for four major species and host races of *Meloidogyne*. *Nematologia Mediterranea*, 27, 123-125.

Jones, D. R. (2000). Diseases of Banana, abaca and enset. CAB International, Wallingford, UK.

Jones, R. K., & Milne, D. L. (1982). Nematode pests of bananas. In: D. P. Keetch and J. Heyns (Eds) *Nematology in Southern Africa*. Pretoria, South Africa: 30-37.

Jones, R. K. (1979). Control of *Helicotylenchus multicinctus* parasitizing bananas using systemic nematicides. *Nematropica*, 9, 147-150.

Kaplan, D. T., & O'Bannon, J. H. (1985). Occurrence of biotypes in *Radopholus citrophilus*. *Journal of Nematology*, 17, 158-162.

Kaplan, D. T., & Opperman, C. H. (1997). Genome similarity implies that citrus-parasitic burrowing nematodes do not represent a unique species. *Journal of Nematology*, 29, 430-440.

Kashaija, I. N., Speijer, P.R., Gold, C. S., & Gowen, S. (1994). Occurrence, distribution and abundance of plant parasitic nematodes of bananas in Uganda. Preliminary result of a diagnostic survey. *African Crop Science Journal*, 2, 99-104.

Kashaija, I. N., McIntyre, B. D., Ssali, H., & Kizito, F. (2004). Spatial distribution of roots, nematode populations and root necrosis in Highland banana in Uganda. *Nematology*, 6, 7-12.

Keetch, D. P. (1972). Some host plants of the burrowing eelworm, *Radopholus similis* (Cobb) in Natal. *Phytophylactica*, 4, 51-58.

Khan, R. M. (1999). Distribution of *Radopholus similis* in India, its spread in new regions and an analysis of the nematofauna of banana crop pathosystem. *Nematologia Mediterranea*, 27, 239-245.

Kirby, M. F, Kirby, M. E., Siddiqui, M. R. & Loof, P. A. A. (1980). Fiji nematode survey report: Plant parasitic nematode distributions and host association. Ministry of Agriculture and Fisheries, Fiji, Bulletin No. 68.

Kloepper, J.W., Leong, J. Teintze, M. & Schroth, M.N. (1980). Enhanced plant growth by siderophores produced by plant growth-promoting rhizobacteria. *Nature*, 286, 885-886.

Koshy, P.K., Eapen, S.J. & Pandey, R. (2005). Nematode Parasites of Spices, Condiments and Medicinal Plants. In : Luc, M., Sikora, R.A., and Bridge, J. (Eds), *Plant Parasitic Nematodes in Subtropical and Tropical Agriculture*, 2nd Edition. CAB International, Wallingford, UK: 751-791

Krishnamoorthy, V. & Kumar, N. (2005). Preliminary evaluation of diploid banana hybrids for yield potential, male fertility and reaction to *Radopholus similis*. *Plant Genetic Resources Newletters*, 141, 39-43.

Kumar, A. C., Viswanathan, P. R. K., & D'Souza, G. I. (1971). A study of plant-parasitic nematodes of certain commercial crops in coffee tracts of south India. *Indian Coffee*, 35, 222-224.

Kwa, M. (2003). Activation de bourgeons latents et utilisation de fragments de tige de bananier pour la propagation en masse de plants en conditions horticole *in vivo*. *Fruits*, 58, 315-328.

Larter, L. N., & Allen, E. F. (1953). Notes on current investigations, Oct. to Dec. 1952. *Malayan Agriculture Journal*, 36, 36-43.

Leach, R. (1958). Blackhead toppling disease of bananas. *Nature*, 181, 204-205.

Lehman, P. S. (1980). Weeds as reservoirs for nematodes that threaten field crops and nursery plants. *Nematology circular of the Florida Department of Agriculture and Consumer Services*, n° 66.

Lejju, B. J., Robertshaw, P., & Taylor, D. (2006). Africa's earliest bananas ? *Journal of Archeological Science*, 33, 102-113.

Lescot, T. (2004). Banane: Production, commerce et variétés. *FruiTrop*, 118, 5-9.

Lin, Y. Y., & Tsay, T. T. (1985). Studies on banana root knot nematode disease in central area of Taiwan. *Journal of the Chinese Society for Horticultural Science*, 31, 44-46.

Linbing, X., Hu, Y., Bingzhi, H., & Yuerong, W. (2004). Production and R&D of banana in China. In: *Advancing banana and plantain R&D in Asia and the Pacific*. Vol. 12. INIBAP–AP, Los Baños, Laguna, Philippines: 49-60.

Loof, P. A. A. (1964). Free-living and plant-parasitic nematodes from Venezuela. *Nematologica*, 10, 201-300.

Loos, C. A., & Loos, S. B. (1960b). The black-head disease of bananas (*Musa acuminata*). *Proc. Helm. Soc. Wash.*, 27, 189-193.

Loos, C. A. (1959). Symptoms expression of *Fusarium* wilt disease of the Gros Michel banana in the presence of *Radopholus similis* (Cobb, 1893) Thorne, 1949 and *Meloidogyne incognita acrita* Chitwood, 1949. *Proc. Helminth. Soc. Wash.*, 26,103-111.

Loos, C. A. (1961). Eradication of the burrowing nematode *Radopholus similis* from bananas. *Plant Disease Reporter*, 45, 457-461.

Loos, C. A. (1962). Studies on the life-history and habits of the burrowing nematode, *Radopholus similis*, the cause of black-head disease of banana. *Proc. Helminth. Soc. Wash.*, 29,43-52

Loos, C. A., & Loos, S. B. (1960a). Preparing nematode fruit banana 'seed'. *Phytopathology*, 50, 383-386.

Lopez, L., Torrez, J. L., Rodriguez, J. L., Morales, S. R., & Martin, J. (2000). Emploi d'un nouveau nématicide biologique pour la protection racinaire du bananier plantain (*Musa* AAB) multiplié par micropropagation. *InfoMusa*, 9(2), 8-9.

Loridat, P. (1989). Etude de la microflore fongique et des nématodes associées aux nécroses de l'appareil souterrain du bananier en Martinique. Mise en évidence du pouvoir pathogène du genre *Cylindrocladium*. *Fruits*, 44, 587-597.

Luc, M., & De Guiran, G. (1960). Les nématodes associés aux plantes de l'Ouest Africain. *Agron. Trop. Nogent*, 15, 434-449.

Luc, M., & Vilardebo, A. (1961). Les nématodes associés aux bananiers dans l'Ouest Africain. 1. Espèces parasites, dommages causés. *Fruits*, 16, 205-219.

Luc, M., Sikora, R. A., & Bridge, J. (2005). Plant Parasitic Nematodes in Subtropical and Tropical Agriculture, 2nd Edition, CAB International, Wallingford, UK: 890 pp.

Luis, J. G. (1998). Phenylphenalenone-type phytoalexin and phytoanticipins from susceptible and resistant cultivars of *Musa* species, its potential for engineering resistance to fungi and nematodes into banana. *Acta Horticulturae*, 490, 425-430.

Maas, P. W. T. (1969). Two important cases of nematode infestation in Surinam. In Peachey, J.E. (Ed) *Nematodes of Tropical Crops*. Commonwealth Agricultural Bureaux, Slough, UK: 148-154.

Machon, J. E., & Hunt, D. J. (1985). *Pratylenchus goodeyi*. C.I.H. Descriptions of plant-parasitic nematodes. Commonwealth Agricultural Bureaux, Farnham Royal, UK, Set 8, n° 120.

Mallamaire, A. (1939). La pourriture vermiculaire du bananier de Chine causée par *Anguillulina similis* Goodey en Afrique Occidentale Française. *L'Agronomie Coloniale, Année 28*, (254)32-42, (255)65-75.

Marcelino, L. L., Viquez, M., & Tarté, R. (1978). Fluctuaciones estacionales de la densidades de poblacion de *Radopholus similis* en raices de banano Valery (*Musa acuminata* AAA) en la zona Pacifica de Panama. *Nematropica*, 8, 52-55.

Marin, D. H., Sutton, T. B., & Barker, K. R. (1998a). Dissemination of bananas in Latin America and the Caribbean and its relationship to the occurrence of *Radopholus similis*. *Plant Disease* , 82, 964-974.

Marin, D. H., Sutton, T. B., Barker, K. R., Kaplan, D. T., & Opperman, C.H. (1998b). Burrowing-nematode resistance in black Sigatoka resistant banana hybrids. *Nematropica*, 28, 241-247.

Marin, D. H., Barker, K. R., Kaplan, D. T., Sutton, T. B., & Opperman, C. H. (1999). Aggressiveness and damage potential of Central American and Caribbean population of *Radopholus* spp. in banana. *Journal of Nematology*, 31, 377-385.

Marin, D. H., Barker, K. R., Kaplan, D. T., Sutton, T. B., & Opperman, C. H. (2000). Development and evaluation of a standard method fort screening for resistance to *Radopholus similis* in bananas. *Plant disease*, 84, 689-693

Martin, G. C., James, G. L., Bisset, J. L., & Way, J. I. (1969). Trials with field crops and *Radopholus similis* with observations on *Pratylenchus* sp., *Meloidogyne* sp. and others plant parasitic nematodes. *Rhod. J. agric. Res.*, 7, 149-157.

Mateille, T. (1990). Monoxenic culture of banana parasitic nematodes on *Musa acuminata* cv. Poyo shoots. *Journal of Nematology*, 22, 608-611.

Mateille, T. (1992). Comparative development of three banana-parasitic nematodes on *Musa acuminata* (AAA group) cvs Poyo and Gros-Michel vitro-plants. *Nematologica*, 38, 203-214.

Mateille, T. (1993). Effects of banana-parasitic nematodes on *Musa acuminata* (AAA group-cvs Poyo and Gros-Michel vitro-plants. *Tropical Agriculture (Trinidad)*, 70, 325-331.

Mateille, T. (1994.) Réactions biochimiques provoquées par trois nematodes phytoparasites dans les racines de *Musa acuminata* (groupe AAA) variétés Poyo et Gros Michel. *Fundamental and Applied Nematology*, 17, 283-290.

Mateille, T., Cadet, P., & Quénéhervé, P. (1984). Influence du recépage du bananier Poyo sur le développement des populations de *Radopholus similis* et d'*Helicotylebnchus multicinctus*. *Revue Nématol.*, 7, 355-361.

Mateille, T., Quénéhervé, P., & Hugon, R. (1994). The development of plant-parasistic nematode infestation on micropropagated banana plants following field control measures in Côte d'Ivoire. *Annals of Applied Biology*, 125, 147-149.

McIntyre, B. D., Speijer, P. R., Riha, S. J., & Kizito, F. (2000). Effects of mulching on biomass, nutrients and soil water in bananas inoculated with nematodes. *Agronomy Journal*, 92, 1081-1085.

McSorley, R. (1979). Plant parasitic nematodes associated with bananas and plantains in southern Florida. *Plant Disease Reporter*, 63, 663-665.

McSorley, R., & Parrado, J. L. (1981). Population fluctuations of plant-parasitic nematodes on bananas in Florida and its control. *Proceeding of the Florida State Horticultural Society*, 96, 201-207.

McSorley, R., & Parrado, J. L. (1986). *Helicotylenchus multicinctus* on bananas: an international problem. *Nematropica*; 16, 73-91.

Melin, P., & Vilardebo, A. (1973). Nématicide et désinfection à l'eau chaude dans la lutte contre *Radopholus similis* en bananeraie. *Fruits*, 28, 843-849.

Merny, G., Fortuner, R., & Luc, M. (1974). Les nématodes phytoparasites de Gambie. *L'Agronomie Tropicale*, 29, 702-707.

Mian, I. H. (1986). Plant-parasitic nematode with some crop species in Bangladesh. *Bangladesh J. Plant Pathology*, 2(1), 7-13.

Minz, G., Ziv, D., & Strich-Harari, D. (1960). Decline of banana plantations caused by spiral nematodes in the Jordan valley and its control by DBCP. *Ktavim*, 10, 147-157.

Moens, T. Araya, M., & De Waele, D. (2001). Correlations between nematode numbers and damage to banana (Musa AAA) roots under commercial conditions. *Nematropica*, 31, 55-65.

Moens, T. Araya, M., Swennen, R., & De Waele, D. (2004). Enhanced biodegradation of nematicides after repetitive applications and its effect on root and yield parameter in commercial banana plantations. *Biology and Fertility of Soils*, 39, 407-414.

Moens, T., Araya, M., Swennen, R., & De Waele, D. (2005). Screening of Musa cultivars for resistance to *Helicotylenchus multicinctus*, *Meloidogyne incognita*, *Pratylenchus coffeae* and *Radopholus similis*. *Australasian Plant Pathology*, 34, 299-309.

Moens, T., Araya, M., Swennen, R., & De Waele, D. (2006). Reproduction and pathogenicity of *Helicotylenchus multicinctus*, *Meloidogyne incognita* and *Pratylenchus coffeae*, and their interaction with *Radopholus similis* on Musa. *Nematology*, 8, 45-58.

Moens, T., Araya, M., Swennen, R., De Waele, D., & Sandoval, J. (2003). Effect of growing medium, inoculum density, exposure time and pot volume : Factors affecting the resistance screening for *Radopholus similis* in banana (*Musa* spp.) *Nematropica*, 33, 9-26.

Nair, K. K. R. (1979). Studies on the chemical control of banana nematodes. *Agric. Res. J. Kerala*, 17, 232-235.

Nair, M. R. G. K., Das, N. M., & Menon, M. R. (1966). On the occurrence of the burrowing nematode *Radopholus similis* (Cobb, 1893) Thorne, 1949, on banana in Kerala. *Indian Journal of Entomology*, 28, 553-554.

Newhall, A.G. (1958). The incidence of Panama disease of bananas in presence of the root knot and burrowing nematodes (*Meloidogyne* and *Radopholus*). *Plant Disease Reporter*, 42, 853-856.

Nguyet, D. T. M., Elsen, A., Tuyet, N. T., & De Waele, D. (2002). Réponse de plantes-hôtes de bananiers Pisang jari buaya et Mysore à *Radopholus similis*. *InfoMusa*, 11, 19-21.

Nguyet, D. T. M., Thuy, T. T. T., Tuyet, N. T., Tu, D. M., Yen, N. T., Thanh, D. T., & Nhi, H. H. (2003). Occurrence of *Pratylenchus coffeae* and occurrence, damage and reproduction of *Radopholus similis* in the Northen and Central Niglands of Vietnam. In: *Advancing banana and plantain R&D in Asia and the Pacific*. Vol. 11. INIBAP–AP, Los Baños, Laguna, Philippines: 65-78

O'Bannon, J. H. (1977). Worldwide dissemination of *Radopholus similis* and its importance in crop production. *Journal of Nematology*, 9,16-25.

Ogier, J. P., & Merry, C. A. F. (1970). Yield decline of plantain *Musa paradisiaca* in Trinidad, associated with the nematode *Pratylenchus* sp. *Turrialba*, 20, 407-412.

Orion, D., Levy, Y., Israeli, Y., & Fisher. E. (1999). Scanning electron microscope observations on spiral nematodes (*Helicotylenchus multicintus*). *Nematropica*, 29, 179-183.

Orton William, K. J. (1980). Plant-parasitic nematodes of the Pacific. Technical report, Vol.8, UNDP/FAO-SPEC Survey of Agricultural Pests and Diseases in the South Pacific. Commonwealth Institute of Helminthology, St Albans, England, 192 pp.

Osborne, R. E. (1962). Bodles Altafort. A new banana for Jamaica. *Banana Board Research Department, Occ. Bull.*, n° 3.

Oteifa, B. A. (1962). Species of root-lesion nematodes commonly associated with economic crops in the delta of the U.A.R. *Plant Disease Reporter*, 46, 572-575.

Paparu, P., Dubois, T., Gold, C. S., Niere, B., Adipala, E., & Coyne, D. (2006). Colonisation pattern of non-pathogenic *Fusarium oxysporum*, a potential biological control agent, in roots and rhizomes of tissue cultured *Musa* plantlets. *Annals of Applied Biology*, 149, 1-8.

Peregrine, W. T. H., & Bridge, J. (1992). The lesion nematode, *Pratylenchus goodeyi*, an important pest of *Ensete* in Ethiopia. *Tropical Pest Management*, 38, 325-326.

Perrier, X., & Tezenas du Montcel, H. (1990). Musaid: a computerized determination system. In R. L. Jarret, (Ed.), *Identification of Genetic diversity in the genus Musa*. INIBAP, Monferriez sur Lez, France: 76-91.

Pinochet, J. (1977). Occurrence and spatial distribution of root-knot nematodes on bananas and plantains in Honduras. *Plant Disease Reporter*, 61, 518-520.

Pinochet, J. (1978). Histopathology of the root lesion nematode, *Pratylenchus coffeae*, on plantains, Musa AAB. *Nematologica*, 24, 331-340.
Pinochet, J. (1979). Comparaison of four isolates of *Radopholus similis* from Central America on Valery bananas. *Nematropica*, 9, 40-43.
Pinochet, J. (1986). A note on nematode control practices on bananas in Central America. *Nematropica*, 16, 197-203.
Pinochet, J. (1987). La variabilidad de *Radopholus similis* en banane en las differentes regiones productoras del mundo. In: *Memorias VII Reunion ACORBAT* . CATIE, Turrialba, Costa Rica: 175-182.
Pinochet, J. (1988a). Comments on the difficulty in breeding bananas and plantains for the resistance to nematode. *Revue de Nématologie*, 11, 3-5.
Pinochet, J. (1988b). A method for screening bananas and plantains to lesion forming nematodes, In : *Nematode and the borer weevil in bananas : Present status of research and outlook*. Bujumbura, Burundi: 62-65.
Pinochet, J., & Rowe, P. R. (1979). Progress in breeding for resistance to *Radopholus similis* in banana. *Nematropica*, 9, 76-79.
Pinochet, J., & Stover, R. H. (1980). Fungi associated with nematode lesions on plantains in Honduras. *Nematropica*, 10, 112-115.
Pinochet, J., & Stover, R. H. (1980). Fungi in lesions caused by burrowing nematodes on bananas and their root and rhizome rotting potential. *Trop. Agric. Trinidad*, 57, 227-232.
Pinochet, J., & Ventura, O. (1977). Plant parasitic nematodes associated with bananas in Belize. *Tropical Agriculture (Trinidad)*, 54, 349-352.
Pinochet, J., & Ventura, O. (1980). Nematodes associated with agricultural crops in Honduras. *Turrialba*, 30, 43-47.
Pinochet, J., Fernandez, C., Jaizme, M., & Tenoury, P. (1997). Micropropagated banana infected with *Meloidogyne javanica* responds to *Glomus intraradices* and phosphorus. *HortScience*, 32, 35-49.
Pinochet, J., Jaizme, M., Fernandez, C., Jaumot, M., & De Waele, D. (1998). Screening bananas for root-knot (*Meloidogyne* spp) and lesion nematode (*Pratylenchus goodeyi*) resistance for the Canary Islands. *Fundamental and Applied Nematology*, 21, 17-23.
Pocasangre, L., Sikora, R. A., Vilich, V., & Schuster, R. P. (2000). Survey of banana endophytic fungi from Central America and screening for biological control of *Radopholus similis*. *Acta Horticulturae*, 531, 283-289.
Prachasaisoradej, S., Chinnasri, B., Tungjitsomkid, N., & Chiemchaisri, Y. (1994). Status of nematode and weevil borer problems affecting banana in Thailand. In: R.V. Valmayor, R. G. Davide, J. M. Stanton, N. L. Treverrow and V. N. Rao (Eds), *Banana Nematodes and Weevil Bores in Asia and the Pacific*. INIBAP/ASPNET, Los Baños, Philippines: 115-121
Price, N. (1999). Les bananiers d'altitude en Colombie. *InfoMusa*, 8(2), 26-28.
Price, N. S. (1994a). Alternate cropping in the management of *Radopholus similis* and *Cosmopolites sordidus*, two important pests of banana and plantain. *International Journal of Pest Management*, 40, 237-241.
Price, N. S. (1994b). Field trial evaluation of *Musa* varieties and of other crops as hosts of *Pratylenchus goodeyi* in Cameroon. *Afro-Asian Journal of Nematology*, 4, 11-16.
Price, N. S. (1995). The origin and development of banana and plantain cultivars. In: S. Gowen (Ed.), *Bananas and Plantains*. Chapman and Hall, London, UK: 1-12.
Price, N. S. (2006). The banana burrowing nematode, Radopholus similis (Cobb) Thorne, in the Lake Victoria region of East Africa: its introduction, spread and impact. *Nematology*, 8, 801-817.
Price, N. S., & Bridge, J. (1995). *Pratylenchus goodeyi* (Nematoda: Pratylenchidae) a plant parasitic nematode from the montane highlands of Africa. *Journal of African Zoology*, 109, 435-442.
Quénéhervé, P. (1988). Populations of nematodes in soils under banana cv Poyo in the Ivory Coast. 2. Influence of soil texture, pH, and organic matter on nematode populations. *Revue de Nématologie*, 11, 245-251.
Quénéhervé, P. (1989a). Populations of nematodes in soils under banana, cv. Poyo, in the Ivory Coast. 3. Seasonal dynamics of populations in mineral soil. *Revue de Nématologie*, 12, 149-160.
Quénéhervé, P. (1989b). Populations of nematodes in soils under banana, cv. Poyo, in the Ivory Coast. 4. Seasonal dynamics of populations in organic soil. *Revue de Nématologie*, 12, 161-170.

Quénéhervé, P. (1990). Spatial arrangement of nematodes around the banana plant in the Ivory Coast, related comments on the interaction among concomitant phytophagous nematodes. *Acta Oecologica*, 11, 875-886.

Quénéhervé, P. (1993a). Banana phenology in relation to phytophagous nematodes. In : *Biological and Integrated Control of Higland Banana and Plantain Pests and Diseases*. Gold, C.S. and Gemmil, B. (Eds), IITA, Cotonou, Benin: 218-230.

Quénéhervé, P. (1993b). Nematode management in banana agroecosystems, comments and outlook from the Côte d'Ivoire experience. *Crop Protection*, 12, 164-172.

Quénéhervé, P., & Cadet, P. (1985a). Localisation des nematodes dans les rhizomes du bananier cv 'Poyo'. *Revue Nématol.*, 8, 3-8.

Quénéhervé, P. & Cadet, P. (1985b). Etude de la dynamique de l'infestation en nématodes transmis par les rhizomes du bananier cv. Poyo en Côte d'Ivoire. *Revue Nématol.*, 8, 257-263.

Quénéhervé, P., & Cadet, P. (1986). Une nouvelle technique d'échantillonnage pour l'étude des nématodes endoparasites du bananier. *Revue Nématol.*, 9, 95-97.

Quénéhervé, P., Cadet, P., & Mateille, T. (1991). New approaches to chemical control of nematodes on bananas: field experiments in the Ivory Coast. *Revue de Nématologie*, 14, 543-549.

Quénéhervé, P., Marie-Luce, S., Barout, B., & Grosdemange, F. (2006). Une technique de criblage variétal précoce des bananiers envers les nématodes phytoparasites. *Nematology*, 8, 147-152.

Quénéhervé, P., Van den Berg, E., Topart, P., & Hostachy, B. (1997). Analyse écologique de la spécificité parasitaire des nématodes phytoparasites associés à quelques plantes ornementales cultivées à la Martinique. *Nematologica*, 43, 214-227..

Quénéhervé, P., Chabrier, C., Auwerkerken, A., Topart, P., Martiny, B., & Marie-Luce, S. (2006). Status of weeds as reservoirs of plant parasitic nematodes in banana fields in Martinique. *Crop Protection*, 25, 860-867.

Quimi, V. H., & Villacis, J. (1977). Estudio comparativo de dos methodos de extraccion del nematodo *Radopholus similis* de la raices de banano. *Nematropica*, 7, 44-47.

Rajendran, G., & Sivakumar, C. V. (1996). Pathogenicity of the spiral nematode, *Helicotylenchus multicinctus* and effect on yield and nutrient status of banana. *Pest Management in Horticultural Ecosystems*, 2, 23-27.

Rajendran, G., Naganathan, T. G., & Vadivelu, S. (1979). Studies on banana nematodes. *Indian Journal of Nematology*, 9, 54.

Ramclam, W., & Araya, M. (2006). Frequency of occurrence and abundance of root nematodes on banana (Musa AA) in Belize. *International Journal of Pest Management*, 52, 71-77.

Razak, A. R. (1994). Plant parasitic nematodes, a potential threat to commercial cultivation of banana in Malaysia. In: Valmayor, R. V., Davide, R. G., Stanton, J. M., Treverrow, N. L. and Rao, V. N. (Eds), *Banana Nematodes and Weevil Borers in Asia and the Pacific*. INIBAP/ASPNET, Los Baños, Philippines: 34-45.

Reversat, G., & Soriano, I. (2002). The potential role of bananas in spreading rice root-knot nematode, *Meloidogyne graminicola*. *International Rice Research Notes*, 27(2), 23-24.

Rivas, X., & Roman, J. (1985). Estudio sobre la gama de hospederos de una poblacion de *Radopholus similis* de Puerto Rico. *Nematropica*, 15, 165-170.

Rodriguez-Romero, A. S., Guerra, M. S. P., & Jaizme-Vega, M. D. (2005). Effect of arbuscular mycorrhizal fungi and rhizobacteria on banana growth and nutrition. *Agronomy for Sustainable Development*, 25, 395-399.

Roman, J. (1978). Nematodos del bananero y el platanero. In: J. Roman, (Ed), *Fitonematologia tropical*. Universidad de Puerto Rico, Mayaguez, Puerto Rico: 93-110

Roman, J., Oramas, D., Green, J., & Torres, A. (1983). Control of nematodes and black weevils in plantains. *Journal of Agriculture of the University of Puerto Rico*, 67, 270-273.

Roman, J., Rivas, X., Rodriguez, J., & Oramas, D. (1976). Chemical control of nematodes in plantains (AAB). *Journal of Agriculture of the University of Puerto Rico*, 60, 36-44.

Roman, J., Rivas, X., Rodriguez, J., & Oramas, D. (1977). Further experiments on the chemical control of nematodes in plantains (AAB). *Journal of Agriculture of the University of Puerto Rico*, 61, 192-199.

Rowe, P., & Rosales, F. (1994). Musa breeding at FHIA. In: Jones, D.R. (Ed) *The Improvement and Testing of Musa: a Global Partnership*. INIBAP, Montpellier, France: 117-129.

Salas, J. A., Oyela, R., & Stover, R. H. (1976). Effet of fallow on the burrowing nematode (*Radopholus similis*) of bananas. *Plant Disease Reporter*, 60, 863-866.

Santor, W. & Davide, R.G. (1982). Interrelationships of *Radopholus similis* and *Meloidogyne incognita* in banana. *Phil. Phytopath.*, 18, 22-23.

Sarah, J-L. (1989). Banana nematodes and their control in Africa. *Nematropica*, 19, 199-216.

Sarah, J-L., Lassoudière, A., & Guérout, R. (1983). La jachère nue et l'immersion du sol: deux méthodes intéressantes de lutte intégrée contre *Radopholus similis* (Cobb) dans les bananeraies de sol tourbeux de Côte d'Ivoire. *Fruits*, 38, 35-42.

Sarah, J-L., Sabatini, C., & Boisseau, M. (1993). Differences in pathogenicity to banana (Musa sp Poyo) among isolates of *Radopholus similis* from different production area of the world. *Nematropica*, 23,75-79.

Sarah, J-L., Blavignac, F., Sabatini, C., & Boisseau, M. (1992). Une méthode de laboratoire pour le criblage variétal des bananiers vis-à-vis de la résistance aux nématodes. *Fruits* 47, 559-564.

Scotto la Massèse, C. (1968). Plant nematodes of crops in the French West Indies. In : Peachey, J.E. (Ed), *Nematodes of Tropical Crops*. Technical Communication Commonwealth Bureau of Helminthology, No 40. CAB, St. Albans, UK: 164-183.

Sher, S. A. (1954). Observations on plant-parasitic nematodes in Hawaii. *Plant Disease Reporter*, 38, 687-689.

Sikora, R. A. (1979). Observations on *Meloidogyne* with emphasis on disease complexes, and the effect of host on morphometrics. Proceeding of the second research planning conference on root knot nematodes, *Meloidogyne* spp. Athens, Greece: 93-104

Sikora, R. A. (1992). Management of antagonistic potential in agricultural ecosystems for the control of plantlet parasitic nematode. *Annual Review of Phytopathology*, 12, 245-270.

Sikora, R. A. & Pocasangre, L. (2006). The concept of a suppressive banana plant: Root health management with a biological approach. Proceeding of the XVII ACORBAT Meeting, Santa Catarina, Brasil: 241-248.

Sikora, R. A., & Schloesser, E. (1973). Nematodes and fungi associated with root system of banana in a state of decline in Lebanon. *Plant Disease Reporter*, 57, 615-618.

Sikora, R. A., & Schuster, R. P. (1998). Novel approaches to nematode IPM. In: E. A.Frison, C. S. Gold, E. B.Karamura and R. A. Sikora (Eds), *Mobilizing IPM for sustainable banana production in Africa* . INIBAP, Montpellier, France: 127-136.

Sikora, R. A., Bridge, J., & Starr, J .L. (2005). Management practices: an overview of integrated nematode management technologies. In: M. Luc, R. A. Sikora and J. Bridge (Eds), *Plant Parasitic Nematodes in Subtropical and Tropical Agriculture*, 2nd Edition. CAB International, Wallingford, UK: 793-825.

Simmonds, N. W. (1960). Evolution of the bananas. Longmans, London, UK.

Simmonds, N. W. (1966). Bananas. 2nd Ed. Longmans, London, UK.

Simmonds, N. W., & Shepherd, K. (1955). The taxonomy and origin of the cultivated bananas. *Journal of the Linnean Society of Botany*, 55, 302-312.

Sipes, B., & Litchy, J. S. (2002). *Radopholus similis* damage to *Anthurium andreanum*. *Nematropica*, 32, 77-81.

Sipes, B., Caswell-Chen, E. P., Sarah, J-L., & Apt, W. J. (2005). Nematode parasites of Pineapple. In M. Luc, R. A. Sikora and J. Bridge (Eds), *Plant Parasitic Nematodes in Subtropical and Tropical Agriculture*, 2nd Edition. CAB International, Wallingford, UK: 709-731.

Smith, L. R., & Thames, W. H. (1969). Plantain nematodes and preliminary crop rotation studies fot their control in the Dominican Republic. Proceedings of the Symposium on Tropical Nematology, Rio Piedras, Puerto Rico: 84-89.

Smithson, P. C., McIntyre, B. D., Gold, C. S., Ssali, H. & Kashaija, I. N. (2001). Nitrogen and potassium fertilizer vs. nematode and weevil effects on yield and foliar nutrient status of banana in Uganda. *Nutrient Cycling in Agroecosystems*, 59, 239-250.

Speijer, P. R., & Bosch, C. (1996). Susceptibility of Musa cultivars to nematodes in Kagera Region, Tanzania. *Fruits*, 51, 217-222.

Speijer, P. R., & De Waele, D. (1997). *Screening of* Musa *germplasm for resistance and tolerance to nematodes*. INIBAP Technical Guidelines 1. IPGRI, Rome, Italy; INIBAP, Montpellier, France, IITA, Ibadan, Nigeria.

Speijer, P. R., & De Waele, D. (2001). Nematodes associated with East African Highland cooking bananas and cv. Pisang awak (*Musa* spp.) in Central Uganda. *Nematology*, 3, 535-541.

Speijer, P. R., & Gold, C. S. (1996). Musa root health assessment, a technique for the evaluation of *Musa* germplasm for nematode resistance. In: Frison, E.A., Horry, J.P. & De Waele, D. (Eds), *New*

Frontiers in Resistance Breeding for Nematodes, Fusarium *and Sigatoka.* INIBAP, Montpellier, France: 62-67.

Speijer, P. R., & Fogain, R. (1999). *Musa* and *Ensete* nematode pest status in selected African countries. In: Frison, E.A., Gold, C.S., Karamura, E.B. & Sikora, R.A. (Eds). *Mobilizing IPM for sustainable banana production in Africa.* INIBAP, Montpellier, France: 99-108.

Speijer, P. R., Kajumba, C., & Ssango, F. (1999). East African Highland banana production as influenced by nematodes and crop management in Uganda. *International Journal of Pest Management,* 45, 41-49.

Speijer, P. R., Rotimi, M. O., & De Waele, D. (2001). Plant parasitic nematodes associated with plantain (*Musa* spp., AAB-group) in southern Nigeria and their relative importance compared to other biotic constraints. *Nematology,* 2001, 3, 423-436.

Ssango, F., Speijer, P. R., Coyne, D. L., & De Waele, D. (2004). Path analysis: a novel approach to determine the contribution of nematode damage to East African Highland banana (*Musa* spp., AAA) yield loss under two crop management practices in Uganda. *Field Crops Research,* 90, 177-187.

Stanton, J. M. (1994). Status of nematode and weevil borer problems affecting banana in Australia. In: R.V. Valmayor, R.G. Davide, J. M.Stanton, N. L. Treverrow and V. N. Rao (Eds), *Banana Nematodes and Weevil Borers in Asia and the Pacific.* INIBAP/ASPNET, Los Baños, Philippines: 48-56.

Stanton, J. M. (1999). Assesment of resistance and tolerance of in vitro propagated banana plants to the burrowing nematode, *Radopholus similis. Australian Journal of Experimental Agriculture,* 39, 891-895.

Stanton, J. M., Pattison, A. B., & Kopittke, R. A. (2001). A sampling strategy to assess banana crops for damage by *Radopholus similis* and *Pratylenchus goodeyi. Australasian Journal of Experimental Agriculture,* 41, 675-679.

Sticher, L., Mauch-Mani, B., & Métreaux, J.P. (1997). Systemic acquired resistance. *Annual Review of Phytopathology,* 35, 235-270.

Stirling, G. R. (1991). Biological control of plant-parasitic nematodes: Progress, problems and prospects, CAB International, Wallingford, UK.

Stoffelen, R., Verlinden, R., Pinochet, J., Swennen, R. L., & De Waele, D. (2000b). Host plant response of *Fusarium* wilt resistant *Musa* genotypes to *Radopholus similis* and *Pratylenchus coffeae. International Journal of Pest Management,* 46, 289-293.

Stoffelen, R., Verlinden, R., Xuyen, N. T., Swennen, R., & De Waele, D. (2000a). Host plant response of Eumusa and Australimusa bananas (*Musa* spp.) to migratory endoparasitic and root-knot nematodes. *Nematology,* 2, 907-916.

Stover, R. H. (1966). Fungi associated with nematodes and non-nematode lesions on banana roots. *Canadian Journal of Botany,* 44, 1703-1710.

Stover, R. H. (1972). Banana, plantain, and abaca diseases. Commonwealth Mycological Institute, Kew, England: 316 pp.

Stover, R. H., & Fielding, M. J. (1958). Nematodes associated with root injury of *Musa* spp in Honduran banana soils. *Plant Diseases Reporter,* 42, 938-940.

Stoyanov, D. (1967). Especies de nematodos parasitos del platano en Cuba y posibilidades de control. *Revista de Agricultura,* 1, 9-47.

Suit, R. F., & DuCharme, E. P. (1953). The burrowing nematode and other parasitic nematodes in relation to spreading decline. *Citrus Leaves,* 33:8-9, 32-33.

Sundararaju, P., & Cannayane, I. (2003). Evaluation of different biopesticides against major nematode pathogens infesting banana cv. Nendran. In: *National Symposium on Organic Farming in Horticulture for Sustainable Production.*CISH, Lucknow, India: 66-67.

Sundararaju, P., Shanthi, A., & Sathiamoorthy, S. (2005). Status report on Musa nematodes problems and their management in India. In: F. S. De la Cruz, I. Van den Bergh, D. De Waele, D. M. Hautea, and A. B.Molina (Eds), *Towards management of Musa nematodes in Asia and the Pacific.* INIBAP-AP, Los Baños, Laguna, Phillipines: 21-42.

Swennen, R., de Langhe, E., Janssen, J., & Decoene, D. (1986). Study of the root development of some *Musa* cultivars in hydroponics. *Fruits,* 41, 515-524.

Tandigan, I. C., & Davide, R. G. (1986). Biological control of *Tylenchulus semipenetrans* on citrus and *Radopholus similis* on banana with *Paecylomyces lilacinus* and *Penicillium anatalicum. Philippine Phytopathology,* 22, 42-48.

Tarjan, A. C. (1961). Longevity of *Radopholus similis* (Cobb) in host free soil. *Nematologica,* 6, 170-175.

Tarté, R., & Pinochet, J. (1981). Problemas nematologicos del banano. Contribuciones recientes a su conocimiento y combate. Union Paises Exportadores Banano (UPEB), Panama.

Tarté, R., Pinochet, J., Gabrielli, C., & Ventura, O. (1981). Differences in population increase, host preferences and frequency of mophological variants among isolates of the banana race of *Radopholus similis*. *Nematropica*, 11, 43-52.

Taylor, A. L., & Loegering, W. Q. (1953). Nematodes associated with root lesions in abaca. *Turrialba*, 3, 8-13.

Tenkouano, A., Hauser, S., Coyne, D., & Coulibaly, O. (2006). Clean planting material and management practices for sustained production of banana and plantain in Africa. *Chronica Horticulturae*, 46, 14-18.

Ternisien, E., & Melin, P. (1989). Etude des rotations culturales en bananeraie. II. Impact des cultures de rotations sur la production bananière et l'état sanitaire du sol. *Fruits*, 44, 445-454.

Timm, R. W. (1965). A preliminary survey of plant parasitic nematodes of Thailand and the Philippines. Report of South East Asia Treaty Organization.

Tixier, P., Risède, J. M., Dorel, M., & Malézieux, E. (2006). Modelling population dynamics of banana plant-parasitic nematodes: a contribution to the design of sustainable cropping systems. *Ecological Modelling*, 198, 321-331.

Torrealba, P. A. (1969). Survey of plant-parasitic nematode genera from Venezuela. In: J. E. Peachey (Ed), *Nematodes of Tropical Crops*. Technical Commununication No 40. Commonwealth Bureau of Helminthology, Albans, Herts, England: 257-263.

Treutter, D. (2006). Significance of flavonoids in plant resistance : a review. *Environmental Chemistry Letters*, 4, 147-157.

Tripathi, L. (2003). Genetic engineering for improvement of *Musa* production in Africa. *African Journal of Biotechnology*, 2, 503-508.

Umesh, K. C., Krishnappa, K., & Bagyaraj, D. J. (1988). Interaction of burrowing nematode, *Radopholus similis* (Cobb, 1893) Thorne, 1949 and VA mycorrhiza, *Glomus fasciculatum* (Thaxt.) Gerd. & Trappe, in banana. *Indian Journal of Nematology*, 18, 6-11.

Urwin, P. E., Leversley, A., McPherson, M. J., & Atkinson, H. J. (2000). Transgenic resistance to the nematode *Rotylenchulus reniformis* conferred by *Arabidopsis thaliana* plants expressing proteinase inhibitors. *Molecular Breeding*, 6, 257-264.

Urwin, P. E., Lilley, C. J., McPherson, M. J., & Atkinson, H. J. (1997.) Resistance to both cyst and root-knot nematodes conferred by transgenic *Arabidopsis* expressing a modified plant cystatin. *Plant Journal*, 12, 455-461.

Urwin, P. E., Troth, K. M., Zubko, E. I., & Atkinson, H. J. (2001). Effective transgenic resistance to *Globodera pallida* in potato field trials. *Molecular Breeding*, 8, 95-101.

Valette, C., Mounport, D., Nicole, M., Sarah, J-L., & Baujard, P. (1998). Scanning electron microscope study of two African populations of *Radopholus similis* (Nematoda: Pratylenchidae) and proposal of *R. citrophilus* as a junior synonym of *R. similis*. *Fundamental and Applied Nematology*, 21, 137-144.

Valette, C., Nicole, M., Sarah, J. L., Boisseau, M., Boher, B., Fargette, M., & Geiger, J. P. (1997). Ultrastructure and cytochemistry of interactions between banana and the nematode *Radopholus similis*. *Fundamental and Applied Nematology*, 20, 65-77.

Valle-Lamboy, S., & Ayala, A. (1976). Control of plantain nematodes with contact nematicides. *Nematropica*, 6(2), 55-59.

Valmayor, R. V. (1990). Banana and Plantain R&D in Asia and the Pacific. Proceedings of a regional consultation on banana and plantain R&D networking, 20-24 November, Manila, INIBAP, Montpellier, France.

Van den Bergh, I., Nguyet, D. T. M., Tuyet, N. T., Nhi, H. H., & De Waele, D. (2002). Screening of Vietnamese *Musa* germplasm for resistance to root knot and root lesion nematodes in the greenhouse. *Australasian Plant Pathology*, 31, 363-371.

Van der Vecht, J. (1950). Plant parasitic nematodes. In: L. G. E. Karshoven and J. V. D. Vecht (Eds), *Diseases of cultivated plants in Indonesia Colonies*. W. van Woeve, The Hague, The Netherlands: 16-41.

Veech, J. A. (1982). Phytoalexins and their role in the resistance of plants to nematodes. *Journal of Nematology*, 14, 2-9.

Viaene, N., Duran, L. F., Rivera, J. M., Duenas, J., Rowe, P., & De Waele, D. (2003). Responses of banana and plantain cultivars, lines and hybrids to the burrowing nematode *Radopholus similis*. *Nematology*, 5, 85-98.

Vilardebo, A. (1959). Note sur la lutte contre les nematodes en Guinée. *Fruits*, 14, 125-126.

Vilardebo, A. (1970). Perspectives d'utilisation de nouveaux nématicides en bananeraie. *Fruits*, 25, 371-378.

Vilardebo, A. (1974). Méthode d'essai d'efficacité pratique de nématicides étudiés sur *Radopholus similis* Cobb en bananeraies. *Société Française de Phytiatrie* et *de Phytopharmacie. Commission des essais biologiques*, N° 49.

Vilardebo, A. (1976). Populations dynamics of *Radopholus similis* in relation to climatic factors and the physiology of the plant. *Nematropica*, 6, 54-55.

Vilardebo, A., & Guérout, R. (1976). Nematode species in West Africa, Madagascar, and La Réunion, with some comments on their biology. *Nematropica*, 6,53-54.

Vilardebo, A., & Robin, J. (1969). Nematicidal treatment of banana planting material. In: J. E. Peachey, (Ed), *Nematodes of Tropical Crops*. Technical Commununication No 40. Commonwealth Bureau of Helminthology, Albans, Herts, England: 133-141.

Villanueva, L. M. (2004). Status of nematode problem affecting banana in the Philippines. In: F. S. De la Cruz, I. Van den Bergh, D. De Waele, D. M. Hautea, and A. B. Molina (Eds,) *Towards management of Musa nematodes in Asia and the Pacific* . INIBAP, Los Baños, Philippines: 51-59.

Vovlas, N., Avgelis, A., Goumas, D., & Frisullo, D. (1994). A survey of banana diseases in sucker propagated plantations in Crete. *Nematologia Mediterranea*, 22, 101-107.

Vu, T., Hauschild, R., & Sikora, R. A. (2006). *Fusarium oxysporum* endophytes induced systemic resistance against *Radopholus similis* on banana. *Nematology*, 8, 847-852.

Walker, P. T., Hebblethwaite, M. J., & Bridge, J. (1984). Project for banana pest control and improvement in Tanzania. EEC Report for the government of Tanzania. Tropical Development and Research Institute, London, UK.

Wardlaw, C. W. (1961). Banana diseases including Plantains and Abaca. Longman, Green, London, UK.

Wehunt, E. J., & Edwards, E. I. (1968). *Radopholus similis* and other nematodes species on bananas. In: G. C. Smart and V. G. Perry (Eds), *Tropical Nematology*. University of Florida Press, Gainesville, USA: 1-19.

Wehunt, E. J., Hutchinson, D.J. & Edwards, D.I. (1978). Reactions of banana cultivars to the burrowing nematode *Radopholus similis*. *Journal of Nematology*, 10, 368-370.

Whyte, E. B., & Gowen, S. R. (1974). Recovery of nematodes from banana roots and soil samples. *Nematropica*, 4, 27-31.

Williams, J. R. (1969). Nematodes attacking surgarcane. In: J. E. Peachey, (Ed), *Nematodes of Tropical Crops*. Technical Commununication No 40. Commonwealth Bureau of Helminthology, Albans, Herts, England: 84-203.

Winoto, S. R. (1976). Nematological problems in Malaysian Agriculture. In: Proceedings of the Asian regional planning on root-knot nematode Research Program. University Press, Los Baños, Philippines: 17-24.

Winoto, S. R., & Sauer, M. R. (1982). Plant parasitic nematodes associated with cultivated plants in Peninsular Malaysia. *Malaysian Applied Biology*, 11, 5-17.

Wuyts, N., Swennen, R., & De Waele, D. (2006). Effect of plant phenylpropanoid pathway products and selected terpenoids and alkaloids on the behaviour of the plant-parasitic nematodes *Radopholus similis*, *Pratylenchus penetrans* and *Meloidogyne incognita*. *Nematology*, 8, 89-101.

Wuyts, N., Lognay, G., Sagi, L., De Waele, D., & Swennen, R. (2003). Secondary metabolites in roots and implications for nematode resistance in banana (*Musa* spp.). In: D. W. Turner and F. E. Rosales (Eds), *Banana root system: towards a better understanding for its productive management*. INIBAP, Montpellier, France: 238-246.

Yepez, G., Meredith, J., & Perez, A. (1972). Nematodos de banano y platano (*Musa* sp.) en Venezuela. *Nematropica*, 2(2), 47-51.

Zem, A. C., & Alves, E. J. (1978.) Nematoides associados a bananeiras no Estado do Maranhao. *Ciencia e Cultura, Sao Paulo*, 30(7), 13.

Zem, A. C., & Alves, E. J. (1983). Efeito de diferentes praticas sobre a populaçao de *Radopholus similis*. Reuniao Brasileira de Nematologia, Piracicaba (BR). *Sociedade Brasileira de Nematologia* Publicaçao, Brazil, 7, 215-225.

Zem, A. C., & Lordello, L. G. E. (1983). Estudos sobre hospedeiros de *Radopholus similis* e *Helicotylenchus multicinctus*. Reuniao Brasileira de Nematologia, Piracicaba (BR). *Sociedade Brasileira de Nematologia* Publicaçao, Brazil, 7, 175-187.

Zimmerman, A. W. P. (1898). Die nematoden der koffiewortels. *Mededelingen's Lands Plantentuin (Buitenzorg)*, 27, 1-64.
Zuckerman, B. M., & Strich-Harari, D. (1963). The life stage of *Helicotylenchus multicinctus* (Cobb) in banana roots. *Nematologica*, 9, 347-353.
Zum Felde, A., Pocasangre, L. E., & Sikora, R.A. (2004). Use of microbial communities inside suppressive banana plants to increase biocontrol of burrowing nematode, *Radopholus similis*. In: D. W. Turner and F. E. Rosales (Eds), *Banana root system: towards a better understanding for its productive management*. INIBAP, Montpellier, France: 169-177.

2

RENATO CROZZOLI

NEMATODES OF TROPICAL FRUIT CROPS IN VENEZUELA

Universidad Central de Venezuela,
Facultad de Agronomía, Instituto de Zoología Agrícola, Laboratorio de Nematología Agrícola, Apdo. 4579, Maracay, Venezuela

Abstract. Data on nematodes of main fruit crops in Venezuela are reviewed, including acerola, avocado, banana and plantain, breadfruit, cashew, citrus, coconut, date palm, fig, grapevine, guava, mango, papaya, passionfruit, peach, pineapple, sapodilla and tamarind. For each crop, main nematode species are reviewed, with dataon their distribution, damage and management.

1. INTRODUCTION

Venezuela possesses an area of around 912050 km^2, and is located totally in the American tropics of the north hemisphere (0°38'53" and 12°11'46" LN). It benefits by uniform and moderate temperatures all over the year, which is typical of those regions, with the exception of the mountainous areas of more than 2000 msl. Being the temperature average of about 27°C with few fluctuations, the daily variation (10-15°C) is higher than the annual variation. This little temperature variability is accompanied by the uniformity in the day length, e.g. the difference between the longest (12h 42') and shortest days is only 1h 10' for Maracay (10°10'N).

In such a way the temperatures, the day length and the high radiation allow the growth of plants during all the year, being water, and the oxygen deficit in the soil, the two main limiting factors. If no water restriction is present, the growth rates of plants are very high and the interval between sowing and harvest is often shorter than in other subtropical or temperate areas.

In tropical areas the rain is irregular in distribution, falling in two defined periods: dry or "summer" and rainy or "winter" seasons, whose durations vary depending on distance from Ecuador. Also orography often affects plowing. Between the 0 and 5° lat. N, the rainy period lasts almost the whole year, with 2 or 3 months less humid than the remaining.

Venezuela has a low proportion of soils without limitations (2%), the remaining being affected by excessive relief (44%), low natural fertility (32%), drainage lack

A. Ciancio & K. G. Mukerji (eds.), *Integrated Management of Fruit Crops and Forest Nematodes*, 63-83.

(18%) and aridity (4%) (Comerma & Walls, 1978). One of the most complex agronomic practices in fruit crop production is the selection of soil and the choice of best fertilization practices, which forces the development of appropriate technologies for areas with few deep soils, high clay percentages, low natural fertility, little capacity of gas exchange and poor internal drainages (Avilán & Leal, 1990).

The most important fruit crops in Venezuela are: banana (aprox. 55000 ha), plantain (aprox. 62000 ha), mango (aprox. 4700 ha), avocado (aprox. 6500 ha) and orange (aprox. 43950 ha), pineapple (16000 ha), papaya (7500 ha) which represent 76.16 % of the fruit crop area cultivated. Grape represents 6.15 %, whereas annona, guava, passion fruit, acerola, fig, cashew and others cover almost 15.2 % of the cultivated area.

The banana and plantain production represents more than 50% of the total fruit crops production (Leal & Avilan, 1997; FAO, 2006). Yields are extremely low if compare with the optima of other areas: banana (20 ton/ha), plantain (8,5 ton/ha), orange (13 ton/ha), lime (20 ton/ha), mango (15 ton/ha), papaya (12 ton/ha), pineapple (18.5 ton/ha). In general, the low productivity of the fruit crops are due to: *i)* lack of an appropriate crops zonification; *ii)* inadequate agronomic management; *iii)* lack of genetic material well adapted to the different domestic ecological conditions; *iv)* problems of insects pests, mites, weeds, diseases and nematodes; and *v)* lacks of appropriate post harvest management and trading (Leal & Avilan, 1997; Crozzoli, 2002).

Nematodes constitute a major pests in many fruit crops. General signs of nematode damage include stunting, premature wilting, leaf yellowing, root malformation and related signs typical of nutrient deficiencies, frequently very evident. Stunting and poor stand development tend to occur in patches throughout the field as a result of the irregular distribution of nematodes in soil. Unfortunately, producers often consider nematodes only when the problem is very serious and frequently not easy to solve.

With the exception of banana, citrus, coconut and plantain, informations concerning nematode damage is relatively scarce in Venezuela. Included here are some fruit crops for which nematological information exists. The fruit trees crops are reviewed in alphabetical order of their common names within each section.

2. NEMATODES BY CROP

2.1. Acerola (Malpighia glabra)

Acerola is still very limited in production, but it is enjoying increasing interest as a commercial product rich in vitamin C. The most important nematodes associated with this crop in Zulia (Castellano et al., 2004) and Aragua States are: *Tylenchorhynchus annulatus*, *Helicotylenchus dihystera*, *Hoplolaimus seinhorsti*, *Hemicriconemoides strictathecatus* (syn. *H. mangiferae*), *Xiphinema brasiliense*, *Meloidogyne incognita* and *Monotrichodorus monohystera*. However, recently, in Lara State *Meloidogyne mayaguensis* was identified as associated with acerola

(Lugo et al., 2005). Damage caused by *M. incognita*, *M. javanica* and *M. arenaria* are severe, especially in sandy soil or in nursery stock (Costa et al., 1999).

2.2. Avocado (Persea americana)

Species of *Aphelenchus*, *Criconemella*, *Helicotylenchus*, *Meloidogyne*, *Pratylenchus*, *Tylenchus*, *Xiphinema* as well as *Rotylenchulus reniformis* and *Radopholus similis* are the most important plant parasitic nematodes associated with avocado in Venezuela (Petit, 1990; Suárez & Rosales, 1998). However, there is no evidence that these nematodes cause significant economic damages to avocado plants. The presence of *Rotylenchulus* and *Helicotylenchus* was observed to be associated with higher incidence of the fungus *Phytophthora cinnamomi*, causal agent of a severe avocado root disease (Suárez et al., 1984; 1992).

*2.3. Banana and Plantain (*Musa *AAA and AAB)*

All of the most cultivated clones are growing in Venezuela: *Musa* AA, AAA, AAB and ABB. The most important are *Musa* AAA (banana) and *Musa* AAB (plantain). For a general treatment of banana nematodes see Chapter 1, this volume.

The most widespread and important nematode species occurring on banana and plantain in Venezuela are: *R. similis*, *Helicotylenchus multicinctus*, *M. incognita*, *R. reniformis* and *Pratylenchus coffeae* (Yépez et al., 1972, Crozzoli et al., 1993; 1995; Crozzoli, 2002; Suárez & Rosales, 2004).

Radopholus similis is widely distributed in the western regions (Lake Maracaibo, Zulia, Mérida and Trujillo States) and central western region (Yaracuy, Barinas and Portuguesa States) associated principally with plantain. In the central region, it has been detected only in the Carabobo and Aragua States on plantain and banana, respectively (Haddad et al., 1975; Petit, 1990; Crozzoli et al., 1993; Montiel et al., 1997).

Helicotylenchus multicinctus is probably the most widespread nematode in banana and plantain in Venezuela, and is present in all growing areas (Haddad et al., 1975; Petit, 1990; Crozzoli et al., 1993; Crozzoli 2002; Suárez & Rosales, 2004).

High populations of *P. coffeae* have been detected on banana and plantain in Aragua, Carabobo, Sucre, Barinas, Yaracuy and Zulia States (Haddad et al., 1975; Petit, 1990; Crozzoli et al., 1993; Montiel et al., 1997; Suárez y Rosales, 1998).

Meloidogyne incognita is also distributed in all the growing areas of banana and plantain. The highest populations has been detected in Aragua State (Haddad et al., 1975; Crozzoli et al., 1993). This polyphagous species is very pathogenic on plantain in Cojedes State.

Rotylenchulus reniformis is widely distributed in the central region (Aragua and Carabobo States) (Crozzoli et al., 1993; Crozzoli, 2002). Recently *Hoplolaimus seinhorsti* has been detected in association with roots of Giant Cavendish banana in Trujillo State at densities higher than 24 nematodes per g of roots.

However, nematode parasitism on banana and plantain roots is characterized by the simultaneous infestations by several species. In Aragua State, combination of

H. multicinctus and *M. incognita* severely reduced banana yields. Treatment with ethoprop on Giant Cavendish banana reduced the nematodes population and increased the weight of banana bunches in the second year after treatments compared with untreated control (Crozzoli et al., 1995). In the Zulia State simultaneous infestations by *H. multicinctus*, *R. similis* and *P. coffeae* are common. The highest population densities on roots, in these cases, are reached by *H. multicinctus* or *P. coffeae*, and rarely for *R. similis* (Montiel et al., 1997), suggesting that the burrowing nematode is not the more important pest on these cultivations in Venezuela.

In Aragua State, an important producer of bananas, as well as in the Zulia State, the most important producer both of banana and plantain, the use of nematicides is a common practice in production growing and in the planting hole. However, in Aragua State, the practice of supporting plants with bamboo poles to prevent plant toppling, is widespread, as is the rotation with alternative crops (maize and leguminous) for one-two years. The use of tissue cultured plants and selection of disease-free suckers is also becoming more and more common. Positive experiences exist in the control of *H. multicinctus* with the use of *Trichoderma* in grower producing fruit for the international export trade.

Unfortunately no study exists in order to determine the damage threshold of banana and plantain to nematodes in Venezuela, as well as no data are available on the possible detection of genetic resistance on local cultivars.

2.4. Breadfruit (Artocarpus altilis)

Little information exists about nematode problems of the breadfruit tree. In Venezuela, prominent galls induced by *M. incognita* have been observed in breadfruit tree used for shade of cocoa in Cumboto (Aragua State). Evidence of damage is unclear.

2.5. Cashew (Anacardium occidentale)

The cashew nut plant is native of Brazil and it is widely cultivated in Venezuela, principally in Oriental States and Zulia State. Limited information on nematodes attacking cashew exists. The nematological situation in the Oriental States is unknown, but the species associated with this crop in the Zulia State are: *Tylenchorhynchus annulatus*, *Helicotylenchus dihystera*, *H. seinhorsti*, *H. stricthatechatus*, *X. brasiliense*, *M. monohystera* and *R. reniformis*. Evidence of damage is unclear. It is important to emphasize that cashew (cvs. Rojo and Amarillo) is not attacked by the root knot nematode *M. incognita* (Castellano et al., 2004).

2.6. Citrus (Citrus *spp.*)

After plantain, citrus is the most important fruit crop in Venezuela with 40000 ha cultivated and a production of 52000 ton. The principal citrus crops are orange, mandarins, lemon, grapefruit and tangelo, that are cultivated in almost all regions. The main producing States are: Carabobo, Yaracuy, Monagas, Aragua (oranges), Miranda (mandarins), Zulia and Monagas (lemons and grape fruit). The principal rootstocks used are: *Citrus volkameriana*, *C. reshnii*, Citrange Carrizo and Troyer (*C. sinensis* × *Poncirus trifoliata*) and Citrumelo Swingle (*C. paradisi* × *P. trifoliata*) that are resistant to 'Tristeza virus'.

Many species of plant parasitic nematodes has been detected associated with citrus crops (Table 1) (Crozzoli et al., 1997; 1998).

Figure 1. Growth differences between Citrus volkameriana *8 years old healthy plants (A) and plants infested by the citrus nematode,* Tylenchulus semipenetrans *(B). Typical symptom of micronutrients deficiency on* C. volkameriana *leaves of a nematode infested plant (C).*

The citrus nematode, *Tylenchulus semipenetrans*, is the causal agent of the 'slow decline' of citrus. This is the most important nematode on this crop in Venezuela and the world. McBeth observed for first time *T. semipenetrans* in 1955 and considered it a limiting factor of the crop in Maracay (Aragua State). Later, Dao (1961) found the nematode in the Carabobo State and Yépez (1965) in Monagas, Nueva Esparta and other Oriental States. At the moment it is broadly disseminated in all the production areas, causing yield reductions between 25 and 32% (Dao, 1961; Yépez & Meredith, 1970; Petit, 1991). Aragua and Zulia States are the most affected, with 100% of the orchards infested. The main infestation source is the use of infested propagation material from nurseries (Crozzoli & Fúnes, 1992).

Symptoms are similar to those associated with poor root development. Leaves are smaller and may become chlorotic. Micronutrients deficiency (Mn, Cu, Fe) symptoms are frequent. Wilting occurs earlier during periods of water stress and leaf drop is more pronounced, producing exposed branch terminals (Fig. 1).

Infected feeder roots have a dirty appearance due to soil particles that adhere to gelatinous eggs masses on the root surface (Fig. 2). Feeder roots decay, losing the epidermis integrity and the cortex feeding efficiency. Numerous organisms i.e. fungi and bacteria can also invade roots.

Table 1. Nematode species associated with Citrus *spp. in Venezuela*

Aorolaimus holdemani
Criconema demani
Gracilacus aculenta
Helicotylenchus crenacauda, H. dihystera, H. erythrinae, H. multicinctus
Hemicriconemoides communis
H. strictathechatus
Hoplolaimus seinhorsti
Meloidogyne exigua, M. incognita
Criconemoides (=Mesocriconema) onoense, C. ornatum, C. sphaerocephala,
Monotrichodorus monohystera
Paratrichodorus minor
Paratylenchus elachistus, P. minutus
Pratylenchus brachyurus, P. zeae
Rotylenchus caudaphasmidius
Scutellonema brachyurum
Tylenchorhynchus annulatus, T. capitatus
Tylenchulus semipenetrans
Xiphinema brasiliense, X. brevicollum, X. krugi, X. peruvianum, X. simillimum, X. vulgare

Physiological races or biotypes of *T. semipenetrans* exist based on host suitability: three biotypes are commonly recognized. The 'Citrus' biotype that reproduces poorly on *P. trifoliata* but well on *Citrus* spp. and the hybrids 'Carrizo' and 'Troyer' (*C. sinensis* × *P. trifoliata*), on olive (*Olea europaea*), grape (*Vitis vinifera*) and persimmon (*Diospyros* spp.). The 'Poncirus' biotype that reproduces on *Citrus*, *P. trifoliata* and grape, but not olive, and the 'Mediterranean' biotype, similar to the 'Citrus' biotype except that it does not reproduce on olive (Duncan et al., 2005). For an extensive review of the citrus nematode biology, damage and management, see chapter 6 of this volume.

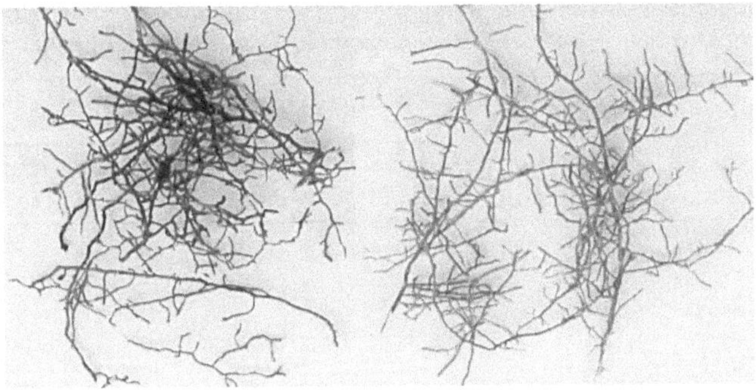

Fig. 2. Differences between roots of Citrus volkameriana *infested (left) by the citrus nematode,* Tylenchulus semipenetrans *and uninfested control (right). Parasitised feeder roots show a dirty appearance due to soil particles adhering to the gelatinous eggs masses on the root surface.*

In a study carried out in Venezuela with populations of *T. semipenetrans* proceeding from the main citrus producing areas, the Venezuelan populations appeared uniform since their reproductive behavior was similar on all rootstocks tested. Also, it could be demonstrated that the Citrange Carrizo and Troyer, and Citrumelo Swingle are resistant to the nematode (Table 2).

When patogenicity tests were carried out with a population of *T. semipenetrans* proceeding from the Aragua State, *C. reshnii* appeared as the most susceptible rootstock, with *C. volkameriana* as best host. In none of the rootstocks derived from *P. trifoliata* the nematode reproduced. These rootstocks can be hence considered as an effective control measure, with a broader protection range, given that rootstocks derived from *P. trifoliata* are also tolerant to *Phytophthora* (Table 3).

Fluctuation in population levels of the citrus nematode are observed during the year and are related mainly to the plants phenological stages and precipitation. In plants of Tahiti lime grafted on *C. volkameriana* rootstock with or without flooding irrigation, the first population increase of *T. semipenetrans* densities corresponds to the month of February, with a second increase observed in October. Both corresponded to the start of the period of major vegetative activity of plants (December-January for the first and September-October for the second peack, respectively). The nematode populations were also inversely correlated with the precipitation, with no effect observed for irrigation (Fig. 3A and 3B).

In plants of Tahiti lime grafted on *C. amblycarpa* rootstock with irrigation (by flooding), the first population increase of *T. semipenetrans* corresponded to the month of January and the second one was observed in March. Both corresponded to the major vegetative activity of plants (December-January and March, respectively).

In plants of Tahiti lime grafted on *C. reshnii* with flooding irrigation, the first population increase of *T. semipenetrans* was observed in January–February and the second one in October. Both corresponded to the periods of major vegetative activity of plants (December-January and September–October, respectively) (Fig. 4A and 4B).

Table 2. Relation between final population (Pf) and initial population (Pi) of different citrus rootstocks to the Citrus nematode, Tylenchulus semipenetrans.

Populations	Rootstocks				
	Citrus volkameriana [a]	*Citrus reshnii*	Citrange Carrizo	Citrange Troyer	Citrumelo Swingle
Aragua	++	+	-	-	-
Miranda	++	+	-	-	-
Monagas	++	+	-	-	-
Zulia	++	+	-	-	-
Carabobo, Valles altos	++	+	-	-	-
Carabobo, Valles bajos	++	+	-	-	-
Yaracuy, Valles altos	++	+	-	-	-
Yaracuy, Valles bajos	++	+	-	-	-

[a] ++: Pf / Pi >10, +: Pf / Pi = 2-10, - : Pf / Pi < 1

Table 3. Populations of Tylenchulus semipenetrans, *relationship between final and initial population densities (Pf/Pi) and effect on aerial fresh (AFW) and dry weights (ADW) on different citrus rootstocks (from* Crozzoli & González, 1989)

Rootstock				AFW (g)		ADW (g)	
	♀♀ / g roots[1]	J2 + ♂♂ / g roots	Pf/Pi	Io	Ii	Io	Ii
C. volkameriana	20705 A	31444 A	46	25,3	17,6	10	6,7*
C. aurantium	586 C	1151 C	4,1	20	19,3	8,8	9,1
C. reshnii	845 BC	5754 B	6,8	18,7	11,7	7,9	5,6
C. reticulata	484 C	1994 C	2,3	15,9	14,6	6,4	5,9
C. amblicarpa	1426 B	5330 B	5,2	9,8	9,1	4,3	4
Citrange Carrizo	30 D	3 D	0,04	11,1	10,9	4,8	4,6
Citrange Troyer	67 D	134 D	0,1	11,2	11	5	4,9
Citrumelo Swingle	1 D	0 D	0,001	7,8	9,3	3,4	3,7
P. trifoliata Rubidouix	3 D	6 D	0,002	4,3	3,3	2,2	1,9
P. trifoliata FL strain	7 D	3 D	0,008	4,4	5,6	2,2	2,2
P. trifoliata Argentina	3 D	1 D	0,009	3,6	3,5	2,1	1,8

[1] Data flanked in colums by the same letter are not significantly different according to Duncan's Multiple Range Test (P=0.01).

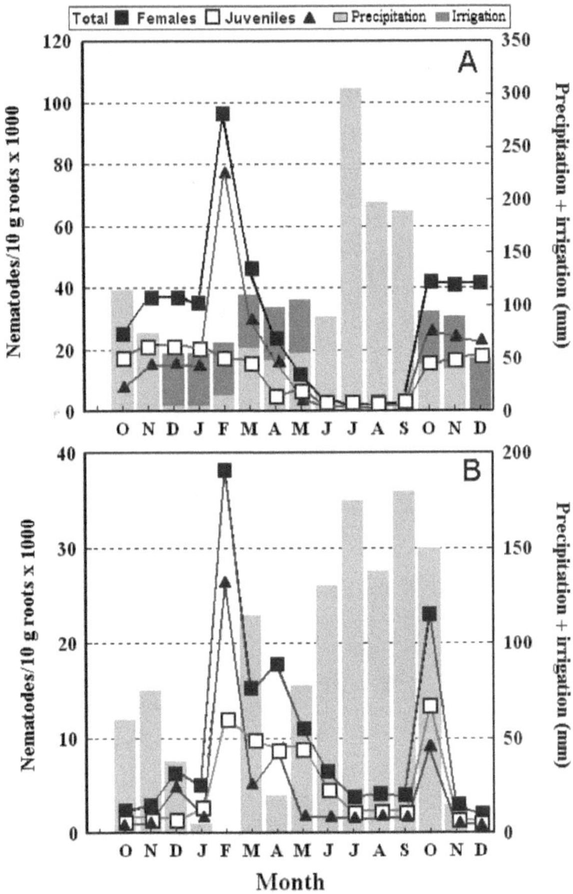

Figure 3. Populations of Tylenchulus semipenetrans *on roots of six years old* Citrus volkameriana, *related to precipitation and irrigation (A) or precipitation only (B) in Aragua State, Venezuela.*

Rainfall hence appears as an important factor, since the phenology of plants affects the subsequent population fluctuation levels. Although the principal periods of vegetative activity are definite, during the rainy season (May–November) vegetative growth also occurs, but no increments of *T. semipenetrans* densities were evidenced.

2.7. Coconut (Cocos nucifera)

The major nematodes affecting this crop in Venezuela is the causal agent of the red ring disease, caused by *Bursaphelenchus cocophilus*. This disease was first reported as occurring in the East (Sucre State), near Trinidad, by Salazar (1934) and was the first disease caused by nematodes to be investigated in Venezuela. Malaguti (1953) pointed out the same disease in oil palm (*Elaeis guineensis*) in the State of Yaracuy.

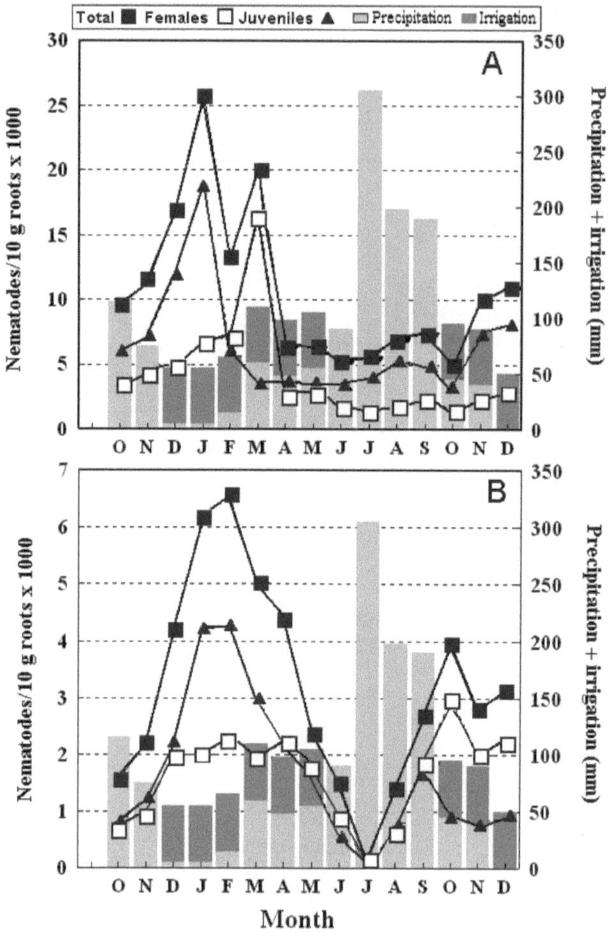

Fig. 4. Populations of Tylenchulus semipenetrans *on roots of six years old* Citrus amblycarpa *(A)*, and C. reshnii *(B)*, *related to precipitation and irrigation in Aragua State, Venezuela.*

In 1953, 10% of palms in production had died but in 1950 mortality was 30% of 20000 growing plants. In 1955 the red ring disease was considered the most important disease of coconut trees. In 1962 it was also detected in the Falcon State and posteriorly in Aragua and Carabobo States. In 1970 the percentage of coconut trees affected by red ring was 20% and in 1973, approximately 40% of the 4-10 years old coconut trees were affected by the disease (Servicio Shell para el Agricultor, 1958; 1973; Webster & González, 1959; 1960; González & Webster, 1960; Yépez & Martínez, 1969).

In 1977 red ring was considered the disease causing the biggest damage to the domestic coconut palms. In that year, FONCOPAL carried out a recognition in the cultivations of the coast in the Distrito Federal and Miranda State and detected red ring disease. At the moment the situation is less dramatic since appropriate control measures are adopted. The management measures for red ring disease in coconut is based on prevention rather than cure, by the destruction of infested palm material, and by the trapping and killing of the weevil vectors, principally *Rhyncophorus palmarum*, before they can spread the nematodes.

In Venezuela, the following control measures are applied: *i)* periodical and careful revision of the plantations to detect sick plants; *ii)* elimination of these plants; *iii)* use of traps impregnated with insecticides with the purpose of eliminating the highest quantities of possible vectors; *iv)* maintainance of plantations free of overgrowths and trunks, to facilitate inspections.

2.8. Date Palm (Phoenix dactylifera)

Although dates palm is not cultivated by its fruit in Venezuela, it is very common as an ornamental plant. Little is known about plant parasitic nematodes in the crop, however, high populations of *Pratylenchus* sp. have been observed in Margarita Island, Nueva Esparta State. The nematode produces small, elongate, brown lesions, that subsequently coalesce causing extensive root rotting. Large numbers of nematodes and their eggs may be observed in these cavities. *Pratylenchus penetrans* has been associated with root damage in Algeria (Lamberti, 1973) but no data are available on incidence and damage in Venezuela.

2.9. Fig (Ficus carica)

In Venezuela the cultivation of fig was very important in the years sixties-seventies in the Aragua state (about 400-500 msl). Unfortunately, *Meloidogyne* spp. destroyed all plantations. At the moment, orchards exist in the Andean, central (> 1200 msl) and Zulia states. In this last one, attacks by root-knot nematodes, putatively identified as *M. mayaguensis*, are frequent. *Xiphinema index* also has been detected in El Jarillo (Miranda state) associated with fig, but no evidence of damage were reported (Renaud, 1990).

2.10. Grapevine (Vitis *sp.*)

Grapevine cultivation in Venezuela began in 1960 in Aragua, Lara and Zulia States. Actually, 90% of all cultivated grapes are grown in Lara and Zulia States, where total acreage reached 1000 ha in 1997. In addition to root-knot nematodes (*Meloidogyne* spp.) many other nematode species attack grapes.

Nine genera of plant parasitic nematodes were identified from samples of *Vitis vinifera* cvs. Cardinal, Italia, Alphonse Lavalle, Rossetti, Violeta and Tucupita in Lara State. *Helicotylenchus*, *Meloidogyne* and *Pratylenchus* were the genera most frecuently observed. Other genera included *Rotylenchulus*, *Tylenchorhynchus*,

Aphelenchus, *Tylenchus*, *Paratylenchus* and *Xiphinema* (Petit, 1978). Later, *R. reniformis*, *T. semipenetrans* and *Trichodorus* sp. were identified in Zulia and Lara States (Petit, 1990). The most important nematodes, however, are *M. incognita* and *M. javanica.* Both species are common in vineyards of Anzoátegui, Guárico, Aragua, Lara and Zulia States (Renaud, 1978). Stunted growth, changes in color, and increased sensitivity to stress usually are associated to root-knot nematode attacks, with worse symptoms observed in sandy soils.

A study to evaluate the combined effect of *M. incognita* and *M. javanica* was carried out on cvs. Cardinal, Criolla Negra, Tucupita and Villa Nueva. Results indicated that Criolla Negra is resistant to the nematode penetration and can be used as rootstock. For Villa Nueva, the combination of both species did not significantly affect development or characteristics of the plants. Cardinal and Tucupita appeared as good hosts, with gall indexes of 4 and 5, respectively, in a 0-5 scale (Petit, 1980). In a greenhouse experiment, *M. incognita* reduced height, stem diameter, fresh and dry weights of Criolla Negra and Italia compared to the uninoculated controls. Dry and top weight of Italia were severely suppressed by the nematode whereas Criolla Negra was only slightly affected. Histological examination of infested roots of Italia showed giant cells in the central cylinder where the nematode established its feeding sites. On Criolla Negra hypersensitive response, although giant cells were found, a low (2) root gall index was observed, suggesting that this rootstock is a less suitable host to the *M. incognita* population used, than Italia (Petit, 1993).

Tylenchulus semipenetrans has been detected in Anzoategui State associated with grape planting in soil previously planted with citrus. Symptoms, however, did not appear on grapes infested with citrus nematode (Crozzoli, 1990; unpublished). The first level of management aims at the exclusion of nematodes from uninfested areas wherever possible. It is important that nematode-free rootings are used for new vineyards or replants. However, chemical control has been also applied. To control mixed populations of *Pratylenchus*, *Helicotylenchus*, *Rotylenchulus*, *Criconemoides*, *Meloidogyne* and *Xiphinema*, three rates of systemic nematicides (carbofuran), were applied to determine their efficacy during two cycles in a eight year-old vineyard with grape variety Cardinal grafted on native rootstock Criolla Negra. The rates were 2.7; 1.35 and 0.68 g a.i. per vine. For the first cycle there were no significant differences in yield among treatments. In the second cycle carbofuran at 2.7 and 1.35 g a.i. per vine increased yields more effectively than at rates of 0.68 g a.i. Soil samples showed a reduction of the nematode population during the first cycle, but at the end of the second cycle the nematode populations had increased considerably (Petit, 1982).

2.11. Guava (Psidium guajava)

The nematodes associated with guava in Venezuela are: *Meloidogyne* spp., *Pratylenchus brachyurus*, *Tylenchorhynchus contractus*, *Xiphinema americanum sensu lato*, *Aorolaimus levicaudatus*, *R. reniformis*, *H. seinhorsti*, *Helicotylenchus dihystera*, *Criconemoides sphaerocephala* (= *Mesocriconema sphaerocephalus*) and *C. onoense* (= *M. onoense*) (Crozzoli et al., 1991; Crozzoli, 2002). Guava production has declined steadily during the past 10 years due to increase pressure from *Meloidogyne* spp. Dieback and reduced yields have been reported on sandy soil, also in relation to toxicity due to the use of water with high contents of salts (Fig. 5).

Crozzoli et al., (1991) identified the species of the genus *Meloidogyne* as a mixed population of *M. incognita* and *M. arenaria*, addressing the nematode problem by screening other *Psidium* species for presence of possible resistant rootstocks. *Psidium fiedrichsthalianum* showed a high degree of resistance to *M. incognita* and its tolerance limit was 60-fold lower than that of *P. guajava* cultivars (Casassa et al., 1997; 1998; Matheus et al., 1999). Treatments with nematicides and manure did not give satisfactory results (Casassa et al., 1996).

Association between *M. incognita* race 1 (Crozzoli & Casassa, 1998), *Macrophomina phaseolina* and *Fusarium oxysporum* are common. These fungi inoculated alone invaded cortical parenchyma and vascular parenchyma tissues, respectively, but in combination with *M. incognita* they invaded also tissues modified by the nematode feeding (Suárez et al., 1998).

Recently, *M. mayaguensis* was identified by means of isozyme analyses (Molinari et al., 2005). This nematode may cause a severe guava decline in Brazil (Carneiro et al., 2001) and Venezuela it was detected in States of Zulia and Lara (Lugo et al., 2005).

Fig. 5. Guava trees (A) infested (Ni) or not infested (No) by Meloidogyne *spp. and galls (B) induced on roots.*

2.12. Mango (Mangifera indica)

In successive samplings carried out in mango plantations located in the States Carabobo, Portuguesa, Yaracuy, Cojedes and Lara, the associated nematodes were:

Helicotylenchus, *R. reniformis*, *Tylenchorhynchus* and *Paratylenchus* spp. Other nematodes recovered at a lower scale were: *Hemicriconemoides*, *Criconemoides*, *Pratylenchus* and *Xiphinema*. In the visited plantations, however, clear symptoms related to the nematodes species associated to the fruit crop were not observed (Petit, 1990). *Hemicriconemoides strictathecatus* is widely distributed in association with mango throughout the worlds and its pathogenicity to mango has been demonstrated (McSorley, 1992). In Aragua State, mango populations were as high a 1000 nematodes/100 cm^3 of soil. However, no damage to plants has been reported thus far, but due to its wide distribution in Venezuela, pathogenicity studies appear necessary (Crozzoli et al., 1995). This species is very frequent at high densities in plants proceeding from nurseries.

2.13. Papaya (Carica papaya)

Many plant parasitic nematodes have been found in association with papaya roots in Venezuela. They are: *Aphelenchus*, *Criconemoides*, *Ditylenchus*, *Helicotylenchus erythrinae*, *Helicotylenchus* sp., *Hemicriconemoides*, *Meloidogyne*, *Paratylenchus*, *Pratylenchus*, *Rotylenchulus*, *Tylenchorhynchus*, *Tylenchus*, *Xiphinema americanum sensu lato* and *Xiphinema* sp. (Yépez & Meredith, 1970; Petit, 1992; Suarez & Rosales, 2001). A survey of papaya in seven Venezuelan States detected high incidence of *Meloidogyne* and *Rotylenchulus reniformis.* Up to 85% of samples from Yaracuy and Falcon States showed *Meloidogyne* spp. and 59% from Monagas State showed *R. reniformis* (Suárez & Rosales, 2001). Root galls caused by *Meloidogyne* spp. have been reported in nursery seedlings and young papaya plants in Yaracuy and Portuguesa States (Petit, 1990). Only three species, however, appear economically significant in papaya plantations in Venezuela, namely *M. incognita*, *M. javanica* and *R. reniformis* (Bustillo et al., 2000; Crozzoli et al., 2005; Crozzoli, 2006, unpublished).

Meloidogyne incognita is very common in papaya orchards of Paraguaná Peninsula (Falcon State). In pot experiments the nematode tolerance limit for fresh and dry top weights of papaya plants (Paraguanera type), were 0.16 and 0.25 eggs, juveniles and young females/cm^3 of soil. The reduction were 22.5% for fresh and dry top weight and 18% for stem diameter and height increase at Pi $\geq$ 16 eggs, juveniles and young females/cm^3 of soil (Bustillo et al., 2000) (Fig. 6A).

Rotylenchulus reniformis causes severe plant damage and yield reductions (Fig. 7). In pot experiment tolerance limit to the nematode for fresh and dry top weights of papaya (Paraguanera type) were 0.25 and 0.18 eggs, juveniles and young females/cm^3 of soil, respectively. The reduction were 33% and 35% at Pi $\geq$ 16 eggs, juveniles and young females/cm^3 of soil, for fresh and dry top plant weights, respectively (Crozzoli et al., 2005).

Rotylenchulus reniformis appears more pathogenic than *M. incognita* in Venezuela. Recently, *M. javanica* has also been detected in plants proceedings from Aragua State nurseries (Fig. 6B and C).

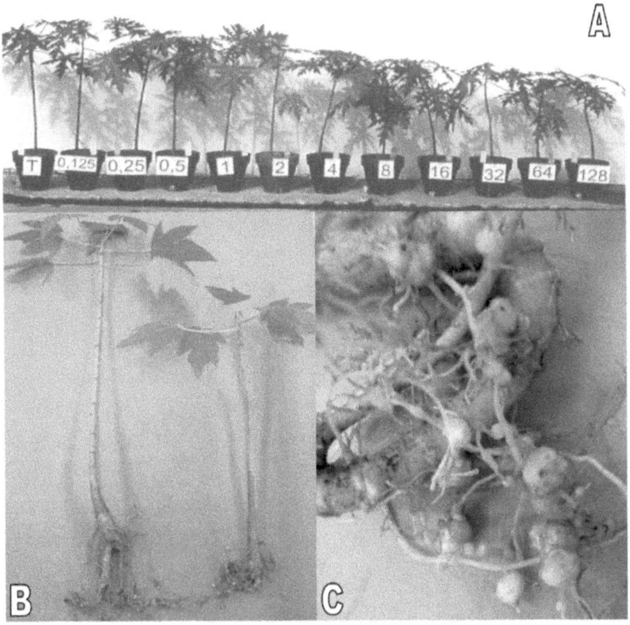

Fig. 6. A) Papaya plants (Paraguanera-type) inoculated with different initial populations of M. incognita. *B) Young plants of papaya infected by* M. javanica. *C) Detail of galls induced by* M. javanica *on papaya roots.*

2.14. Passionfruit (Passiflora edulis *f. sp.* flavicarpa)

Two varieties of *P. edulis* are known as purple passion fruit (*P. edulis*) and yellow passionfruit (*P. edulis* f. sp. *flavicarpa*). In Venezuela the last one only is cultivated, in the States of Carabobo, Lara, Mérida, Yaracuy and Zulia (Suárez & Rosales, 1998).

Although a number of plant parasitic nematodes are reported associated with passionfruit, including *Helicotylenchus* spp., *Meloidogyne* spp., *R. reniformis*, *Paratylenchus* and *Xiphinema* sp. (Petit, 1990), only *R. reniformis* is reported as the causal agent of economic damages (Suárez & Rosales, 1998). *Rotylenchulus reniformis* and *Phytophthora* spp. are the most important pathogens on this crop, severely reducing either fruit production and plant longevity (Suárez & Rosales, 2003). Under controlled conditions, *R. reniformis* reduced the plant growth and increased its susceptibility to soil fungi attacks, as much in nurseries as in fields, during the growth phase (Suárez et al., 1999). An interaction between *R. reniformis* and *Fusarium solani* also occurs, but resistant cultivars can break this relationship (Pernía et al., 2002).

Suárez et al. (2004), determined the reaction of nine *Passiflora* accessions to the attack of *R. reniformis*, using the scheme proposed by Cook (1974). Results showed

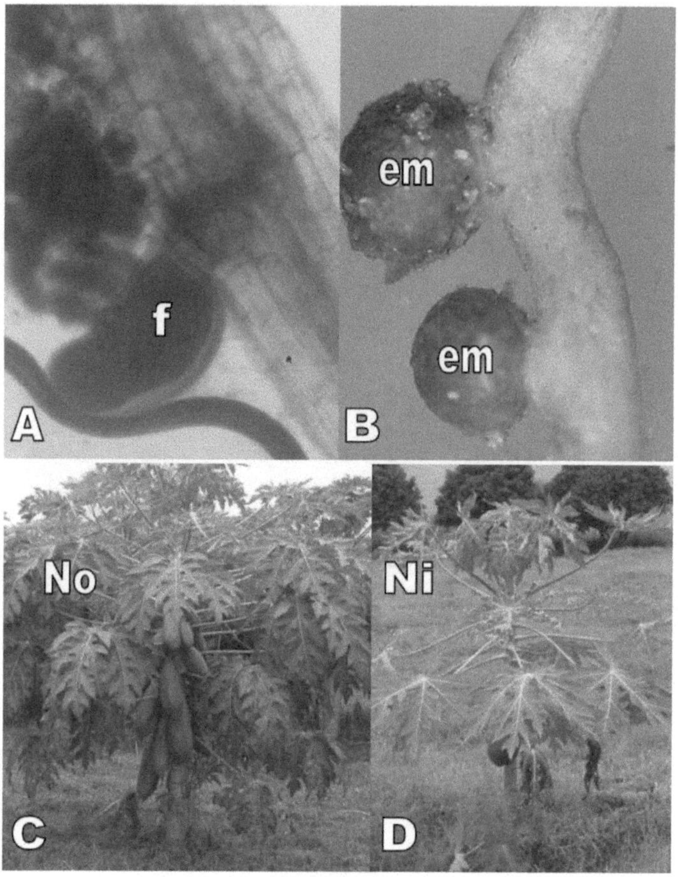

Figure 7. A) Female (f) of Rotylenchulus reniformis *with egg mass (em); B) Egg mass (em) of the nematode; one year old plants of papaya* Paraguanera-*type in absence of* R. reniformis *(C, No) and infested the nematode (D, Ni)*

that accessions 219-2, 221, 232-1, 241, 247-2 of *P. edulis* f. sp. *flavicarpa* and accessions 214-1 and 236-2 resulted resistant-tolerant, with 219-1 as susceptible. Accession 231 of *Passiflora quadrangularis* was also evaluated and should be considered resistant-tolerant. Resistant materials is promising for future breeding programs or as rootstocks.

2.15. Peach (Prunus persica)

Although peach is not a tropical crop, this cultivation in Venezuela is developed in agroecological areas with special characteristic: altitude between 900 and 2.400 msl, temperature from 12 to 23 °C, annual precipitations from 1.000 to 2000 mm, lands of irregular topography with moderate to high slope declivities. Peach is generally cultivated in small properties (0.5 to 10 ha) using the varieties Criollo amarillo (Yellow Creole) and Jarillazo. It is cultivated mainly in Aragua and Miranda States. However, small orchards exist in Merida, Lara, Táchira and Trujillo States, that are not considered in production statistics. For the year 1999 the harvested surface reached 1850 ha with a production of 28070 tons and a yield national average of 15 tons/ha (MPC, 1999).

Plant parasitic nematodes associated with peach are: *M. javanica*, *M. incognita* and some unidentified species of *Pratylenchus*, *Paratylenchus* and *Helicotylenchus* (Vargas, 1981; Crozzoli et al., 1987); However, patogenicity has been demonstrated only for *M. javanica*.

Decline or sudden death is one of the most important problems in peach orchards. Even when the cause or etiology of this disease has not been clarified completely, roots and soil of declining plants revealed the consistently presence of *Phytophthora cinnamomi*, *Pythium* sp. and *Fusarium* spp., generally associated with *Meloidogyne* spp., which is broadly diffused in the orchards (Crozzoli & Vargas, 1989); Suárez et al., 1999; González, 1993; Rondón, 1990). The symptoms of the disease consist on to dry rot and roots death, accompanied by foliage yellowings, swelling loss (flaccidity) of leaves and dried of branches, leading to the sudden death of the plant. Yellowing of leaves is also associatet to the lack of macronutrients in the plant, although they are present in soil. The use of resistant rootstocks to *Meloidogyne* spp. is not common in Venezuela although the resistance of Okinawa stock has been demonstrated (Crozzoli & Vargas, 1989). The main control method is the application of granulated nematicides. For further data on peach nematodes and their management, see chapter 7 in this volume.

2.16. Pineapple (Ananas comosus)

In Venezuela there are about 8000 ha cultivated in the Lara (6000 ha), Trujillo (1500 ha) and Sucre-Anzoátegui (500 ha) States. The cultivated groups are Red Spanish in Lara State, Perolera (Valera Roja and Valera Amarilla) in Trujillo State and a local cultivar, Cumanesa in Sucre-Anzoátegui States (Jiménez et al., 2001).

Many plant parasitic nematodes are associated with pineapple in Venezuela: genera include *Aphelenchus*, *Ditylenchus*, *Helicotylenchus*, *Meloidogyne*, *Paratylenchus*, *Pratylenchus*, *Rotylenchulus*, *Tylenchorhynchus*, *Tylenchus* and *Xiphinema* which occur in Aragua, Carabobo, Lara and Trujillo States (Yépez & Meredith, 1970; Petit, 1992). *Peltamigratus macbethi* and *M. incognita* are reported from Anzoátegui State. *Pratylenchus* spp. are reported from Táchira State (Suárez, 1977), and *P. brachyurus* from Bolivar State (Renaud, 1985). Jiménez et al., (2001) identified *Aorolaimus holdemani*, *Criconema demani*, *Ditylenchus acutus*, *Criconemoides (=Mesocriconema) ornatum*, *M. incognita*, *Paratylenchus nawadus*, *P. brachyurus*, *R. reniformis* and *X. dimidiatum* associated with pineapple in Lara and Trujillo State.

Suárez and Rosales (1998) and Jiménez et al., (2001) reported in roots infected by *M. incognita* terminal club-shaped small galls, with non-terminal fusiform galls causing brooming of the root system. Severe infections result in a stunted root system, poor anchorage and plants that are more susceptible to moisture and nutrient stress. *Pratylenchus brachyurus* causes necrosis that extend progressively over the whole surface of the root, separating the cortex from the central cylinder. Foliar symptoms result from deficient water and mineral supply. In both *M. incognita* and *P. brachyurus* attacks, the fruit size is reduced.

Red Spanish group and Valera Roja, the most cultivated varieties in Venezuela, are susceptible either to *M. incognita* and *P. brachyurus* (Suárez & Rosales, 1998).

2.17. Sapodilla (Manilkara zapota)

Sapodilla is grown principally in the Zulia State (about 1500 ha). The nematode fauna of sapodilla in Venezuela has been investigated by Petit (1990) and Crozzoli et al., (1997). *Hemicriconemoides strictathecatus* was the most common plant parasitic nematode observed, occurring in 40% of sapodilla soil samples proceeding from the Zulia State. *Hoplolaimus seinhorsti*, *Criconemoides (=Mesocriconema) ornatum*, *Xiphinema simillimum* and *Paratylenchus* sp. were also common (Crozzoli et al., 1997). Other nematodes observed include *Aphelenchus*, *Ditylenchus*, *Helicotylenchus*, *R. reniformis*, *Pratylenchus*, *Tylenchorhynchus* (Petit, 1990). The patogenicity of *H. strictathecatus* has been demonstrated (Saeed, 1974), but no evidence of damage is available.

2.18. Tamarind (Tamarindus indica)

In Venezuela, this crop is cultivated in small areas, scattered in the whole country. Parasitic nematodes associated are: *H. strictathecatus*, *X. brasiliense* and *R. reniformis* detected in the Zulia State (Castellano et al., 2004), and *Xiphidorus amazonensis* detected in the Apure State (Lamberti et al., 1999). *Hemicriconemoides stricthatechatus*, has been considered as pathogenic to tamarind (Saeed, 1974), but no evidence of damage is available.

REFERENCES

Avilán, L., & Leal, F. (1990). *Suelos, fertilizantes y encalado para frutales*. Editorial América, Caracas: 459 pp.

Bustillo, Y., Crozzoli, R., Greco, N., & Lamberti, F. (2000). Efecto del nematodo agallador *Meloidogyne incognita* sobre el crecimiento de la lechosa (*Carica papaya*) en vivero. *Nematologia Mediterranea*, 28, 163-170.

Carneiro, R. M. D. G., Moreira, W. A., Alves, A. M. R., & Gomes, A. C. M. M. (2001). First record of *Meloiodgyne mayaguensis* on guava in Brazil. *Nematología Brasileira*, 25, 223-238.

Casassa, A. M., Crozzoli, R., Matheus, J., Bravo, V., & Marín, M. (1998). Efecto del nematodo agallador *Meloidogyne incognita* sobre el crecimiento del guayabo (*Psidium* spp.) en vivero. *Nematologia Mediterranea*, 26, 237-242.

Casassa, A. M., Matheus, J., Crozzoli, R., Bravo, V., & González, C. (1997). Respuesta de algunas selecciones de guayabo al nematodo *Melidodgyne incognita* en el Municipio Mara del estado Zulia. *Fitopatologia Venezolana*, 10, 5-8.

Casassa, A. M., Matheus, J., Crozzoli, R., & Casanova, A. (1996). Control químico de *Meloidogyne* spp. en el cultivo del guayabo (*Psidium gujava* L.) en el municipio Mara del estado Zulia, Venezuela. *Revista de la Facultad de Agronomia*, (LUZ), 13, 303-312.

Castellano, G., Quijada, O., Jiménez-Pérez, N., & Briceño, E. (2004). Nematodos fitoparasíticos asociados con merey, tamarindo y semeruco en el estado Zulia y respuesta de dos cultivares de merey ante el nematodo agallador *Meloidogyne incognita*. *Fitopatología Venezolana*, 17, 6-8.

Comerma, J., & Paredes, R. (1978). Principales limitaciones y potencial agrícola de las tierras de Venezuela. *Agronomía Tropical,* 28, 71-85.

Cook, R. (1974). Nature and inheritance of nematodo resistance in cereals. *Journal of Nematology*, 6, 165-174.

Crozzoli, R. (2002). Especies de nematodos fitoparasíticos en Venezuela. *Interciencia*, 27, 354-364.

Crozzoli, R., & Casassa, A. M. (1998). Identificación de especies y razas de *Meloidogyne* asociadas al cultivo del guayabo en el municipio Mara del estado Zulia. *Revista de la Facultad de Agronomia*, (LUZ), 15, 107-108.

Crozzoli, R., Casassa, A. M., Bravo, V., & Matheus, J. (1997). Nematodos fitoparásitos asociados a cultivos del níspero y de los cítricos en el estado Zulia, Venezuela. *Nematropica*, 27,106 (Abstract).

Crozzoli, R., Casassa, A. M., Rivas, D., & Matheus, J. (1991). Nematodos fitoparásitos asociados al cultivo del guayabo en el estado Zulia. *Fitopatología Venezolana*, 4, 2-6.

Crozzoli, R., Chedas, C., & Casassa, A.M. (1995). Two *Hemicriconemoides* (Nemata: Criconematidae) from Venezuela. *Nematologia Mediterranea*, 23, 239-244.

Crozzoli, R., & Fúnes, C. (1992). Presencia del nematodo *Tylenchulus semipenetrans* Cobb,1913 en las principales zonas productoras de cítricos del estado Aragua, Venezuela. *Fitopatología Venezolana,* 5, 17-20.

Crozzoli, R., & González, A. (1989). Evaluación de resistencia a *Tylenchulus semipenetrans* en once patrones de cítricos. *Agronomia Tropical*, 39, 269-279.

Crozzoli, R., Graff, R., & Rivas, D. (1993). Nematodos fitoparásitos asociados al cultivo del banano (*Musa* AAA) en el estado Aragua. *Revista de la Facultad de Agronomia*, (Maracay), 19, 275-287.

Crozzoli, R., Lamberti. F., Greco, N., & Rivas. D. (1998). Nematodos fitoparasíticos asociados con los cítricos en Venezuela. *Nematologia Mediterranea*, 26, 31-58.

Crozzoli, R. Martínez, G., & Rivas, D. (1995). Manejo y fluctuaciones poblacionales de *Helicotylenchus multicinctus* y *Meloidogyne incognita* en banano, en Venezuela. *Nematropica*, 26,61-66.

Crozzoli, R., Perichi, G., Vovlas, N., & Greco, N. (2005). Effect of *Rotylenchulus reniformis* on the growth of papaya in pots. *Nematropica*, 35, 53-58.

Crozzoli, R., Rivas, D., Greco, N., Montes, L., & Gómez, K. (1997). Presencia de *Tylenchulus semipenetrans* en las principales zonas productoras de cítricos de los valles altos de los estados Carabobo y Yaracuy, Venezuela. *Nematologia Mediterranea*, 25, 151-154..

Crozzoli, R. & Vargas, G. (1989). Reacción de trece patrones de durazno a infestaciones de *Meloidogyne javanica*. *Fitopatología Venezolana*, 2, 16-18.

Dao, F. (1961). Los nematodos en relación a los cultivos y la producción. Caracas: Shell, 49-51.

Duncan, L. (2005). Nematode parasites of citrus. Pp. 437-466 *in* M. Luc, R.A. Sikora and J. Bridge, eds. Plant Parasitic Nematodes in Subtropical and Tropical Agriculture. CAB International, Wallingford, U.K.

FAO. (2006). Estadísticas de la FAO. On line at http://faostat.fao.org/ (30/04/2006).

González, J. A., & Webster, B. N. (1960). Investigaciones sobre la enfermedad "anillo rojo" en cocos y palma de aceite. Shell, Cagua: 14 pp.

Haddad, O., Meredith, J., & Martínez, G. (1975). Nematodos fitoparasitos asociados a cultivares y clones de banano en Venezuela. *Nematropica*, 5, 33-39.

Jiménez, N., Crozzoli, R., Petit, P., & Greco, N. (2001). Nematodos fitoparasíticos asociados con al cultivo de la piña, *Ananas comosus*, en los estados Lara y Trujillo, Venezuela. *Nematologia Mediterranea*, 29, 13-17.

Lamberti, F. (1973). Presenza di *Pratylenchus penetrans* in Algeria su palme da dattero in deperimento. *Nematologia Mediterranea*, 1, 63-65.

Lamberti, F., Crozzoli, R., Molinari, S., De Luca, F., Agostinelli, A., & Greco, N. (1999). Two species of *Xiphidorus* Monteiro (Nematoda: Dorylaimida): New records for Venezuela. *Nematologia Mediterranea*, 27, 83-93.

Leal, F. y Avilán, L. (1997). Situación de la fruticultura en Venezuela: un análisis. *Revista de la Facultad de Agronomia*, (Maracay), 23, 1-30.

Lugo, Z., Molinari, S., Crozzoli, R., Perichi, G., Greco, N., Castellano, G., & Jiménez-Pérez, N. (2005). *Meloidogyne mayaguensis* (Nematoda: Tylenchida) en Venezuela. Resumenes XIX Congreso Venezolano de Fitopatología. Barquisimeto, 14 - 17 de noviembre 2005.

Malaguti, G. (1953). "Pudrición del cogollo" de la palmera de aceite africana (*Eleaeis guineensis* Jacq.) en Venezuela. *Agronomía Tropical*, 3, 13-32.
Matheus, J., Suárez, H. Z., Rosales, L. C., Tong, F., Casassa, A. M., Bravo, V., & Nava, A. (1999). Histological reaction of *Psidium* spp. selections to *Meloidogyne incognita* in Venezuela. *Nematologia Mediterranea*, 27, 247-251.
McSorley, R. (1992). Nematological problems on tropical and subtropical fruit tree crops. *Nematropica*, 22, 106-116.
Molinari, S., Lamberti, F., Crozzoli, R., Sharma, S. B., & Sánchez-Portales, L. (2005). Isozyme patterns of exotic *Meloidogyne* spp. populations. *Nematologia Mediterranea*, 33, 61-65.
Montiel C., A., Sosa, L., Medrano, C., & Romero, D. (1997). Nematodos fitoparásitos en plantaciones de plátano (*Musa* AAB) de la margen izquierda del río Chama, Estado Zulia,Venezuela. *Revista de la Facultad de Agronomia* (LUZ), 14, 245-251.
MPC (Ministerio de Producción y Comercio). (1999). Plan de producción y disponibilidad. Subsector Agrícola Vegetal. Cultivos permanentes y semipermanentes (Superficie, Rendimiento y Producción). D.G.S.P. División de Frutales, Café y Cacao. Caracas.
Pernía, A., Zambrano, C., Rivero, R., & Bracho, M. (2002). Manejo integrado de plagas en el cultivo de parchita, *Passiflora edulis* Sims f. *flavicarpa* Degener en la agropecuaria "El Chamito". Memorias de la primera reunión venezolana sobre investigación y producción de Passifloras. UCLA-Postgrado en Horticultura. Barquisimeto, Venezuela, 59-60.
Petit, P. (1978). Estudio preliminar sobre los nematodo fitoparásitos asociados al cultivo de la vid (*Vitis vinifera* L.) en Venezuela. *Nematropica*, 8, 66-68.
Petit, P. (1980). Respuesta de cuatro variedades de vid al nematodo nodulador (*Meloidogyne* spp.). *Nematropica*, 10,103-106.
Petit, P. (1982). Resultados preliminares sobre control químico de nematodos fitoparasitos en la variedad de vid Cardinal. *Agronomía Tropical*, 33, 23-32.
Petit, P. (1990). Reconocimiento de nematodos fitoparasitos asociados a frutales de importancia económica en Venezuela. *Fitopatologia Venezolana*, 3, 2-5.
Petit, P. (1991). Presencia del nematodo de las cítricas (*Tylenchulus semipenetrans*) en la zona citrícola del centro de Venezuela. *Fitopatología Venezolana*, 4, 10-12.
Petit, P. (1993). Respuesta del portainjerto híbrido Criolla negra y el cultivar de vid Italia a la infección de *Meloidogyne incognita*. *Fitopatología Venezolana*, 6, 18-21.
Renaud, J. (1978). Nematodos asociados al cultivo de la vid en Venezuela. Proceeding of the 6° conference on virus and virus disease of the grapevine. Ministerio de Agricultura. INIA. España: 301.
Renaud, J. (1985). Consideraciones sobre *Pratylenchus*, Filipjev, 1936 (Nematoda: Pratylenchidae), contribución a especies encontradas en Venezuela. Trabajo de Ascenso, Universidad Centrooccidental Lisandro Alvarado. Barquisimeto, Venezuela: 83 pp.
Renaud, J. (1996). El género *Xiphinema* (Nematoda: Longidoridae) en Venezuela. Trabajo de ascenso, Barquisimeto, Venezuela: Universidad Centroccidental Lisandro Alvarado. Barquisimeto, Venezuela: 126 pp.
Rondón, A. (1990). Enfermedades de los frutales en Venezuela. Instituto de Investigaciones Agronómicas, Maracay: FONAIAP-CENIAP, Serie B, N. 9.
Rosales, L. C., & Suárez, Z. (2001). Nematodos fitoparasíticos asociados al lechoso y distribución geográfica en Venezuela. *Fitopatología Venezolana*, 14, 21-23.
Saeed, M., (1974). Studies on some stylet-bearing nematodos associated with sapodilla (*Achras zapota* L.) with special referente to *Hemicriconemoides mangiferae* Siddiqi, 1961. PhD. Thesis, University of Karachi, Pakistan: 137 pp.
Salazar, C. (1934). Resultados de un viaje de estudio y divulgación en el estado Sucre. *Boletin Ministerio de Salubridad y Agricultura*, 2, 403-457.
Shell (1958). Servicio Shell Para El Agricultor. Hacia el control del anillo rojo en coco. *Noticias Agrícolas*, 1, 137-138.
Shell (1973). Servicio Shell Para El Agricultor. Con el control del coco cigarrón o picudo se evita la incidencia del anillo rojo en los cocotales. *Noticias Agrícolas*, 32,127-129.
Suárez, Z. (1977). Reconocimiento de nematodos fitoparásitos en el cultivo de la piña en Venezuela. Datos preliminares. Memorias IX Jornadas Agronómicas, Maracay, Venezuela, 84 (Abstract).
Suárez, H., Z., Gómez, M. A., & Rosales, L. C. (2004). Reacción de nueve accesiones de *Passiflora* al ataque del nematodo arriñonado, *Rotylenchulus reniformis*. *Fitopatología Venezolana*, 17, 9-11.

Suárez, Z., González, M. S., Rosales, L. C., & Tellechea, V. (1993). Alteraciones histológicas en *Passiflora edulis* f. sp. *flavicarpa* inducidas por *Rotylenchulus reniformis*. *Fitopatología Venezolana*, 6, 11-14.

Suárez, Z., González, María, S., & Tellechea, V. (1994). Distribución geográfica y daño en parchita causado por *Rotylenchulus reniformis* en Venezuela. *Nematropica*, 24, 91-92.

Suárez, Z., Rondon, A., Tellechea, V., Solorzano, R., & Navas, R. (1984). Seis años de estudio sobre la asociación de hongos telúricos del suelo y nematodos fitoparásitos en aguacate. Memorias XI Jornadas Agronómicas, Maracaibo, Venezuela, 36 (Abstract).

Suárez, Z., Rondon, A., Tellechea, V., Solorzano, R., & Navas, R.(1992). Asociación de hongos del suelo con nematodos fitoparásitos en aguacatero. *Agronomía Tropical*, 42, 321-328.

Suárez, Z., & Rosales, L. C., (1998). Nematodos asociados a los frutales de importancia y su control. I. Frutales perennes. *FONAIAP Divulga*, 59, 13-18.

Suárez, Z., & Rosales, L. C., (1998). Nematodos asociados a los frutales de importancia y su control. II. Frutales anuales. *FONAIAP Divulga*, 60, 38-41.

Suárez, H. Z., & Rosales, L. C. (2004). Problemas nematológicos en Musáceas. *Revista Digital del Centro Nacional de Investigaciones Agropecuarias de Venezuela, CENIAP HOY, 6.* http://www.ceniap.gov.ve/ceniaphoy/articulos/n6/arti/suarez_z/arti/suare.

Suárez, Z., Rosales, L. C., Rondón, A., & González, M. S. (1999). Nematodos fitoparasíticos en el complejo de enfermedades de algunos frutales en Venezuela. *Nematropica,* 29, 135 (Abstract).

Suárez, Z., Rosales, L. C., Rondón, A., & González, M. S. (1998). Histopatología de raíces de *Psidium guajava* atacadas por el nematodo *Meloidogyne incognita* raza 1 y los hongos *Macrophomina phaseolina* y *Fusarium oxysporum*. *Fitopatología Venezolana*, 11, 44-47.

Torrealba, P. (1968). Control de enfermedades. Nematodos en Hortalizas. Noticias agrícolas. *Servicio Shell para el Agricultor*, 13, 34-41.

Vargas, G. (1981). Censo de nematodos en huertos de durazneros en la zona montañosa central de Venezuela. Facultad de Agronomía, Universidad Central de Venezuela, Estación Experimental "Bajo seco", Boletín 1, 17 pp.

Webster, B. N., & González, J. A. (1959). Investigaciones sobre la enfermedad "anillo rojo" en cocos y palma de aceite. Memorias IX Convención de la Asociación Venezolana para el Avance de la Ciencia. Caracas, Venezuela: 14 pp.

Webster, B. N., & González, J. A. (1960). Investigaciones sobre la enfermedad "anillo rojo" en cocos y palma de aceite. Memorias X Convención de la Asociación Venezolana para el Avance de la Ciencia. Caracas, Venezuela: 13 pp.

Yépez, G. (1965). Presencia del "nematodo de las cítricas" (*Tylenchulus semipenetrans* Cobb, 1913) en la zona de Caripe, estado Monagas. *Revista de la Facultad de Agronomia*, (Maracay), 3, 111- 116.

Yépez, G., & Martínez, G. (1969). El "anillo rojo" enfermedad del cocotero y la palma africana. *Natura*, 38, 46-50.

Yépez, G., & Meredith, J. (1970). Nematodos fitoparásitos en cultivos de Venezuela. *Revista de la Facultad de Agronomia*, (Maracay), 5, 33-80.

Yépez, G., Meredith, J., & Pérez, A. (1972). Nematodos del banano y plátano (*Musa* sp.) en Venezuela. *Nematropica,* 2, 47-51.

3

JUAN CARLOS MAGUNACELAYA

CONCEPTS IN MANAGEMENT OF TREE CROPS NEMATODES IN FRUIT PRODUCTION SYSTEMS

Universidad Católica de Valparaíso, Facultad de Ciencias, Avda. Brasil 2950, Valparaíso, Chile

Abstract. The effects of irrigation and other techniques applied in intensive fruit crop productions in Chile are discussed, with emphasis on roots protection and nematode management strategies. Concepts on resistance and tolerance are given and the important role of the interactions among rootstocks, irrigation practices and nematodes is highlighted. Nematode management options include prevention, chemical treatments and use of plant resistance elicitors. Soil conditions and diagnosis are then reviewed focusing on plant vigor, root system analysis by means of test pits inspections, sampling, application of minirhizotrons and other management strategies.

1. INTRODUCTION

Professionals dedicated to the agricultural sector today focus on obtaining an orchard efficient from the agronomic point of view, which is reflected by better fruit quality and competitiveness in terms of export potentials. Under this perspective, the greatest importance should be given to the tre root system. Progress was made from an "ancient" vision in which all plant need had to be supplied through the plant leaf surfaces, to a new vision in which researchers and producers dedicate time to study the underground condition of the plant, a research topic considered for years as the less studied side of agronomy.

The process of economic globalization takes us to the opening of new markets, consequently competition emerges and producers are forced to achieve higher levels of efficiency to handle their fields. A series of elements which producers were not used to consider must now be integrated. Today, strategical associations must be searched to get new and better markets, but these proposals require products of good quality, and success depends on plants in good shape. As a consequence, fruit must show best qualities in taste, nutrition and aspect. During the post-harvest phases, the best quality of fruit must be lenghtened in order to reach consumers in best conditions, although this process can take even weeks or months.

At this regard, fruit producers must know how to handle their orchards in order of increasing their commercial expectatives, being also efficient as concerns investments and costs in relation to returns. To achieve this goal, trees must be

A. Ciancio & K. G. Mukerji (eds.), *Integrated Management of Fruit Crops and Forest Nematodes*, 85-99.

balanced, well handled, as well as correctly located and adequate to the properties of each region and climate. Even though soils are not always the most appropriate, efficient watering, fertilizing and handling of sanitary problems are needed to obtain products of better quality.

To achieve these objectives, many specialists and producers in the world have a strong tendency to consider more solid, stable and sustainable procedures to be applied to plants. A good balance between aerial and underground or radical development is fundamental at this regard, since the aerial part and hence production depend on better nutrition conditions. Consequently, to improve fruit production quality and quantity, producers must to aim at identifying those cultural conditions allowing the orchards to produce in a continuous and balanced way, sustainable in time.

When our objective is the permanent quality, phytoparasitic nematodes must be incorporated into the "agricultural equation" as a limiting factor affecting the productive balance of the trees. Nematode damage must be suspected when trees do not react to the cares deployed by producers, and no increment in productivity is achieved. Good practices, hence, require producers to worry about the radical problems caused by phytoparasitic nematodes. To improve their production technology, producers should try to control at least some of the several variables affecting the agricultural system represented by the cultivated field. No system other than crops works with a greater diversity of living beings, and agricultural practices try, indeed, to make homogeneous something that, by nature, is mostly heterogeneous. Diversity must be considered, in general terms, as a positive factor, and producers should look for the genetic variability since it provides the basis for increasing the possibilities of adaptation to new or unknown conditions (Magunacelaya & Dagnino, 1999).

From a genetic point of view, phytoparasitic nematodes are often in a favorable situation since plants, many times, have a monoclonal origin with identical strengths and weaknesses characterising all individuals in a field. It is easy to estimate the disadvantage that plants have in this equation, since the radical health (on which the production cycle relies) is what nematodes damage first. This damage is produced by their stylets which cause physiological as well as productive alterations, symptoms that usually begin with reductions of vigour, then lowering the plant conversion efficiency to finally affect quality. If nematodes act in association with other pathogens, their action can be expressed (or measured) directly on the aerial parts of the plants. As in the case of nematodes transmitting viruses, or of those species that weaken the plant growth, parasitism exposes plants to damage by other pathogens i.e. fungi or bacteria, as well.

All strategies of handling nematodes on fruit crops must involve diverse actions to produce a satisfactory result, and these must be worked out as a whole, with the only objective of reducing the incidence of the nematodes attack, especially when dealing with sedentary species. In this case, feeding activity permanently affects the root cell structure, due to the permanent feeding site. In susceptible plants, parasites depend on the viability of the cells which nourish them, and the main management goal is to keep the lodging plant alive for many years, searching for the conditions leading to a tolerance of the parasite population in roots.

In Chile, the increase and diversification of watering techniques has been accompanied by an increase of nematodes problems. Watering technology has allowed the establishment of crops in soils which were difficult or impossible to cultivate before, when traditional watering methods were used. However, too frequent waterings expose these soils to unbalanced soils-to-water volumes. In these conditions the plants are forced to survive in a "plant pot" of different size, depending on the technology applied. The roots can try to evade from the watered zone only during the most humid season, but later, during the dry season, they die due to the lack of humidity. Many producers have forgotten that fruit trees (and their roots too) require, to thrieve normally, relatively high levels of oxygen (variable among species), usually between 18 and 20%, and that soils with a higher content of clay or mud can require many days after the irrigation to regain these levels (Allen et al., 1969) .

In conditions of dry land or traditional laying or furrow irrigation, nematodes do not benefit of favourable conditions to parasitise or damage the roots, and when attacking, their damage may be unnoticed. However, the variability among plants affects their reaction with differences in vigour, precociousness, productivity and other characteristics which, in a more primitive agriculture, were accepted as "normal".

Roots asphyxia is often observed by producers in many countries, even in sandy, light soils with high porosity. Some groups of nematodes are favoured by this high level of permanent humidity, which allows them to travel and locate growing roots. Fruit trees with perennial leaves differ from deciduous fruit trees for their morphologic and physiological qualities. In relation to the management of nematode parasites, they require the same care as for other fruit trees. The particular adaptation of keping leaves during the winter season is related to physiological aspects, and this brings a series of significant differences in the managing of plagues and air diseases, since leaves are present all over the year. Significant problems may be experienced in winter with frosts, but with soil submersion, roots growth will be limited to very defined periods.

In the management strategies that will be exposed later, it is necessary to check out a few concepts, which relate directly to the characteristics of nematodes. As soon as nematodes colonize a soil we must recognize that it will last forever since, being structurally and functionally simple, nematodes are very resistant, either as individuals and population, to almost any change applied to soil. For this reason we must introduce the concepts of "living together" and "defending roots from the nematodes damage" rather than "killing all individuals". The major or minor efficiency of a nematicide should be considered in relation to the major or minor degree of protection given by the radical growth. When root tissues are sclerotised, the root cannot be considered any more vulnerable to the nematodes assault or to other pathogens, and a protection level may be considered achieved. At this moment, indeed, roots are more vulnerable to bad irrigation practices.

In field conditions, nematologists rely on holes or test pits to bring together precedents of the radical systems of an orchard, to develop decisions concerning products application or other actions. The radical system observation pits can result more important and informative than the analyses of nematodes in soil, which inform us only about the nematode species and its population levels. What definitively must be known concerns the dimension of the nematodes assault to the roots, which may be deduced by observing them all along the pit profile.

In general, nematodes are unnoticed at the eyes of farmers and move or disperse easily by means of several agricultural activities. As a consequence, the first and best form of control is to prevent their dispersal. In the last 20 years many progresses in this respect were obtained in Chile, but nothing can be done yet in the more ancient valleys, where nematodes managed to be already established (Magunacelaya, 2003).

2. GOOD QUALITY OF ROOT, AND BALANCED PLANTS

As mentionned before, when nematodes appear the most suitable action is trying to live with them, which means using all the possible, available and most efficient tools, in the right moment. The depletion of soil or fatigue is a common phenomenon in agriculture, and today it is possible to find abundant technical notes concerning the negative outcomes of a crop repetition. In many situations, the determinant factors of this depletion or fatigue can be related to nematodes. The introduction of new technical production systems requires the use of selected plant material and accentuates situations of soil roots variability. When the orchards begin to be watered technically, their response to the requirements of the producer is not always possible when unsuspected enemies as nematodes are present. Consequently, a production failure, in the different growing conditions, may be eventually observed.

3. CONCEPTS OF RESISTANCE AND TOLERANCE

Plants resistance or tolerance to nematodes should be considered as two different reactions. Resistance considers nematodes as the reference entities, observing the variation of the population reproductive index (increases/decrease or Pf/Pi). In general terms, when the population increases the plant resistance is lacking, whereas when nematodes diminish we can consider that a given level of resistance is present. To reduce the population it is necessary to rely on some kind of interference affecting nematodes reproduction. In general, this can be a barrier or obstacle that the plant deploys, in any moment during the interaction established with the parasite. Nematodes use a complex series of chemoreceptors to detect minimal amounts of exudates released during active root growth, specially from the elongation zone immediately located after the root apex. This zone also presents thinner cellular or walls in the process of suberification.

A later level of resistance is given by the walls of the radical cells which sclerotise more rapidly, impeding the nematodes stylet penetration towards the root interior zone, or in case of endoparasitic species, which penetrate with all their body to establish themselves into the radical tissue. When nematodes enter completely the roots, their parasitic bahaviour can be of two types: mobile and sedentary. Mobile nematodes will move in the interior of the root causing destructions, which they will exploit to feed. This is the case of injuring nematodes i.e. lesion nematodes, *Pratylenchus* spp.

In sedentary nematodes, i.e. *Tylenchulus semipenetrans* (Fig. 1), after its establishment in the feeding site the parasite is not able to move any more, and a permanent site of nourishment which feeds the adult for all its life, is formed. To achieve this goal, the juvenile establishs itself inside the cell, releasing secretions of its esophageal glands that, once reaching the plant cell cytoplasm, alter its metabolism, starting a cascade of biochemical changes leading to cells hyperplasia and hypertrophy. The growing plant cells form the so-called nurse cells, active during the whole life-cycle of the parasite which exclusively feeds on them. This very narrow relationship between nematodes and roots evolved during many million years, and developed in parasites a kind of host dependence on the activity of these cells on which their life depends. The apical end of the stylect also acts as a kind of sieve, due to its tiny orifice, thus preventing the suction of cytoplasm structures or organelles, vital for the cells. A barrier of resistance to this level is given by the obstacles that the plant puts in forming nurse or giant cells, for what the nematode in the root cannot be fed.

Tolerance is different from resistance, and refers to a plant condition regarding the nematode assault, being admitted as very tolerant the plant whose roots grow in spite of the fact that nematodes are attacking, eating, and entering the radical tissue. The populations of nematodes arise resting on these roots, but the plant can grow and produce anyway, without falling down (González, 1984; 1987).

Figure 1. Damage caused on citrus trees parasitised by a chilean population of Tylenchulus semipenetrans: *note defoliation (A) and stunted growth with fruit loss (B).*

4. ROOTSTOCKS AND NEMATODES

In the genetic diversity that characterised tree crops in the era of pre-industrialised agriculture, a greater variability of resistance or tolerance to nematodes acted as a factor of sustainability, no more available today or largely unknown. This seems a valid hypothesis explaining the actual levels of plants susceptibility, since plants grafted on seed produced rootstocks differ physiologically and morphologically, and that resistance indeed depends on morphologic and physiological factors. The modern procedures of clones reproduction or spread of this standard material reduce significantly the basic roots variability. However, it is necessary that the nematode factor be considered in this selection process, to get major guarantees of success.

Nematodes are today an important factor recognized in soils of the whole world, and must be present in the equation that allows for decision making, in any agricultural project. Since they can increase the costs of production, their preventive evaluation is needed, to avoid any kind of future "surprise". The rootstock radical systems, as well as the plants grafted on seed produced rootstocks, respond in function of their aptitude to a series of soil characteristics i.e. irrigation, soil type (sandy, clay, muddy), compaction, depth, presence of salts, conductivity and pH, among others. The plants grafted on seed produced rootstocks are more sensitive to phenomena of asphyxia or anoxia, and the roots surviving in soils watered and compacted with excessive frequencies, are the structural ones. The thinnest radical tissues, more sensitive to anoxia, die and disappear in situations of unbalanced irrigation. Any excesses in the irrigation frequency can, in some situations, affect the roots negatively, but they affect also the parasitic nematodes, i.e. the citrus nematodes, whose development may be favored by permanent humidity conditions (Inserra et al., 1994).

More rustic soils, naturally selected during long periods of time, are difficult to handle in productive systems with technical irrigation. It is difficult to determine the times of irrigation, the frequencies, or to relate with the phenologic states, the climatic conditions, the variations of soil or the presence of uneven soils, as well as of groundwater and its quality. At this regard knowledge-based agriculture is also costly, and not always an improvement in the management of technical parameters can reduce the complexity of the crop system. Trying to make fewer mistakes or to give good conditions to the plant, we fall down in situations that favor the main root enemies, phytoparasitic nematodes. On the other hands, the rootstocks used to accommodate better to the shallowest and compacted soils show greater sensibility to asphyxia or other diseases. A balance must hence be searched for, based on knowledge of previous field history and experience (Magunacelaya, 1996a; 1996b).

5. NEMATODES MANAGEMENT STRATEGIES

5.1. Virgin Soils

The first and most efficient way of controlling nematodes, even though very basic, is to prevent them from coming to our orchard. For this, it is essential that whenever "virgin" soils are going to be cultivated, healthy plant material be used. Quarantine and certification schemes also apply at regional scales and certified material free of pathogens and/or parasites must always be used. Althouh more expensive, this precaution may save future costs of differnet orders of magnitude, including removal of trees before time (Magunacelaya, 2004). It requires however, skilled personnel for the identification of species(Molinari et al. 2004), knowledge about their endemism and epidemiology, as well as the deployment of eventual inspection and monitoring efforts to avoid their introduction or spreading.

5.2. Fumigations

Soil fumigation with products like methyl bromide or 1,3 dichloropropene, represents an alternative when soils are highly infested by nematodes, generally present in situations of replant in which the new plants inherit plagues and diseases from the previous ones. In the sequence of control actions, however, it is possible to set a series of other less aggressive alternatives. For fumigation to be effective in nematode control, a temperature of more than 20 °C is needed to activate the chemical fumigating product. Low levels of humidity are also needed to allow the product to move well in soil (Magunacelaya, 2005). With the fumigation, the correct concept that should be handled is helping the new plants that will be established to grow as rapidly as possible in infested soils, so that they can start colonizing volumes of soil with minor nematode loads.

5.3. Replants

In replant situations, the nematode distribution in the soil volume is in relation to the roots distribution of the previous plantation. The zones in which the previous fruit trees had major root concentrations is where the highest concentrations of nematodes can be found. If the management actions do not include fumigation as part of the protocol, this information will be relevant at the moment of establishing the frame of plantation for the new orchard.

When replanting without fumigating, and depending on the legislation locally applied, it may be necessary to use other nematicides, i.e. granular carbamates, at the moment of establishing the plantation. The best effects of these nematicides can be obtained when they are added in the first layers of soil in which the plant is installed. In this way, the product remains in the soil top part and later irrigations will dissolve, activate and spread the active molecules in soil.

The application of composted organic materials also appears useful, which contributes to a good quality soil, rich in nutrients and without parasites. It is possible to use fresher, or less composted manure. In the latter case the risk occurs of some phytotoxics effects on plants roots, for it is convenient that these contributions be spread in the volume of soil in places that the root will explore several months later, after transplant.

The last action concerns the possible use of nematicides through the irrigation systems. However, in orchards whose plantation frame is less than 1000 plants · ha^{-1}, this type of treatments is not recommendable, preference being given to the applications with a reservoir, allowing the nematicide solution to come directly on the plants. It is not suitable to make the application through the dropper system, since the nematicide will fall down in volumes of soil which the roots do not reach yet and therefore there is nothing to protect. Also, considering that in newly planted orchards only one out of every three droppers will get directly to the roots of a young newly transplanted tree, two thirds of the volume of the applied solution or product will be lost. This will hence increase the cost of application. Let's remember, however, that the nematicide products must be used to protect the radical system of the plants, rather than killing nematodes.

In the initial plant growth phases it is necessary to achieve the maximum potential growth of roots per volumes of soil. This needs appropriate irrigations to promote the roots to develop a suitable radical volume, and stimulants, i.e. products that generally contain hormones like auxines. Today, a new generation of products strengthening the innate response of plants in any affected organ are appearing, i.e. the products inducing a systemic acquired resistance (SAR), which are showing promising results. Very encouraging results are observed in plants when, in presence of parasitic nematodes, the nematicides are used applying, after a few days, SAR products or elicitors.

In conclusion, although these practices may sustain the farmer when replanting a fruit orchard, when the nematodes are already in soil, farmers must learn how to live with them as best as possible. This conceptual change in roots protection by phytoparasitic nematodes implies a great change in the form in which these phytoparasites must be managed. Later on some ideas of "living together" with the nematodes are developed.

6. TREATMENTS

As stated before, chemical treatments have the advantage of a quick response and appear useful in emergency situations. These products are nematicides and also root products, elicitors and SAR promoters. In this group we also include the soil fumigating products, useful in conditions of extreme sanitary problems with nematodes and/or fungi. Generally, they are toxic compounds that affect the operator and plant too, so they cannot be used in established plantations and must be applied with care by protected operators.

6.1. Fumigation

Minimal temperatures (around 20° C) are needed for a successful fumigation. An appropriate preparation of subsoil to 70 cm of depth as minimum is also needed, to release and open spaces that promote aeration and allow a good distribution of the gas-phase products. The ideal humidity is between 30 to 50 % of the field capacity, in order to fumigate with better results when the soil surface has humidity as for sowing, and it is drier towards the deepest layer. The dose of fumigant depends on the type of soil, the product and the pathogen/parasite density (Magunacelaya et al., 2004a; 2004b).

6.1.1. Doses and Dates

The dates of the product application are important, together with the doses per ha or volume of soil or plant, as well as the concentrations of application. The moment of products application must be based on the condition (state) of the root that we want to protect. In fruit trees, it is not useful to posticipate the moment of application since the nematodes will not be present or will be in a more resistant condition, making the early application useless. A too early nematicide application can be peformed if the majority of the nematode population is in the egg stage, which makes it very resistant to any application, since the juveniles that we want to control are protected within the egg and, according to the species, very difficult to penetrate. The suitable choice is to expect the nematodes to be active to apply the product, by monitoring in time the densities changes of vermiform stages in soil.

6.1.2. Application

Attention must be given to the activity of the plant roots, which in some cases may activates the nematodes out of their condition of latency. For some nematode groups (mainly cyst species) some root exudates stimulate the eggs hatching. The best way to select the product application date is to know the time of roots growth and to protect them during those periods. In general, perennial plants have two key moments for root growth, the first in spring, when the fruit has thickened and is growing, and the second in autumn, when the plant works for itself accumulating reserves.

Recognizing these two dates of maximum root growth intensity allows us to apply the nematicides, the root products and/or resistance elicitors, in a timely manner. A too early application will not last enough to protect all new roots, due to the characteristics of main nematicides of worldwide use, since in these periods (specially in spring), farmers irrigate more intensively, diluting the products and making them prematurely inactive.

6.1.3. Concentration

All the nematicidal products are toxic at certain concentrations, and every much diluted nematicide stops being toxic. The root products and elicitors of resistance also depend on the concentrations of application though their action does not depend so directly as the nematicides. If the products are applied much diluted they will not have effect of any type. Every nematocidal product that is commercialized should display, following the local legislation and rules, a dose per ha and a concentration of application. It can be complicated for a producer to be adviced that a certain dose at a certain concentration must be applied, but it is easy, if a technician performs the calculations and the producer is given a time of application, which is calculated according to the particular characteristics of the irrigation system, or the time that the product must be dropping through it. The number of application represents, in the systems of technified irrigation, known volumes of water. By knowing the volumes of the product and the water of the irrigation systems applied, the concentrations can be exactly determined. This is one of the several factors which may give erratic results, when products are applied incorrectly. However, a few liters of nematicide per ha cannot eradicate nematodes from a great volume of soil. Several tests made in the last years suggested that it is very difficult to locate nematicide products, with plants established in more than 300 or 400 cubic meters of soil per ha, aiming at protecting the plant roots growth. Finally, it is obligation of the specialists to evaluate the results of any application with chemical products or with with another type of treatment applied (Magunacelaya, 2005; San Martín & Magunacelaya, 2005.).

6.1.4. Volume of Treated Soil

Also in relation to chemical products, it is important to determine where to make the application, that is to say, to determine which will be the volume of soil (out of the total) to which the product solution will be applied. This is due to the fact that it will be impossible to cover the whole field volume of soil with a few liters of solution. It is frequent that under the irrigation hoses a precarious radical quality is present, due to the inadequate use of the irrigation system. Then, the roots grow at the edges of this asphyctic bulb, which is where they find an optimal level of humidity and oxygen. In a condition such as this, it is not wise to apply the nematicide or root products in this volume of anoxic soil, since it will be impossible that in this place the roots will grow, until the strategy of irrigation does not change. For a suitable application of the nematicide, root products or SAR elicitor, we need to know very well our irrigation system, as regards how much water the system precipitates per unit of surface and time, in order to be able to correctly determine the time in which the solution must precipitate, and the volume of soil to cover with the application.

6.1.5. Elicitors of Resistance

The elicitors or SAR activators are very useful in the management of fungal or bacterial diseases, as well as of phytoparasitic nematodes. SAR is a system of wide spectrum of plant resistance, which produces certain natural compounds or so-called activators, inductors or elicitors, and that does not possess direct action on the pathogen or parasites.

When a pathogen gets in contact with any part of a plant, two local and further systemic reactions are produced. Normally, the plant puts in place some barriers to the pathogens, i.e. the cellular wall and waxes. Also, it can generate local defenses as the death of some cells. The local hypersensitivity reaction produces the synthesis of phytoalexins, accumulation of salicylic acid, and/or a major resistance of the cellular wall. All these elements restrict the development of the invading parasite or pathogen. SAR is a secondary and wide response through the whole plant that goes beyond the exact place where the stimulation was produced, leading to the systemic formation of resistance proteins and the stimulation of acids like the salicylic and jasmonic. Among the proteins or enzymes which are increased, there are the chitinases, suberin and lignin.

The salicylic acid is capable of inducing genes of defense in not infected leaves, though some works suggest that it is necessary to establish the defense response in local and systemic leaves, but it does not represent the mobile sign itself. In response to the pathogen assault, SAR activates numerous genes whose products degrade the cellular wall of bacteria or fungi, destroy infected cells, etc. The induction does not occur only in the tissue initially infected, but also in leaves and other tissue exposed to the pathogen, due to the signs that are transported through the phloem. With the activated SAR the plant has the defenses ready to be used in case of need. Chitosan is considered in Chile as an elicitor of nematode resistance in plants, whose active ingredient (chitin) after a configuration change, allows for solubility (Magunacelaya, et al., 2003; 2004a).

7. SOIL CONDITIONS

A requirement of vital importance to get best perfomance with any product in soil is optimal soil humidity, since dry or flooded soil should be avoided. Nematodes live in the soil pores, and in this space they must be reached by the chemicals applied. However, these products flow through the soil microspaces. A dry soil will in fact absorb the product solution whereas a flooded one will prevent its movement, specially the vertical one, down to deeper levels.

When a nematicide is applied, the basic concept is that this is a product application rather than an irrigation or a fertilization with a nematicide added. A previous calibration of the dripping irrigation system is also important to apply chemical products. If possible, a known volume of water must be stained in the reservoir of injection, (i.e. 500 liters), measuring how long the dye leaks in one of the central droppers of every irrigation sector. This strategy will allow the nematicide application at the suitable concentration or in the adequate time, regulating the water volume of the reservoir in which the product will be dissolved.

8. DIAGNOSIS

The diagnosis of phytoparasitic nematodes, according to their importance, should proceed in the order herein shown.

8.1. Plant Vigor

The first step concerns the detection of reduction in the plant vigor, specially if it does not appear to be related to a known reason. The observer must recognize some problems due to the loss of vigor, specially in the periods in which the trees need to use their radical systems. In general, it is detected when in spring, with suitable temperature conditions, adequate soil humidity, the plants need that the radical system works to generate new outbreaks and to make the fruit grow, but this does not happen. A reduced plant vigor can represent the initial symptom of an increasing population of nematodes in the roots. It can also be related to a lower fertility or induction of fruit loss or small size. When doubts exist in the identification of the reasons of these deficiencies, the actions that must be made, in order of priority or utility will be the test pits confection or root pits observation, followed by the sampling of soil for nematode analysis, and finally the use of minirhizotrons.

8.2. Test Pits Analysis

The test pits, or pits of observation, must be wide enough to allow the observation of the soil profile at the major possible magnitude, in vertical and horizontal directions. The most standard dimension of a test pit for root study must be up to the depth where roots exist, laterally including at least half of the distance among the plantation rows. It is necessary to visualize both the sector watered more directly, the bulb of irrigation, and the space between rows, as well as the driest sectors of soil profile.

The test pits must be made constantly, and some producers have the habit of prepearing test pits with periodicity. In a test pit, the quality of the roots can be determined if these are affected or not by nematodes. It is also possible to identify the nematodes in the roots based on their symptoms (i.e. galls), or to determine the roots distribution along the soil profile, as well as where the roots are found (under or far from the dropper), or in the whole profile, characterizing the type of soil, and the presence of layers. Test pits hence are used for evaluating the present and past irrigation conditions, the need of nematode control, the efficiency of already applied treatments and also deciding for analysis of another type.

In a test pit, it is possible to determine the results of a tested strategy of nematodes handling, as well as the use of chemicals, fertilizer or compost, and the needs of subsoils (Fig. 2). The pit must be made in the good sectors and bad sectors as well, studying both situations, examining first the pit of the good sector and then the deficit one. The complement of the test pit actions is the soil sampling, to analyze the presence of nematode species and their population levels.

8.3. Sampling Strategies

For statistical samplings, 33 subsamples are usually collected from a 0.5 ha surface. This strategy is useful when there is still nothing established in the area. When plants are already established, the sampling should be directed to the plant roots, since they attract nematodes.

The most interesting sampling when establishing variations in the nematode populations in time is monitoring, which consists in sampling in certain fixed places marked in such a way to avoid the variability due to the spatial distribution of the nematodes in soil. At least, two samples or two monitoring stations per irrigation sector should be made, and every sample should consider at least 4 or 5 subsamples of similar volume from the zone, to get a whole sample with approx. 0.5 kg or 500 cm^3 of soil. The samples can be made in any period of the year. Sampling before spring allows the farmer to plan decisions about necessary treatments, or in function of better prices expected. The samples must have an appropriate humidity, neither flooded nor dry, and must be obtained from good or regular plants, never from the worst plants, unless a specific decision on them has to be taken.

8.4. Treatments

After test pit analyses and soil sampling, the third action concerns the treatments. To decide if nematode control treatments are needed there must be several complemented conditions: the roots condition, the condition of the aerial part of the plants, their growing capacity, the rate of aerial spring growth and the results of the nematode analyses from the rizosphere. All these precedents allow to decide the treatments in any of three possible moments: at post-crop, winter treatment, or spring treatment. The treatments may be applied in different period of the year, depending on the different objectives that must be adapted to the special conditions of every orchard.

8.5. Minirhizotrons

The minirhizotrons are a useful tool in diagnosis. They are small glasses that cover a section of a soil and roots cut that allow the inspection and measure of the radical growth, through the glass. The construction technique of the minirhizotron consists of digging a hole of approx. 60 cm depth × 50 cm width × 70 cm length, at not more than 50 cm from the base of the chosen plant, leveling the smallest wall carefully and fastening the glass to it with stakes in place. The hole must be dug along the row, and finally the glass will remain perpendicular to the rows and to the step of the tractors. This glass position helps the step of the tractors not to break it, and the spaces that can remain between the glass and the soil wall are refilled by sieved soil. This minirhizotron has the only objective of determining the radical growth, yielding data on the root growth to produce technically correct applications. Roots photographies may be made periodically through the crystals of the small minirhizotrons, to measure their growth.

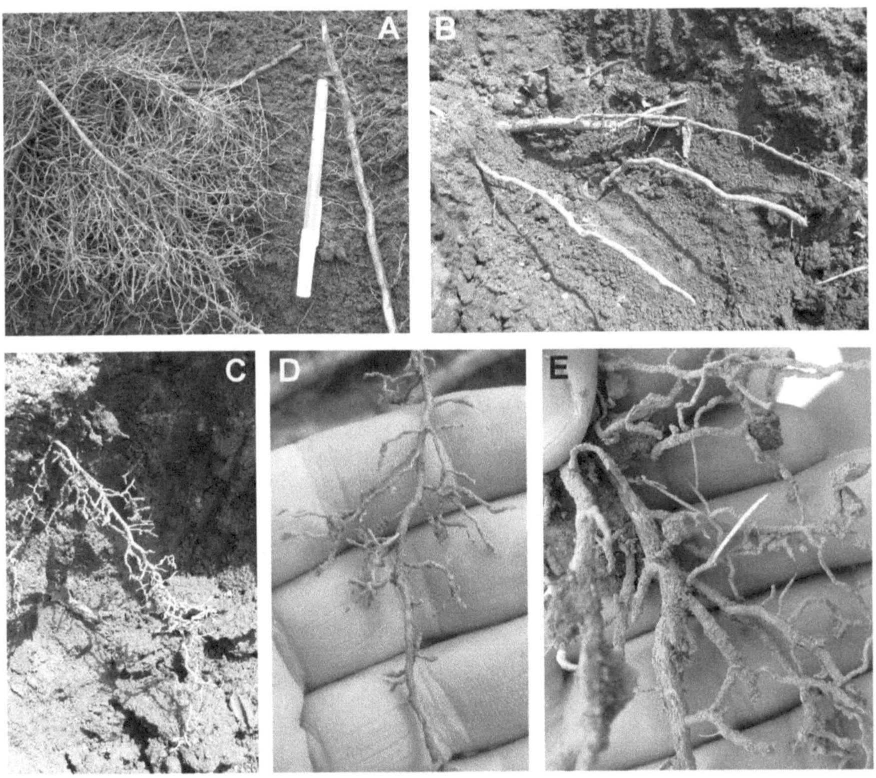

Figure 2. Inspection of damage induced by Tylenchulus semipenetrans *on citrus roots: control roots treated with chemicals with secondary growth (A) and untreated nematode infested roots (B); infested roots may be recognized by the adhesion of soil aggregates (C-E).*

9. OTHER MANAGEMENT STRATEGIES

When nematodes are already in the orchard, best pre-plantation and postplantation actions should be identified. To live with nematodes, a series of tools are useful, including rootstocks, chemical products, irrigation, choice of fallows, rotations, use of fresh organic and/or composted matter, solar light and cover cultivation.

Fallow is a very effective strategy, if the drying up of the soil is rapid, due to a good movement of the soil. The change or rotation of cultivation cannot be effective with all the parasites, since some nematode species (*Meloidogyne* spp.) are very polyphagus. However, other species which present a narrow host spectrum can be controlled. Handling of nematodes by organic matter may be useful, but if the material is not completely degraded or composted (Magunacelaya, 2005), the fermentation in the field may yield to a degradation of the material with emission of one or more organic molecules such as the methane, ethane, propane, butane, metanoic acid, etanoic, acetic, and butanoic acid. The degradation of the aminoacids produces ammonia. The fermentation causes significant increases of the temperature that can reach more than 70ºC, which is translated into a nematocidal, fungicidal and bactericidal action. Once the fermentation is finished, the colonization of this new environment is produced by microorganisms, in general beneficial, which work as unspecific biological control agents for the nematodes entering the space they occupy.

REFERENCES

Allen, M. W., Noffsinger, E. M., Hart, W. H., & Valenzuela, A.(1969). El nemátodo de los cítricos en Chile (*Tylenchulus semipenetrans*). *El Campesino*, 100, 26-31.

González, H. (1984). El nematodo de los cítricos en Chile. *Investigación y Progreso Agropecuario*, 25, 14-16.

González, H. (1987). El nematodo de los cítricos (*Tylenchulus semipenetrans*) y la importancia de su Estudio, en Chile. *Aconex*, 17, 5-8.

Inserra, R. N., Duncan, L. W., O'Bannon, J. H. & Fuller, S .A. (1994). Citrus nematode biotypes and resistant citrus rootstocks in Florida. Institute of Food and Agricultural Sciences (IFAS), FL. *Nematology Circular*, 205: 5 pp.

Magunacelaya, J. C. (1996a). Nemátodos vectores de virus en Chile. En Avances de sanidad vegetal en frutales y vides. Universidad de Chile, Facultad de Ciencias Agrarias y Forestales. Santiago de Chile: 147-154.

Magunacelaya, J. C. (1996b). *Pratylenchus* y nemátodos ectoparásitos en Chile. En: Avances de sanidad vegetal en frutales y vides. Universidad de Chile, Facultad de Ciencias Agrarias y Forestales. Santiago de Chile: 133-138.

Magunacelaya, J. C. (2003). Nemátodos de importancia en la agricultura chilena, situación actual y perspectivas. *Nematropica*, 33, 108.

Magunacelaya, J. C. (2004). Manejo de nemátodos / Nemátodo Management. Cómo cortar el problema de raíz/Solving the problema of the root. *Vitivinicultura*, 3 (13), 12-18.

Magunacelaya, J. C. (2005). Aspectos generales de manejo de nemátodos fitoparásitos de importancia agrícola en viñedos en Chile. *Red Agrícola*, 2, (8), 17-20.

Magunacelaya, J. C. (2005a). Uso de extracto de quillay para el control de nemátodos. At *http://mazinger.sisib.uchile.cl/repositorio/lb/ciencias_agronomicas/monteale.*

Magunacelaya, J. C., & Dagnino, E. (1999). *Nematología Agrícola en Chile*. Serie Ciencias Agronómicas N.° 2, Facultad de Ciencias Agronómicas Universidad de Chile, Santiago de Chile: 288 pp.

Magunacelaya, J. C. (2005b). Control de nemátodos mediante uso de materia orgánica. At *http://mazinger.sisib.uchile.cl/repositorio/lb/ciencias_agronomicas/monteale.*

Magunacelaya, J. C., Abogabir, P., & Pacheco, H. (2003). Acción enraizante de biorend (quitosano) y extracto de quillay, en plantas de tomate en bolsas de polietileno con suelo, y en "rizotrones". *Nematropica*, 33, 109.

Magunacelaya, J. C., Abogabir, P. & Pacheco, H. (2004a). Estudio de Quitosano (BioRend) como complemento a la acción de productos nematicidas, en Vid de mesa y Vinífera en Chile. *Nematropica*, 34, 134-135.

Magunacelaya, J. C., Pierce, J., & Ahumada, M. T. (2004b). Acción nematicida y beneficios para la planta (Var.Chardonnay), del uso de 1,3 Dicloropropeno (Triform) en suelos altamente infestados con *Meloidogyne ethiopica*, entre las temporadas 2001 y 2004 en la zona central de Chile. *Nematropica*, 34, 135-136.

Molinari, S., Lamberti, F., Duncan, L. W., Halbrendt, J., McKenry, M., Abawi, G. S., et al. (2004). SOD polymorphism in the *Xiphinema americanum*-group (Nematoda: Longidoridae). *Nematology*, 6, 867-876.

San Martín, R., & Magunacelaya, J. C. (2005.) Control of plant parasitic nematodes with extracts of *Quillaja saponaria*. *Nematology*, 7, 577-585.

4

NAHUM MARBÁN-MENDOZA

NEMATODES MANAGEMENT IN COFFEE PRODUCTION SYSTEMS

Universidad Autónoma Chapingo, Posgrado en Protección Vegetal, Chapingo, Edo. de México, CP 56230, México

Abstract. Coffee production in Mexico with particular reference to the diversity of cropping systems and nematode parasites are reviewed. Different cropping systems, including natural or mountain systems, traditional polyculture, specialized or commercial polyculture and sunlight systems are described. Nematodes affecting coffee include root-knot and lesion species, and their interactions with other pathogens, including fungal diseases, are then reviewed. Control strategies and tactics coffee nematodes are revised. The development of programs in the management of nematodes is then proposed, based on different tactics including prevention through quarantine, cultural management, development of clean planting systems, solarisation, use of antagonistic plants, soil amendments, weed host control, inter and intracropping (shade coffee), resistance, applications of chemical nematicides and organic amendments, biological control or use of natural products.

1. INTRODUCTION

Coffee production and use in Mexico is considered to have begun by the end of the XVIII century, when the plant was first introduced into the region of Córdoba, Veracruz, in 1796. It was then reported from the state of Michoacán in 1823 and later in the region of Tuxtla Chico, Chiapas, in 1847. However, it was not until 1882 that Mexico began to export this grain which was continuously exported, since that time (Regalado, 1996; Castillo et al., 1997).

Since first crops, the coffee cropping systems showed some variations and adaptations to either climate and environment. For example, the production zone on the Gulf of Mexico slope is humid but presents cold fronts known as “nortes”, whereas in the Pacific region, the environmental conditions are characterized by a prolonged drought from October to May. The composition and structure of the shade conditions thus vary from one zone to another, representing the main variability source among these areas (Escamilla & Díaz, 2002).

At the present time, Mexico ranks in the seventh place in the list of producing countries, with an invested surface of 801656 ha for the year 2006, reporting a national average yield of 1.9 ton · ha $^{-1}$. The yields are relatively lower than those

A. Ciancio & K. G. Mukerji (eds.), *Integrated Management of Fruit Crops and Forest Nematodes*, 101-118.

known prior to 2005, in which average yields of over 2 ton · ha $^{-1}$ were reported, with a gradual decrease observed since 1996, when a maximum of 2.6 ton · ha^{-1} were reported. This value decreased in the following decade despite the fact that coffee cultivation and production extended to 15 states of the Mexican Republic (SIAP, 2005).

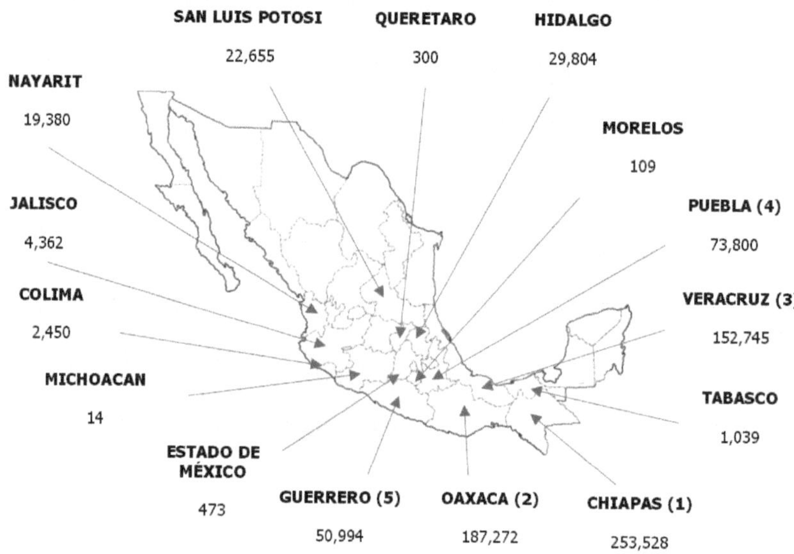

*Figure 1. States of the Mexican Republic where coffee (*Coffea arabica*) is produced. The total area, for each state, is shown in hectares (SIAP, 2005).*

The production zones can be grouped into four large regions ranging from the southern border with Guatemala, up to the state of Nayarit in the North Pacific (Fig. 1). Each region presents its own characteristics, but mostly they show conditions that are suitable for the production of quality coffee. They are:

a) Gulf of Mexico Slope. This includes the states of San Luis Potosí, Querétaro, Hidalgo, Puebla, Veracruz, part of Oaxaca and Tabasco. The period of intense rains begins in June and is interrupted in August, to begin again in September and finally in October or November. Annual rainfalls vary from 1300 to 3000 mm.
b) Pacific Slope. This region includes the states of Colima, Guerrero, Jalisco, Nayarit, Michoacán and part of Oaxaca. This region is characterized by long periods of drought, from November until May. The winter season is dry and hot, which facilitates the fruits harvesting and the drying process as well.
c) Soconusco Region. This region is formed by a large part of the state of Chiapas, and although it is geographically located on the Pacific slope, it differs in climate, since rainfalls vary from 2500 to 5000 mm throughout

the year, without significant droughts. In this region, an important volume of organic coffee (which is in great demand on the american and european markets) is produced.

d) North-Central Region of Chiapas. This area is characterized by prolonged periods of drought from November to April, due to the direct influence of moist winds proceeding from the Gulf of Mexico.

e) Another two states which were not traditionally characterized for coffee growing, were recently annexed to the producing states. They are: the State of Mexico and Morelos. As expected, the cultivated surface is small, but these areas are outstanding for impressive yields, higher than 4 ton $\cdot$ ha $^{-1}$.

Of the 15 producing states (Fig. 1), the highest production is reported in Chiapas, which represents 39.8 % of total production, followed by Puebla, (21.5 %) and Veracruz (20.2 %). Together, these states concentrate 81.6 % of the mexican coffee production. In this group, Puebla presents the highest yields, with 4.67 ton $\cdot$ ha $^{-1}$ (SIAP, 2005).

The coffee producing farms, as well as the highest number of sown surfaces, are mostly concentrated in small productive units, and the cultivation systems present a marked difference as concerns the size of the plots where this crop is grown. For example, the average size of the plots is 2.7 ha, with 92 % of the coffee growers of the country located in surfaces of 5 ha or less, whereas a small minority of plantations possess over 50 ha (Table 2). This fact, which appears as a factor of fragility because of plots fragmentation, may hide an advantage, especially if we consider the type of manual care given to the production, which offers an added value.

Table 1. Surface and varieties in coffee plantations in Mexico.

Variety	Area (Ha)	(%)
Coffea arabica	738,330	97
Typica	205,515	27
Caturra	152,233	20
Mundo Novo	129,398	17
Garnica	121,786	16
Catual	53,282	7
Bourbon	30,445	4
Maragogype	22,835	3
Catimor	15,224	2
Pacamar	7,612	1
Coffea canephora	22,835	3
Total	761,165	100

Source: Instituto Mexicano del Café. Census, 1992.

Considering coffee types, Mexico is an outstanding producer of *C. arabica*, given that 97% of the total area pertains to this species. The varieties Typica, Mundo Novo and Caturra are of particular importance, as they were the first to be introduced during the decade of the 1950's. These are followed by the variety Garnica, a hybrid which showed adaptability and higher productivity. This variety was the result of breeding investigations carried out by Mexican scientists of what was formerly the Mexican Institute of Coffee (Instituto Mexicano del Café).

Presently, Mexico is the fourth coffee exporter of the world and ranks in the seventh place as producer, and first place in the production of organic coffee. This crop generates over 350000 jobs, due to the fact that in many plantations and states, the production process is almost completely manual, from sowing to harvest. If we also consider the families and personnel linked to the transformation and commercialization of the crop, then approximately 3 million Mexicans depend on the cultivation and production of coffee (Regalado, 1993; Santoyo et al., 1994; Castillo et al., 1997).

Table 2. Characterization per production surface of the coffee plantations in Mexico.

Stratification	Farmers		Area	
(ha)	Number	%	ha	%
0.01-2.0	194718	68.9	247483	32.5
2.01-5.0	64617	22.9	229623	30.2
5.01-20.0	22017	7.8	199519	26.2
20.01-50.0	815	0.3	25991	3.4
> de 50.0	425	0.1	58548	7.7
Total	282592	100.0	761164	100.0

Source: Instituto Mexicano del Café. Census, 1992.

The economic and social importance of coffee is due to the fact that this grain is included among the first six crops for cultivation surface and value of its production, about USD 361.5 million for 2005 (INEGI, 2006). Furthermore, it contributes substantially over the capital generated by the agricultural sector (close to 270 million USD exported in 2005).

As for its social relevance, it should be considered that due to international conventions, like the regulation of prices until the latter part of the 1980's and the state support, many producers were led to convert their production areas to coffee cultivations. This was convenient at first because of the support and prices that guaranteed the production, along with the geographic characteristics of some areas with agro-climatic conditions favourable for coffee growing. However, this led to specialization and the consequent economic dependence of many family units on the cultivation of this grain, and even whole communities dedicated to coffee production.

In the environmental aspect, nearly 98% of the coffee producing areas in Mexico are shaded, providing important wood masses that favour aspects such as the protection of native fauna, with an important influence on the water cycle. Under the premises of agroecosystem conservation, certain practices were implemented aiming at a more efficient use of chemical products for fertilizing and pests and diseases control. Studies of agro-ecological zoning were carried out with the purpose of establishing, according to the environmental conditions, polycultivation systems and to obtain the greatest advantages by the conditions of the diverse coffee production regions (Pérez & Geissert, 2006). Furthermore, the production of organic coffee was established in Mexico for over 30 years, allowing an added price in the international market to be achieved (Sosa & González, 1995). This production is mainly (90%) finalized to export towards the United States, Europe and Asia.

Despite the variation in price, coffee is a very important crop in the Mexican agricultural sector, due to its economic, social and ecological aspects. It adapts well to topographic conditions where it would be difficult to produce any other crop. The state of Chiapas reached first place in production and quality due to the altitude at which it is produced, and because of late maturation, with higher qualities obtained in the product as concerns flavour and aroma (Hernández, 2000).

2. CROPPING SYSTEMS

Escamilla and Díaz (2002) provided a description of the coffee cropping systems in Mexico. These are mainly defined by the composition of the shade trees, as well as by the incorporation of labor and agricultural goods aimed at the sustainability of the whole agro-forest system.

2.1. Natural or Mountain System

This cultivation system is the first used in Mexico for coffee production, originating in the zone of Córdoba, Veracruz. It is characterized by the diversity of the natural vegetation, with no particular species predominating, but useful to provide shade. This system naturally simulates the conditions of growth and development of *C. arabica*, as it can be found in the middle of the natural vegetation, with the upper stratus providing shade and the lower strata replaced by coffee trees.

This production system is more frequent in the coffee regions of the Pacific slope, in response to the long period of drought. In states such as Nayarit, Guerrero and Oaxaca, it is the predominant system with 60-70% of the total of areas dedicated to its production. However, the importance of this system decreases toward the Gulf slope, due to the presence of cold fronts from October til February, and the higher relative humidity of this zone, which reduces the period of drought and its effects. For this reason, the plantations that maintain this system represent only 5% of the land dedicated to coffee in this zone of Mexico, especially in the state of Veracruz, although in the states of Tabasco and Chiapas, it may represent 30 % of the total area.

According to Escamilla and Díaz (2002), in Mexico this cultivation system is characterized by the predominance among the plantations of the 'criolla' variety, with most of the plants of 30-80 years of age. However, it is possible to find tall varieties such as Bourbon, Mundo Novo, Pluma Hidalgo, and even improved low varieties such as Caturra and Garnica. Even in the state of Puebla in the region of Tlacotepec de Díaz, the species *C. canephora* is cultivated, known as 'robust' coffee.

The propagation of new coffee plants is carried out through seedlings emerging within the coffee grove, sown with this purpose. The plantation density is 800-1200 trees · ha^{-1}, planted at distances of 3 × 4 or 3 × 3 m, without a clearly defined distribution or arrangement, but responding to the obstacles present in the plots and at times fostering the growth of new plants, that are born within the coffee grove.

Sometimes the shade conditions are excessive, surpassing 300 shade trees per hectare. This may favor the appearance of fungal diseases, one of the most common in these zones being known as 'ojo de gallo' (*Mycena citricolor*). In this production system, conditions of abandonment are common in plantations, with little or no agronomic management. It can be observed that there is regulation of shade, no fertilization and phytosanitary management are applied as in general, agrochemicals. Occasional pruning takes place, and the weeds and underbrush are only eliminated prior to harvest with the use of a machete, given that excessive shade conditions limit growth.

As it would be expected, yields under these conditions are low, reporting an average of 2-6 quintals · ha^{-1}, which practically places these producers in the category of gatherers. However, the people who tend this crop, mainly indigenous groups in remote mountain zones, do not see coffee as the only source, nor as the principal source of income, given that they use the natural resources to obtain wood for fuel and for the construction of homes and furniture and the collection of edible plants and mushrooms. The hunting of birds or some mammals is possible because of the diversity of plants and abundance of wild fauna in the coffee producing zones located towards the Mexican Pacific coast. Some of the indigenous groups which are reported as being associated to this system of coffee production are Totonacas, Nahuas, Otomis, Chinantecos, Mixes, Mazatecos, Mixtecos, Tsotziles, Tzelzales, Choles, Tojolobales, Zoques and other groups (Escamilla & Díaz, 2002).

2.2. Traditional Polyculture System

The aforementioned mountain system and the polycultivation system are considered to be 'ecological coffee cultivation', and the distinguishing characteristic of the latter system is the predominance of natural vegetation, but with the incorporation of some introduced or native cultivated species. This is the cultivation system that is most widely distributed in the coffee growing regions in Mexico, especially towards the Pacific region. In this system, there is a presence of small scale producers, but with an elevated participation of indigenous producers and laborers.

Plant diversity consists of species with defined uses, such as wood, fruit, ornamental, vegetable or medicinal species. However, plants of the genus *Inga* are used for the arboreal stratus (which never surpasses 50% of the total) and whose purpose is to provide shade. In this system there is a high presence of fruit trees such as diverse species of native avocado, *Calocarpum mammosum*, *Carica papaya*,

Annona spp., *Acanthocereus pentagonus*, *Psidium guajava*, *Pimienta dioica*, *Spondias purpurea*, *Theobroma cacao*, and *Ananas comosus*. Introduced fruit species are also found, such as some varieties of sweet orange and other species of the genus *Citrus*, *Musa* spp., *Mangifera indica*, *Prunus persica*, *Macadamia* spp., to name a few, along with native and introduced wood and ornamental species, vegetables and medicinal plants. Although the products of this system can be used at the family level, there are species and products which are commercialized. Therefore, plant diversity corresponds to an economic strategy in which diverse products are obtained per surface unit.

Despite its importance, a spatial ordering is not observed in most of the plantations that produce under this system, which may respond to efficient strategies of utilization, although ecological crop management of the products assumes great relevance.

In this system, tall varieties can be found such as Typica, Bourbon, Mundo Novo, Pluma Hidalgo and Maragogipe. Low varieties are also grown, such as Caturra Rojo, Amarillo, Garnica and Pacamara. It is also common to find more than one variety in a plantation, but almost never more than five varieties, one of which is predominant. In some areas of warm climates of the states of Veracruz, Chiapas, Oaxaca and Puebla, it is common to find the Robusta coffee variety.

The population densities found in this system are 800-1600 plants $\cdot$ ha^{-1}, although they may reach 2500 plants $\cdot$ ha^{-1}. The planting is generally carried out in squares of 2 × 2 or 3 × 3 m, but there are also plantations of 3 × 4 or 2 × 3 m. In some plots planting is set in the slope direction, which causes serious erosion problems, and some producers intercalate young plants with old ones, due to the fact that some seeds germinate and are cultivated rather than eliminated. In some cases densities of up to 4000 plants $\cdot$ ha^{-1} were reported. These plantations vary in age from 20 to 40 years, and very young or up to 80 years old plants can be found.

In this system, certain practices are carried out such as weeding, fertilization, shade regulation, and depending on the coffee prices, weed control can be applied one or more times. Nearly 20% of producers apply fertilizations once or twice, and nursery plants are sometimes acquired. Phytosanitary control generally is carried out by means of pruning, but not by chemical applications. The problems that are most frequently reported are those caused by insects (*Hypothenemus hampei*, *Plagiohammus maculosos*, *Leucoptera coffeella*), nematodes (*Meloidogyne* spp.), and fungi (*Mycena citricolor*, *Hemileia vastatrix* and *Corticium koleroga*).

Some producers, especially in Chiapas and Oaxaca, occasionally apply organic fertilizers and carry out soil conservation practices by means of terracing or live walls, whereas the control of broca is sometimes carried out with the fungus *Beauveria bassiana*. Yields under this system vary and may reach 14 quintals $\cdot$ ha^{-1} in case of high performances, or 3 quintals $\cdot$ ha^{-1} when they are low. Under the organic cultivation system, 8-10 quintals $\cdot$ ha^{-1} are reported. However, these diversified systems present high productivity that is not well quantified, due to the fact that the destination of production is for auto-consumption, and some of the remaining production is destined for commercialization or barter, thus the strategy of the producers is to insure food, shelter and health.

2.3. Specialized System

This system is oriented towards monocultivation, given that only coffee is produced under shade and uses leguminous plants almost exclusively to provide shade. This specialized system was the product of a technological package developed by the 'Instituto Mexicano del Café' (INMECAFE) in 1970. However, due to the low price crisis, many of these systems have been converted to polycultivation.

Some data reported for the state of Veracruz, Mexico, indicate that the system is more accepted as the altitude over sea level increases, and becomes the predominant system (58%) in zones located over 900 msl. Some other important characteristics of this system are the propagation and distribution of coffee seedlings, the impulse toward the renovation and rehabilitation of coffee plantations introducing improved varieties, the increase in plantation density, the application of substances such as fertilizers and fungicides, and management of pruning. The principal variety promoted was Garnica. The plantation design presents a defined arrangement and, depending on the conditions of altitude, radiation and edaphic characteristics, the population density of the shade trees can range from 40 to 400 trees per hectare.

Other states in which this system was widely adopted were Puebla and Chiapas, where in some regions it can represent up to 80% of the plots dedicated to the crop. In Oaxaca and Nayarit, acceptance was much lower, but some plantations can be found. In general, this system represented the principal factor explaining the high production indices registered in the final decades of the twentieth century in Mexico, although due to the present situation of coffee prices, it does not result very profitable.

Presently, the plantations which maintain this system are characterized by the predominance of improved varieties of low height such as Garnica and Caturra. However, tall varieties are also frequent, such as Mundo Novo and Bourbon. Recently, higher productivity has been sought with improved varieties such as Catual, and varieties that are resistant to blight and which are also precocious, such as Oro Azteca, Colombia and Costa Rica 95, and some variants of Catimor. Another aspect is the search for quality through the varieties Typica, Maragogipe and Pacamara, especially applied for production of 'organic coffee'.

The age of these plots is almost never over 20 years, although older plantations may be found. The coffee plantation densities vary from 1000 to 3300 plants $\cdot$ ha^{-1}. The recommendations of INMECAFE varied from 1600 to 2000 plants $\cdot$ ha^{-1} for tall varieties and 2500 plants $\cdot$ ha^{-1} for low varieties. The planting was carried out in real mark, rectangular or triangular or 'tresbolillo' arrangement, often established in the direction of the slope, but this practice is now being substituted by the implementation of curved rows that are level or contoured. Soil conservation practices include the use of terraces and live walls.

Weed control is carried out frequently and varies from one to four times, and is usually carried out manually with the use of some implement, although herbicides are sometimes applied. Pruning is carried out, principally the so-called 'Veracruz pruning', which considers criteria of productive tissue, phytosanitation, rejuvenation with deep pruning and complemented with the elimination of suckers. Shade is also regulated, with an umbrella shape to the crown of the coffee plants at a 5 m height.

The application of chemical fertilizers is carried out from one to three times a year. The fertilizers most often applied have the NPK formula 18-12-06 or 17-17-17. Organic products are occasionally applied such as coffee pulp or chicken manure. Some producers make compost and apply it at the moment of planting or every two or three years. Under this system, a high production is obtained, reporting from eight to eighty quintals · ha^{-1} or more in the municipality of Amatlán de los Reyes in Veracruz, but production varies according to the regions and the management of plantations.

The pests and diseases are almost always the same as with the other systems, and chemical control is almost never applied, except in the control of *H. hampei* with *B. bassiana* or with endosulfan combined with other practices under the principles of integrated management.

2.4. Commercial Polyculture System

In this system, two to four species coexist in association with coffee. These are destined for commercialization, in a production diversification strategy and are distributed according to a spatial order, with a more intensive soil use. As a result of the coffee prices crisis, the polycultivation system gained importance, as producers were motivated to adopt this method (Escamilla et al., 1994; Escamilla, 1997).

The species that are often present in the polycultivation system are fruit trees such as: *Persea americana*, *Annona muricata*, *Citrus latifolia*, *C. reticulata*, *C. sinensis*, *Calocarpum mammosum*, *Litchi chinensis*, *Macadamia integrifolia*, *Macadamia tetraphylla*, *Carica papaya*, *Musa acuminata* and *Eriobotrya japonica*. Ornamental foliage species are also found, such as *Chamaedorea elegans*, *Musa acuminata* and *Chamaedorea tepejilote*, among others. Vegetable species include *Capsicum annum*, *C. pubescens*, *Lycopersicum esculentum*, *Mangifera indica*, *Xanthosema* spp. and *Physalis ixocarpa*. Wood species include *Cedrella mexicana*, *Inga* spp., *Juglans* spp., *Swiettenia* spp. and *Acrocarpus fraxinifolius*. Other species found are *Pimienta dioica*, *Vanilla planifolia*, *Hibiscus sabdariffa*, *Hevea brasiliensis*, *Phaseolus vulgaris* and *Zea mays*. The species found vary from one region to another and even within the same state (Licona et al., 1995; Baltazar, 1999; Santacruz, 2000).

Other aspects characterizing this system are the homogeneity of plantations as concerns the coffee variety or the associated crops exploited, and the specific tasks which are carried out for each species. Under this system of polycultivation, producers insure their income throughout the year from the commercialization of products that are harvested according to the crop cycle. Due to the products diversification, economic stability is increased and jobs are generated during most of the year, and this system is expanding. However, its success depends on the producers technical implementation and administrative capacity.

2.5. Sunlight System

This is the most intensive system found in Mexico, and is referred to as "sunlight system" or "open sky system". In this system the coffee plants are grown without shade and are maintained in monocultivation. It is used mainly in the states of Puebla, Veracruz and Chiapas, by the large scale producers, who use higher technification to increase production and reduce costs. The technology transfer occurred through the hiring of technicians and foreign advisors, as well as through the exchange of experiences with other countries that use the sunlight system.

At one point, this system represented nearly 50% of the coffee producing regions of Puebla. However, due to the heavy frosts that occur and which in some years are extremely severe, damages are caused in the unshaded coffee groves. As a result, many producers and large plantations returned to the shade system, and at present, the unshaded system represents less than 10% of the surface of the lands that produce coffee in the state of Puebla.

In the state of Veracruz, its introduction is very recent and covers almost 1% of the lands of this state, but in altitudes close to 900 msl. There have been also attempts in Chiapas, but on a much smaller scale than in Puebla and Veracruz. Other states such as Oaxaca, Guerrero and Nayarit have had extremely unfortunate results due to the lack of moisture of this region of the Pacific slope and the high cost in fuel and economy. However, the unshaded system exposes the coffee plants to different conditions, forcing them to increase growth and production as a survival response. This system demands cultivation practices such as fertilization, weed control, pruning and phytosanitary control, high plant density and the use of improved varieties of low height, such as Caturra, Catual, Garnica and, recently, varieties resistant to blight, for example Costa Rica 95, Colombia, Oro Azteca and Catimores. One characteristic of these systems is the high population density, with up to two or three plants per rootstock, which is called, such as the case may be, "double or triple posture". Densities vary from 4000 to 15000 plants · ha^{-1}, the rectangular planting system of 2 × 1 m being the most frequent. Under this system, soil and foliar analyses are carried out frequently in order to establish fertilization programs in which micronutrients are included.

With the unshaded system, many of the pests found in the other systems are reduced, but others appear or increase their density, such as *Colletotrichum gloesporioides*, *Cercospora coffeicola*, *Oligonychus coffeae*, *Saissetia* spp., *Coccus* spp., *Toxoptera aurantii*, *Planococcus citri*, *Pseudococcus cryptus*, *Meloidogyne* and *Pratylenchus* spp., as well as *H. hampei*, which represent all together the most important diseases and pests reported for this system.

In spite of the above, yields are as high and up to 40 - 80 quintals, sometimes surpassing 100 quintals · ha^{-1}. This system and its productivity tend to report yields that do not offer the same quality as the coffee grown under shade. Therefore it is not possible to verify whether the system justifies the economic and environmental costs that it demands.

3. PHYTOSANITARY ASPECTS OF THE COFFEE CROP

3.1. Nematodes Affecting Coffee

Villanueva et al. (1990) made a general description of phytosanitary status of coffee and mentionned among them some pests like the coffee grain weevil *H. hampei*, the coffee leaf borer *Leucoptera coffeella*, the flour lice *Planococcus citri*, the stem and leaf borer *Plagiohammus maculosos*, *P. mexicanus*, *P. spinipensis*, the Red spider *Oligonychus coffeae*, the Chacuatete *Idiarthron subquadratum*, the scales *Saissetia coffea*, *S. oleae*, *Coccus viridis*, *C. hesperidum*, *Selenaspidus articulatus*, some aphids (*Toxoptera aurantii*), and ants (*Atta fervens*, *A. mexicana*, *A. cephalotes*, *Pogonomyrmex barbatus*).

According to Castillo (1993), diseases represent one of the limiting factors reducing the production of coffee plantations, and the most important pathogens, due to their frequency and damage they cause, are fungi and nematodes. Among the diseases caused by fungi are coffee blight *Hemileia vastatrix*, *Corticium koleroga*, *Mycena citricolor*, antracnosis *Colletotrichum coffeanum*, the rust spot *Cercospora coffeicola*, the black root rot *Rossellinia bunodes*, *Phoma costarricensis* and pink blight *Corticium salmonicolor*.

Nematodes represent one of the principal limiting factors of coffee. Since 1878, Jobert, cited by García (1993), pointed out the problems they caused in coffee, but did not attribute them importance. In Mexico, one of the first reports was given by Alcocer and Gottwald (1963) indicating that *Meloidogyne*, among other genera, were associated to yield losses. Sasser (1979) pointed out that *Meloidogyne* spp. were the cause of 10% of coffee yield losses in Mexico and Central America. Anzueto et al., (1995), studied nematodes damages on coffee mainly due to *Meloidogyne* spp. and, to a lesser extent, to *Pratylenchus* spp., in different farms of Mesoamerica countries. They indicated that yield losses were in the range of 10-25%, depending upon farm management.

Although the damages caused by nematodes associated to coffee were reported in Mexico for several years, they received little relevance. Longer and sytematic studies on the issue are lacking. The data available deal mostly with nematode distribution and identification at the genera level, and with chemical and non chemical evaluation for nematode control. However, in the organic production system of some regions in Chiapas, coffee plants presented a wilting syndrome that began in the apex and descended until provoking the death of the plant. The disease began to cause concern among producers, who observed that these symptoms were spatially distributed in patchy patterns. The sampled soil and roots of some farms showed that the damages were caused by a high population of nematodes, which caused direct and indirect damage to the coffee plants (Ventura et al., 2005). This motivated further investigations, mainly for the identification of associated genera and control methods.

Presently, diverse genera are recognized in Mexico such as: *Meloidogyne*, *Pratylenchus*, *Paratylenchus*, *Tylenchorhynchus*, *Radopholus*, *Xiphinema*, *Criconemella*, *Peltamigratus*, *Tylenchus*, *Aphelenchoides*, *Hemiciclyophora* and *Helicotylenchus*. The list includes the genera that are more frequently associated (Topete, 1966; Vázquez et al., 1992; Castillo et al., 1992), but the most frequent are *Meloidogyne* and *Helicotylenchus* (Regalado, 1993). Species from these genera were reported mainly in farms localized in specialized systems, commercial polycultivation systems and sunlight systems, where damage and severe losses are also reported, particularly in patchy patterns.

On the contrary, nearly no report exists on nematodes in both the natural or mountain system or the traditional policultivated system. This might be explained, in part, by the differences between the two subgroups such as plant diversity, coffee varieties, shade degree, farm management, etc. However, it would be desirable to study this in deph for better understanding.

In general, nematodes did not receive attention, since it was considered that the symptoms they produced responded to several factors, such as soil fertility. In addition, the presence of the pathogenic complex *M. incognita* and *Pratylenchus* spp. has been detected, with *Fusarium oxysporum* and *Trichoderma* causing severe losses, especially in the central region of Veracruz, where it may cause up to 40% of these losses (Vargas, 1992; Castillo, 1995; Paz & Escamilla, 1996).

Also reported was the presence of symptoms similar to those of "corchosis del cafeto", a pathogenic complex involving *Meloidogyne arabicida*, *F. oxysporum* f. sp. *coffeae*, as well as *Verticillium*, *Cylindrocladium*, *Phialophora* and *Gonytrichum* spp. This syndrome was reported in Juan Viñas, Costa Rica (Marbán-Mendoza et al., 1989). Replanting Caturra or Catuai varieties in a patch severly infested without any treatment normally caused total failure of the crop. Furthermore, the symptoms were observed also in the central region of Veracruz, Mexico but were induced by *M. incognita* and *Pratylenchus* spp., in combination with *F. oxysporum* and *Trichoderma* (Vargas, 1992). Téliz et al. (1993) also refer *Verticillium* and *Helicotylenchus* spp. to this complex.

Proposals have been made for the management of nematodes. Marbán-Mendoza (2002) proposed a plan (Fig. 2; Table 3) that can be adjusted to any crop, human resources, budgetary constrains and time available. In brief, it consists in an agroecosystem diagnostic produced ideally by experts and growers to define hierarchically problems and aiming at setting up a time table, containing specific actions to deal with.

This theoretical framework has some constraints for practical applications. In practice, a single expert with strong commitment on Integrated Pest Management (IPM) practices can start working on some actions. So, an independent nematology team might be involved in a project seeking to solve nematode problems in a given crop-region under the principles of Integrated Nematode Management (INM) (Fig.2, left column), and /or they can be part of a major project including further pest experts trying to solve complex pathosistems in a given crop(s) growing in a given region(s). This must be done also under the phylosophy of IPM, (Fig. 2, right column).

The INM approach might take longer times, due to the nature of nematode problems and the low impact generated to obtain monetary resources for research and developments, like in the coffee case. However, under some circumstances many factors may align to coincide, causing a major impact on the society. This is the case of the phasing out of Methyl Bromide (MB) under the Montreal Protocol, a major achievement of the Enviromental Program of the United Nations. Here, from 1990 up to the present time, the world has been spending in plant protection, million USD on Research and Development to look for reliable alternatives to MB.

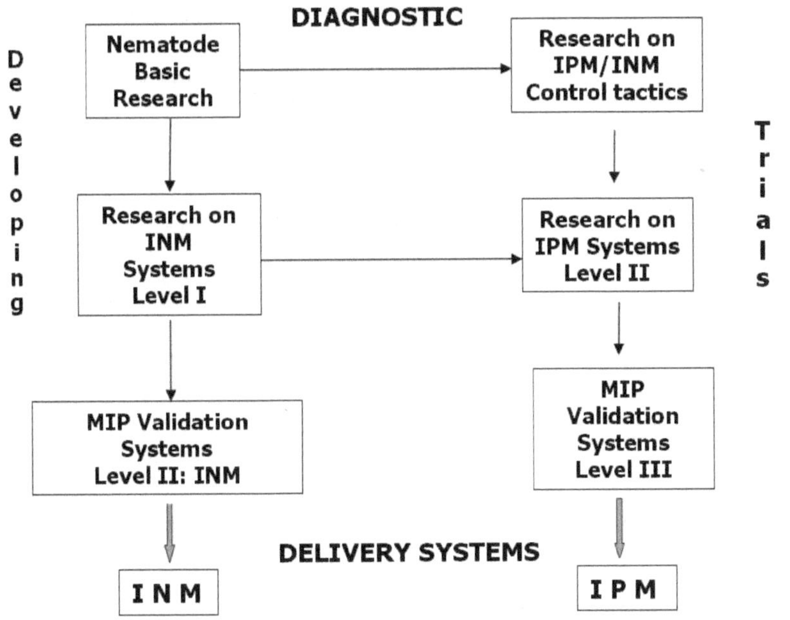

Figure 2. Proposal for the development of programs in the management of nematodes affecting coffee plants.

Althought soil-borne problems include nematodes as part of a complex of diseases limiting crops yields, fungi are indeed the key pests in soil fumigation problems. Nevertheless, nematologists across the world working on high cash crops like tomato, bell pepper, strawberries, melon and watermelon, tobacco and cutting flowers, had the chance to seek for more ecological alternatives, under the phylosophy of IPM. The net result, in about two decades, is the elimination of MB use in developed countries (restricted to critical use only). Close to 50 thousand metric tonnes were eliminated for annual use in soil fumigations (http://ozone.unep.org/teap/Reports/MBTOC/index.shtml). Knowledge generated in this effort might be applicable to coffee and other crops affected by nematodes with time.

Given the problems caused by nematodes, some practices have been implemented to combat their damaging effects, one of which was the evaluation of varieties, testing *C. arabica* varieties grafted on *C. canephora* in their response to the nematodes attack (Rivera, 1990). Grafted plants presented greater resistance to nematodes than plants propagated by seed (not grafted). Plant development and production of coffee beans increased in grafted plants. This practice appeared as an efficient management alternative, given that *C. canephora* turned out to be less susceptible to nematodes. In general, grafted plants present higher growth rates as concerns height, stem thickness and number of branches.

Table 3. Strategies and tactics for the control of coffee nematodes in Mexico.

Strategy	*Tactics*	*Comments*
Exclusion	Quarantine (legal)	Pending
	Cultural	Research
Reduction of initial population density	Developing clean planting systems (solarisation)	Research
	Antagonistic plants (coffee)	Research, some adoption
	Soil amendments	Research, some adoption
	Weed host control	Research
	Inter-Intracropping (shade in coffee)	Research, some adoption
Suppression	Resistance	Research, some adoption
	Chemical nematicide	On going progress
	Organic amendments	Research, some adoption
	Biological (natural/amendments)	Research

3.2. Coffee Nematodes Investigations

Bautista (2001) tested the varieties of *C. arabica*: Colombia Green Shoot, Colombia Brown Shoot, Pacarama, Costa Rica 95, Garnica F5 and Oro Azteca, grafted on *C. canephora* in different environments (localities at different altitudes) evaluating the population fluctuations of *Meloidogyne* and *Pratylenchus* spp. and the plant response in time. *Meloidogyne* and *Pratylenchus* spp. were the most abundant, with no significant differences in their population densities up to 20 months after transplanting (TAT). However, at 25 TAT in the Huatusco locality, grafted plants showed much less nematodes, although not clear differences could be established among grafted varieties (Bautista, 2001). Nevertheless, 36 months later root galling was so severe in the non-grafted plants, that some of them started to show dieback symptoms leading to full mortality 60 months later (Marbán-Mendoza, 2006, pers. comm.) (Table 4).

Significant differences were found in the plant response variables when grafted on *C. canephora* var. *robusta* under the commercial polycultivation and sunlight systems conditions, in the coffee region of Veracruz (Bautista, 2001).

Among the varieties, the direct sowing treatment (no grafted) showed higher damages than the grafted plants. Similarly, grafted plants presented higher height, number of branches and first production (Table 5), although the stem diameter did not show significant differences. Among environments, the best conditions for plant development were obtained at Maromilla, due in part to the relatively warmer climate (lower altitude) than the other areas studied (Bautista, 2001). The variety Colombia Green Shoot showed lower yields at first harvest (1810 kg) as compared with Costa Rica 95 (2520 kg) (Table 5).

Table 4. Population densities of Meloidogyne incognita *and* Pratylenchus coffeae *in grafted and ungrafted coffee plants (*Coffea arabica*) in three localities with different altitudes, in the central region of Veracruz, Mexico.*

Locality (msl)	Plants	Nematode densities [a]					
		P. coffeae		*M. incognita*			
		20 [b]	25 [b]	20 [b]	25 [b]	36 [c]	60 [d]
Huatusco (1300)	Grafted	1.5 a	0.8 a	12.2 a	8.6 b	1	0
	No graft	6.3 a	1.0 a	21.9 a	37.1 a	5	100
Manuel González (900)	Grafted	0.5 a	0.5 a	2.0 a	13.0 a	0.5	0
	No graft	2.0 a	2.7 a	3.0 a	29.4 a	5	100
Maromilla (700)	Grafted	0.4 a	0.4 a	2.0 a	3.6 a	0	0
	No graft	1.0 a	1.5 a	5.3 a	25.7 a	5	100

[a] Means within a column with the same letter do not differ significantly (Tukey, $\alpha = 0.05$).
[b] Mean density (vermiform stages $\cdot$ g^{-1} soil), after transplanting (months)
[c] Gall index (0-5).
[d] Plant mortality (%).

In a field study with *C. arabica* varieties grafted on *C. canephora* in the region of Tuxtla Chico, Chiapas, under sunlight conditions, Colmenares (2001) found that best yields were obtained with grafted plants, being outstanding the varieties Colombia Brown Shoot, Colombia Green Shoot and Oro Azteca. As for the plant height, although the best result was observed in plants grafted with Pacamara, a clear effect on the other varieties could not be established, even 30 months after transplanting.

In a field trial evaluating some nematode control tactics in *Coffea arabica* var. Garnica F5 in a sunlight system in Huatusco, Veracruz, Hernández (2000) reported no significant differences among the treatments, which consisted of chemical products such as aldicarb, ethroprop, a bio-controller or a coffee barry pulp compost. However, this author observed a higher vigour in plants treated with the combination of compost and chemicals.

More recently, evaluations were made for some management practices, especially considering the point of view of "organic coffee". Ventura (2005) studied alternatives of control and evaluated biological products that do not release chemical residues having an effect on the consumers' health. The products evaluated were a Quillay (Chilean tree) formulated extract, a fermentation solids and soluble extracts of *Myrothecium-verrucaria* (Ditera™ DF 90%), ground hen eggshells and aldicarb, compared as a traditional nematocidal treatment. The author reported, apart of *M. incognita*, presence of *Criconemella*, *Helicotylenchus*, *Paratylenchus*, *Pratylenchus* and *Tylenchorhynchus* spp., all associated to different levels of damage to coffee plants (var. Caturra, Catuai). The trial was carried out in sunlight systems at Chicomuselo, Chiapas.

*Table 5. Average fruit production at first harvest (2001) for coffee plants (*C. arabica*), grafted and ungrafted, in three localities of the central region of Veracruz, Mexico.*

	Yields [1]	Total yield [2]
Locality/ Altitude (m)		
Huatusco (1300)	0.875 c [3]	2916
Manuel González (900)	1.902 b	6276
Maromilla (700)	3.392 a	11193
Propagation		
Grafting	2.366 a	7885
No grafting	2.050 b	6862
Varieties		
'Colombia brote verde'	1.810 d	6032
'Colombia brote café'	2.366 a	8865
'Pacamara'	1.830 cd	6099
'Costa Rica 95'	2.520 ab	8865
'Garnica F5'	2.280 ab	7632
'Oro Azteca'	2.210 bcd	7365

[1] Mean production per plant (kg).
[2] Mean total production (kg · ha^{-1}) estimated for 3333 coffee plants.
[3] Means with the same letter in any column are not significantly different, according to Duncan's Multiple Range Test (P=0.05).

Table 6. Root-knot nematodes at each sampling date in the rhizosphere of coffee plants growing at Chicomuselo, Chiapas.

Date	Treatments				
(months)	egg shell 250 g · plant $^{-1}$	Quillay® 7 ml · plant $^{-1}$	Ditera® 10 g · plant $^{-1}$	aldicarb 7 g · plant $^{-1}$	control
0 [1]	1181 a [2]	1098 a	1183 a	1354 a	1417 a
6 [1]	720 a	288 b	266 b	55 c	955 a
8	1002 a	83 c	353 b	4 c	1174 a
12	960 a	370 b	430 b	20 c	1220 a

[1] Applications carried out at each treatment.
[2] J_2 · 100 cc^{-1} soil. Means flanked by the same letter within a column do not differ significantly according to Duncan's Multiple range Test (P = 0.05).

The first sampling was carried in May 2004 and the following samplings followed an average of every two months during the first year (Table 6). After the first sampling (initial population), a trend reducing nematode population was observed in plants treated with the Quillay formulated extract, Ditera and aldicarb during approx. one year. However, the effect became less evident as compared with aldicarb treated plants, on which nematode suppression lasted until the end of the experiment. Nematode suppression was more evident during the first days after treatments except in the treatments with eggshell and the untreated control plants. Additionally, qualitative evaluations were made of the foliar area and of the root system, observing that the plants treated with the Quillay formulated extract, Ditera and aldicarb showed higher foliar area development and secondary roots growth, and lower root galling. A laboratory work with gradient concentrations of the Quillay extract showed that *Meloidogyne* (J2) incubated for 12 h at 3500 ppm become immobile and did not recover during 72 hours after successive washings with distilled water. Aldicarb incubated nematodes behave about the same but at 15 ppm. Hoever, none of the products appeared economically viable for use in coffee plantations affected by nematodes under the present circumstances of the crop in Mexico.

REFERENCES

Alcocer, G. L., & Gottwald, C. (1963). Determinación de nematodos fitoparásitos en México. In: Resúmenes del primer congreso nacional de la ciencia del suelo. Sociedad Mexicana de la Ciencia del Suelo, México, D .F., 208-213.

Anzueto, F., Bertrand, B., Peña, M., & Marbán-Mendoza, N. (1995). Desarrollo de una variedad porta-injerto resistente a los principales nematodos de América. *In*: Resúmenes del XVII Simposio de Cafeticultura Latinoamericana. IICA-PROMECAFE. San Salvador, El Salvador.

Bautista, O. U. (2001). Respuesta de seis variedades de café (*Coffea arabica* L.), injertados sobre café robusta (*Coffea canephora*) y no injertadas, a nematodos asociados en Huatusco, Veracruz, México. Universidad Autónoma Chapingo. Departamento de Parasitología Agrícola, Chapingo, México, 56 pp.

Baltazar, H. J. (1999). Diagnóstico del cultivo de vainilla (*Vanilla planifolia* A.) como alternativa de diversificación en comunidades cafetaleras de Oaxaca, Puebla y Veracruz. Professional Thesis. Universidad Autónoma Chapingo. Huatusco, Veracruz, México: 153 pp.

Castillo, P. G., Hernández, E. E., Téliz, O. D., Nieto, A. D., Obregón, D. E., Castillo, H. J., & Ruiz, B. R. (1992). La "corchosis" del cafeto en México. *In*: Resúmenes del XV Simposio de Cafeticultura Latinoamericana. IICA-PROMECAFE. Xalapa, Veracruz, México: 90 pp.

Castillo, P. G., Contreras, A., Zamarrita, C. A., Méndez, L. I., Vázquez, M., Olgui, M. F., & Fernández, R. A. (1997). Tecnología para la producción de café en México. Folleto técnico N. 8, División Agrícola. SARH. INIFAP. CIRGOC. Xalapa, Veracruz, México. 90 p.

Castillo, P. G. (1995). Manejo integrado de la corchosis en Veracruz, México. *In*: Resúmenes del XVII Simposio de Cafeticultura Latinoamericana. IICA-PROMECAFE. San Salvador, El Salvador.

Castillo, P. G. (1993). Enfermedades del cafeto y su control en México. Folleto técnico # 4. División Agrícola. SARH. INIFAP. CIRGOC. Xalapa, Veracruz, México. 30 p.

Colmenares, A. D. (2001). Respuesta de seis variedades de café (*Coffea arabica*), injertadas y no injertadas, a fitonematodos asociados, en Tuxtla chico, Chiapas, México. Universidad Autónoma Chapingo. Departamento de Parasitología Agrícola: 67 pp.

Escamilla, P. E., & Díaz, C. S. (2002). Sistemas de cultivo de café en México. Universidad Autónoma Chapingo. CRUO-CENIDERCAFE. Huatusco, Veracruz. México: 64 pp.

Escamilla, P. E. (1997). Evaluación técnica-económica de plantaciones de café en el sistema de policultivo comercial en Veracruz. Colegio de Postgraduados. Manlio Fabio Altamirano, Veracruz. México: 180 pp.

Escamilla, P. E., Licona, V. A. L., Díaz, C. S., Santoyo, C. V. H., Sosa, R., & Rodríguez, L. (1994). Los sistemas de producción de café en el centro de Veracruz, México. Un análisis tecnológico. Centro de Investigaciones Históricas, Universidad de Costa Rica. *Revista de Historia*, 30, 41-67

García, P. P. (1993). Distribución y niveles poblacionales de fitonematodos de importancia económica asociados al cafeto. *In*: Resúmenes del XVI Simposio de Cafeticultura Latinoamericana. IICA-PROMECAFE. Managua, Nicaragua.

Hernández, B. M. A. (2000). Identificación, dinámica poblacional y tácticas de control de nematodos en café (*Coffea arabica* var. Garnica F5) en la región de Huatusco, Veracruz, México. Tesis profesional. Universidad Autónoma Chapingo. Departamento de Parasitología Agrícola. Chapingo, México.

Licona, V. A., Escamilla, P. E., Díaz, C. S., & Pérez, J. R. (1995). Diversificación productiva en regiones cafetaleras de México. *In*: III Simposio Internacional del Café. Confederación Mexicana de Productores de Café. Xicotepec de Juárez, Puebla, México.

Marbán-Mendoza, N. (2002). Nematode management practices in MesoAmerica. *Nematology*, 4 (Part 2), 131.

Marbán-Mendoza, N., Torres, O. M., & Calderón, V. (1989). Etiología de la corchosis del cafeto en Costa Rica. *In*: Resúmenes del XII Simposio de Cafeticultura Latinoamericana. IICA-PROMECAFE. San Pedro Sula, Honduras. 425 pp.

Paz, A. G., & Escamilla, P. E. (1996). Manual de propagación de injertos. Folleto Chapingo, México: 42 pp.

Pérez, P. E., & Geissert, K. D. (2006). Zonificación agroecológica de sistemas agroforestales: el caso café (*Coffea arabica* L.)-palma camedor (*Chamaedorea elegans* Mart.). *Interciencia*, 31, 556-562

Regalado, O. A. (1993). Problemática fitosanitaria de la cafeticultura en México. Tesis profesional. Universidad Autónoma Chapingo. Departamento de Parasitología Agrícola. Chapingo, México: 210 pp.

Regalado, O. A. (1996). Manual para la cafeticultura mexicana. SAGARPA. INCA RURAL. Consejo Mexicano del Café. D. F. México: 156 pp.

Rivera, F. A. (1990). Variedades de café cultivadas en México. *In*: El cultivo del cafeto en México. Instituto Mexicano del Café. INMECAFE-CIA. NESTLE. Ed. La fuente. D. F. México, 35-40.

Santacruz, R. C. A. (2000). La cafeticultura del estado de Nayarit: Situación actual y problemática. Tesis profesional. Universidad Autónoma Chapingo. Huatusco, Veracruz, México: 86 pp.

Santoyo, C. V. M., Díaz, C. S., & Rodríguez, B. (1994). Sistema agroindustrial café en México: diagnóstico, problemática y alternativas. SAGAR. INCA RURAL. CIESTAAM. Universidad Autónoma Chapingo. Chapingo, México. México:157 pp.

SIAP. (2005). Resumen Nacional por Producto. Avance Comparativo de Siembras y Cosechas Perennes. Café cereza. Secretaria de Agricultura Ganadería Pesca y Alimentación. Servicio de Información y Estadística Agroalimentaria, D. F. México.

Sosa, M. L., & González, V. J. (1995). El cultivo de café orgánico en México. Universidad Autónoma Chapingo. Dirección de Centros Regionales. Chapingo, México.

Téliz, O. D., Castillo, P. G., & Nieto, A. D. (1993). La corchosis del cafeto en México. *Revista Mexicana de Fitopatología*, 11, 5-12.

Topete, P. E. (1966). Plagas y enfermedades del cafeto. Instituto Mexicano del Café. México: 70 pp.

Vargas, H. E. E. (1992). Determinación y cuantificación de los nematodos asociados a las raíces del cafeto (*Coffea arabica* L.) en la cabecera municipal de Tlaltetela, Veracruz, México. Professional Thesis, Universidad Veracruzana. Xalapa, Veracruz, México: 43 pp.

Vázquez, V. M., López, de L. E., & Sánchez, de L. A. (1992). Evaluación de selecciones de *Coffea arabica* con tolerancia a roya y con cierto nivel de tolerancia a nematodos. *In*: Resúmenes del XV Simposio de Cafeticultura Latinoamericana. IICA-PROMECAFE. Xalapa, Veracruz, México.

Ventura, A. J. (2005). Dinámica poblacional, identificación y combate de nematodos en café orgánico en la colonia Monte Sinaí, Municipio de Chicomuselo Chiapas, México. Professional Thesis, Universidad Autónoma Chapingo, Departamento de Parasitología Agrícola. Chapingo, México: 69 pp.

Villanueva, M. A. E., Aranda, D. E., & Regalado, O. A. (1990). Plagas del cafeto. *In*: El cultivo del cafeto en México. INMECAFE-CIA NESTLE. Ed. La fuente, México, 165-177.

5

SAMUEL B. ORISAJO

NEMATODES OF CACAO AND THEIR INTEGRATED MANAGEMENT

Plant Pathology Division
Cocoa Research Institute of Nigeria
P. M. B. 5244, Ibadan, Oyo State, Nigeria

Abstract. Geographic distribution of nematodes parasitic or associated with cocoa crops are revised. Damage symptoms are described for root-knot nematodes, "sudden death" of plants and disease complexes associated with other species. Integrated management approaches to nematode control are discussed, with reference to occurrence and use of resistant planting materials, production of nematode-free seedlings in nurseries, use of nematicides, organic amendments and biological control.

1. INTRODUCTION

Cacao (*Theobroma cacao*) is a small tree originated in South America on the edges of Upper Amazon region of Latin America. The genus *Theobroma*, of the family Sterculiaceae, includes about twenty species of which only *T. cacao* is of economic interest. The three main recognized groups are Criollo, Amazonian Forastero and Trinitario. The fruit of cacao (cocoa) is the part of the tree that is most visible and, together with its component parts (seeds), it is most closely connected with the commercial activities (Bartley, 2005). Cocoa is one of the most flavored rich food, containing cathechin and epicathechin, which are responsible for some of its protective properties. As a result of this dual role, it ranks third as a beverage after tea and coffee (Lass & Wood, 1985) and it has wider application and usages. The cocoa mass is used for chocolate, biscuit and confectionary whereas the butter is used in making sweet, chocolate, perfumes and in pharmacy. The pod husk, a by-product, is being utilized in livestock feeds and as a fertilizer.

1.1. Production

The cacao plant is cultivated in many countries of South and Central America, Africa, Asia and Oceania, located mostly between 10° North and South of the Equator (Figure 1). Since the end of the First World War, West Africa has dominated the world cocoa market. The principal cacao growing countries of West

A. Ciancio & K. G. Mukerji (eds.), *Integrated Management of Fruit Crops and Forest Nematodes*, 119-134.

Africa, namely Côte d'Ivoire, Ghana, Nigeria, Cameroon and Togo make up about 70 % of the worldwide cocoa production. Asia accounts for approximately 17 %, with Indonesia being the largest producer in the region, whereas and Central and South America produce approximately 13 %.

New cacao growing regions have yet to be proven, especially Vietnam and India. It may be possible to produce cocoa beans in volume in other regions, but to duplicate the unique flavour, as well as the chemical and physical characteristics of the West African cocoa may prove to be very difficult (Taylor & Taylor, 2006). In general, main crop West African beans are used primarily for chocolate production, whereas the Indonesian and Malaysian beans are used for cocoa butter.

1.2. Climatic Requirements

Cacao is a low altitude crop growing best from sea level up to an altitude of 700 msl. In West Africa, the best crops have been produced within altitudes of 100-300 msl (Opeke, 2005). Temperature is the determining factor in choosing the maximum altitude for cacao, which needs a high temperature with no great variations. A mean temperature of 24 - 28 °C is favourable with a daily range of less than 10 °C, which is necessary to start bud bursting. However, bud busting and flushing become excessive when daily temperature ranges are in excess of 10 °C. Low temperatures adversely affect cacao trees. Cacao is fairly demanding and requires at least 1500 mm of appropriately distributed rainfall and optimum relative humidity of 80% (Van Himmer & Snoek, 2001). Excessive humidity may lead to mossy trunks and be at the origin of aggressive fungal diseases.

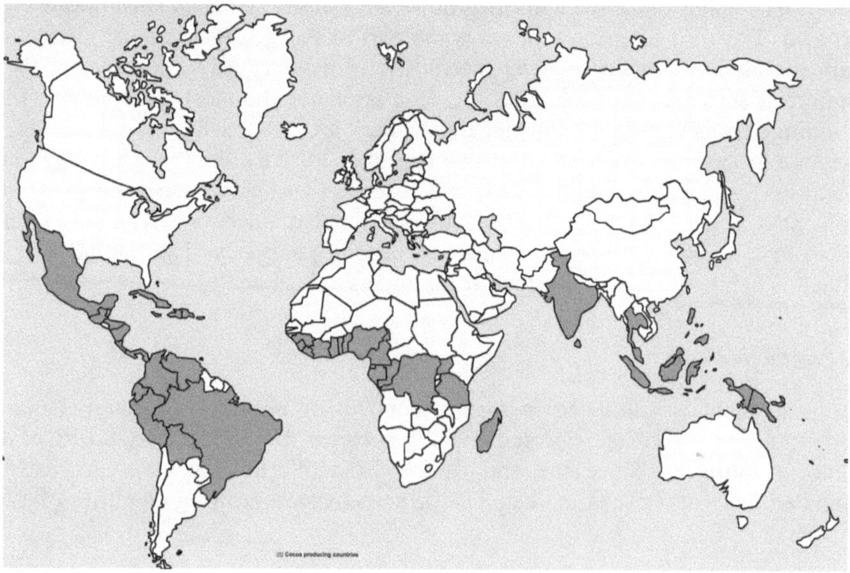

Figure 1. World map showing main cacao-growing countries (shaded).

1.3. Cultivation Techniques and Practices

The propagation of cacao can be by seed and vegetative, and the former is the cheapest. As there are many disadvantages to sowing directly in the field, the seeds are usually sown in nursery seedbeds, black polythene bags or woven baskets, as cacao seedlings do not withstand transplanting, especially with naked roots. Germination takes 7 to 10 days and seedlings are transplanted to the fields when 5-6 months old. Vegetative propagation is by cuttings, graftings and buddings. The establishment of clonal plantations has already become current practice. This method is used particularly to develop germplasm and establish seed gardens. The cuttings or grafts are usually taken in a budwood garden. Vegetative propagation by grafting has the advantage of not requiring heavy investments. Side grafting, cleft grafting and especially chip budding are still favoured in several American and Southeast Asian countries, where a skilled workforce is available.

Biotechnology has been explored in recent years in cacao propagation through the application of plant tissue culture techniques. Approaches to crop improvement have been taking principally through the avenues offered by somatic embryogenesis (SE). The first report of cacao SE by Esan (1975) described a method using immature zygotic embryo tissue explants, followed up by other reports using similar methods (Pence et al., 1979; Tan et al., 2000).

Several other scientists had since used other cacao parts through sporolytic tissues including leaves (Litz, 1986), nuccellus (Chatelet et al., 1992; Figueira & Janick 1993; Sondahl et al., 1993) and floral explants such as petals and staminodes (Lopez-Baez et al., 1993, Alemanno et al., 1996a, b; Alemanno et al., 1997). Other explants used include cambial tissues and sexual embryos, but few plantlets have been regenerated in vitro. As a result, application of tissue culture to cacao have concentrated on thc formation of the callus for the induction of somatic embryos and eventual regeneration of complete plants.

A more recent effort focused on the use of staminodes and petals of cacao flower parts, which are capable of producing propagules for propagating a wide variety of cocoa genotypes (Li et al., 1998). This has led to several breakthroughs in cacao tissue culture (Alemanno et al., 2000; Guiltinan et al., 2000; Lambert et al., 2000; Maximova et al., 2000).

Spacing for cacao varies between areas. Optimum spacing should provide the maximum yield per unit area over a given period. Closer spacing is used in Africa such as 3 × 2.5, 3 × 3, 3.1 × 3.1 and 4.5 × 4.5 m. In America and Asia, spacing is predominantly 4 × 4, 3.6 × 3.6 and 3 × 3 m.

Shading is indispensable for young cacao plants during the first few years in the field. Shade adjustment is one of the most important maintenance tasks in a young plantation. Regular interventions should not only aim at selecting specific shade trees, but also at reconciling the favourable effects of shade with the harm caused by competition between the roots of the shade trees and those of the crop. Any necessary thinning and pruning of low branches, potentially harmful to crown development, should be carried out in due time.

The relatively dense shading of the early years should be gradually removed to allow 50 - 70 % of the total light to pass. This gradual elimination of shade, should only take place when the cacao trees are full-grown and when their crowns join to form a continuous cover. Plantain (*Musa* spp.) for shading predominates mostly in

Africa, while in America, Asia and Oceania the shade trees planted are mostly *Gliricidia*, *Albizia*, *Erythrina*, *Calliandra*, *Leucaena* and *Pithecolobium* spp.

2. NEMATODES OF CACAO

Nematodes are unique in their ubiquity and variety of types as soil borne pests. Almost every crop has its complement of nematode parasite. Nematodes injure plants directly by their feeding which disturbs the host tissue mechanically and chemically, and indirectly by transmitting soil-borne viruses or by increasing the host susceptibility to bacterial and fungal pathogens.

Various surveys conducted over the years by different workers have shown the association of plant parasitic nematodes with cacao (Table 1). Damage potentials depend on the type and age of cacao attacked, with the root-knot nematode, *Meloidogyne* spp., as the most damaging genus. Damage to cacao range from growth retardation in the nursery and stunted growth characterized by small-sized leaves, a factor that is closely correlated with growth and yield, to yield losses in fruiting trees and, in severe cases, sudden death of trees. Transplanting nematode-infested seedlings constitute a major source of intra- and inter-farm transfer of nematode inoculum.

2.1. Root-Knot Nematodes

Root knot nematodes, *Meloidogyne* spp., are the most important nematodes parasitic on cacao because of the damage they cause and their wide distribution in cocoa producing regions.

2.1.1. Geographic Distribution

The earliest report of root-knot nematodes on cacao was that of Ritzema-Bos (1900). Later *Meloidogyne* spp. have been reported in Congo (Ghesquière, 1921), São Tomé (Cotterel, 1930), Ghana (Edwards, 1955; Gerard, 1962), Côte d'Ivoire (Luc & de Guiran, 1960), Malawi (Corbett, 1961; Martin, 1961), Nigeria (Caveness, 1967; Afolami & Caveness, 1983), Brazil (Lordello, 1968), Venezuela (Torrealba, 1969), India (Sosamma et al., 1980) and Bolivia (Bridge et al., 1982).

2.1.2. Species

Meloidogyne incognita appears to be the most comon nematode of cacao. It is a common pest in West Africa (Whitehead, 1969; Asare-Nyako & Owusu, 1979; Orisajo & Dongo, 2005; Fademi et al., 2006), India (Sosamma et al., 1980) and is widespread in cacao regions of Brazil (Sharma & Sher, 1974; Sharma, 1982). A variety, *M. acrita*, has been reported on cacao in Nigeria (Caveness, 1967) and on cacao in the coffee areas of South India (Kumar et al., 1971).

Other species of *Meloidogyne* reported on cacao include: *M. javanica* in Malawi (Corbett, 1961) and Central Africa (Martin, 1961), *M. arenaria* and *M. thamesi* in Brazil (Manço, 1969; Sharma, 1979) and *M. exigua* in Bolivia (Bridge et al., 1982).

2.1.3. Damage Symptoms

The histological changes observed on cocoa roots have a pattern similar to the damage reported for several other host plants. Following the invasion of roots by juveniles, the root cells swell and even coalesce to form giant cells leading to the formation of galls. Egg masses are deposited on the root surface through rupture of the cortex, but some eggs may be laid and hatch within root tissues.

Symptoms of damage on cacao seedlings associated with *M. incognita* are dieback, stunting, wilting, chlorosis of the leaves and reduction in size of leaves (Figure 2). In the field, when the dieback conditions occur, trees die down to their roots, which remain alive developing shoots in the following growing season and when the dead terminals are pruned off, as well (Fig. 3). On roots, tiny galls and females with egg masses can be observed (Afolami, 1981; Afolami & Ojo, 1984; Orisajo & Fademi, 2005). A definitive study in Brazil (Sharma & Maia, 1976) showed that *M. incognita* caused small, rounded and elongated galls with conspicuous egg masses. Considerable growth reductions of seedlings were evident with decreased stem girth, shoot and dry root weights.

In Ghana, *M. incognita* infection of cacao seedlings resulted in above ground hypocotyls swellings in addition to large galls on tap roots and smaller galls on feeding roots (Asare-Nyako & Owusu, 1979). The head end of the swollen female nematode was in the cambial region in both hypocotyl and root infections, and the typical hyperplasia and hypertrophy of parenchyma cells associated with root-knot diseases were present. Also, significant reductions in root weight and seedling height were observed. Asare-Nyako and Owusu suggested that the roots reduction in was responsible for the height reduction observed, rather than the extent of gall formation. It was also suggested that roots galling is not a sufficient criterion to assess parasitism by *Meloidogyne*, since on some crops infections can be overlooked (Whitehead, 1969).

Cacao roots infested by *M. javanica* forms galls in the environmental conditions of Central Africa (Martin, 1961). In Malawi slow growth rate of cacao and failure of young seedlings to get established in the field have been associated with soils heavily infested with *M. javanica* (Corbett, 1961). Damage symptoms were also observed on cacao roots infested by *M. exigua* in Bolivia (Bridge et al., 1982). General galling of secondary roots and atrophy of primary roots were reported as caused by *M. arenaria* in Brazil (Manço, 1969).

2.1.4. "Morte Subite" or Sudden Death

This sudden death of cacao is a condition in which green leaves suddenly turn yellow, then brown, and finally dry up, but remain hanging intact on the tree for a long time, even after the tree death (Fig. 4). This is a common occurrence on cacao fields in Nigeria, Brazil and other cacao growing regions of the world. The syndrome of the sudden death disease is a permanent wilting. Jimenez-Saenz (1971) and Sharma and Sher (1973) associated the occurrence of sudden death with root-knot nematodes. In Congo, Ghesquiere (1921) showed that *Meloidogyne* caused a form of 'morte subite' as a consequence of intensive attack on roots, and in Brazil a combination of attack by root-knot and root-lesion nematodes was considered as possible cause of the disease (Saenz, 1969).

2.2. Other Nematodes Parasitic on Cacao

The root lesion nematode, *Pratylenchus brachyurus* was found around cacao roots in Côte d'Ivoire (Luc & de Guiran, 1960), Nigeria (Caveness, 1967) and Brazil (Sharma & Sher, 1973). *Pratylenchus coffeae* has been reported to infect cacao roots in Java (Fluitter & Mulholland, 1941). In India, *P. coffeae* and *Rotylenchulus reniformis*, which is also known in association with cacao in Jamaica (Dixon, 1961), multiplied on cacao (Kumar et al., 1971). In pot experiments in Brazil, an inoculation of 1000 *Helicotylenchus dihystera* led to stunting and reduction of the root system of cacao seedlings (Campelo & Galli, 1980).

Dieback and death of cacao seedlings in the nursery has been attributed to infection by *Dolichodorus minor* in Brazil (Sharma, 1971). The entire root system was reduced, blackened and showed disintegrated cortex and beadlike gall formation, characterized by a reddish-brown color and hardness.

2.3. Disease Complexes

The fungus *Mycoleptodiscus terrestris*, was found to be associated with sudden death in Bahia, Brazil (Ram, 1973) and it possibily interacts with root-knot nematodes. Nematodes have been found with spores of *Phytophthora megakarya* in an extract from black pod of cocoa in Nigeria (unpublished). Whereas disease complex involving nematodes and fungi have been documented, not much can be said of disease complexes involving nematodes and bacteria, or nematodes and viruses. This is particularly true for tropical tree crops. Nevertheless, and in spite of the fact that this aspect has not been investigated as it relates to cacao, the possibility of a real threat cannot be excluded.

Presently, longidorid and trichodorid nematodes have been reported as vectors of plant viruses (Lamberti & Roca, 1987). In Longidoridae, species of the genera *Xiphinema*, *Longidorus* and one *Paralongidorus* are listed as vectors. Among the Trichodoridae, vector species belong to the genera *Trichodorus* and *Paratrichodorus*. Nematodes belonging to the above mentioned genera were found associated with cacao in Brazil, Côte d'Ivoire, Ghana and Nigeria.

Although no concerted effort has been made as regards the possibility of nematode-virus associations in cacao, apart of the mealybug-transmitted Cocoa Swollen Shoot Virus (CSSV), a nematode-transmitted virus, Cacao necrosis nepovirus, a serotype of tomato black ring virus reported in Ghana (Kenten, 1972), was documented (Brunt et al., 1996).

3. NEMATODE CONTROL IN CACAO

It has been reported earlier that nematodes in the nursery can retard the seedlings growth. The transplant of nematode infested seedlings carries nematodes to the plantations, where the plant may die. Replanting infested plantations is expensive and difficult, hence the need for control of nematodes both in the nursery and the field. Nematode control could be achieved through the use of resistant varieties/planting materials, chemicals (nematicides), and lately use of organic amendments. However, control of nematodes in a perennial crop like cacao is more difficult than in annual or herbaceous crops. The long-time nature of perennial crops makes rotation schemes, which are readily adoptable for annual crops, impractical in tree (perennial) crops, since nematodes that survive initial control practice have time to recover and build up again to destructive levels.

Table 1. Nematode associated with cacao roots in the cacao-growing world

Nematode	Country [a]
Criconemella goodeyi	Côte d'Ivoire
Criconemoides onoense	Brazil
Discocriconemella limitanea	Côte d'Ivoire
Dolichodorus minor	Brazil
Helichotylenchus cavenessi	Nigeria
H. dihystera	Brazil, Nigeria
H. erythirinae	Jamaica
Hemicriconemoides cocophilus	Nigeria
Hemicycliophora paradox	Côte d'Ivoire
H. loofi	Brazil
H. oostenbrinki	Nigeria
Heterodera spp.	Brazil
Hoplolaimus galeatus	Costa Rica
H. pararobustus	Nigeria
Longidorus spp.	Brazil, Côte d'Ivoire, Ghana, Nigeria
Meloidogyne arenaria	Brazil
M. exigua	Bolivia
M. incognita	Brazil, Côte d'Ivoire, Ghana, India, Nigeria
M. javanica	Malawi
M. thamesi	Brazil
Neodiplogaster tropica	Guatemala
Paratrichodorus christiei	Brazil
Paratylenchus arculatus	Côte d'Ivoire
Peltamigratus holdemani	Brazil
Pratylenchus brachyurus	Brazil, Côte d'Ivoire, Nigeria
P. coffeae	India, Java, Jamaica
Radopholus similis	Côte d'Ivoire, Jamaica, Nigeria
Rotylenchulus reniformis	Brazil, India, Jamaica
Rotylenchus microstriatus	Brazil
Scutellonema brachyurum	Brazil, Jamaica, Nigeria
S. clathricaudatum	Brazil, Jamaica, Nigeria
Trichodorus monohystera	Brazil, Côte d'Ivoire, Ghana, Nigeria
Trophurus imperialis	Côte d'Ivoire
Tylenchorhynchus martini	Nigeria
Tylenchus coffeae	Brazil, Nigeria
Xiphinema attorodorum	Nigeria
X. brevicollum	Brazil
X. ebriense	Ghana, Nigeria
X. elongatum	Phillipines
X. ifacolum	Brazil, Ghana, Nigeria
X. insigne	Phillipines
X. longicaudatus	Nigeria
X. nigeriense	Ghana, Nigeria
X. setariae	Brazil, Nigeria

[a] Source: Afolami & Caveness, 1983; Bridge et al., 1982; Caveness, 1967; Fademi & Orisajo, 2005; Martin, 1961; Sharma, 1982; Sharma & Loof, 1974; Sharma & Sher, 1973, 1974; Sosamma et al., 1980a,b; Thorold, 1975; Whitehead, 1969.

Figure 2. Stunted growth, chlorosis, reduction in leaf size, wilting and dieback of cacao seedlings infested with M. incognita *(left) compared with similar age plant (right) free of nematodes.*

Figure 3. Dieback conditions on cacao seedlings infested by Meloidogyne incognita.

Figure 4. Sudden death of cacao seedlings caused by root-knot nematodes

3.1. Integrated Management Approach to Nematode Control

Integrated Pest Management (IPM), which aims at utilizing a range of control mechanisms and replace broad-spectrum pesticides, brings exciting prospects for the whole cacao production chain. Improved planting material and plant husbandry, biocontrol and rational pesticide use can all play a role in IPM. The objective is to combine the available control methods with other potential methods in an integrated management approach. The potential control methods are revised in the following sections.

3.1.1. Resistant Planting Materials

Resistant cultivars hold out most promises for effective and economic control of nematodes. A resistant cultivar without nematode treatment often yields as much as high yielding susceptible cultivars treated with nematicides (Epps et al., 1981). Conveniently, the use of resistant planting materials is the most economic approach to nematode control in plants. However, this method is unfortunately not very applicable to cacao. Although cacao was not listed among the eight nut/fruit trees which have locally available nematode-resistant or nematode-tolerant seeds or rootstocks (Sasser & Frekman, 1987), assessment of materials in the Nigerian cacao germplasm has shown that only six clones possessed some degree of resistance (Afolami & Ojo, 1984). These, unfortunately, do not fall within the categories selected by breeders for broad-based multiplication for distribution to farmers. Nematodes hence pose a threat to the cacao production that cannot be addressed through resistance or tolerance of planting materials, although there is every possibility of unlocking the lack of tolerance with modern biotechnology tools.

3.1.2. Production of Nematode-Free Seedlings in the Nursery

This method involves raising of cacao seedlings in pre-treated soils in which all fauna (pathogenic and non-pathogenic) has been eliminated. Steam sterilization and chemical treatment (as partial sterilization procedure) are conventional means of sterilizing soil for raising cacao seedlings in baskets or polythene bags, where the cost is justified by the returns.

In Nigeria, Afolami (1993) controlled nematodes in bagged nursery soils with Basamid with some degree of effectiveness as steam sterilization of nursery soil. In the report, 2 g Basamid granules in 2-litre nursery bag placed at 0.5 and 0.25 inch depth completely eliminated plant-parasitic nematodes, but preserved free-living *Rhabditis* spp. and predatory Mononchids. A higher rate of 4 g per 2-litre bag, though equally effective, was highly phytotoxic.

3.1.3. Use of Nematicides in the Field

Historically, nematicides have been highly effective in controlling a wide range of nematode species in crops with quick acting, leading to spectacular increases in yields, specially of high-valued crops. Sosamma et al., (1980) reported increase in yield of cacao by the application of fenamiphos, fensulfothion and ethoprop. DBCP (dibromo-choropropane) a nematicide actually banned because of its toxicity to humans anf higher animals, injected into the soil around infected trees, controlled nematode damage temporarily but increased crop yields (Entwistle & Caveness, 1963; Ichinohe, 1967). The response of nematodes and cacao to applications of some nematicides to both potted plants and field trees is summarized in Table 2.

Since the first use of carbon disulfide (a fumigant) as a soil nematicide, several fumigant and non-fumigant nematicides have been developed. The more recently developed nematicides are the organophosphates or carbamates, all with dual action as nematicides and insecticides.

Many of these chemicals ended up unregistered because they were phytotoxic, while others, though initially registered, had to be withdrawn. In Nigeria, experiment is being conducted on the low dosage of carbofuran (nematicide and insecticide) with organic amendments for the control of nematodes in both the nursery and in the field.

3.1.4. Organic Amendments

The incidence of pesticide poisoning and mortality in some countries (Kottengoda, 1985) serves as a grim warning about the risks that arise when pesticides are widely used under poor management. Increased social and legislative pressure to restrict the use of methyl bromide, an effective soil fumigant used extensively to control a broad spectrum of pests including nematodes, has created the impetus to evaluate alternative approaches for management of soil borne diseases (Chellemi et al., 1994; Gullino et al., 2003). Hence the more urgent need derives, to develop new management tools that are environmentally and toxicologically safe.

Table 2. Responses of cacao to nematicides[*]

Nematicide a.i.	*Control*	*Effect on seedlings growth*	*Effect on mature trees*	*Reference*
Aldicarb	effective	good response	-	Sharma & Ferraz, 1977
Basamid granular	effective	good response	-	Afolami, 1993
DBCP	effective in field	no response	increased yields	Martin, 1961 Sharma & Ferraz, 1977 Entwistle & Caveness, 1963
DBCP	poor in pot tests	high concentrations cause damage	phytotoxic, decreased yields at 78 kg / ha	Martin, 1961 Sharma & Ferraz, 1977 Jimenez & Bonates, 1971
Ethoprophos	-	-	increased yield	Tarjan et al., 1971 Tarjan et al., 1972 Sosamma et al., 1980
Fenamiphos	very effective	growth much improved	increased yield	Sharma & Ferraz, 1977 Tarjan et al., 1971 Tarjan et al., 1972 Sosamma et al., 1980
Fensulfothion	effective	growth much improved	increased yields	Sharma & Ferraz, 1977 Martin, 1961 Tarjan et al., 1971 Tarjan et al., 1972 Sosamma et al., 1980
Fosthietan	poor	some response	-	Sharma & Ferraz, 1977
Lannate	poor	poor response	-	Sharma & Ferraz, 1977
Oxamyl	no control	poor response	-	Sharma & Ferraz , 1977

* Mention does not constitute an endorsement.

Organic amendments that are generally used for increasing agricultural productivity have been shown to have a suppressive effect on plant parasitic nematodes (Kaplan et al., 1992; Mehta et al., 1994; Widner & Abawi, 1998, 2002; Walker, 2004). Consequently, in Nigeria, use of organic amendments is being evaluated for suppression of nematode population in cacao. Recent reports (Orisajo & Fademi, 2005; Orisajo et al., 2005, 2008) showed that root-knot nematode populations were suppressed by cocoa pod husks, neem-fortified cocoa pod husks and poultry litter applied as soil amendment, with significant increases in plant height, stem girth, leaf area and dry shoot/root weights of cacao seedlings. Leaf extracts of *Ocimum gratissimum*, *Carica papaya*, *Azadirachta indica*, *Vernonia amygdalina Bixa orellana, Acalypha ciliate, Jatropha gossypifolia* and *Allium ascalonicum* reduced nematode populations and enhanced the growth of cacao seedlings in the nursery (Orisajo & Dongo, 2005; Orisajo et al., 2007).

The use of organic amendments is suggested to cocoa farmers in Nigeria as a good substitute for nematicide use in the management of root-knot disease of cacao seedlings. This will reduce the current level of frustration faced by resource-poor farmers in establishment of cacao seedlings in the field.

3.1.5. Biological Control

Biological control using antagonistic microorganisms, alone or in combination with other control methods in IPM programs, may provide a possible solution to root-knot nematodes. The most studied group among the nematode-antagonistic organisms is given by nematophagous fungi.

Vesicular-arbuscular mycorrhiza (VAM) fungi have been reported to increase plant tolerance and offset the growth reductions caused by *M. arenaria* in groundnut (Carling et al., 1996). Inoculation of tomato plants with *Glomus mosseae* (mycorrhizal fungus) suppressed gall index and the average number of galls per root system by 52% and 66%, respectively, compared with seedlings inoculated with *M. javanica* alone (Al Raddad, 1995). The association of mycorrhiza with cacao roots have been reported in Malaysia (Nadarajah, 1980), Brazil (Ezeta & Santos, 1981), Ecuador and Indonesia (Kramadibrata & Hedger, 1987). It has been reported that mycorrhiza likely do play a role in early cacao seedling growth, when the proportion of infected roots to total tree biomass is much greater than in established trees (Kramadibrata & Hedger, 1987).

Alternative management measures of nematodes in cacao using early mycorrhizal infection that would confer protection against root-knot nematode at a seedling stage when plants are most vulnerable are currently being explored in Nigeria. These measures are considered important, taking into account a widespread change towards production systems that use in vitro material propagated in treated substrates, free of mycorrhiza and other beneficial microorganisms. Research on the use of *Trichoderma* spp. as a possible biocontrol agent of nematodes in cacao seedlings is also being carried out in Nigeria.

4. CONCLUSIONS AND OUTLOOK FOR FUTURE CONTAINMENT

Association of nematodes with cacao has been often reported. Whereas the nature of pathogenic relationship between nematodes and other crops (arable and tree crops) have been investigated and reported, it is only recently that studies on pathogenic relationship between nematodes and cacao have being carried out. Even then, the studies appear to be restricted to effects of the root-knot nematodes on cacao, and are exclusive of interactions between the nematode and other soil pathogens.

However, the economic importance of cocoa justifies that a wide range of approaches be adapted to breeding and plant pathology researches. Therefore, it is pertinent to acknowledge the fact that breeding programs (especially for disease resistance) in cacao can no longer neglect the effect of nematodes, particularly the root-knot species.

It is also very important to note that Research and Development agencies have not paid, for long time, due attention to other biotic agents concentrating their efforts only on black pod disease and mirid problems. It will, therefore, be expedient if a network approach can be adapted for a regenerated research on role of nematodes in cacao, especially with regards to estimation of yield losses, distribution, pathogenicity, and interaction with other biotic constraints.

Although the present status has identified the possibility of an integrated use of chemicals, organic amendments and biocontrol agents in nurseries towards an early protection of cacao seedlings before transplant, a concerted awareness program for farmers on relevance of this procedure will be necessary. Until breeding programs manage to produce truly resistant varieties, implementation of the above protection strategy will reduce the current level of frustration faced by cocoa farmers in the establishment of new plantings, rehabilitation of old farms and the 'hard-to-explain' (by farmers) declining productivity of existing farms.

REFERENCES

Afolami, S. O. (1981). Symptoms of root–knot nematode infection on *Theobroma cacao* seedlings in Nigeria – a preliminary investigation. In Proceedings 3rd Regional Conference of the International Meloidogyne Project, IITA, Ibadan, Nigeria.

Afolami, S. O. (1993). The Effect of Basamid Granular (Dazomet) on nematodes parasitic on cacao seedlings in the nursery. In Proceedings 11th International Cocoa Research Conference, Bahia, Brazil.

Afolami, S. O., & Caveness, F. E. (1983). The Frequency of occurrence and geographical distribution of plant parasitic nematodes associated with *Theobroma cacao* in Nigeria. *Turrialba*, 33, 97-100.

Afolami, S. O., & Ojo, A. A. (1984). Screening of *Theobroma cacao* germplasm for resistance against a root- knot nematode – *Meloidogyne incognita* in Nigeria. In Proceedings 9th International Cocoa Research Conference, Lome, Togo.

Alemanno, L., Berthouly, M., & Michaux-Ferriere, N. (1996a). Embryogenèse somatique du cacaoyer a partir de pièces florales. *Plantations Recherche Developement*, 3, 225-237.

Alemanno, L., Berthouly, M., & Michaux-Ferriere, N. (1996b). Histology of somatic embryogenesis from floral tissues cocoa. *Plant Cell, Tissue and Organ Culture*, 46, 187-194.

Alemanno, L., Berthouly, M., & Michaux-Ferrière, N. (1997). A comparison between *Theobroma cacao* L. zygotic embryogenesis and somatic embryogenesis from floral explants. *In Vitro Cellular Development and Biology - Plant*, 33, 163-172.

Alemanno, L., Maximova, S., Michaux-Ferrière, N., & Guiltinan, M. J. (2000). Comparison de l'embryogenese somatique primaire et secondaire: evaluation de l'efficacité et ontogeny des embryons somatiques. In Proceedings of 13th International Cocoa Research Conference, Kota Kinabalu, Sabah, Malaysia.

Al Raddad, A. M. (1995). Interaction of *Glomus mosseae* and *Paecilomyces lilacinus* on *Meloidogyne javanica* of tomato. *Mycorrhiza*, 5 (3), 233-236.

Asare-Nyako, A., & Owusu, G. K. (1979). *Meloidogyne incognita* infection of cocoa seedlings. In Proceedings 7th International Cocoa Research Conference, Douala, Cameroun.

Bartley, B. G. D. (2005). The genetic diversity of cacao and its utilization. CABI Publishing.Wallingford, UK: 341 pp.

Bridge, J., Page, S. L. J., & Waller, J. M. (1982). Plant parasitic nematodes and diseases of crops in the Santa Cruz Department of Bolivia. U.K. Overseas Development Administraion Report: 60 pp.

Brunt, A. A., Crabtree, K., Dallwitz, M. J., Gibbs, A. J., & Watson, L. (1996). Viruses of plants. Descriptions and lists from the VIDE database. CABI International. Oxon, UK: 1484 pp.

Campelo, A. M. F. L., & Galli, F. (1980). Patogenicidade de *Helicotylenchus dihystera* (Cobb) Sher em *Theobroma cacao* L. *Revista Theobroma*, 10 (1), 5-14.
Carling, D. E., Roncadori, R. W., & Hussey, R. S. (1996). Interactions of arbuscular mycorrhizae, *Meloidogyne arenaria*, and phosphorus fertilization on peanut. *Mycorrhiza*, 6 (1), 9-13.
Caveness, F. E. (1967). End of tour progress report on the Nematology project, Ibadan, Ministry of Agriculture and Natural Resources: USAID / Nigeria project 620-11-110-050, 135p.
Chatelet P., Michaux-Ferrière, N., & Dublin, P. (1992). Potentialites embryogenes du nucelle et du tegument mnterne de graines immatures de cacaoyer (*Theobroma cacao* L.). *CR AcM Scm Paris*, 315, 55-62.
Chellemi, D. O., Olsen, S. M., & Mitchell, D. J. (1994). Effects of soil solarization and fumigation on survival of soilborne pathogens of tomato in northern Florida. *Plant Disease*, 78, 1167-1172.
Corbett, D. C. M. (1961). Report of Department of Agriculture 1959/1960. Part II. Nyasaland: 157-158.
Cotterel, G. S. (1930). Bulletin Department of Agriculture, Accra Gold Coast, 22p.
Dixon, W. B. (1961). Nematological investigations 1958-61. Jamaica: Bulletin Ministry of Agriculture, 59 (New series), 1-35.
Edwards, E. E. (1955). Further observations in the occurrence of nematodes of the genus *Meloidogyne* in Gold Coast. *Journal of Helminthology*, 29, 153-170.
Entwistle, P. F., & Caveness, F. E. (1963). Nematology. Annual Report, West Africa Cocoa Research Institute, 1961-62, 113-114.
Epps, J. M., Young, L. D., & Hartwig, E. E. (1981). Evaluation of nematicides and resistant cultivar for control of soybean cyst nematode race 4. *Plant Disease*, 65, 665-666.
Esan, E. S. (1975). Tissue culture studies on cacao (*Theobroma cacao* L.). In: Proceedings of 5th National Cocoa Research Conference, Ibadan, Nigeria.
Ezeta, F. N., & Santos, O. M. (1981). Importancia da endomicorriza na nutricao mineral do cacauiero. *Revista Brasileira de Ciencia do Solo*, 5, 22-27.
Fademi, O. A., Orisajo, S. B., & Afolami, S. O. (2006). Impact of plant parasitic nematodes on cocoa production (in Nigeria) and outlook for future containment of the problem. In: Proceedings 15th International Cocoa Research Conference, San Jose, Costa Rica.
Figueira, A., & Janick, J. (1993). Development of nucellar somatic embryos of *Theobroma cacao*. *Acta Horticulturae*, 336, 231-238.
Fluitter, H. J., & Mulholland, J. J. (1941). Gegevens, Verkregen biji het onderzoek naar de waardplanten van *Tylenchus coffeae*. *Bergcultures*, 15, 1588-1593.
Gerard, P. M. (1962). Nematode studies. Annual Report West Africa Cocoa Research Institute, 1961-65: 32-33.
Ghesquière, J. (1921). Nouveaux parasites du cacaoyer. Maladie vermiculaire du cacaoyer (*Tylenchus* (*Heterodera*) *radicola*. Greef.) et sa relation avec la maladie de *Diplodia* (Coup de soleil, Die back). *Bullentin de Agriculture*, 12, 709-718.
Guiltinan M., Miller, C., Traore, A. & Maximova, S. N. (2000). Greenhouse and field evaluation of orthotropic cacao plants produced via somatic embryogenesis, micro and macro-propagation. In Proceedings of 13th National Cocoa Research Conference, Kota Kinabalu, Sabah, Malaysia.
Gullino, M. L., Camponogara, A., Gasparrini, G., Rizzo, V., Clini, C., & Garibaldi, A. (2003). Replacing methyl bromide for soil disinfestations: the Italian experience and implications for other countries. *Plant Disease*, 87, 1012-1021.
Ichinohe, M. (1967). Plant protection in Japan. VII. Control of plant nematodes. Agriculture. Asia. Special issue No. 5, 93-98.
Jimenez, S. E., & Bonates, J. (1971). The tolerance of cacao, *Theobroma cacao* L. to the nematicide 1, 2-dibromo-3-chloropropane. *Revista Theobroma*, 1, 30-36.
Kaplan, M., Noe, J. P., & Hartel, P. G. (1992). The role of microbes associated with chicken litter in suppression of *Meloidogyne arenaria*. *Journal of Nematology*, 24, 522-527.
Kenten, K. H. (1972). The purification and some properties of cocoa necrosis virus, a serotype of tomato black ring virus. *Annals of Applied Biology*, 71, 119-126.
Kumar, A. C., Viswanathan, P. R. K., & D'Sousa, G. I. (1971). A study of plant parasitic nematodes of certain commercial crops in coffee tracts of South India. *Indian Coffee*, 35, 222-224.
Kramadibrata, K., & Hedger, J. N. (1987). Comparative studies on the mycorrhizal symbionts of cocoa in Ecuador and Indonesia. In Proceedings 10th International Cocoa Research Conference, Santo Domingo, Dominican Republic.
Lass, R. A., & Wood, C. A. R. (1985). Cocoa production, present constraints and priorities for research. World Bank. Washington, DC, USA.

Lamberti, F., & Roca, F. (1987). Present status of nematodes as vectors of plant viruses. In: J. A. Veech and D. W. Dickson (Ed.), Vistas on Nematology. Society of Nematologists. Hyattsville, Maryland, USA, 321-328.

Lambert, S. V., Guiltinan, M., Maximova, S., & Aitken, W. M. (2000). Ex vitro propagation of acclimated somatic embryo derived cocoa plants. In 13th International Cocoa Research Conference, Kota Kinabalu, Sabah, Malaysia.

Li, Z., Traore, A., Maximova, S., & Guiltinan, M. J. (1998). Somatic embryogenesis and plant regeneration from floral explants of cacao (*Theobroma cacao* L.) using thidiazuron. *In Vitro Cellular Development and Biology - Plant*, 34, 293-299.

Litz, R. E. (1986). Tissue culture studies with *Theobroma cacao*. In P. S. Dimick (Ed.), Cacao biotechnology symposium (pp. 111-120). Florida: The Pennsylvania State University.

Lopez-Baez, O., Boron, H., Eskes, A. B., & Petiard, V. (1993). Embryogenese somatique de cacaoyer *Theobroma cacao* L., a partir de pièces florales. *Comptes Rendus de l'Académie des Sciences de Paris*, 316, 579-584.

Lordello, L. G. E. (1968). Nematoides associados a uma doença do cacaueiro. *Revista de Agricultura Piracicaba (Brazil)*, 43, 154.

Luc, M., & de Guiran, G. (1960). Les nematodes associes aux plantes de l'ouest Africain. Liste preliminaire. *L'Agronomie Tropicale Nogeut*, 15, 434-449.

Manço, G. R. (1969). Nematoides em plantidas de cacau. In: Informe Technico 1968-69. Itabuna, Bahia, Brazil, 99-101

Martin, G. C. (1961). Plant species attacked by root-knot nematodes (*Meloidogyne* spp.) in the Federation of Rhodesia and Nyasaland. *Nematologica*, 6, 130-134.

Maximova, S. N., Alemanno, L., Young, A., Traore, A., Michaux-Ferrière, N., & Guiltinan, M. (2000). Efficiency, origin and quality of cacao somatic embryogenesis. In: Proceedings of 13th International Cocoa Research Conference, Kota Kinabalu, Sabah, Malaysia.

Mehta, U. K., Sundararaju, P., & Natesan, N. (1994). Effect of five oil cakes on control of *Pratylenchus zeae* in sugarcane. *Nematologia Mediterranea*, 22, 219-220.

Nadarajah, P. (1980). Species of Endoganaceae and mycorrhizal association of *Elaesis guineensis* and *Theobroma cacao*. In: P. Nikola (Ed.), Tropical Mycorrhiza Research. Oxford, UK: Clarendon Press, 232-237

Opeke, L. N. (2005). Tropical commodity tree crops. Spectrum Books Limited. Ibadan, Nigeria:

Orisajo, S. B., & Dongo, L. N. (2005). Nematicidal potential of some indigenous plant extracts against root-knot nematode on cacao. *African Scientist*, 6, 129-134.

Orisajo, S. B., & Fademi, O. A. (2005). Influence of neem-fortified cocoa pod husks soil amendment on *Meloidogyne incognita* in cocoa. *African Scientist*, 6, 125-128.

Orisajo, S. B., Fademi, O. A., & Afolami, S. O. (2005). Influence of cocoa pod husks as soil amendment on *Meloidogyne incognita* in cocoa. In Abstract Malaysian International Cocoa Conference, Kuala Lumpur, Malaysia.

Orisajo, S. B., Okeniyi, M. O., Fademi, O. A., & Dongo, L. N. (2007). Nematicidal effects of water leaf extracts of *Acalypha ciliata, Jatropha gossypifolia, Azadirachta indica* and *Allium ascalonicum* on *Meloidogyne incognita* infection on cacao seedlings. *Journal of Research in Biosciences*, 3, 49-53.

Orisajo, S. B., Afolami S. O., Fademi O. A., & Atungwu J. J. (2008). Effects of poultry litter and carbofuran soil amendments on *Meloidogyne incognita* attacks on cacao. *Journal of Applied Biosciences*, 7, 214-221

Pence, V. C., Hasegawa, P. M., & Janick, J. (1979). Asexual embryogenesis in *Theobroma cacao* L. *Journal of the American Society of Horticultural Science*, 104, 145-148.

Ram, A. (1973). Relatorio Annual de Fitopatologia. Ilheus, Bahia, Brazil: Centro de Pesquisao do Cacau, 24-26.

Ritzema-Bos, J. (1900). Les nematodes parasites des plantes cultivees. 6th Congrés Internationale d'Agriculture, Paris II: 306-313.

Sasser, J. N., & Freckman, D. W. (1987). A world perspective on nematology: the role of society. IIn: J. A. Veech and D. W. Dickson (Ed.), Vistas on Nematology. Society of Nematologists: Hyattsville, Maryland, USA, 7-14.

Saenz, E. J. (1969). Relacion entre el ataque de nematodes y la muerte subita del cacau (*Theobroma cacao* L.) en Bahia, Brazil. *Turrialba*, 19, 255-260.

Sharma, R. D. (1971). Nematode associated with cacao and rubber in Bahia, Brazil. *Revista Theobroma*, 1, 43-45.

Sharma, R. D. (1979). Informe técnico – Setor de nematologia. CEPEC-CEPLAC, Itabuna, Bahia, Brazil 56-60.

Sharma, R. D. (1982). Nematodes associated with cocoa hybrids and clones in Bahia, Brazil. *Nematologia Brasileira*, 6, 85-91.

Sharma, R. D., & Ferraz, E. C. A. (1977). Efficacia de nematicidas sistemicos no controle aos nematóides fitoparasitas associados a mudas de cacaueiro (*Theobroma cacao* L.) *Revista Theobroma*, 7(1), 3-12.

Sharma, R. D., & Loof, P. A. A. (1974). Nematodes of the cocoa region of Bahia, Brazil. III – Plant Parasitic and free living nematodes in the rhizospheres of six different plant species. *Revista Theobroma*, 4, 39-43.

Sharma, R. D., & Maia, M. A. Z. (1976). Pathogenicity of the root-knot nematode *Meloidogyne incognita* on cocoa. *Revista Theobroma*, 6, 56-65.

Sharma, R. D., & Sher, S. A. (1973). Nematodes associated with coffee in Bahia, Brazil. *Archivos do Instituto Biologico de Sao Paulo*, 40, 131-135.

Sharma, R. D., & Sher, S. A. (1974). Nematoides da regiao cacaueira do Espirito Santos, Brazil. I. Nematoides associates ao cacaueiro (*Theobroma cacao* L.). *Revista Theobroma*, 4, 26-31.

Sondahl, M. R., Liu, S., Bellato, C., & Bragmn, A. (1993). Cacao somatic embryogenesis. *Acta Horticulturae*, 336, 245-248.

Sosamma V. K , Koshy , P. K. & Sundararaju, P. (1980a). Nematodes associated with cacao – A review. Plant Protection Committee for the South East Asia and Pacific region, Technical Document N. 123: 14 pp.

Sosamma V. K., Koshy, P. K., & Sundararaju, P. (1980b). Plant parasitic nematodes associated with cacao. *Cocoa Growers' Bulletin*, 29, 27-30.

Tan C . L., Davey, M. R., Lowe, K. C., Power, J. B., & Furtek, D. B. (2000). A study on the factors affecting somatic embryogenesis in *Theobroma cacao* L. In Proceedings of 13th International Cocoa Research Conference, Kota Kinabalu, Sabah, Malaysia.

Tarjan, A. C., Jimenez, M. F., & Soria, V. J. (1971). Reactions of nematized cacao to chemical treatment. *Nematropica*, 1, 16.

Tarjan, A. C., Jimenez, M. F., & Soria, V. J. (1972). Improving yields from nematode infested cacao trees (*Theobroma cacao*) in Costa Rica by use of nematicides. *Nematropica*, 2, 10-11.

Tarjan, A. C., Jimenez, M. F. & Soria, V. J. (1973). Increasing yields of cacao by application of nematicides. *Turrialba*, 23, 138-142.

Taylor, C., & Taylor, L. (2006). Future trends in cocoa industry - a perspective. *Coffee & Cocoa International*, 33, 39-41.

Thorold, C. A. (1975). Diseases of Cocoa. Clarendon Press. Oxford, UK: 423 pp.

Torrealba, P. A. (1969). Survey of plant parasitic and free-living nematode genera from Venezuela. In J. E. Peachey (Ed.). Nematodes of Tropical Crops. Commonwealth Agricultural Bureaux, St. Albans, UK: 257-263

Van Himmer, M., & Snoek, J. (2001). Beverage and stimulant crops. In R. H. Raemaekers (Ed.), Crop production in tropical Africa. Directorate General for International Co-operation (DGIC). Brussels, Belgium: 889-921.

Walker, G. E. (2004). Effects of *Meloidogyne javanica* and organic amendments, inorganic fertilizers and nematicides on carrot growth and nematode abundance. *Nematologia Mediterranea*, 32(2), 181-188.

Whitehead, A. G. (1969). The distribution of root-knot nematodes *Meloidogyne* spp. in Tropical Africa. *Nematologica*, 15, 315-333.

Widmer, T. L., & Abawi, G. S. (1998). Marketable yields of carrots in *Meloidogyne hapla* infested soils as affected by a green manure of Sudan grass. *Journal of Nematology*, 30, 522.

Widmer, T. L., & Abawi, G. S. (2002). Relationship between levels of cyanide in sudangrass hybrids incorporated into soil and suppression of *Meloidogyne hapla*. *Journal of Nematology*, 34, 16-22.

6

LARRY W. DUNCAN

MANAGING NEMATODES IN CITRUS ORCHARDS

University of Florida, IFAS, Citrus Research and Education Center, 700 Experiment Station Road, Lake Alfred, FL 33850 USA

Abstract. Citrus trees are damaged by several nematode parasites, all but one of which are of limited distribution worldwide. The most economically important species is *Tylenchulus semipenetrans* by virtue of its presence in all citrus producing regions. The nematode is noteworthy for its intimate association with citrus and the ability of trees to support very large populations before damage becomes evident. Several of the remaining species are among the most damaging parasites in the citrus rhizosphere. This chapter reviews the biology, ecology and economic importance of these nematodes from the standpoint of pest management.

1. INTRODUCTION

Most commercial citrus species are in the genera *Citrus* (oranges, mandarins, pomelos, grapefruit, lemons, limes and citrons), *Fortunella* (kumquats) and *Poncirus* (trifoliate oranges), all in the family Rutaceae (Swingle & Reese, 1967). Citrus production worldwide exceeded 72 million tons in 2006 (Anon., 2007). *Citrus* spp. evolved as understory plants and are naturally deep rooted (Ford, 1954a,b). Trees grow best in well-drained soils because roots will not grow into or remain in saturated zones for more than a few days without permanent damage. In areas with high water tables, soil is bedded to provide adequate rooting volume and conditions. Citrus grows well in either humid or arid zones provided that adequate soil moisture can be maintained. Irrigation of citrus is commonly practiced by a variety of methods that range from orchard flooding to low-volume drip or microsprinkler systems. In areas with sporadic rainfall, the ability to manage soil moisture is critical for good production, particularly during the period when fruit are set after the first seasonal flower bloom (Sites et al., 1951). There is a tendency at present in the United States and elsewhere to increase early returns by planting higher density orchards with shorter life expectancies due to such diseases as citrus blight, tristeza and greening (Hearn, 1986; Spyke and Castle, 2007).

Numerous nematode species are associated with the citrus rhizosphere (Cohn, 1972; Duncan 1999). Few, however, have been shown to be of economic importance. With the exception of *Tylenchulus semipenetrans,* nematodes known to be capable of damaging citrus are very limited in distribution, due either to edaphic conditions or to the natural distribution of a particular nematode. Relatively little is

A. Ciancio & K. G. Mukerji (eds.), *Integrated Management of Fruit Crops and Forest Nematodes*, 135-173.

known about the relationships between citrus and several of these species. Although regional in scope, nematode parasites of citrus are varied in their habits. Migratory endoparasites (lesion and burrowing nematodes), sedentary endoparasites (citrus and root-knot nematodes), and several species of ectoparasitic nematodes can damage citrus. Additionally, there are nematode species commonly found in the citrus rhizosphere for which insufficient information exists to determine their pathogenic potential.

2. TYLENCHULUS SEMIPENETRANS

The "citrus nematode", *T. semipenetrans*, is a parasite of several woody plant species. It is aptly named because it is ubiquitous in the citrus producing regions of the world. More than 75 rutaceous species (mainly citrus, their hybrids and close relatives) are suitable hosts, but only a few non-rutaceous hosts are known, most notably grape, olive, lilac and persimmon. Due to its narrow host range, *T. semipenetrans* can be readily excluded from new citrus plantings through nursery sanitation. Unfortunately, regulatory exclusion is rare and surveys estimating infestation of 50-90% of orchards are common in many parts of the world (Van Gundy & Meagher, 1977 ; Heald & O'Bannon, 1987; Esser et al., 1993; Sorribas et al., 2000; de Campos et al., 2002; Iqbal et al., 2006; Maafi & Damadzadeh, 2008; Sorribas et al., 2008).

The nematode was first detected on citrus roots in California in 1912, and was named and described during the next two years (Cobb 1913; 1914; Thomas 1923). It causes the disease "slow decline of citrus", so named because tree debilitation by *T. semipenetrans* is gradual in newly infested sites. Population growth is slow and the nematode is well-adapted to citrus, with very high numbers required to significantly affect the growth and health of its host (Cohen et al., 1965).

2.1. Slow Decline Symptoms

Aboveground symptoms of slow decline are those associated with poor root development, drought or lack of nutrition. Leaves and fruit are smaller and chlorosis may be evident (Philis, 1989; McClure & Schmitt, 1996; Kallel et al., 2004). Fibrous roots are less abundant so that wilting occurs earlier during periods of water stress and leaf drop is more pronounced causing thinner tree canopies in heavily infested trees (Heald & O'Bannon, 1987; Hamid et al., 1988; Duncan et al., 1995). Symptoms also depend on the suitability of environmental conditions and cultural practices for citrus health (Fig. 1). Infested trees growing under otherwise optimum conditions may appear healthy, but yield somewhat less fruit because of carbon lost to the nematode directly and to the need for more frequent growth of fibrous roots (Hamid et al., 1988; Duncan & Eissenstat, 1993). As conditions become less suitable for tree growth, effects of citrus nematode parasitism are more apparent (Van Gundy & Martin, 1961; Van Gundy et al., 1964; Heald & O'Bannon, 1987). For example, in saline conditions, excessive sodium accumulates in leaves of nematode infected trees (Mashela et al., 1992a). Leaves of heavily infected trees also exhibit reduced concentrations of nutrients such as potassium, which indicates the importance of optimum fertilization (Martin & VanGundy, 1963; Willers, 1979; Mashela & Nthangeni, 2002).

Figure 1. Symptoms associated with slow decline of citrus are affected by management and environmental conditions. Nematode infected trees (A) grown in a site with poor drainage, periodic salinity and high populations of Phytophthora nicotianae *were visually similar to adjacent trees with no or very few* T. semipenetrans *on the roots; however, fruit yield, root mass density and leaf area were significantly less on infected trees (Duncan et al. 1995). Trees heavily infected by* T. semipenetrans *growing in deep sandy soil with optimal management showing few decline symptoms (B); however fruit yield and size on such trees often increase in response to nematode management.*

Citrus nematodes do not cause galling or obvious necrosis to fibrous roots, so that incipient infestations can easily go undetected in nurseries. Soil particles adhere to *T. semipenetrans* gelatinous egg masses on the root surface, and give roots a thicker, dirty appearance even after rinsing (Fig. 2). Secondary organisms can invade the cortex at the nematode infection and feeding sites and the resulting decay can result in cortical sloughing and root death when infections are heavy (Schneider & Baines, 1964; Cohn, 1965b; Hamid et al., 1988).

2.2. Biology and Ecology

Vermiform second stage juvenile *T. semipentrans* hatch and do not immediately penetrate citrus fibrous roots. Second stage female juveniles have been shown to survive more than two years in the absence of roots (Baines, 1950). The nematode is sexually dimorphic with males having the ability develop to the adult stage without feeding. Although amphimictic, females can produce male and female offspring in the absence of males (Van Gundy, 1958; Dalmasso et al., 1972). Female juveniles feed for up to two weeks on epidermal cells before molting in approximately seven days to the third, fourth and adult stages (Van Gundy, 1958). The anterior of young adult females penetrate deep into the fibrous root cortex to initiate a permanent feeding site at which time the nematode becomes increasingly swollen (Cohn, 1964). The female feeds repeatedly from several transfer or *nurse cells* that surround the head (Kallel et al., 2005; Fig. 3). The posterior of the female remains exposed on the root surface (Fig 2). The adult female has no rectum or anus and the excretory pore is just anterior to the vulva (Cobb, 1914; Gutierrez, 1947). The rectum and anus is discernable in live, second-stage juvenile females (Duncan & Inserra, 2005). Approximately six weeks after hatching at 25 °C, females lay eggs on the root surface in a gelatinous egg mass secreted from the excretory pore (Van Gundy, 1958; Cohn, 1964).

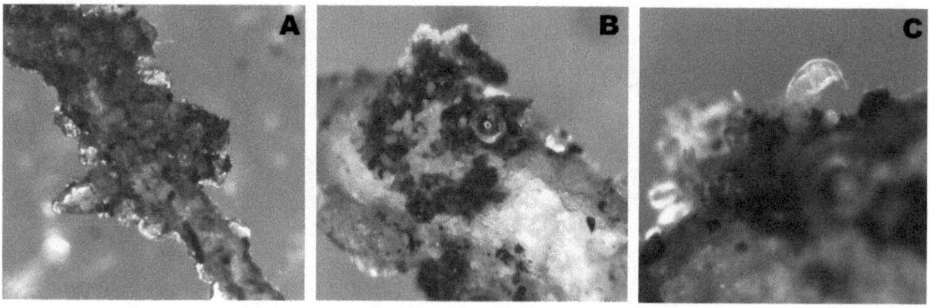

Figure 2. Tylenchulus semipenetrans *life stages viewed with a dissecting microscope on citrus fibrous roots. Soil particles and organic films adhere to gelatinous egg masses (A), giving a dirty appearance to roots even after rinsing (note actual root diameter in lower portion of figure). An egg mass (B) uncovered by removing sand and organic matter. The posterior of a female (C) adjacent to an egg mass deposited by a neighboring female.*

The ecology of *T. semipenetrans* reflects an intimate coevolution with citrus and other deep-rooted understory trees and vines. The nematode is able to induce an intricate series of nurse cells in the cortex of just a few plant species. Nutrients are continuously transferred to the nematode via nurse cells with little damage to the host, which can consequently support large numbers of the parasite for years. The nematode develops fastest at the generally moderate temperatures typical of soil in the shade of the tree canopies. Unlike many nematodes, *T. semipenetrans* is unable to survive at very low soil water potential, which occurs less frequently in the surface rhizosphere of deep-rooted trees (see below) compared to shallow-rooted herbaceous plants.

Development of *T. semipenetrans* is regulated by temperature as it interacts with geographic and temporal variability in the soil environment and the host phenology. The nematode displays distinct and often predictable patterns of annual population growth. Depending on the region, one (Prasad & Chawla, 1966; Bello et al., 1986; Sorribas et al., 2000; Maafi & Damadzadeh, 2008), two (Vilardebo, 1964; O'Bannon et al., 1972; Salem, 1980; Baghel & Bhatti, 1982; Duncan et al., 1993; Al Hinai & Mani, 1998; Sorribas et al., 2000; Galeano 2002), or three (Hamid et al., 1988) distinct periods of active population development per year are reported, although no seasonality was evident during a survey in Israel (Cohn, 1966). Maximum development of *T. semipenetrans* occurred at 25°C with slower rates as temperatures approached upper (31°C) and lower (20°C) limits for population growth (O'Bannon et al., 1966). In most regions, low winter temperatures can regulate population size (Duncan et al., 1993; Maafi & Damadzadeh, 2008) and high summer soil temperatures are associated with seasonally low populations in warmer regions such as Egypt, Texas, Oman, and Spain (Salem, 1980; Davis, 1984; Al Hinai & Mani, 1998; Sorribas et al., 2000; Korayem & Hasabo, 2005). Reynolds and O'Bannon (1963a) speculated that population growth was minimal on young trees in Florida and Arizona until the canopies developed enough to provide shade and reduce soil temperature.

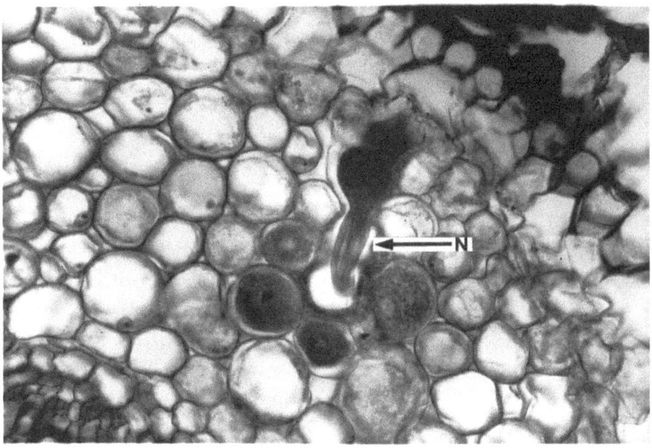

Figure 3. Tylenchulus semipenetrans *feeding site in the root cortex showing the nematode anterior (N) with the head in a cavity surrounded by dark colored nurse cells that are continually replenished with starch and other nutrients* (from O'Bannon & Esser, 1985).

Tylenchulus semipenetrans has little capacity for desiccation survival compared to many plant parasitic nematodes that must routinely survive soil drying cycles in a state of anhydrobiosis (Tsai & Van Gundy, 1988). Population decline of citrus nematode is pronounced if drought is severe enough to cause wilt in citrus trees (Van Gundy & Martin, 1961; Van Gundy et al., 1964). Nevertheless, soil moisture is frequently inversely related to population growth of *T. semipenetrans* in the field (Tuong, 1963; Duncan et al., 1993; Sorribas et al., 2000; Galeano 2002).

Passive movement of nematodes below the sampled profile due to precipitation is a potential cause of the relationship (Sorribas et al., 2000), although the vertical distribution of some nematodes seems to be unaffected by heavy rainfall events (Chabrier et al., 2008). Alternatively, experiments have shown that populations of *T. semipenetrans* in extremely dry parts of the rhizosphere can either grow very rapidly or decline precipitously, depending on whether part or all of the root system is affected by drought. Hydraulic lift of water deep in soil to drier surface soil horizons via the root xylem (Caldwell et al., 1991) creates a humid zone at the rhizoplane that may not be measurable in the soil, particularly in coarse textured soil. An environment of hydrated roots in dry soil favors population growth of *T. semipenetrans* compared to more humid soil conditions (Duncan & El-Morshedy, 1996). It is unknown whether this is due to increased oxygen in dry soil (Van Gundy et al., 1962; 1964), increased activity of natural enemies in moist soil (Dirac & Menge, 2002), or other factors.

As a parasite of deep-rooted perennials, *T. semipenetrans* likely experienced less selection pressure than many nematodes for anhydrobiotic survival. The response of *T. semipenetrans* to hydraulic lift may affect regional patterns of population density which tend to be reported as higher in arid (Cohn, 1966; Macaron, 1972; Willers, 1979; Davis et al., 1985; Sorribas et al., 2000, 2008) than in sub-tropical regions (Davide, 1971; O'Bannon, et al., 1972; Duncan et al., 1993). Similarly Sorribas et al. (2000) observed that population densities were higher under drip than under flood irrigation. In tropical and subtropical regions, hydraulic lift may also affect seasonal patterns of population change which tend to be bimodal, with peaks in the dry months that precede and follow the summer rainy season (Toung, 1963; Prasad & Chawla, 1966; O'Bannon et al., 1972; Duncan et al., 1993).

The population dynamics of *T. semipenetrans* are also regulated by seasonality in the growth of citrus organs and the availability of nutrients in roots. Citrus fibrous root growth alternates with growth of new leaves during the growing season. Young roots are the most suitable for penetration and development of *T. semipenetrans* and new cohorts of developing nematodes are created during root flushes (Cohn, 1964; O'Bannon et al., 1972). Several peak periods of increased numbers of female *T. semipenetrans* corresponded with root flushes in a California citrus orchard (Hamid, et al., 1986). However, the nematode can alter the normal pattern of carbon allocation. Trees heavily infected by the nematode have less root mass than lightly infected trees, but root growth is initiated more frequently to replace those damaged by the nematode (Hamid, et al., 1986). Starch is a major nutrient requirement of *T. semipentrans* (Cohn, 1965), whereas lignin and phenolic compounds inhibit root infections (Kaplan, 1981). The seasonal concentrations of these compounds in fibrous roots have been shown to be correlated in the expected ways with *T. semipentrans* population growth in field surveys (Van Gundy & Kirkpatrick, 1964; Duncan et al., 1993) and in field experiments in which root carbohydrates were manipulated (Duncan & Eissenstat, 1993).

Unlike plant parasitic nematodes in many crops, damaging levels of *T. semipentrans* can develop in all soils suited to citriculture. Salinity and pH of soil solutions are two factors known to have consistent effects on *T. semipenetrans* and its role in crop loss. The nematode develops poorly in saline soils (Kirkpatrick & Van Gundy, 1966). However, salinity has been shown to be associated with high nematode numbers and increased crop loss in the field (Machmer, 1958; Cohn et al., 1965; Willers & Holmden 1980). The explanation is that soil salinity is seasonal in occurrence, being least during rainy seasons when salt residues from saline irrigation water are leached from surface soils. Citrus that has been exposed to salinity supports greater population growth of *T. semipenetrans* during the periods when salts are washed from the rhizosphere, than does citrus not exposed to salinity or citrus grown continually under saline conditions (Mashela et al., 1992a; 1992b). The mechanism is unknown, but reduced production of phenylalanine ammonia lyase in salt-stressed citrus may limit production of phenolic-based defensive compounds in roots (Dunn et al., 1998). Nematode-infected roots accumulate less sodium and chloride ions, whereas these elements increase in leaves (Willers & Holmden, 1980; Mashela et al., 1992a; 1992b). Increased osmotic pressure from carbohydrate transfer and accumulation in nematode infected roots has been proposed as a mechanism by which salinity elements move through roots to accumulate in the leaves (Mashela & Nthangeni, 2002).

Tylenchulus semipenetrans will develop to damaging levels across a wide range of pH. However population growth is markedly faster and equilibrium density greater at pH 6.0-8.0 (Van Gundy & Martin, 1962; Bello et al., 1986; El-Borai et al., 2003). For this reason, citrus grown in calcareous soil is likely to experience somewhat greater damage from this nematode than is citrus grown in more acidic soils provided that pH is not excessively high. The nematode will also persist in any soil texture in which citrus is grown. Various studies in pots indicate that population growth is faster in moderately fine textured soil than in sand or in very fine textured soil (Van Gundy et al., 1964; Davide, 1971; Bello et al., 1986). However, texture interacts strongly with other factors, especially moisture, and high population densities of the nematode occur in coarse textured soil in the field (Duncan et al., 1993). Moderate levels of organic matter favors population growth of the nematode (Van Gundy, 1958; O'Bannon, 1968).

Tree nutrition influences population levels (Martin & Van Gundy, 1963; Mangat and Sharma, 1981). Similarly, as with most nematode-host combinations, the genetic variety of roots affects population growth of *T. semipenetrans*, even among rootstocks that are not considered to be resistant (Davide, 1971; O'Bannon et al., 1972; O'Bannon & Hutchinson, 1974; Davis, 1984). The scion had no effect on the resistance or susceptibility of a rootstock in some studies, but did influence nematode development and even morphometrics to some extent (Kirkpatrick & Van Gundy, 1966; Scotto La Massese et al., 1975; Das & Mukhopadhyaya, 1985; Bello et al., 1986). However, it was recently reported that resistance through cellular necrosis was conferred to susceptible sour orange rootstocks when resistant trifoliate orange was used as the scion (Kallel et al., 2006).

2.3. Interactions with Soil Organisms

The large numbers of *T. semipenetrans* in the relatively undisturbed citrus rhizosphere support a diverse and often abundant community of natural enemies in most orchards examined (Stirling & Mankau, 1977; Gaspard & Mankau, 1986; Fattah et al., 1989; Roccuzzo et al., 1992). Little is known about the level of natural control of the nematode or the factors that influence it. *Pasteuria* sp. infects and develops inside *T. semipentrans* juveniles, but not in females as is common for this bacterium on *Meloidogyne* spp. (Fattah et al., 1989; Kaplan, 1994). *Pasteuria* was detected on the nematode in just one of 27 orchards in Florida, whereas half of 48 orchards in Spain revealed this association. Although the infection rate by the bacterium is generally low, some soils appear to be very conducive (Walter & Kaplan, 1990). Sorribas et al., (2000) detected up to 47% of juveniles infected by *Pasteuria* in an orchard in which nematode and bacterium incidence were positively correlated over time,which suggests a role for the bacterium in regulating the nematode. Similarly, Gené et al. (2005) found that fungal parasitism of egg masses was directly related to female abundance and inversely related to fecundity in Spanish orchards. While parasitism was generally low, in one orchard the fungus *Paecilomyces lilacinus* infected 75% of females and 10% of eggs. Strong correlations between parasitism of egg masses and the amount of sand, Mg and P content in soil suggests the possibility of identifying edaphic factors that enhance natural control of the nematode (Gené et al., 2005).

Fungi such as *Fusarium oxysporum* and *F. solani* are near-ubiquitous saprophytes in the citrus rhizosphere with the ability to colonize sedentary nematode females and eggs as well as citrus roots (Stirling, 1991). Initial studies revealed no measurable disease synergism between *Fusarium solani* and *T. semipenetrans* on citrus seedlings (Van Gundy & Tsao, 1963); however, additional studies showed that the nematode may increase the pathogenicity of this fungus when conditions are favorable for the fungus (O'Bannon et al., 1967; Labuschagne et al., 1989, Walker & Morey, 1999). *Phytophthora nicotianae* is a virulent pathogen of citrus roots. Levels of *P. nicotianae* in soil were inversely related to those of *T. semipenetrans* that were reduced variously by different nematicides (Graham & Duncan, 1997). Subsequently, it was shown that pre-infection of citrus roots by *T. semipenetrans* can reduce the rate of infection by *P. nicotianae* (El-Borai et al., 2002). There is some evidence that growth of the fungus is inhibited in the vicinity of the nematode eggs, but a mechanism is unknown (El-Borai et al., 2002b).

2.4. Biotypes and Rootstock Resistance

No species in the genus citrus has been found to have resistance to *T. semipenetrans* despite extensive screening. The only source of resistance incorporated into citrus is a dominant and oligogenic trait derived from *Poncirus trifoliata* (Hutchinson, 1985; Verdejo-Lucas and Kaplan, 2002). Eleven RAPD markers associated with this resistance trait were reported and could be used to facilitate identification of resistance in breeding programs (Ling et al., 2000; Deng et al., 2000). Resistant hybrids of *P. trifoliata* provide acceptable rootstocks in many regions (Gottlieb et al., 1986; Spiegel-Roy et al., 1988; Verdejo-Lucas et al., 2000). Swingle citrumelo (*C. paradisi* × *P. trifoliata*) is highly resistant to most populations of *T. semipenetrans*, tristeza virus and *Phytophthora nicotianae*. It is planted extensively in Florida and used to a more limited extent elsewhere, being restricted mainly by an intolerance calcareous soils. Selections of Poorman orange (*Citrus* × hybrid of undetermined origin) × *P. trifoliata* hybrids were similarly resistant to *T. semipenetrans*, *Phytophthora citrophthora* and tristeza virus (Gottlieb et al., 1986; Spiegel-Roy et al., 1988). Recently, *T. semipenetrans* resistant hybrids of *P. trifoliata* × various mandarin (*C. reticulata*) have been identified as acceptable rootstocks for calcareous soils (Verdejo-Lucas et al., 2003). *Severinia buxifolia* is a non-host citrus relative with potential as a source of germplasm for breeding programs. Several citrange cultivars (*C. sinensis* × *P. trifoliata*) once considered resistant or tolerant to citrus nematodes are now apparently susceptible to all biotypes of the nematode.

Resistance to *T. semipenetrans* derived from *P. trifoliata* involves hypersensitivity, wound periderm formation, compounds in root tissues that are toxic to the nematode, and unidentified factors which may repel nematodes and reduce infection (Van Gundy & Kirkpatrick, 1964; Kaplan & O'Bannon, 1981; Galeano et al., 2003; Kallel et al., 2006). Nematodes that manage to develop in resistant roots have low fecundity and produce a higher proportion of males (Verdejo-Lucas et al., 2000).

Soon after widespread use of Troyer citrange rootstock for its reported resistance to *T. semipenetrans*, resistance-breaking populations were detected (Baines et al., 1969a,b). Three main biotypes are now recognized for their differential host ranges on four plant species (Inserra et al., 1980). The "Citrus" biotype reproduces poorly on *P. trifoliata* but will multiply on *Citrus* spp. as well as on olive (*Olea europeae*) grape (*Vitis vinifera*) and persimmon (*Diospyros* spp.). Originally identified in North America and Italy, the biotype has also been detected in South America and the Middle East (Maafi, et al., 2000; Magunacelaya et al., 2004). The "Poncirus" biotype, initially found in California, reproduces on most citrus including *P. trifoliata*, and on grape but not on olive. This biotype is frequently detected in industries that employ Poncirus-based resistance in commercial rootstocks (Duncan et al., 1994; Miller et al., 1996; Murguia, et al., 2005; Kwaye, et al., 2008). A "Mediterranean" biotype is similar to the "Citrus" biotype, except that it does not reproduce on olive. It is found in Mediterranean countries, South Africa and perhaps India (Gottlieb et al., 1986; Verdejo-Lucas et al., 1997; Kallel, et al., 2006). Populations of a reported "Grass" biotype were eventually shown to be new species, *Tylenchulus graminis* and *T. palustris* (Inserra et al., 1988).

2.5. Economic Importance and Crop Loss Prediction

Estimating crop loss and economic thresholds is complex for long-lived perennials compared to annual crops. Preplant population density of nematodes is routinely related to yield in crops that mature in months. However yields of citrus trees are affected by the cumulative stress of nematodes and other factors during many years. At a given point in time, the potential yield of a citrus tree is highly variable compared to that of a corn seed or tomato seedling. Estimating a tree's nematode burden is also complicated by seasonal patterns of population growth. Fruit yield may be inversely related to *T. semipenetrans* abundance during some months and not others (Sorribas et al., 2008). Moreover, management of parasites can cause citrus trees to allocate carbohydrate to vegetative growth before fruit growth (Eissenstat and Duncan, 1992), so that yields may (McClure & Schmitt, 1996) or may not (Le Roux et al., 1991; Duncan, 1989) increase in the first year following nematode management. For reasons such as these, and because of the scarcity of damage functions for different environments and rootstock scion combinations, confidence intervals for loss predictions in citrus tend to be larger than for predictions in annual crops.

A major constraint to crop loss estimation in citrus is the cost of measuring yield in large numbers of trees or blocks of trees. Increased adoption by growers of automated yield mapping technology offers tremendous opportunities to relate nematode density and yield over a broad range of time and sites.

There are no yield loss studies in the field in which *T. semipenetrans* is the only independent variable. Instead experiments manipulated nematode density with nematicides, or surveys related natural patterns of nematode density across orchards to the yield in those sites. In the former approach, direct effects of chemicals on yield or indirect effects on other organisms may be involved in the yield response (Baines et al., 1962, 1966; Mankau, 1968; Cohn et al., 1968; Milne & du Toit, 1976; Milne & De Villiers, 1977; Timmer, 1977; O'Bannon and Nemec, 1978; Wheaton et al., 1985; Childers et al., 1987). In the latter method, effects of unmeasured factors (soil texture, moisture, pH etc.) that co-vary with nematode density and affect yield are likely (Duncan et al., 1995; Sorribas et al., 2008). Nevertheless, results of a large body of research provide reasonably consistent findings that *T. semipenetrans* requires high population density to cause measurable crop loss, but at high densities can seriously affect profitability.

Nematicide treatments are widely reported to increase citrus yield (Baines, 1964; Yokoo, 1964; Cohn et al., 1965; Oteifa et al., 1965; Philis, 1969; O'Bannon & Tarjan, 1973; Vilardebo, et al., 1975; Davide & Dela Rose, 1976; Milne & Willers, 1979; Timmer & Davis, 1982; Childers et al., 1987; Duncan, 1989; Le Roux et al., 1991, 1998; McClure & Schmitt, 1996; Singh, 2004). However, positive yield responses have occurred when treatments failed to reduce *T. semipenetrans* levels (Davis et al., 1982; Childers, et al., 1987) and reduction of populations without measurable yield responses is not uncommon (Davis & Wilhite, 1985; Stirling & Wachtel, 1985). The range of reported yield increases in response to nematicide treatment is wide, but tends to be of the order of 10-30%. Combining chemical management with use of resistant and susceptible rootstocks can partially control for non-target effects of chemical treatments. Sorribas et al. (2003) found no significant difference in the growth of a resistant and a susceptible rootstock during three years in soil fumigated to effectively control *T. semipenetrans*. However, the trunk cross sectional area of resistant trees was 48% greater than those of susceptible trees in non-fumigated plots.

Relationships between nematode density and tree condition show that highest absolute densities are attained on trees before symptoms become severe with the resulting loss of root mass (Reynolds & O'Bannon, 1963b; Davide, 1971; Scotto la Massèse, 1980; Coelho et al., 1983). When tree condition and yield are compared to nematode density per length or weight of roots, the relationships are usually inverse. In Israel, tree condition did not decline until *T. semipenetrans* surpassed 4 000 nematodes per gram of fibrous roots (Cohn et al., 1965). A Florida orchard was identified in which randomly distributed trees were infested or not infested by *T. semipenetrans*. Trees were also damaged by *P. nicotiana* and salinity, but levels of those variables and others such as soil pH, texture, and nutrients, did not differ for infested or non-infested trees. Tree condition was unrelated to presence of *T. semipenetrans*, but leaf area, fibrous root mass density, and fruit yield of infested trees were 32%, 8%, and 22% lower, respectively, than on non-infested trees (Duncan et al., 1995).

Citrus fruit yield has also been negatively correlated with infestation level (Willers, 1979; Timmer & Davis, 1982; Childers et al., 1987; Korayem and Hasabo, 2005). Yield was related to springtime (but not autumn) density of *T. semipenetrans* in roots in two of three Spanish orchards surveyed during two years (Sorribas et al., 2008). The damage functions in these orchards suggested a tolerance limit (below which no loss is measurable) of fewer than three hundred females per gram of root with economic thresholds ranging between 330-710 females per gram of root depending on the cost of the nematicide used and the value of the fruit during the two years. These estimates are similar to those reported in California where greater than 400 or 700 females per gram of root in early spring or early summer, respectively, are considered to merit management in orchards with a history of responding to management (Garabedian et al., 1984). In South Africa, a lower threshold of 100 females per gram of root is recommended (LeRoux et al., 2000).

Because *T. semipenetrans* reduces fruit size, the economic impact of the nematode is greater when fruit are marketed fresh rather than for juice. The greatest profitability from managing the nematode for fresh fruit production can be due to the increased value of larger fruit than to increased yield (Philis, 1989; McClure and Schmitt, 1996).

2.6. Management

Recommendations for managing *T. semipenetrans* vary greatly in different regions. If population levels are not high enough to cause noticeable tree decline, yield loss may not be large enough to be readily noticed. Because the effects of the nematode are similar to those of other biotic and abiotic problems that affect roots, tree decline may not be attributed to the nematode. Nematicides are expensive, not always highly effective, and profitability can be difficult to predict. Environmental concerns, particularly groundwater contamination, can outweigh concerns about the nematode if yield loss is not great. Consequently, managing the nematode should be based on careful consideration of local conditions and the likelihood that practices are profitable and unlikely to have hidden costs to the environment. A variety of tactics exist to characterize the threat from *T. semipenetrans* and to take appropriate actions.

2.6.1. Sampling and Extraction

For advisory purposes, citrus nematodes should be sampled at the same time each year, preferably when population density is likely to be greatest, or when research in a specific region has revealed a relationship between nematode density and yield (Sorribas et al., 2008). Seasonal variation in numbers of nematode life stages in the soil and roots are in the order of 3- to 10-fold (Salem, 1980; Baghel & Bhatti, 1982; Duncan et al., 1993; Sorribas et al., 2000; 2008). Because nematodes are increasingly aggregated in soil as density declines, sample accuracy increases during seasons of peak population size and in locations of highest root and nematode concentration (Nigh, 1981a; Duncan, 1986). Fibrous roots and nematodes are more abundant beneath the tree canopy than at the dripline or in rows between trees (Nigh, 1981b; Davis, 1985; Duncan, 1986, 1989). Low volume irrigation systems concentrate root and nematode populations even further in the wetted zones. Stratification of orchards based on tree health and factors such as texture and moisture that affect nematode density can also improve sample accuracy and management precision (Scotto la Massèse, 1980).

Accurately estimating *T. semipenetrans* population density is expensive. Davis (1984) estimated five samples, each consisting of 12 cores (2.5 × 30 cm) of soil, were needed to estimate population levels to within 20% of the mean in a Texas orchard. Estimates within 40% of the true mean required between 30-75 cores in 2 ha areas of Florida orchards (McSorley & Parrado, 1982b; Duncan et al., 1989, 1994a). The nature of *T. semipenetrans* distribution in most citrus industries increases the value of sampling despite the relatively low level of accuracy that is affordable. The majority of orchards in Florida have no citrus nematodes or have numbers small enough to be easily distinguished from economic threshold levels. In northeastern Spain, it is estimated that 40% of the orchards lose yield to *T. semipenetrans*, with 22% at densities exceeding an economic threshold, and with a majority of populations well above or below the threshold (Sorribas et al, 2008). Some laboratories recommend sampling to a depth of 60 cm (Garabedian et al., 1984), but population levels in the first 30 cm of soil were found to reflect population density to 60 cm (Duncan, 1986). For a given sample size, sample precision for root stages of the nematode is less than that for soil stages (Duncan et al., 1993), although root stages may provide better prediction of yield loss (Sorribas et al., 2008).

Juveniles and males of *T. semipenetrans* can be separated from soil by most conventional methods. Because extraction efficiencies are rarely reported, direct comparison of estimates between laboratories is often not possible. For some soils, techniques based on Baermann funnel principles appear to be similar in efficiency to techniques employing density flotation if the layer of soil extracted is relatively thin (Nigh, 1981b; McSorley & Parrado, 1982a). However, other authors report major differences in efficiency of the two approaches (Galeano, 2002). Because soil populations are usually reported per unit of soil, unless root mass density is also measured, the counts provide no information about the density of nematodes on the roots (Scotto La Massèse, 1980; Duncan, 1986). Nematodes hatching from root samples are easily obtained (Young, 1954; Cohn et al., 1965; McClure and Schmitt, 1996) and females per unit root can also be determined by extraction (Baines et al., 1969b; Duncan et al., 1993; Sorribas et al., 2008) or direct counts on stained roots (Davis and Wilhite, 1985).

2.6.2. Sanitation and Exclusion

Expansion of citrus into new citrus areas presents an important opportunity to reduce the incidence of *T. semipenetrans* because the nematode is rarely encountered outside of citrus orchards or vinyards (Milne, 1982; Lehman, 1996). New infestations usually result from movement of infected planting stock or on contaminated equipment (Tarjan, 1956; Van Gundy & Meagher, 1977). Moreover, *T. semipenetrans* is the only economic nematode pest in most citrus growing regions, which would obviate the need for any nematode control if it can be excluded. Such a condition exists in much of Florida's citrus industry where citrus nurseries are regulated by a state program requiring all commercial nursery stock to be certified free of *T. semipenetrans*, *Radopholus similis*, and *Pratylenchus coffeae* (Inserra et al., 2005). A large portion of Florida's orchards moved southward onto virgin soils following a series of killing freezes in the 1980s. Because all planting stock was nematode-free, a large portion of the industry is now nematode-free. The

Florida nursery certification program was estimated to have saved growers 33 million dollars in 1994 by reducing yield losses from *T. semipenetrans* that would have otherwise occurred from the spread of this nematode (Lehman, 1996). The program is effective because it requires *i)* nursery site certification, followed by continuous monitoring through soil sampling, *ii)* isolating nursery locations to avoid runoff water from infested orchards and *iii)* security to prevent contaminated planting media or equipment from entering the nursery area.

To maintain orchards free of nematodes, separate equipment is needed for use in infested and non-infested orchards; otherwise equipment must be disinfested prior to movement into non-infested orchards, a time-consuming task if done effectively (Esser, 1984). Occasional introduction of *T. semipenetrans* into otherwise clean orchards does not negate the value of sanitation, because the nematode migrates very slowly on its own power (Meagher, 1967; Tarjan, 1971; Baines, 1974; Duncan et al., 1995). Particularly when using low volume irrigation, trees often remain uninfected for years, despite the presence of neighboring infestations. Irrigation with some forms of surface water such as canals and rivers was a source of inter-orchard contamination by *T. semipenetrans* and *Phytophthora nicotianae* in South Africa, spreading the pests widely in a short time (Cohn et al., 1976). The use of settling ponds and filtration systems was suggested, but may be impractical given the volumes of water and maintenance required (Cohn, 1976).

2.6.3. Cultural Practices

Otherwise healthy orchards with large numbers of *T. semipenetrans* are those in which nematode management is most likely to be profitable. The limiting factor principle implies that trees are unable to respond to nematode management if *T. semipenetrans* are but one among other constraints to root growth (Thomason & Caswell, 1987). Although citrus nematode may sometimes exacerbate damage caused by other stresses (Labuschagne & Kotze, 1988; Mashela & Nthangeni, 2002), citrus trees that are damaged by *Phytophthora* spp., poor drainage, salinity, frequent drought or other problems are unlikely to respond consistently to management of just *T. semipenetrans*. Orchards should be managed properly in all respects, before investing in nematode management tactics.

Preplant solarization of soil can promote early growth of citrus, but the reasons are unresolved. Cronje et al., (2002) attributed increased tree growth and yield following solarization in South Africa to early and long-lasting control of *P. nicotianae.* Indeed, the larger, healthier trees in solarized plots supported higher numbers of *T. semipenetrans* up to age 10 years.

The resistant rootstock Swingle citrumelo is now widely planted in Florida and, combined with nursery certification, has appreciably reduced the occurrence of *T. semipenetrans* (Lehman, 1996). New varieties that tolerate calcareous soil will likely prove useful in other regions (Sorribas et al., 2003; Verdejo-Lucas et al., 2003). Resistance management is important to reduce the likelihood of selecting *Poncirus* biotypes that occur in regions with widespread use of *P. trifoliata* rootstocks (Baines et al., 1969b; Duncan et al., 1994; Miller et al., 1996; Murguia, et al., 2005; Kwaye, et al., 2008).Replanting orchards entirely with resistant rootstocks provides a discreet population of nematodes from which to select for resistance breaking individuals. In contrast, replanting a resistant tree adjacent to an infected susceptible tree provides a continuous challenge to the resistance genes (Duncan et al., 1994b; Verdejo-Lucas et al., 2003).

2.6.4. Fumigants and Nematicides

Preplant fumigation of old orchard sites with histories of citrus nematode infestation can be important to prevent the rapid infection of young trees (Baines et al., 1956, 1966; O'Bannon & Tarjan, 1973; Le Roux et al., 1998; Sorribas et al., 2003). Citrus nematodes can survive for up to two years in the absence of plants (Cohn, 1966; Van Gundy et al., 1967) and have been detected in fields that were formerly orchards for as long as 9 years, presumably surviving on root sprouts (Baines et al., 1962; Hannon, 1964). Net income from increased yield during years 4-8 after planting was 46-101% higher in plots fumigated with methyl bromide in South Africa (Le Roux et al., 1998). The most commonly used preplant fumigants in citrus are methyl bromide, 1,3-dichloropropene and metam sodium. Historically, dibromochloropropane (DBCP) was widely used to control citrus nematodes until it was banned for health reasons. Currently methyl bromide is being phased out due to ozone depletion and a variety of use restrictions (residential buffers, soil type-groundwater restrictions) are increasingly imposed on the use of the remaining fumigants (Noling et al., 2007). Fumigants can also be phytotoxic to young trees planted too soon (Cohn et al., 1968; Milne, 1974). In nurseries that experience frequent or very thorough fumigation, mycorrhizal fungi may be nearly eradicated and require reintroduction (O'Bannon & Nemec, 1978). This problem is rare in orchards because young plants are already mycorrhizal or are quickly invaded by fungi from adjacent soil (Graham, 1988).

Post-plant nematicides in citrus are primarily carbamate or organophosphate, acetylcholinesterase inhibitors. Efficacy against *T. semipenetrans* varies considerably among these compounds (Le Roux et al., 1998; McClure & Schmitt, 1996). The effectiveness of a nematicide cannot be assessed from studies of two or three years duration because continuous use has resulted in accelerated microbial degradation and loss of efficacy for some of them (Smelt et al., 1996; Johnson, 1998).

Several post-plant nematicides are translocated systemically within the tree and suppress insects and mites (both pest and beneficial species) in addition to nematodes. Some also have basipetal movement from the point of application to provide greater control of nematodes in the deeper soil profiles (O'Bannon & Tarjan, 1979). All of the nematicides used in citrus are incorporated in the soil either mechanically or with irrigation for efficacy and safety. They are inappropriate for small farms that lack proper application equipment and safety apparel. As with fumigants, nematicides are being continually deregistered or restricted for environmental concerns, particularly groundwater contamination. In Florida, treatment of nematode pests in citrus orchards resulted in contamination of large numbers of drinking water wells with fumigants and nematicides, which were subsequently banned or severely restricted in their use (Kaplan, 1988).

Nematicide placement and timing are important considerations. Because the abundance of nematodes and fibrous roots in the upper soil horizons decline quickly with distance from the trunk, nematicides - even systemic products - are most effective when applied beneath the tree canopy (Nigh, 1981a; Duncan, 1986; 1989). Application through low volume irrigation systems deliver nematicides to areas of highest root and nematode abundance. Where population levels and root growth are seasonal, treatment should precede periods when nematodes actively invade new roots (Hamid et al., 1988). Splitting the maximum allowable nematicide dose for multiple applications within a season can markedly increase efficacy. The life cycle of the citrus nematode was disrupted by three applications of cadusaphos, made at 60 day intervals, to the extent that nematodes were not detected on roots or in the soil for up to four years (Le Roux, 2000; McClure & Schmitt, 1996).

3. RADOPHOLUS SIMILIS AND R. CITRI

Radopholus similis is commonly called the burrowing nematode because of its extensive tunneling through root tissue as a migratory endoparasite. A race of this nematode causes one of the most economically important diseases of citrus, "spreading decline". Somewhat remarkably, this race of the nematode is only encountered in Florida, almost exclusively on the central ridge of deep sandy soil in the middle of the state. The disease was first described in 1928 and became the foremost citrus problem for several decades, because management tactics were ineffective until the causal organism was identified in 1953 (Suit & DuCharme, 1953). The name of the disease reflects the rapid progression of decline in infested orchard, which can reach 15m/yr (Feldmesser et al., 1960; Poucher et al., 1967; O'Bannon & Tomerlin, 1969a; Tarjan, 1971). The citrus race of *R. similis* also parasitizes banana, but the more widespread banana race is unable to reproduce on citrus (DuCharme & Birchfield, 1956).

The citrus race of *R. similis* was renamed *R. citrophilus* in 1984 and designated as a sibling species to *R. similis* based on putative differences in chromosome number, isozyme patterns, mating behaviour, host preference and morphology (Huettel et al.,1982; 1984; Huettel & Yaegashi, 1988). Independent research failed to confirm the previous work, but provided evidence based on karyotype identity, morphological and genetic similarity and reproductive compatibility that *R. citrophilus* is a junior synonym of *R. similis* (Kaplan & Opperman, 1997; 2000; Kaplan et al., 1997, 2000; Valette et al., 1998; Elbadri et al., 2002). Indeed, the more recent work on *Radopholus* systematics reveals little intraspecific variation, compared to many nematodes, in the DNA sequences of studied genomic regions. The genetic similarity among *R. similis* populations worldwide may result from its wide host range and recent dissemination worldwide on banana from its center of origin in Australasia (Kaplan, 1994b; Fallas et al., 1996; Machon & Bridge, 1996; Marin et al., 1998).

More recently, *R. citri* was discovered in citrus roots in Indonesia (Bridge et al., 1990; Hahn et al., 1994; Machon & Bridge, 1996). The pathogenicity of *R. citri* was demonstrated and the nematode is associated with declining trees in Indonesia, but its economic importance in the region is unknown.

3.1. Spreading Decline Symptoms

Trees infected by *R. similis* have sparse foliage that typically begins high in the canopy during the early stages of disease development (Fig. 4). Leaves and fruit are small and fruit drop is excessive. Branch ends are bare and eventually entire branches die. Affected trees wilt rapidly during periods of water stress that occur regularly during Florida's dry season in winter and spring. It is during these periods that disease progression is most rapid. During the rainy season, trees often recover a more healthy appearance, but decline symptoms become more pronounced with repeated drought cycles. Symptoms of spreading decline can be confused with citrus blight, a major disease of unknown origin. Spreading decline differs from citrus blight in that large contiguous groups of trees are affected as the nematode spreads and expansion of the diseased area is more rapid. The rate of forced water uptake in the trunk of a nematode-infected tree is indistinguishable from normal trees, whereas water cannot be forced into the vessels that are plugged in trees with citrus blight (Graham et al., 1983).

Figure 4. Symptoms of spreading decline of citrus caused by Radopholus similis. *The thin canopy is caused primarily by water deficit during the dry season, due to the massive loss of fibrous roots in the deeper soil horizons.*

The most obvious symptom to the root system is the reduction in the quantity of feeder roots in the deeper soil profiles (Ford, 1952; 1953). At depths up to 20-30 cm, root mass appears normal and without symptoms of damage. However, at depths of 25-50 cm, 75% of the root system may remain and below this level the root system is almost totally destroyed. Mature citrus growing on the deep sands of the ridge may establish as much as half of the feeder roots between 1 and 6 m, enabling access by trees to water deep in soil. Destruction of the deep root system on a large tree accounts for the drought-related aboveground symptoms during the dry season. Fibrous roots develop lesions at the points of nematode entry and activity, which expand and coalesce as secondary pathogens destroy these tissues. Nematodes are gregarious and may burrow in a section of root for several weeks completely destroying the phloem and much of the cortex, and girdling the central cylinder (DuCharme, 1959). On larger roots, the lesions can form callused margins (Feder & Feldmesser, 1956). The nematode penetrates the region of elongation and root tips can become swollen due to hyperplasia and stubby if terminals are penetrated (Feder & Feldmesser, 1956; DuCharme, 1959; 1968).

3.2. Biology and Ecology

Females lay eggs inside the roots at an average rate of nearly two per day. Development is rapid with eggs hatching in 2-3 days, completion of the life cycle in just 18-20 days, and high population growth rates when conditions are favorable (DuCharme & Suit, 1967; DuCharme & Price, 1966). Laboratory colonies initiated with single females attained average population levels of more than 11 000 individuals in less than 3 months, although rhizosphere competitors limit such population growth in the field (DuCharme & Price, 1966). The nematodes normally reproduce sexually; however females that do not mate after a period of time reproduce as hermaphrodites (Kaplan & Opperman, 2000). Mature males do not feed and comprise 0-40% of the population, averaging about 10% (DuCharme & Price, 1966). The nematode migrates from roots only when forced by overcrowding and decay (Fig. 5). It did not survive for longer than 6 months in the absence of host roots in controlled conditions (DuCharme, 1955; Tarjan, 1961), but in the field it was detected after two years of bare fallow, perhaps surviving on larger root fragments (Hannon, 1963; Suit et al., 1967). The nematode is spread in contaminated nursery stock (Poucher et al., 1967), machinery (Tarjan, 1956) and subsoil water (DuCharme, 1955).

Edaphic conditions regulate the biology of *R. similis* on citrus in economically important ways. Soil texture and water table depth are two of the most important factors in disease expression. The nematode migrates best and is more pathogenic to citrus in sandy than loamy soils in pot studies (O'Bannon & Tomerlin, 1971; Tarjan, 1971). In citrus growing regions of Florida *R. similis* occurs very sporadically, other than the central ridge, but populations do not develop to damaging levels. This is probably related to interactions between soil temperature, moisture and patterns of root growth. The cardinal temperature for *R. similis* is 24°C and development occurs between 12 and 32°C. Optimum temperatures occur for the longest periods each year in the deeper soil horizons where highest reproduction and root damage occurs. Increased root growth and carbohydrate availability in the late summer-early autumn period coincides with optimum temperatures to support increased population growth. By late autumn, nematode concentration in roots is high and as the infected roots begin to die absolute numbers of nematodes decline (DuCharme, 1967; 1969). Elevated temperature and periodic water deficit in the surface soil during the major period of root growth may partly explain low population development in surface roots (Tarjan, 1961).

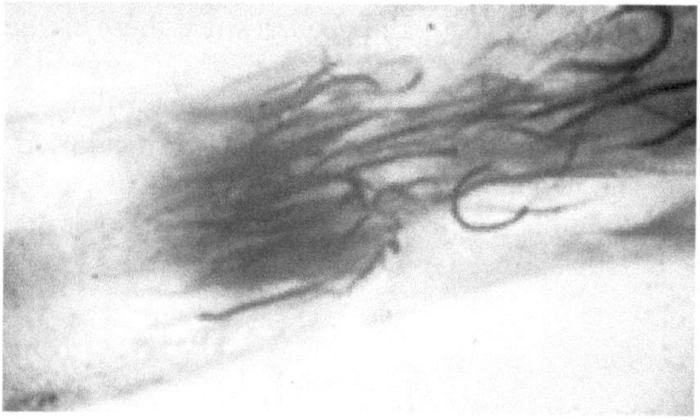

Figure 5. Radopholus similis *in fibrous root cortex* (from Poucher et al., 1967).

3.3. Interactions with Soil Organisms

Secondary fungal invaders play a key role in spreading decline because root lesions are quickly infected by fungi and other rhizosphere inhabitants (Feder et al., 1956; DuCharme, 1968). Fungicide treatment of infected seedlings increased root and plant growth as well as population density of *R. similis* (Feldmesser et al., 1959). Damage to seedlings is greater when *R. similis* occurs with *Fusarium oxysporum* and *F. solani* than when plants are infected by just the nematode or the fungi (Feder & Feldmesser, 1961). Citrus plant tolerance to *R. similis* was enhanced in phosphorus deficient soil by mycorrhizal infection (O'Bannon & Tomerlin, 1971; O'Bannon and Nemec, 1979). Increased phosphorus uptake by citrus induced either with fertilizer or by adding mycorrhizal fungi also reduced *R. similis* population levels (Smith & Kaplan, 1988).

3.4. Biotypes and Resistance

In contrast to *T. semipenetrans*, *Radopholus similis* is able to reproduce on more than 250 plants in 15 families outside of the Rutaceae (Ford et al., 1960). Within the Rutaceae, more than 1200 species, varieties and hybrids have been screened for resistance or tolerance to *R. similis* (Ford & Feder, 1961; O'Bannon & Ford, 1976). Three varieties of citrus, Ridge Pineapple, Estes rough lemon, Milam lemon and Carrizo citrange (*P. trifoliata* × *Citrus sinensis*), have been released as resistant rootstocks since 1958. Considerable variability exists within the progeny of Carrizo citrange for susceptibility to burrowing nematodes (Kaplan, 1986); however, a breeding line known as Kuharski Carrizo has been identified in which resistance is stable. All of the resistant rootstocks have been shown to support biotypes of *R. similis* capable of breaking resistance (Kaplan & O'Bannon, 1985), although the incidence of resistance-breaking populations on these varieties in the field is unknown.

3.5. Economic Importance

Because *R. similis* on citrus is restricted to Florida, the nematode's economic impact is slight on the world market. The nematode was estimated to cause no more than 0.2% yield loss in the world citrus industry (Cohn, 1972). Nevertheless, *Radopholus similis* and a lesion nematode, *Pratylenchus coffeae*, are the most virulent nematode parasites of citrus (O'Bannon et al., 1976). Losses in infested orchards prior to discovery of effective management tactics were estimated at 40-70% (DuCharme, 1968). The damage by spreading decline within orchards has been mitigated in recent years by improved management practices described below. Unfortunately, the discontinuation of programs to prevent migration of burrowing nematode from infested to uninfested orchards has increased the occurrence of this pest.

3.6. Management

Radopholus similis is more easily detected in samples of roots than of soil, and highest densities in roots occur at depths below 30 cm. Laboratories traditionally used expensive mechanized equipment to collect roots to depths of 120 cm to obtain those with large numbers of nematodes (Poucher et al., 1967). Subsequently it was found that using shovels to collect and process large amounts of roots in the surface horizon (<30 cm depth) provides better detection than processing the smaller amount of roots from deep in the soil (Duncan et al., 1994c). Stratification of orchards based on tree decline symptoms is important in sampling for *R. similis*. Random sampling is inappropriate because determination of population levels is generally not the goal of sampling for burrowing nematodes but rather delimiting an area of infestation. Intensive sampling of suspicious trees increases the chance of detecting the nematode, whose population level can be quite low seasonally.

Spreading decline is managed by restricting the spread of the nematode through nursery-stock certification, sanitation, cultural practices, use of resistant rootstocks and use of nematicides. Previously, nursery certification and chemical management was emphasized through state programs that relied on intensive sampling to delimit infested and uninfested parts of orchards. Attempts were made to eradicate the nematode by removing infested trees and a margin of uninfested trees and treating the soil with high rates of fumigants. The soil was maintained under bare fallow for at least 6 months before replanting with resistant rootstocks (Poucher et al., 1967). Alternatively, plant free buffers, 5-18 m wide, were created to prevent the movement of the nematode between infested and non-infested areas. The buffers were periodically fumigated to prevent citrus roots from growth laterally beneath them (Suit & Brooks, 1957; Poucher et al., 1967). These tactics were very expensive (as much as 20 000 dollars/ha in 1977), but they limited the spread of the nematode by more than 90% (O'Bannon, 1977). They were discontinued in 1983 when fumigant residues were detected in local drinking water wells throughout much of the central ridge. Subsequent research to maintain buffers using methyl bromide and mechanical root pruning proved too costly (Duncan et al., 1990). Although local spread of *R. similis* can no longer be prevented, avoiding infestation by the nematode remains a high management priority. Equipment used in infested orchards should be reserved for that purpose when possible or disinfested between operations (Esser, 1984).

Commercial planting stock and soil that is transported into or within citrus areas must be certified as nematode-free. Certification requires regular sampling and inspection of nurseries and soil mines. Lehman (1996) estimated a 14:1 return on investment from the state certification program, which resulted in increased yield worth 40 million dollars/year.

The fact that *R. similis* on the central ridge damages primarily the roots in the deeper (below 45 cm) soil horizon, provides the opportunity to manage spreading decline with cultural practices designed to support a healthy, shallow root system. Practices employed include use of herbicides and mowing rather than cultivation for weed management to avoid cutting surface roots (Tarjan & Simmons, 1966), frequent use of supplemental irrigation to provide sufficient water to the surface root system (Bryan, 1966; 1969), and use of an optimum fertility schedule, preferably through frequent fertigation to maintain nutrients in the shallow rhizosphere. Three rootstocks are recommended for use against spreading decline, Milam lemon, Ridge Pineapple sweet orange and Kuharski Carrizo citrange. The occurrence of resistance-breaking populations of the burrowing nematode have been shown to reproduce well on all resistant cultivars, which indicates a need for rootstocks with additional resistance genes (Kaplan and O'Bannon, 1985).

Systemic nematicides with basipetal movement are used by some growers to suppress *R. similis* in deeper roots and have been demonstrated to increase yield (O'Bannon & Tomerlin, 1977; O'Bannon & Tarjan, 1979).

Tree decline by *R. similis* remains a chronic problem. Nevertheless, infested orchards in which sound practices are employed can remain economically viable and may out-produce state production averages (Bryan, 1966; 1969; Tarjan & O'Bannon, 1977).

4. PRATYLENCHUS SPP.

Pratylenchus coffeae is the most damaging lesion nematode parasite of citrus, but *P. brachyurus* and *P. vulnus* are also known to be economically important. *P. coffeae* has been reported on citrus in the United States (O'Bannon et al., 1972), India (Siddiqi, 1964), Japan (Yokoo & Ikegemi, 1966), Oman (Mani et al., 1997) South Africa (Milne, 1982) and Taiwan (Huang & Chang, 1976). Genetic and morphological variation among *P. coffeae* populations suggests that the group is a species complex (Golden et al., 1992; Duncan et al., 1988; 1999). A lesion nematode identified as *P. coffeae* was found to infest about one percent of the citrus nurseries and orchards in Sao Paulo State, Brazil (Campos et al., 2002). The nematode was redescribed as *P. jaehni* (Inserra et al., 2001). It appears to be very similar to lesion nematodes from coffeae in Sao Paulo (Duncan et al., 1999), although the host ranges differ (Silva & Inomoto, 2002). *Pratylenchus jaehni* is associated with unthrifty citrus trees; however, its virulence on citrus and economic importance remain to be characterized. Within citrus, *P. jaehni* appears to have a very restricted host range (Calzavara et al., 2007). Putative *P. coffeae* on native vegetation in Florida, which threatened the nematode-free certification of some citrus nurseries, were found to be genetically distinct from *P. coffeae*, incapable of reproducing on citrus, and likely represent several undescribed species (Inserra et al., 1996, 1998; Duncan et al., 1999).

In North America, damage by *P. coffeae* occurs in Florida, but is relatively rare due to the nursery certification program there (O'Bannon & Tarjan, 1985). In South Africa, the nematode has not been associated with economic problems (Milne, 1982) as it has in other regions where it is found. Infection occurs in the feeder roots where all stages of the nematode inhabit cortical tissue and where migration is within and between cells. If penetration of the root tip occurs, the meristem is destroyed and lateral roots are often initiated. The nematode is found in vascular tissues only when localized populations are unusually high. Cortical invasion results in extensive cavities, but vascular tissues remain intact until invaded by secondary organisms.

Pratylenchus coffeae is amphimictic with males feeding in the roots and comprising 30-40% of the population (Radewald et al., 1971b; Inserra et al., 2001). Relatively high (26-30°C) soil temperatures are optimum for development with completion of the life cycle in less than one month. Densities as high as 10 000 nematodes/g root have been reported (O'Bannon & Tomerlin, 1969b; Radewald et al., 1971a). The nematode can survive in roots in soil for at least 4 months (Radewald et al., 1971a).

Pratylenchus coffeae reduced root weights by as much as half and plant growth by 38% (Siddiqi, 1964; O'Bannon & Tomerlin, 1969b; Radewald et al., 1971a). In the field, damage by *P. coffeae* causes a pronounced tree decline that requires frequent tree removal and replacement (Fig. 6). A comparison of infected and non-infected young trees during 4 years in the field showed growth reduction ranging from 49-80% and yield loss ranging from 33-95% depending on the rate of growth of the nematode on different rootstocks (O'Bannon & Tomerlin, 1973). Soil types ranging from sands to sandy loams did not affect the pathogenicity of *P. coffeae* to rough lemon roots (O'Bannon et al., 1976). Unlike *R. similis,* reported migration of the nematode through soil is slow, on the order one m/year (Tarjan, 1971; O'Bannon & Tomerlin, 1973; O'Bannon, 1980), although the rate of spread of decline symptoms in orchards is greater. *Pratylenchus coffeae* population growth and spatial pattern may be affected by competition with *T. semipenetrans* because the two species appeared to be mutually exclusive in an orchard, although exclusion of one species by the other was not observed in experiments (Kaplan & Timmer, 1982). *Poncirus trifoliata* and some selections of a Microcitrus hybrid appear to have some resistance, but none exists in commercial rootstocks (O'Bannon & Esser, 1975). Cleopatra mandarin was somewhat more tolerant of the nematode than were rough lemon or sour orange rootstocks (O'Bannon & Tomerlin, 1973).

The biology of *Pratylenchus brachyurus* is similar to *P. coffeae. Pratylenchus brachyurus* is widely distributed, but its distribution in citrus varies with region. In Florida, the nematode was present in 90% of orchards sampled (Tarjan & O'Bannon, 1969), whereas it has not been reported from citrus orchards in South Africa, even though it is widespread in that country (Milne, 1982). Citrus is a less favorable host for *P. brachyurus* than for *P. coffeae*, with root population densities generally 10-fold lower for the former species (Radewald et al., 1971a). Thus, while it is a pathogen of seedlings and young trees (Brooks & Perry, 1967; Tarjan & O'Bannon,

1969; Radewald et al., 1971a; Tomerlin & O'Bannon, 1974; Frederick & Tarjan, 1975), *P. brachyurus* does not greatly affect mature citrus in the absence of other problems such as severe drought (O'Bannon et al., 1974). When populations of *P. brachyurus* in mature Valencia orange trees on rough lemon rootstock were controlled with aldicarb, trees suffered less frost damage during a severe winter and subsequent yields were increased (Wheaton et al., 1985; Childers et al., 1987). It is unclear, however, what other factors may have been affected by the systemic pesticide. Like *P. coffeae, P. brachyurus* reproduces best at temperatures above 25°C and can affect seedling growth in coarse and medium textured soils. Movement of *P. brachyurus* through soil is not as rapid as that of *P. coffeae* (O'Bannon, 1980).

Figure 6. Symptoms caused by Pratylenchus coffeae *infecting citrus in Florida. Note the extensive loss of trees damaged by the nematode.*

Pratylenchus vulnus has been found associated with citrus only in Italy (Inserra & Vovlas, 1974) and California (Siddiqui et al., 1973). It was shown to be capable of causing severe damage to nursery seedlings, but is not reported to damage mature trees (Inserra & Vovlas, 1977). Spanish populations of the nematode did not increase on citrus cultivars tested (Pinochet et al., 1992). As with other species of

Pratylenchus, the nematode is pathogenic in a range of soils from sand to sandy clay loam. Biology, population growth rates and root damage are similar to those described for *P. coffeae*.

5. BELONOLAIMUS LONGICAUDATUS

Belonolaimus longicaudatus, the "sting nematode", was detected in fewer than 10% of Florida citrus orchards (Esser et al., 1993), but in the coarse sandy soils of the central ridge it was encountered in 64% of orchards surveyed (Duncan et al., 1996). *B. longicaudatus* is widely distributed on a number of cultivated and non-cultivated host plants in the southeastern United States and several other species in the genus are reported from this region and from Australia, Venezuela, Puerto Rico and Brazil (Gozel et al., 2006). The nematode is not known to be a problem to citrus outside of Florida although it was reported to be associated with unhealthy citrus in Costa Rica (Lopez, 1978). In sandy soil, *B. longicaudatus* causes severe damage, especially to young trees, by greatly reducing the quantity and quality of roots. Although ectoparasitic, the nematode can be spread on infested planting stock, even when the soil is washed from roots (Kaplan, 1985).

In nurseries, as few as 40 nematodes/dm^3 soil can cause aboveground symptoms of stunted, chlorotic plants (Kaplan, 1985). The nematode feeds at the root tip, giving root systems of infested trees a coarse appearance due to a reduction in the number of lateral roots, swollen fibrous roots, swellings at terminals due to hyperplasia and multiple apices (Standifer & Perry, 1960; Kaplan, 1985; Fig. 7). Heavily infested roots rot from secondary infection. The host range varies for different populations of the nematode, which suggests the existence of races (Abu-Gharbieh & Perry, 1970; Robbins & Hirschmann, 1974); and comparison of rDNA sequences of populations in Florida revealed that *B. longicaudatus* is likely a species complex (Gozel et al., 2006).

Sting nematodes cause severe stunting of trees on all tested rootstocks in the field and in controlled greenhouse studies (Standifer & Perry, 1960; Abu-Gharbieh & Perry, 1970; Esser & Simpson, 1984; Kaplan, 1985; Duncan et al., 1996). The economic importance of sting nematodes appears to have increased when growers provided additional plant hosts by adopting mowing rather than disking or use of herbicides to manage vegetation between rows.

Newly planted orchards often contain extensive patches of stunted trees. Surprisingly, stunted trees support larger numbers of the nematode than do healthy trees with larger root systems. Soil water potential beneath heavily infested young trees with few roots remains higher than that beneath lightly infested trees with more roots and greater transpiration. Presumably young trees planted in locations with numerous sting nematodes suffer continuous loss of roots, which maintains a wetter, more favorable habitat for the nematode. Such trees remain stunted for several years until they manage enough root growth to periodically dry the surface soil horizon, forcing the nematode deeper in the soil, and allowing tree growth to resume (Duncan et al., 1996).

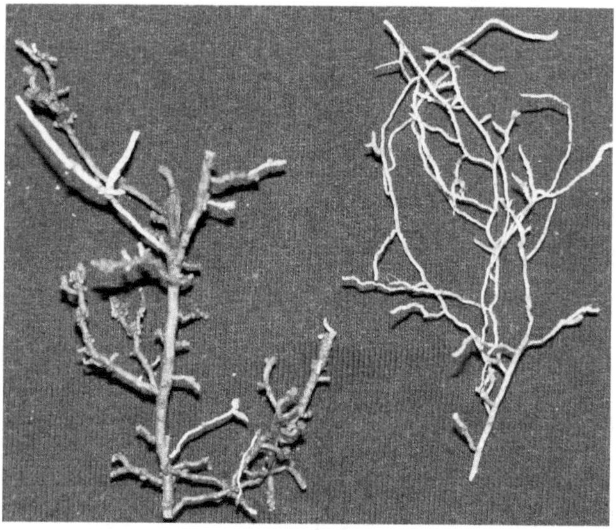

Figure 7. Restricted growth of citrus fibrous roots and multiple root initiation caused Belonolaimus longicaudatus *feeding on meristematic tissue at the root apex (left) compared to healthy roots (right).*

Soil fumigation and post-plant nematicide treatments have alleviated symptoms of sting nematode parasitism (Bistline et al., 1967; Author, unpublished). Hot water treatment of rootstocks for 5 min at 49°C eliminates *B. longicaudatus,* providing a method to disinfest nurserystock (Kaplan, 1985).

6. MELOIDOGYNE SPP.

Root-knot nematodes *(Meloidogyne* spp.*)* can be very damaging to citrus, but are limited in distribution and, in many cases, little studied. A pathogenic species of root knot nematode given the common name "Asiatic pyroid citrus nematode" was reported from Taiwan and New Delhi where it caused elongated galls on citrus roots. The nematode can reproduce on several citrus and other plant species including corn and sweet potato. Non-hosts included *Crotalaria* sp., strawberry, peanut and soybean, which were recommended as trap crops because the nematode could invade but not develop in the roots (Chitwood & Toung, 1960). *Meloidogyne fujianensis* (Pan, 1985) and *M. oteifae* (Pan, 1984) have been reported from citrus in China with the former species parasitizing up to 60% of citrus trees surveyed. A California population of *M. javanica* was eradicated when it was discovered reproducing on a dooryard citrus tree (Gill, 1971).

The most common species of root knot nematodes (*M. incognita, M. javanica* and *M. arenaria*) are rarely encountered reproducing on citrus outside of a few localized regions in China, the Indian subcontinent and the Far East (Minz, 1956; Den Ouden, 1965; Whitehead, 1968; Scotto la Massèse, 1969; Gill, 1971; De Brito et al., 2000; Rao, 2005; Musarrat et al., 2006). Nevertheless, root knot nematodes may cause problems in citrus trees adjacent to good hosts. Van Gundy et al. (1959) reported *M. incognita, M. javanica* and *M. arenaria* infected roots of unthrifty Troyer citrange and sour orange. Roots had galls but lacked nematode egg masses. The nematodes reproduced abundantly on weeds in the field. Inserra et al. (1978) reported extensive root damage due to invasion of citrus roots by *M. javanica* even though no reproduction occurred, and Orion and Cohn (1975) showed that infection of citrus by *M. javanica* resulted in a hypersensitive response and failure to establish giant cells.

7. HEMICYCLIOPHORA SPP.

A number of species of *Hemicycliophora* have been identified from the citrus rhizosphere. *H. arenaria* is a species native to plants in the desert valleys of southern California that causes damage in citrus nurseries (McElroy et al., 1966). The nematode was closely studied (Van Gundy, 1959) and quarantined to prevent its spread to other areas of that state. It appears to have a wide host range (ten of nineteen hosts tested) although the rutaceous host status is variable. *Citrus limon*, *C. aurantifolia*, *C. reticulata* and *Severinia buxifolia* are susceptible, while *Poncirus trifoliata*, *C. aurantium*, *C. paradisi* and *C. sinensis* are resistant (Van Gundy & Rackham, 1961). The nematode feeds in large numbers at root tips whose roots typically develop round galls arising from hyperplasia (Fig. 8). Seedling growth in pot studies was reduced by 35%. *Hemicycliophora nudata* causes similar symptoms on citrus in Australia (Colbran, 1963). *H. arenaria* can be eradicated from root systems with hot water dips (10 min at 46°C), preplant soil fumigation is very effective and a number of rootstocks resistant to the nematode are available (Van Gundy and McElroy, 1969).

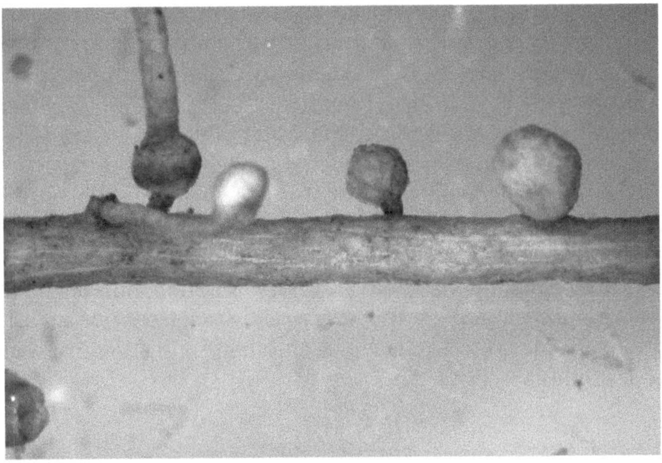

Figure 8. Citrus fibrous root tip galls caused by sheath nematode, Hemicycliophora *spp. (*courtesy of M. C. Pretorius*).*

8. DORILAIMID SPECIES

A large number of plant parasitic species in the order Dorylaimida are vectors of plant viruses. However, none of the several longidorid and trichodorid species known to damage citrus has been implicated in vectoring virus disease agents.

Species in the genus *Xiphinema* (dagger nematodes) are extremely common inhabitants of thecitrus rhizosphere (Baines et al., 1978). Damage by *Xiphinema* is primarily to epidermal and outer cortical cells, which become necrotic and give a typically dark appearance to damaged roots (Cohn, 1970; Cohn & Orion, 1970; Baines et al., 1978). Very little research has been done regarding the pathogenicity to citrus of these ectoparasitc nematodes even though high populations of some species have been consistently associated with citrus in North America, Africa and the Middle East. Most species of *Xiphinema* predominate in lighter textured soils (Cohn, 1969). *Xiphinema brevicollum* and *X. index* are associated with unthrifty trees in localized areas of Israel and were shown to be pathogenic to sour orange seedlings in pot studies (Cohn & Orion, 1970). Similarly, high populations of *X. brevicollum* in Sudan were associated with declining grapefruit trees and pot studies with the nematode revealed similar symptoms of stubby, swollen roots and reduced root abundance (Yassin, 1974). In South Africa, control of *X. brevicollum* with DBCP did not improve tree quality (Milne, 1982). High populations of *X. vulgare* are associated with declining citrus trees in Florida and caused necrosis and severe reduction of the root systems of seedlings in pots (Duncan et al., 1994; Leone et al., 1997). The nematode is very long-lived, requiring 274 days at 24 °C to complete its life cycle on citrus (Coiro, et al., 2002).

Trichodorus and *Paratrichodorus* spp. (stubby root nematodes) are also commonly detected in citrus orchards (Baines et al., 1959; Malo, 1961; Colbran, 1965). Population density in groves is often low, but the nematode often responds to soil fumigation with strong population resurgence (Perry, 1953; Standifer & Perry, 1960). *Paratrichodorus porosus*, *P. lobatus* and *P. minor* feed at root tips, reducing root elongation and cause stubby root symptoms on citrus similar to those of sting nematodes (Baines et al., 1978; Standifer & Perry, 1960; Stirling, 1976). *Paratrichodorus lobatus* is widespread and often abundant in citrus nurseries and orchards in Australia, but its importance is unclear (Stirling, 1976). Despite decreasing feeder root weight in a pot study, *P. lobatus* did not affect taproot or seedling weights (Stirling, 1976). Nursery trees infested with the nematode at levels of 1500/500 cm^3 soil had reduced root systems, poor leaf colour and tended to wilt during the day, but population levels in a nursery were not correlated with tree size. Meagher (1969) reported that *P. lobatus* was associated with unthrifty trees in orchards and nurseries in New South Wales and that fumigation markedly improved the growth of seedlings.

REFERENCES

Abu-Gharbieh, W. I., & Perry, V. G. (1970). Host differences among Florida populations of *Belonolaimus longicaudatus* Rau. *Journal of Nematology*, 2, 209-216.

Al Hinai, M. S., & Mani, A. (1998). Seasonal population changes and management of *Tylenchulus semipenetrans* using organic amendments and fenamiphos. *Nematologia Mediterranea*, 26, 179-184.

Anonymous. (2007). *Citrus Summary 2006-07*. Florida Agricultural Statistics Service, Orlando, FL, USA, 48 pp.
Baghel, P. P. S., & Bhatti, D. S. (1982). Vertical and horizontal distribution of phytonematodes associated with citrus. *Indian Journal of Nematology*, 12, 339-344.
Baines, R. C. (1950). Nematodes on citrus. *California Agriculture*, 4, 7.
Baines, R. C. (1950). Citrus-root nematode investigations. *California Citrograph*, 35, 344
Baines, R. C. (1964). Controlling citrus nematode with DBCP increases yields. *California Citrograph*, 49, 222-233.
Baines, R. C. (1974). The effect of soil type on movement and infection rate of larvae of *Tylenchulus semipenetrans*. *Journal of Nematology*, 6, 60-62.
Baines, R. C., Foote, F. J., & Martin, J. P. (1956). Fumigate soil before replanting citrus for control of the citrus nematode. *Citrus Leaves*, 36, (6-8), 24, 27.
Baines, R. C., Van Gundy, S. D., & Sher, S. A. (1959). Citrus and avocado nematodes. *California Agriculture*, 13, 16-18.
Baines, R. C., Martin, J. P., DeWolfe, T. A., Boswell, S. B., & Garber, M. J. (1962). Effect of high doses of D-D on soil organisms and the growth and yield of lemon trees. *Phytopathology*, 52, 723.
Baines, R. C., Klotz, L. J., DeWolfe, T. A., Small, R. H., & Turner, G. O. (1966). Nematocidal and fungicidal properties of some soil fumigants. *Phytopathology*, 56, 691-698.
Baines, R. C., DeWolfe, T. A., Klotz, L. J., Bitters, W. P., Small, R. H., & Garber, M. J. (1969a). Susceptibility of six *Poncirus trifoliata* selections and Troyer citrange to a biotype of the citrus nematode and growth response on fumigated soil. *Phytopathology*, 59, 1016-1017.
Baines, R. C., Miyakawa, T., Cameron, J. W., & Small, R. H. (1969b). Biotypes of the citrus nematode. *Proceedings of the 1st International Citrus Symposium*. Riverside, California: *2*, 955-956.
Baines, R. C., Van Gundy, S. D., & DuCharme, E. P. (1978). Nematodes attacking citrus. Chapter 7. In: W. Reuther, E. C.Calavan & G. E. Carman (eds.). *The Citrus Industry*, Volume IV. University of California, Division of Agricultural Science, 321-345.
Bello, A., Navas, A., & Belart, C. (1986). Nematodes of citrus-groves in the Spanish Levante Ecological study focused to their control. In: R. Cavalloro & E. Di Martino, (eds.). Proceedings of the Expert's Meeting, Acireale, March 26-29, 1985. *Integrated Pest Control in Citrus Groves*. A.A. Balkema Publ. Co., Rotterdam, Boston: 217-226.
Bistline, F. W., Collier, B. L., & Dieter, C. E. (1967). Tree and yield response to control of a nematode complex including *Belonolaimus longicaudatus* in replanted citrus. *Nematologica*, 13, 137-138.
Bridge, J., Machon, D., & Djatmiadi, D. (1990). A new nematode problem of citrus caused by *Radopholus* sp. in Java. *Nematologica*, 36, 336.
de Brito, J. A., de Lourdes Mendes, M., & Rodrigues, R. (2000). ReaHao de Porta-enxerto de *Citrus* spp. a *Meloidogyne incognita* RaHas 2 e 4 e *M. javanica*. *Nematologia Brasileira*, 24, 253-256.
Brooks, T. L. & Perry, V. G. (1962). Apparent parthenogenetic reproduction of the burrowing nematode *Radopholus similis*. *Soil Crop Science Society of Florida Proceedings*, 22, 160-162.
Brooks, T. L. & Perry, V. G. (1967). Pathogenicity of *Pratylenchus brachyurus* to citrus. *Plant Disease Reporter 51*, 569-573.
Bryan, O. C. (1966). Soil moisture - the key factor in the production of nematode-infested groves. *Citrus and Vegetable Magazine 29*, 39-40.
Bryan, O.C. (1969) Living with the burrowing nematode. *Citrus and Vegetable Magazine 33*, 29-38.
Caldwell, M. M., Richards, J. H., & Beyschlag, W. (1991). Hydraulic lift: Ecological implications of water efflux from roots. In: D. Atkinson (ed.). *Plant Root Growth: An Ecological Perspective*. Blackwell Scientific, Oxford: 423-436.
Calzavara, S. A., Santos, J. M., & Favoreto, L. (2007). Citrus rootstocks resistance to *Pratylenchus Jaehni* (Nematoda: Pratylenchidae). *Nematologia brasileira*, 31, 7-11.
de Campos, A. S., dos Santos, J. M., & Duncan, L. W. (2002). Nematodes of citrus in open nurseries and orchards in Sao Paulo State, Brazil. *Nematology*, 4, 263-264.
Chabrier, C., Carles, C., Quénéhervé, P. & Yves-Marie C. (2008). Nematode dissemination by water leached in soil: Case study of *Radopholus similis* (Cobb) Thorne on nitisol under simulated rainfall. *Applied Soil Ecology*, 40, 299-308.
Childers, C. C., Duncan, L. W., Wheaton, T. A., & Timmer, L. W. (1987). Arthropod and nematode control with Aldicarb on Florida citrus. *Journal of Economic Entomology*, 80, 1064-1071.
Chitwood, B. G. & Toung, M. C. (1960). Host-parasite interactions of the Asiatic pyroid citrus nema. *Plant Disease Reporter*, 44, 848-854.

Cobb, N. A. (1913). Notes on *Mononchus* and *Tylenchulus*. *Journal of the Washington Academy of Science*, 3, 287-288.

Cobb, N. A. (1914). Citrus-root nematode. *Journal of Agricultural Research*, 2, 217-230.

Coelho, Y. D. S., Paguio, O. D. L. R., Zem, A. C., & Filho, H. P. S. (1983) Citrus decline and nematode population on citrus. *Fitopatologia Brasileira*, 8, 367-370.

Cohn, E. (1964). Penetration of the citrus nematode in relation to root development. *Nematologica*, 10, 594-600.

Cohn, E. (1965a). On the feeding and histopathology of the citrus nematode. *Nematologica*, 11, 47-54.

Cohn, E. (1965b). The development of the citrus nematode on some of its hosts. *Nematologica*, 11, 593-600.

Cohn, E. (1966). Observations on the survival of free-living stages of the citrus nematode. *Nematologica*, 12, 321-327.

Cohn, E. (1969). The occurrence and distribution of species of *Xiphinema* and *Longidorus* in Israel. *Nematologica*, 15, 179-192.

Cohn, E. (1970). Observations on the feeding and symptomatology of *Xiphinema* and *Longidorus* on selected host roots. *Journal of Nematology*, 2, 167-173.

Cohn, E. (1972). Nematode diseases of citrus. In: Webster, J.M. (ed.). *Economic Nematology*, Academic Press, London: 215-244.

Cohn, E. (1976). *Report of investigations on nematodes of citrus and subtropical fruit crops in South Africa*. Citrus and Subtropical Fruit Research Institute, Nelspruit, SA: 41 pp.

Cohn, E. & Orion, D. (1970). The pathological effect of representative *Xiphinema* and *Longidorus* species on selected host plants. *Nematologica*, 16, 423-428.

Cohn, E., Minz, G., & Monselise, S. P. (1965). The distribution, ecology and pathogenicity of the citrus nematode in Israel. *Israel Journal of Agricultural Research*, 15, 187-200.

Cohn, E., Feder, W. A., & Mordechai, M. (1968). The growth response of citrus to nematicide treatments. *Israel Journal of Agricultural Research*, 18, 19-24.

Cohn, E., Hough, A., & Mulder, N. (1976) Elimination of nematodes and other plant pathogens from irrigation water by filtration techniques. *13th International Nematology Symposium*, Dublin, Ireland, Sept. 5-11, 16-17.

Coiro, M. I., Lamberti, F., Sasanelli, N., Duncan, L. W., & Agostinelli, A. (2002). Reproduction and longevity of *Xiphinema vulgare* (Nematoda). *Nematologia Mediterranea*, 30, 91-95.

Colbran, R. C. (1963). Studies of plant and soil nematodes. 6. Two new species from citrus orchards. *Queensland Journal of Agricultural Science*, 20, 469-474.

Colbran, R. C. (1965). Studies of plant and soil nematodes. 9. *Trichodorus lobatus* n. sp. (Nematoda: Trichodoridae), a stubby-root nematode associated with citrus and peach trees. *Queensland Journal of Agricultural and Animal Sciences*, 22, 273-276.

Cronjé, C., Le Roux, H. F., Truter, M., Van Heerden, I., & Phillips, H. (2002). Long-term effect of preplant soil solarisation on growth of replant citrus trees in South Africa. *African Plant Protection*, 8, 41-49.

Dalmasso, A., Macaron, J., & Berge, J. B. (1972). Details of reproduction in *Tylenchulus semipenetrans* and *Cacopaurus pestis* (Nematoda-Criconematoidea). *Nematologica*, 18, 423

Das, T. K. & Mukhopadhyaya, M. C. (1985). Influence of citrus species on the morphological variations of *Tylenchulus semipenetrans*. *Indian Journal of Nematology*, 15, 114-116.

Davide, R. G. (1971). *Survey of the distribution of different plant parasitic nematodes associated with the citrus decline in the Philippines*. A report of NSDB project No. 2203, University of Philippines, College of Agriculture, Laguna: 73 pp.

Davide, R. G. & Dela Rose, A. G. (1976). Host-parasite relationships and control of the citrus nematode in the Philippines. II. Evaluation of nematode damage on citrus. University of Philippines, College of Agriculture, Laguna, *NSDB Technical Journal*, 1, 8-15.

Davis, R. M. (1984). Comparison of numbers of *Tylenchulus semipenetrans* Cobb on six citrus rootstocks in south Texas. *Journal of Rico Grande Valley Horticultural*, 37, 95-98.

Davis, R. M. (1984). Distribution of *Tylenchulus semipenetrans* in a Texas grapefruit orchard. *Journal of Nematology*, 16, 313-317.

Davis, R. M. (1985). Citrus nematode control in Texas. *Citrograph*, 70, 212-213.

Davis, R. M., & Wilhite, H. S. (1985). Control of *Tylenchulus semipenetrans* on citrus with fenamiphos and oxamyl. *Plant Disease*, 69, 974-976.

Davis, R. M., Heald, C. M., & Timmer, L. W. (1982). Chemical control of the citrus nematode on grapefruit. *Journal of the Rio Grande Valley Horticultural Society*, 35, 59-63.

Den Ouden, H. (1965). An infestation on citrus in Surinam caused by *Meloidogyne exigua*. *Surinamese Landbouw*, 13, 34.
Deng, Z., Huang, S., Ling, P., Chen, C., Yu, C., Weber, C. A., Moore, G. A., & Gmitter, Jr., F. G. (2000). Cloning and characterization of NBS-LRR class resistance-gene candidate sequences in citrus. *Theoretical and Applied Genetics*, 101, 814–822.
Dirac, M. F. & Menge, J. A. (2002). High temperature are not responsible for lack of infection of citrus roots by *Phytophthora citrophthora* during the summer, but suppressive soil microorganisms may inhibit infection by *P. citrophthora*. *Plant and Soil*, 241, 243-249.
DuCharme, E. P. (1955). Sub-soil drainage as a factor in the spread of the burrowing nematode. *Proceedings of the Florida State Horticultural Society*, 68, 29-31.
DuCharme, E. P. (1959). Morphogenesis and histopathology of lesions induced on citrus roots by *Radopholus similis*. *Phytopathology*, 49, 388-395.
DuCharme, E. P. (1967). Annual population periodicity of *Radopholus similis* in Florida citrus groves. *Plant Disease Reporter*, 51, 1031-1034.
DuCharme, E. P. (1968). Burrowing nematode decline of citrus, a review. In: Smart, G. C. & Perry, V. G. (eds.). *Tropical Nematology*. University of Florida Press, Gainesville: 20-37.
DuCharme, E. P. (1969). Temperature in relation to *Radopholus similis* (Nematoda) spreading decline of citrus. *Proceedings of the 1st International Citrus Symposium*, 2, 979-983.
DuCharme, E. P. & Birchfield, W. (1956). Physiologic races of the burrowing nematode. *Phytopathology*, 46, 615-616.
DuCharme, E.P. & Price, W.C. (1966). Dynamics of multiplication of *Radopholus similis*. *Nematologica*, 12, 113-121.
DuCharme, E. P. & Suit, R. F. (1967). Population fluctuations of burrowing nematodes in Florida citrus groves. *Proceedings of the Florida State Horticultural Society*, 80, 63-67.
Duncan, L. W. (1986). The spatial distribution of citrus feeder roots and of the citrus nematode, *Tylenchulus semipenetrans*. *Revue de Nématologie*, 9, 233-240.
Duncan, L.W. (1989). Effect of Fenamiphos placement on *Tylenchulus semipenetrans* and yield in a Florida citrus orchard. *Supplement to the Journal of Nematology*, 21(4S), 703-706.
Duncan, L. W. (1999). Nematode diseases of Citrus. In: Timmer, L. W. & Duncan, L. W. (eds.). *Citrus Health Management*, APS Press, St. Paul, MN, USA: 136-148.
Duncan, L. W. & Eissenstat, D. M. (1993). Responses of *Tylenchulus semipenetrans* to citrus fruit removal: Implications for carbohydrate competition. *Journal of Nematology*, 25, 7-14.
Duncan, L. W., & Inserra, R. N. (2005) Datasheet for *Tylenchulus semipenetrans*. In: Crop Protection Compendium. CAB International, Wallingford, UK.
Duncan, L. W. & El-Morshedy, M. M. (1996). Population changes of *Tylenchulus semipenetrans* under localized versus uniform drought in the citrus root zone. *Journal of Nematology*, 28, 360-368.
Duncan, L. W., Ferguson, J. J., Dunn, R. A., & Noling, J. W. (1989). Application of Taylor's Power Law to sample statistics of *Tylenchulus semipenetrans* in Florida Citrus. *Supplement to the Journal of Nematology*, 21 (4S), 707-711.
Duncan, L. W., Kaplan, D. T., & Noling, J. W. (1990). Maintaining barriers to the spread of *Radopholus citrophilus* in Florida citrus orchards. *Nematropica*, 20, 71-88.
Duncan, L. W., Graham, J. H., & Timmer, L. W. (1993). Seasonal patterns associated with *Tylenchulus semipenetrans* and *Phytophthora parasitica* in the citrus rhizosphere. *Phytopathology*, 83, 573-581.
Duncan, L. W., El-Morshedy, M. M., & McSorley, R. (1994a). Sampling citrus fibrous roots and *Tylenchulus semipenetrans*. *Journal of Nematology*, 26, 442-451.
Duncan, L. W., Inserra, R. N., O'Bannon, J. H., & El-Morshedy, M. M. (1994b). Reproduction of a Florida population of *Tylenchulus semipenetrans* on resistant citrus rootstocks. *Plant Disease*, 78, 1067-1071.
Duncan, L. W., Toole, J. D., Inserra, R. N., Castle, W. S., & O'Bannon, J. H. (1994c). Comparing two sampling methods to detect *Radopholus citrophilus* in Florida citrus. *Proceedings of the Soil Crop Science Society of Florida*, 53, 42-45.
Duncan, L. W., Mashela, P., Ferguson, J., Graham, J., Abou-Setta, M. M., & El-Morshedy, M. M. (1995). Estimating crop loss in orchards with patches of mature citrus trees infected by *Tylenchulus semipenetrans*. *Nematropica*, 25, 43-51.
Duncan, L. W., Noling, J. W., Inserra, R. N., & Dunn, D. (1996). Spatial patterns of *Belonolaimus* spp. among and within citrus orchards on Florida's central ridge. *Journal of Nematology*, 28, 352-359.
Duncan, L. W., Inserra, R. N., & Dunn, D. (1998). Seasonal changes in citrus fibrous root starch concentration and body length of female *Pratylenchus coffeae*. *Nematropica*, 28, 263-266.

Duncan, L. W., Inserra, R. N., Thomas, W. K., Dunn, D., Mustika, I., Frisse, L. M., et al. (1999). Molecular and morphological analysis of isolates of *Pratylenchus coffeae* and closely related species. *Nematropica*, 29, 61-80.

Dunn, D. C., Duncan, L. W., & Romeo, J. T. (1998). Changes in argenine, PAL activity, and nematode behavior in salinity-stressed citrus. *Phytochemistry*, 49, 413-417.

Eissenstat, D. M. & Duncan, L. W. (1992). Root growth and carbohydrate responses in bearing citrus trees following partial canopy removal. *Tree Physiology*, 10, 245-257.

Elbadri, G. A. A., De Ley, P., Waeyenberge, L., Vierstraete, A., Moens, M., & Vanfleteren, J. (2002). Intraspecific variation in *Radopholus similis* isolates assessed with restriction fragment length polymorphism and DNA sequencing of the internal transcribed spacer region of the ribosomal RNA cistron. *International Journal for Parasitology*, 32, 199-205.

El-Borai, F. E., Duncan, L. W., & Graham, J. H. (2002a). Infection of citrus roots reduces root infection by *Phytophthora nicotianae*. *Journal of Nematology*, 34, 384-389.

El-Borai, F. E., Duncan, L. W., Graham, J. H., & Dickstein, E. (2003). *Tylenchulus semipenetrans* alters the microbial community in the citrus rhizosphere. *Journal of Nematology*, 35, 167-177.

El-Borai, F. E., L. W. Duncan, & J. H. Graham. (2002b). Eggs of *Tylenchulus semipenetrans* inhibit growth of *Phytophthora nicotianae* and *Fusarium solani in vitro*. *Journal of Nematology*, 34, 267-272.

Esser, R. P. (1984). How nematodes enter and disperse in Florida nurseries via vehicles. *Florida Department of Agriculture and Consumer Services, Division of Plant Industry, Nematology Circular*, 109: 2 pp.

Esser, R. P. & Simpson, S. E. (1984). Sting nematode on citrus. *Florida Department of Agriculture and Consumer Services, Division of Plant Industry, Nematology Circular*, 106: 2 pp.

Esser, R. P., Smith, G. T., & O'Bannon, J. H. (1993). An eleven-year phytoparasitic nematode survey of Florida citrus groves and their environs. *Bulletin 15, Florida Department of Agriculture and Consumer Services,* Gainesville, FL.

Fallas, G. A., Hahn, M. L., Fargette, M., Burrows, P. R., & Sarah, J. L. (1996). Molecular and biochemical diversity among isolates of *Radopholus* spp. from different areas of the world. *Journal of Nematology 28*, 422-430.

Fattah, F. A., Saleh, H. M., & Aboud, H. M. (1989). Parasitism of the citrus nematode, *Tylenchulus semipenetrans*, by *Pasteuria penetrans* in Iraq. *Journal of Nematology*, 21, 431-433.

Feder, W. A. & Feldmesser, J. (1956). Root abnormalities caused by burrowing nematode infections. *Phytopathology*, 46, 11.

Feder, W. A. & Feldmesser, J. (1961). The spreading decline complex: The separate and combined effects of *Fusarium* spp. and *Radopholus similis* on the growth of Duncan grapefruit seedlings in the greenhouse. *Phytopathology*, 51, 724-726.

Feder, W. A., Feldmesser, J., & Walkinshaw, C. H. (1956) Microorganisms isolated from feeder roots of citrus seedlings affected by spreading decline. *Soil Crop Science Society of Florida Proceedings 16*, 127-129.

Feldmesser, J., Rebois, R. V., & Taylor, A. L. (1959). Progress report on growth response of burrowing nematode infected citrus following chemical treatments under greenhouse conditions. *Plant Disease Reporter*, 43, 261-263.

Feldmesser, J., Cetas, R. C., Grimm, G. R., Rebois, R. V., & Widden, R. (1960). Movement of *Radopholus similis* into rough lemon feeder roots and in soil, and its relation to *Fusarium* in the roots. *Phytopathology*, 50, 635.

Ford, H. W. (1952). The effect of spreading decline on the root distribution of citrus. *Proceedings of the Florida State Horticultural Society*, 65, 47-50.

Ford, H. W. (1953). Effect of spreading decline disease on the distribution of feeder roots of orange and grapefruit trees on rough lemon rootstocks. *Journal of the American Society for Horticultural Science*, 61, 68-72.

Ford, H. W. (1954a). The influence of rootstock and tree age on root distribution of citrus. *Proceedings of the American Society for Horticultural Science*, 63, 137-172.

Ford, H. W. (1954b). Root distribution in relation to the water table. *Proceedings of the Florida State Horticultural Society*, 67, 30-33.

Ford, H. W. & Feder, W. A. (1961). Additional citrus rootstock selections that tolerate the burrowing nematode. *Proceedings of the Florida State Horticultural Society*, 74, 50-53.

Ford, H. W., Feder, W. A., & Hutchins, P. C. (1960). Citrus varieties, hybrids, species, and relatives evaluated for resistance to the burrowing nematode *Radopholus similis*. *Citrus Experiment Station Mimeo Series 60-13*: 26 pp.

Fortucci-Marongiu, P. (1988). *Three decades of the world citrus economy.* Intergovernmental Group on Citrus Fruit, Commodities and Trade Division, FAO, Rome, Italy: 9 pp.

Frederick, J. J., & Tarjan, A. C. (1975). Control of *Pratylenchus brachyurus* in rough lemon seedlings with Dowco 275 (Diethylfluropyridyl phosphorothioate). *Nematropica*, 5, 10-13.

Galeano, M. (2002). Dinamica di popolazione di *Tylenchulus semipenetrans* e della nematofauna di vita libera nella rizosfera di agrumi in Spagna. *Nematologia Mediterranea*, 30 (*Supplemento*), 49-53.

Galeano, M., Verdejo-Lucas, S., Sorribas, F. J., Ornat, C., Forner, J. B., & Alcaide, A. (2003). New selections from Cleopatra mandarin × *Poncirus trifoliata* with resistance to *Tylenchulus semipenetrans* Cobb. *Nematology*, 5, 227-234.

Garabedian, S., Van Gundy, S. D., Mankau, R. & Radewald, J. D. (1984). Nematodes. In: *Integrated Pest Management for Citrus.* University of California, Riverside, 129-131.

Gene, J., Verdejo-Lucas, S., & Stchigel, A. M. (2005). Microbial parasites associated with *Tylenchulus semipenetrans* in citrus orchards of Catalonia, Spain. *Biocontrol Science and Technology*, 15, 721-731.

Gill, H. S. (1971). Occurrence and reproduction of *Meloidogyne javanica* on three species of citrus in California. *Plant Disease Reporter*, 55, 607-608.

Golden, M. A., Lopez, C. H. R., & Vilchez, R. H. (1992). Description of *Pratylenchus gutierrezi* n. sp. (Nematoda: Pratylenchidae) from coffee in Costa Rica. *Journal of Nematology*, 24, 298-304.

Gottlieb, Y., Cohn, E., & Spiegel-Roy, P. (1986). Biotypes of the citrus nematode (*Tylenchulus semipenetrans* Cobb) in Israel. *Phytoparasitica*, 14, 193-198.

Graham, J. H. (1988). Fumigation induced stunting. In: J. O. Whiteside, S. M. Garnsey & L. W. Timmer, (eds.). *Compendium of Citrus Diseases*: 61.

Graham, J. H., & Duncan, L. W. (1997). Suppression of *Phytophthora nicotianae* in citrus roots by the citrus nematode (*Tylenchulus semipenetrans*). *Phytopathology*, (Suppl.) 87, S35.

Graham, J. H., Timmer, L. W., & Lee, R. F. (1983). Comparison of zinc, water uptake by gravity infusion and syringe injection tests for diagnosis of citrus blight. *Proceedings of the Florida State Horticultural Society*, 96, 45-47.

Gutierrez, R. O. (1947). A nematode attacking the roots of citrus in Argentina. *Revenue Investment in Agriculture, Buenos Aires*, 1, 119-146.

Hahn, M. L., Burrows, P. R., Gnanapragasam, N. C., Bridge, J., Vines, N. J., & Wright, D. J. (1994). Molecular diversity amongst *Radopholus similis* populations from Sri Lanka detected by RAPD analysis. *Fundamental and Applied Nematology*, 17, 275-281.

Hamid, G. A., Van Gundy, S. D., & Lovatt, C. J. (1988). Phenologies of the citrus nematode and citrus roots treated with oxamyl. *Proceedings of the International Society of Citriculture*, 2, 993-1004.

Hamid, G. A., Vangundy, S. D., Lovatt, C. J. (1985). citrus nematode alters carbohydrate partitioning in the washington navel orange. *Journal of the American Society for Horticultural Science*, 110, 642-646.

Hannon, C. I. (1963). Longevity of *Radopholus similis* under field conditions. *Plant Disease Reporter*, 47, 812-816.

Hannon, C. I. (1964). Longevity of the citrus-root nematode in Florida. *Soil Crop Science Society of Florida Proceedings*, 24, 158-161.

Heald, C. M. & O'Bannon, J. H. (1987). Citrus declines caused by nematodes. V. Slow decline. *Florida Department of Agriculture and Consumer Services, Division of Plant Industry, Nematology Circular*, 143: 4 pp.

Hearn, C. J. (1986). Production trends around the world. In: W.F. Wardowski, S. Nagy & W. Grierson, (eds.). *Fresh Citrus Fruits.* The AVI Publishing Co., Westport, CT: 127-132.

Huang, C. S. & Chang, Y. C. (1976). Pathogenicity of *Pratylenchus coffeae* on sunki orange. *Plant Disease Reporter*, 60, 957-960.

Huettel, R. N. & Yaegashi, T. (1988). Morphological differences between *Radopholus citrophilus* and *similis*. *Journal of Nematology*, 20, 150-157.

Huettel, R. N., Dickson, D. W., & Kaplan, D. T. (1982). Sex attractants and behavior in the two races of *Radopholus similis*. *Nematologica 28*, 360-369.

Huettel, R. N., Dickson, D. W., & Kaplan, D. T. (1984). *Radopholus citrophilus* n. sp. (Nematoda), a sibling species of *Radopholus similis*. *Proceedings of the Helminthological Society of Washington*, 51, 32-35.

Hutchinson, D. J. (1985). Rootstock development, screening and selection for disease tolerance and horticultural characteristics. *Fruit Varieties Journal*, 39, 21-25.

Inserra, R. N. & Vovlas, N. (1974). Danni da *Pratylenchus vulnus* su arancio amaro in Puglia. *Nematologia Mediterranea*, 2, 183-185.

Inserra, R. N. & Vovlas, N. (1977). Effects of *Pratylenchus vulnus* on the growth of sour orange. *Journal of Nematology*, 9, 154-157.

Inserra, R. N., Perotta, G., Vovlas, N., & Catara, A. (1978). Reaction of citrus rootstocks to *Meloidogyne javanica*. *Journal of Nematology*, 10, 181-184.

Inserra, R. N., Vovlas, N., & O'Bannon, J. H. (1980). A classification of *Tylenchulus semipenetrans* biotypes. *Journal of Nematology 12*, 283-287.

Inserra, R. N., Vovlas, N., O'Bannon, J. H., & Esser, R. P. (1988). *Tylenchulus graminis* n. sp. and *T. palustris* n. sp. (Tylenchulidae), from native flora of Florida, with notes on *T. semipenetrans* and *T. furcus*. *Journal of Nematology*, 20, 266-287.

Inserra, R. N., Duncan, L. W., Vovlas, N., & Loof, P. A. A. (1996). *Pratylenchus loosi* from pasture grasses in Central Florida. *Nematologica*, 42, 159-172.

Inserra, R. N., Duncan, L. W., Dunn, D. C., Kaplan, D. T., & Porazinska, D. (1998). *Pratylenchus pseudocoffeae* from Florida and its relationship with *P. gutierrezi* and *P. coffeae*. *Nematologica*, 44, 683-712.

Inserra, R. N., Duncan, L. W., Troccoli, A., Dunn, D., Maia Sos Santos, J., & Vovlas, N. (2001). *Pratylenchus jaehni* sp. n. from citrus in Brazil and a redescription of *P. coffeae*. *Nematology*, 3, 653-665.

Inserra, R. N., Duncan, L. W., Dunn, D., et al. (2005). *Pratylenchus jordanensis* a junior synonym of *P. zeae*. *Nematropica*, *35*, 161-170.

Inserra, R. N, Stanley, J. D., Obannon, J. H., & Esser, R. P. (2005). Nematode quarantine and certification programmmes implemented in Florida. *Nematologia Mediterranea*, 33, 113-123.

Iqbal, M. A., Mukhtar, T., Ahmad, R., & Khan, H. U. (2006). Ecological prevalence of *Tylenchulus semipenetrans* in four districts of the Punjab Province, Pakistan. *Pakistan Journal of Nematology*, 24, 19-26.

Johnson, A.W. (1998). Degradation of fenamiphos in agricultural production soil. *Supplement to the Journal of Nematology*, 30, 40-44.

Kallel, S., Abdelwahed, A., Ammar, M., & B'Chir, M. M. (2004). Relationship between citrus decline in orchard, population density of *Tylenchulus semipenetrans* and some soil abiotic factors. *International Journal of Nematology*, 14, 236-245.

Kallel, S., Louhichi, A., B'Chir, M. M., & van Oostveldt, P. (2005). Structure du site trophique induit par Tylenchulus semipenetrans sur bigaradier observé en microscopie photonique, confocale et électronique à transmission. *Nematologia Mediterranea*, 33, 171-178

Kallel, S., Louhichi, A., & B'Chir, M. M. (2006). Résistance de *Citrus aurantium* induite par le *Poncirus trifoliata* vis-à-vis de *Tylenchulus semipenetrans* Cobb. *Nematology*, 8, 671-679.

Kaplan, D. T. (1981). Characterization of citrus rootstock responses to *Tylenchulus semipenetrans* (Cobb). *Journal of Nematology*, 13, 492-498.

Kaplan, D. T. (1985). Influence of the sting nematode, *Belonolaimus longicaudatus*, on young citrus trees. *Journal of Nematology*, 17, 408-414.

Kaplan, D. T. (1986). Variation in *Radopholus citrophilus* population densities in the citrus rootstock Carrizo citrange. *Journal of Nematology* 18, 31-34.

Kaplan, D.T. (1988). Future considerations for nematode management in citrus. *Proceedings of the International Society of Citriculture*, 2, 969-975.

Kaplan, D. T. (1994a). Molecular characterization of the burrowing nematode sibling species, *Radopholus citrophilus* and *R. similis*. In: F. Lamberti, C. De Giorgi & D. M. Bird, (eds.). *Advances in Molecular Plant Nematology*. Plenum Press, New York: 77-83.

Kaplan, D. T. (1994b). An assay to estimate citrus rootstock resistance to burrowing nematodes. *Proceedings of the Florida State Horticulture Society*, 107, 85-89.

Kaplan, D. T. & O'Bannon, J. H. (1981). Evaluation and nature of citrus nematode resistance in Swingle citrumelo. *Proceedings of the Florida State Horticultural Society*, 94, 33-36.

Kaplan, D. T. & O'Bannon, J. H. (1985). Occurrence of biotypes in *Radopholus citrophilus*. *Journal of Nematology*, 17, 158-162.

Kaplan, D. T. & Opperman, C. H. (1997). Genome similarity implies that citrus-parasitic burrowing nematodes do not represent a unique species. *Journal of Nematology*, 29, 430-440.

Kaplan, D. T. & Opperman, C. H. (2000). Reproductive strategies and karyotype of the burrowing nematode, *Radopholus similis*. *Journal of Nematology*, 32, 126-133.

Kaplan, D. T. & Timmer, L. W. (1982). Effects of *Pratylenchus coffeae-Tylenchulus semipenetrans* interactions on nematode population dynamics in citrus. *Journal of Nematology*, 14, 368-373.
Kaplan, D. T., Vanderspool, M. C., & Opperman, C. H. (1997). Sequence tag site and host range assays demonstrate that *Radopholus similis* and *R. citrophilus* are not reproductively isolated. *Journal of Nematology*, 29, 421-429.
Kaplan, D. T., Thomas, W. K., Frisse, L. M., Sarah, J. L., Stanton, J. M., Speijer, P. R., et al. (2000). Phylogenetic analysis of geographically diverse *Radopholus similis* via rDNA sequence reveals a monomorphic motif. *Journal of Nematology*, 32, 134-142.
Kirkpatrick, J. C., & Van Gundy, S. D. (1966) Scion and rootstock as factors in the development of citrus nematode populations. *Phytopathology*, 56, 438-441.
Korayem, M., & Hassabo, S. A. A. (2005). Citrus yield in relation to *Tylenchulus semipenetrans* in silty loam soil. *International Journal of Nematology*, 15, 179-182.
Kwaye, R. G., Mashela, P. W., Shimelis, H., & Mapope, N. (2008). Determination of *Tylenchulus semipenetrans* biotype in Zebediela and Champagne, Republic of South Africa. *Plant Disease*, 92, 639-641.
Labuschagne, N. & Kotze, J. M. (1988). Factors affecting feeder root rot of citrus caused by *Fusarium solani*. *Proceedings of the International Society of Citriculture*, Tel-Aviv, Israel: 120 pp.
Labuschagne, N., Van Der Vegte, F. A., & Kotze, J. M. (1989). Interaction between *Fusarium solani* and *Tylenchulus semipenetrans* on citrus roots. *Phytophylactica*, 21, 29-33.
Lehman, P. S. (1996). Role of plant protection organizations in nematode management. *XIX Congress of Brazilian Society of Nematology*. Brazilian Society of Nematology, Rio Quente, Brasil: 137-148.
Lehman, P. S., Smith, W. W., & Inserra, R. N. (1996). A ten-year assessment of benenfits to the citrus nursery and soil pit industries in Florida from regulatory research on *Tylenchulus* species. *Nematology Circular*, 215, 4.
Leone, A., Miano, V., Lamberti, F., Duncan, L. W., Rich, J. R., & Bleve-Zacheo, T. (1997). Cellular changes induced by *Xiphinema vulgare* in the roots of citrumelo and by *Xiphinema intermedium* in the roots of Bermuda grass. *Nematologia Mediterranea*, 25, 199-207.
Le Roux, H. F., Wehner, F. C., Kotzé, J. M., & Grech, N. M. (1991). Combining fosetyl-Al trunk injection or metalaxyl soil drenching with soil application of aldicarb for control of citrus decline. *Plant Disease*, 75, 1233-1236.
Le Roux, H. F., Ware, A. B., & Pretorius, M. C. (1998). Comparative efficacy of preplant fumigation and postplant chemical treatment of replant citrus trees in an orchard infested with *Tylenchulus semipenetrans*. *Plant Disease*, 82, 1323-1327.
Le Roux, H. F., Pretorius, M. C., & Huisman, L. (2000). Citrus nematode IPM in Southern Africa. *Proceedings of the International Society of Citriculture*, Vol. 2, 823-827.
Lopez, C. R. (1978). *Belonolaimus*, a new member of the nematofauna of Costa Rica. *Agronomia Costarricense*, 2, 83-85.
Ling, P., Duncan, L.W., Deng, Z., Dunn, D., Hu, X., Huang, S., & Gmitter, F.G., Jr. (2000). Inheritance of citrus nematode resistance and its linkage with molecular markers. *Theoretical and Applied Genetics*, 100, 1010-1017.
Machmer, J. H. (1958). Effect of soil salinity on nematodes in citrus and papaya plantings. *Journal of the Rio Grande Valley Horticultural Society*, 12, 57-60.
Macaron, J. (1972). Contribution h l'6tude du nematode phytophase *Tylenchulus semipenetrans* Cobb 1913 Nematoda-Tylenchida), Ph.D. thesis, Université des Sciences et Techniques du Languedoc, Montpellier, France.
Machon, J. E. & Bridge, J. (1996). *Radopholus citri* n. sp. (Tylenchida: Pratylenchidae) and its pathogenicity on citrus. *Fundamental and Applied Nematology*, 19, 127-133.
Maafi, Z.T. & Damadzadeh, M. (2008). Incidence and control of the citrus nematode, *Tylenchulus semipenetrans* Cobb, in the north of Iran. *Nematology*, 10, 113-122.
Maafi, Z. T., Ebrahimi, Y., & Anvari, F. (2000). Evaluation of resistance of some citrus rootstocks to *Tylenchulus semipenetrans* in Mazandran Province. *Iranian Journal of Plant Pathology*, 36, 189-196.
Magunacelaya, J. C.,Villegas, C., Lamberti, F., & Ahumada, M. T. (2004). Studies on a population of *Tylenchulus semipenetrans* from Chile. *Nematologia Mediterranea*, 32, 233-234.
Malo, S. E. (1961). Nematode populations associated with citrus roots in central Florida. *Plant Disease Reporter*, 45, 20-23.
Mangat, B. P. S., & Sharma, N. K. (1981). Influence of host nutrition on multiplication and development of citrus nematode. *Indian Phytopathology*, 34, 90-91.

Mani, A., Al Hinai, M. S., & Handoo, Z. A. (1997). Occurrence, population density, and distribution of root-lesion nematodes, *Pratylenchus* spp., in the Sultanate of Oman. *Nematropica*, 27, 209-219.

Mankau, R. (1968). Effect of nematicides on nematode-trapping fungi associated with the citrus nematode. *Plant Disease Reporter*, 52, 851-855.

Marin, D. H., Sutton, T. B., & Barker, K. R. (1998). Dissemination of bananas in Latin America and the Caribbean and its relationship to the occurrence of *Radopholus similis*. *Plant Disease*, 82, 964-974.

Martin, J. P., & Van Gundy, S. D. (1963). Influence of soil phosphorous level on the growth of sweet orange seedlings and the activity of the citrus nematode (*Tylenchulus semipenetrans*). *Soil Science*, 96, 128-135.

Mashela, P. W., & Nthangeni, M. E. (2002). Osmolyte allocation in response to *Tylenchulus semipenetrans* infection, stem girdling, and root pruning in citrus. *Journal of Nematology*, 34, 273-277.

Mashela, P., Duncan, L. W., Graham, J. H., & McSorley, R. (1992a). Leaching soluble salts increases population densities of *Tylenchulus semipenetrans*. *Journal of Nematology*, 24, 103-109.

Mashela, P., Duncan, L. W., & McSorley, R. (1992b). Salinity reduces resistance to *Tylenchulus semipenetrans* in citrus rootstocks. *Nematropica*, 22, 7-12.

McClure, M. A. & Schmitt, M. E. (1996). Control of citrus nematode, *Tylenchulus semipenetrans*, with cadusafos. *Supplement to the Journal of Nematology*, 28(4S), 624-628.

McElroy, F. D., Sher, S. A., & van Gundy, S. D. (1966). The sheath nematode *Hemicycliophora arenaria,* a native to California soils. *Plant Disease Reporter*, 40, 581-583.

McSorley, R., & Parrado, J. L. (1982a). Plans for the collection of nematode soil samples from fruit groves. *Nematropica, 12*, 257-267.

McSorley, R., & Parrado, J. L. (1982b). Relationship between two nematode extraction techniques on two Florida soils. *Soil Crop Science Society of Florida Proceedings*, 41, 30-36.

Meagher, J. W. (1967). Observations on the transport of nematodes in subsoil drainage and irrigation water. *Australian Journal of Experimental Agricultural Animal Husbandry*, 7, 577-579.

Meagher, J. W. (1969). Nematodes as a factor in citrus production in Australia. *Proceedings of the 1st International Citrus Symposium*, Riverside, California, 2: 999-1006.

Miller, J., Miller, D., & Bird, P. (1996). Does a new biotype of citrus nematode exist in Sundays River Valley? *Inligtingsbulletin - Instituut vir Tropiese en Subtropiese Gewasse*, 284, 8.

Milne, D. L. (1974). Citrus seedbeds: Methyl bromide and mycorrhizae. Citrus and Subtropical Fruit Research Institute, Nelspruit: 9-11.

Milne, D. L. (1982). Nematode pests of citrus. In: Keetch, D. P. & Heyns, J. (eds.). *Nematology in Southern Africa. Republic of South Africa, Department of Agriculture and Fisheries Science Bulletin*, 400, 12-18.

Milne, D. L., & De Villiers, E. A. (1977). Soil application of systemic pesticides for control of thrips and nematodes on citrus. *Citrus and Subtropical Fruit Journal*, 518, 9-18.

Milne, D. L., & du Toit, W. (1976). The effect of citrus nematicides on the earthworm population in the soil. *Citrus and Subtropical Fruit Research Institute*, Nelspruit: 13-15.

Milne, D. L., & Willers, P. (1979). Yield and nutritional responses to phenamiphos treatment of citrus infested with citrus nematodes. *Subtropica*, 1, 11-14.

Milne, D. L., & Willers, P. (1980). Yield and fruit size increases due to control of citrus nematode with phenamiphos. *Citrus and Subtropical Fruit Research Institute, Information Bulletin*, 90, 11-14.

Minz, G. (1956) The root-knot nematode, *Meloidogyne* spp. in Israel. *Plant Disease Reporter*, 40, 798-801.

Murguia, C., Abad, P., Jorda, C., & Bello, A. (2005). Short communication. Identififcaton of the Poncirus biotype of *Tylenchulus semipenetrans* in Valencia, Spain. *Spanish Journal of Agricultural Research*, 3(1), 130-133.

Musarrat, A. R., Firoza, K., & Shahina, F. (2006). Study of root-knot nematodes (*Meloidogyne* spp.) in N.W.F.P. and Sindh, Pakistan. *Pakistan Journal of Nematology*, 24, 1-7.

Nigh, E. L., Jr. (1981a) Relation of citrus nematode to root distribution in flood irrigated citrus. *Journal of Nematology*, 13, 451-452.

Nigh, E. L., Jr. (1981b) Evaluation of sampling and extraction techniques to determine citrus nematode, *Tylenchulus semipenetrans*, populations. *Journal of Nematology*, 13, 451.

Noling, J. W., Buchanon, S., & Schumann, A. (2007). Impacts of EPA Proposed Buffer Zone Restrictions on Florida Strawberry Acreage and Production. *2007 Proceedings Annual International Research Conference on Methyl Bromide Alternatives and Emissions Reductions*, October 28-November 1, 2007. San Diego, CA: 451-453.

O'Bannon, J. H. (1968). The influence of an organic soil amendment on infectivity and reproduction of *Tylenchulus semipenetrans* on two citrus rootstocks. *Phytopathology*, 58, 597-601.

O'Bannon, J. H. (1977). Worldwide dissemination of *Radopholus similis* and its importance in crop production. *Journal of Nematology*, 9, 16-25.
O'Bannon, J. H. (1980). Migration of *Pratylenchus coffeae* and *P. brachyurus* on citrus and in soil. *Nematropica*, 10, 70.
O'Bannon, J. H. & Esser, R. P. (1975). Evaluation of citrus, hybrids, and relatives as hosts of the nematode *Pratylenchus coffeae*, with comments on other hosts. *Nematologia Mediterranea*, 3, 113-122.
O'Bannon, J. H., & Esser, R. P. (1985). Citrus declines caused by nematodes in Florida. II. Physiological races. Nematology Circular No. 116. Florida Department of Agriculture and Consumer Services, DPI. Gainesville, Florida, USA: 4 pp.
O'Bannon, J. H., & Ford, H. W. (1976). An evaluation of several *Radopholus similis*-resistant or -tolerant citrus rootstocks. *Plant Disease Reporter*, 60, 620-624.
O'Bannon, J. H., & Hutchinson, D.H. (1974). Development of rootstocks resistant to the citrus nematode, *Tylenchulus semipenetrans*. In: L. K. Jackson, A. H. Krezdorn & J. Soule, (eds.). *Proceedings of the 1st International Citrus Short Course* (Sept. 24-29, 1973). Gainesville, FL: 22-29.
O'Bannon, J. H., & Nemec, S. (1978). Influence of soil pesticides on vesicular-arbuscular mycorrhizae in a citrus soil. *Nematropica*, 8, 56-61.
O'Bannon, J. H., & Nemec, S. (1979). The response of *Citrus limon* seedlings to a symbiont, *Glomus etunicatus*, and a pathogen, *Radopholus similis*. *Journal of Nematology*, 11, 270-274.
O'Bannon, J. H., & Tarjan, A. C. (1973). Preplant fumigation for citrus nematode control in Florida. *Journal of Nematology*, 5, 88-95.
O'Bannon, J. H., & Tarjan, A. C. (1979). Management of *Radopholus similis* infecting citrus with DBCP or phenamiphos. *Plant Disease Reporter*, 63, 456-460.
O'Bannon, J. H., & Tarjan, A. C. (1985). Citrus declines caused by nematodes in Florida. III. Citrus slump. *Florida Department of Agriculture and Consumer Services, Division of Plant Industry, Nematology Circular*, 117: 3 pp.
O'Bannon, J. H., & Tomerlin, A. T. (1969a). Movement of *Radopholus similis* on a weed host (*Solanum nigrum*). *Journal of Nematology*, 1, 21.
O'Bannon, J. H., & Tomerlin, A. T. (1969b). Population studies on two species of *Pratylenchus* on citrus. *Journal of Nematology*, 1, 299-300.
O'Bannon, J. H., & Tomerlin, A. T. (1971). Response of citrus seedlings to *Radopholus similis* in two soils. *Journal of Nematology 3*, 255-259.
O'Bannon, J. H., & Tomerlin, A. T. (1973). Citrus tree decline caused by *Pratylenchus coffeae*. *Journal of Nematology 5*, 311-316.
O'Bannon, J. H., & Tomerlin, A. T. (1977). Control of the burrowing nematode, *Radopholus similis*, with DBCP and oxamyl. *Plant Disease Reporter*, 61, 450-454.
O'Bannon, J. H., Reynolds, H. W., & Leathers, C. R. (1966). Effects of temperature on penetration, development, and reproduction of *Tylenchulus semipenetrans*. *Nematologica*, 12, 483-487.
O'Bannon, J. H., Leather, C. R., & Reynolds, H. W. (1967). Interactions of *Tylenchulus semipenetrans* and *Fusarium* species on rough lemon (*Citrus limon*). *Phytopathology*, 57, 414-417.
O'Bannon, J. H., Radewald, J. D., & Tomerlin, A. T. (1972). Population fluctuation of three parasitic nematodes in Florida citrus. *Journal of Nematology*, 4, 194-199.
O'Bannon, J. H., Tarjan, A. C., & Bistline, F. W. (1974). Control of *Pratylenchus brachyurus* on citrus and tree response to chemical treatment. *Soil Crop Science Society of Florida Proceedings*, 33, 65-67.
O'Bannon, J. H., Radewald, J. D., Tomerlin, A. T., & Inserra, R. N. (1976). Comparative influence of *Radopholus similis* and *Pratylenchus coffeae* on citrus. *Journal of Nematology*, 8, 58-63.
Orion, D. & Cohn, E. (1975). A resistant response of citrus roots to the root-knot nematode, *Meloidogyne javanica*. *Marcellia*, 38, 327-328.
Oteifa, B. A., Shafiee, V. A., & Eissa, F. M. (1965). Efficacy of DBCP flood irrigation in established citrus. *Plant Disease Reporter*, 49, 598-599.
Pan, C. (1984). Studies on plant-parasitic nematodes on economically important crops in Fujian. I. Species of root-knot nematodes (*Meloidogyne* species) and their host-plants. *Acta Zoologica Sinica*, 30, 159-167.
Pan, C. (1985). Studies on plant-parasitic nematodes on economically important crops in Fujian. III. Description of *Meloidogyne fujianensis* n. sp. (Nematoda: Meloidogynidae) infesting Citrus in Najing County. *Acta Zoologica Sinica*, 31, 263-268.

Perry, V. G. (1953). Return of the nematodes following fumigation of Florida soils. *Proceedings of the Florida State Horticultural Society*, 66, 112-114.

Philis, J. (1969). Control of citrus nematode, *Tylenchulus semipenetrans*, with DBCP in established Cyprus citrus groves. *Plant Disease Reporter*, 53, 804-806.

Philis, I. (1989). Yield loss assessment caused by the citrus nematode *Tylenchulus semipenetrans* on Valencia oranges in Cyprus. *Nematologia Mediterranea*, 17, 5-6.

Poucher, C., Ford, H. W., Suit, R. F., & DuCharme, E. P. (1967). Burrowing nematode in citrus. *Florida Department of Agriculture, Division of Plant Industry Bulletin*, 7: 63 pp.

Pinochet, J., Verdejo, S., & Soler, A., et al. (1992). Host range of a population of *Pratylenchus vulnus* in commercial fruit, nut, citrus, and grape rootstocks in Spain. *Journal of Nematology*, 24, 693-698.

Prasad, S. K. & Chawla, M. L. (1966). Observations on the population fluctuations of citrus nematode, *Tylenchulus semipenetrans* Cobb, 1913. *Indian Journal of Entomology 27*, 450-454.

Radewald, J. D., O'Bannon, J. H., & Tomerlin, A. T. (1971a). Temperature effects on reproduction and pathogenicity of *Pratylenchus coffeae* and *P. brachyurus* and survival of *P. coffeae* in roots of *Citrus jambhiri. Journal of Nematology*, 3, 390-394.

Radewald, J. D., O'Bannon, J. H., & Tomerlin, A. T. (1971b). Anatomical studies of *Citrus jambhiri roots* infected by *Pratylenchus coffeae. Journal of Nematology*, 3, 409-416.

Rao, M. S. (2005). Management of *Meloidogyne javanica* on acid lime nursery seedlings by using formulations of *Pochonia chlamydosporia* and *Paecilomyces lilacinus. Nematologia Mediterranea,* 33, 145-148.

Reynolds, H. W. & O'Bannon, J. H. (1963a). Factors influencing the citrus replants in Arizona. *Nematologica*, 9, 337-340.

Reynolds, H. W. & O'Bannon, J. H. (1963b). Decline of grapefruit trees in relation to citrus nematode populations and tree recovery after chemical treatment. *Phytopathology*, 53, 1011-1015.

Robbins, R. T. & Hirschmann, H. (1974). Variation among populations of *Belonolaimus longicaudatus. Journal of Nematology*, 66, 87-94.

Roccuzzo G., Ciancio A., & Lo Guidice V. (1992). Some observations on the ecology of the citrus nematode Tylenchulus semipenetrans Cobb in Southern Italy. *Proceedings of the International Society of Citriculture*, 3, 950-952.

Salem, A. A.-M. (1980). Observations on the population dynamics of the citrus nematode, *Tylenchulus semipenetrans* in Sharkia Governorate. *Egypt Journal of Phytopathology*, 12, 31-34.

Schneider, H., & Baines, R. C. (1964). *Tylenchulus semipenetrans*: Parasitism and injury to orange tree roots. *Phytopathology*, 54, 1202-1206.

Scotto la Massèse, C. (1969). The principal plant nematodes of crops in the French West Indies. In: Peachey, J. E. (ed.). *Nematodes of Tropical Crops.* Commonwealth Bureau of Helminthology (Great Britain) Technical Communication, 40, 168-169.

Scotto la Massèse, C. (1980). Possibilités offertes par les "courbes isopathologiques" dans l'appréciation de la pathogénie des parasites illustrées par deux exemples en nématologie fruitière. *Revue de Zoologie Agricole*, 78, 97-113.

Scotto la Massèse, C., Vassy, R., & Zaouchi, H. (1975). In fluence of thre rootstocks on the yield of two citrus varieties in Algeria and on infestation by *Tyelenchulus semipenetrans. Nematologia Mediterranea*, 3, 29-34.

Sharma, R. D. & Stirling, G. R. (1991). In vivo mass-production systems for *Pasteuria penetrans. Nematologica*, 37, 483-484.

Siddiqi, M. R. (1964) Studies on nematode root-rot of citrus in Uttar Pradesh, India. *Proceedings of the Zoological Society* (Calcutta), 17, 67-75.

Siddiqui, I. A., Sher, S. A., & French, A. M. (1973). *Distribution of plant parasitic nematodes in California.* State of California, Department of Food and Agriculture, Division of Plant Industry: 324 pp.

Silva, R. A., & Inomoto, M. M. (2002). Host-range characterization of two *Pratylenchus coffeae* isolates from Brazil. *Journal of Nematology*, 34, 135-139.

Sing, B. (2004). Control of citrus nematode, *Tylenchulus semipenetrans* in Nagpur mandarin orchard. *Indian Journal of Nematology*, 34, 70-74.

Sites, J. W., Reitz, H. J., & Deszyck, E. J. (1951). Some results of irrigation research with Florida citrus. *Proceedings of the Florida State Horticultural Society*, 64, 71-79.

Smelt, J. H., Van De Peppel-Groen, A. E., Van Der Pas, L. J. T., & Dijksterhuis, A. (1996). Development and duration of accelerated degradation of nematicides in different soils. *Soil Biology and Biochemistry*, 28, 1757-1765.

Smith, G., & Kaplan, D. T. (1988). Phosphorus and burrowing nematode interactions on growth of rough lemon citrus seedlings. *Journal of Nematology*, 20, 539-544.

Sorribas, F. J., Verdejo-Lucas, S., Forner, J. B., Alcaide, A., Pons, J., & Ornat, C. (2000). Seasonality of *Tylenchulus semipenetrans* Cobb and *Pasteuria* sp. in citrus orchards in Spain. *Supplement to the Journal of Nematology*, 32(4S), 622-632.

Sorribas, F. J., Verdejo-Lucas, S., Galeano, M., Pastor, J., & Ornat, C. (2003). Effect of 1,3 dichloropropene and rootstocks alone and in combination on *Tylenchulus semipenetrans* and citrus tree growth in a replant management program. *Nematropica*, 33, 149-158.

Sorribas, F. J., Verdejo-Lucas, S., Pastor, J., et al. (2008). Population densities of *Tylenchulus semipenetrans* related to physicochemical properties of soil and yield of clementine mandarin in Spain. *Plant Disease*, 92, 445-450.

Spiegel-Roy, R., Vardi, A., Elhanati, A., Solel, Z., & Bar-Joseph, M. (1988). Rootstock selection from a Poorman orange x *Poncirus trifoliata* cross. *Proceedings of the International Society of Citriculture*, 1, 195-200.

Spyke, P. & Castle, B. (2007). Advanced production systems: a pathway to disease management and new perspectives on citrus grove value. *Citrus Industry*, September 2007, 10-12.

Standifer, M. S. & Perry, V. G. (1960). Some effects of sting and stubby root nematodes on grapefruit roots. *Phytopathology*, 50, 152-156.

Stirling, G. R. (1976). *Paratrichodorus lobatus* associated with citrus, peach and apricot trees in South Australia. *Nematologica*, 22, 138-144.

Stirling, G. R. (1991). *Biological Control of Plant Parasitic Nematodes: Progress, Problems, and Prospects.* CAB International. Wallingford, UK: 282 pp.

Stirling, G. R. & Wachtel, M. F. (1985). Effects of nematicides on citrus nematode *Tylenchulus semipenetrans* and the yield of citrus in South Australia. *Agricultural Record (South Australia)*, 12, 6-11.

Suit, R. F., & Brooks, T. L. (1957). Current information relating to barriers for the burrowing nematode. *Proceedings of the Florida State Horticultural Society*, 70, 55-57.

Suit, R. F., & DuCharme, E. P. (1953). The burrowing nematode and other parasitic nematodes in relation to spreading decline. *Citrus Leaves*, 33, (8-9), 32-33.

Suit, R. F., Hanks, R. W., Tarjan, A. C., Feldman, A. W., & DuCharme, E. P. (1967). Control of spreading decline of citrus. State Project 773. *Florida Agricultural Experiment Station Annual Report*: 228-229.

Swingle, W. T., & Reese, R. (1967). The botany of citrus and its wild relatives. Chapter 3. In: Reuther, W., Webber, H. J., & Batchelor, L. D. (eds.). *The Citrus Industry*, University of California, Division of Agricultural Science: 190-430.

Tarjan, A.C. (1956). The possibility of mechanical transmission of nematodes in citrus groves. *Proceedings of the Florida State Horticultural Society*, 69, 34-37.

Tarjan, A. C. (1961). Longevity of *Radopholus similis* (Cobb) in host-free soil. *Nematologica*, 6, 170-175.

Tarjan, A. C. (1971). Migration of three pathogenic citrus nematodes through two Florida soils. *Soil Crop Science Society of Florida Proceedings*, 31, 253-255.

Tarjan, A. C., & O'Bannon, J. H. (1969). Observations on meadow nematodes (*Pratylenchus* spp.) and their relation to decline of citrus in Florida. *Plant Disease Reporter 53*, 683-686.

Tarjan, A. C., & O'Bannon, J. H. (1977). Nonpesticidal approaches to nematode control. *Proceedings of the International Society of Citriculture*, 3, 848-853.

Tarjan, A. C., & Simmons, P. N. (1966). The effect of interacting cultural practices of citrus trees with spreading decline. *Soil Crop Science Society of Florida Proceedings*, 26, 22-31.

Thomason, I. J., & Caswell, E. P. (1987). Principles of nematode control. In: R. H. Brown & B. R. Kerry, (eds.). *Principles and Practice of Nematode Control in Crops.* Academic Press, Australia: 87-130.

Thomas, E. E. (1923). The Citrus Nematode, *Tylenchulus semipenetrans*. *California Agricultural Experiment Station*, 2: 34 pp.

Timmer, L. W. (1977). Control of citrus nematode *Tylenchulus semipenetrans* on fine-textured soil with DBCP and oxamyl. *Journal of Nematology*, 9, 45-50.

Timmer, L. W., & Davis, R. D. (1982). Estimate of yield loss from the citrus nematode in Texas grapefruit. *Journal of Nematology*, 14, 582-585.

Tomerlin, A. T., & O'Bannon, J. H. (1974). Effect of *Radopholus similis* and *Pratylenchus brachyurus* on citrus seedlings in three soils. *Soil Crop Science Society of Florida Proceedings 33*, 95-97.

Toung, M.-C. (1963). A study on seasonal influence in quantitative variation of the citrus nema, *Tylenchulus semipenetrans* Cobb. *National Taiwan University Plant Protection Bulletin*, 5, 323-327.

Tsai, B. Y., & Van Gundy, S. D. (1988). Comparison of anhydrobiotic ability of the citrus nematode with other plant parasitic nematodes. *Proceedings of the International Society of Citriculture*, 2, 983-992.

Valette, C., Mounport, D., Nicole, M., Sarah, J. L., & Baujard, P. (1998). Scanning electron microscope studies of two African populations of *Radopholus similis* (Nematoda: Pratylenchidae) and proposal of *R. citrophilus* as a junior synonym of *R. similis*. *Fundamental and Applied Nematology*, 21, 139-146.

Van Gundy, S. D. (1959). The life history of *Hemicycliophora arenaria* Raski (Nematoda: Criconematidae). *Proceedings of the Helminthology Society of Washington*, 26, 67-72.

Vangundy, S. D. (1958). The Pathogenicity of *Hemicycliophora arenaria* on citrus. *Phytopathology*, 48, 399.

Vangundy, S. D. (1980). Nematology - Status and Prospects - Let's take off our blinders and broaden our horizons. *Journal of Nematology 12*, 158-163.

Van Gundy, S. D., & Kirkpatrick, J. D. (1964). Nature of resistance in certain citrus rootstocks to citrus nematodes. *Phytopathology*, 54, 419-427.

Van Gundy, S. D., & Martin, J. P. (1961). Influence of *Tylenchulus semipenetrans* on the growth and chemical composition of sweet orange seedlings in soils of various exchangeable cation ratios. *Phytopathology*, 51, 146-151.

Van Gundy, S. D., & Martin, J. P. (1962). Soil texture, pH and moisture effects on the development of citrus nematode (*Tylenchulus semipenetrans*). *Phytopathology*, 52, 31.

Van Gundy, S. D., & McElroy, F. D. (1969). Sheath nematode: Its biology and control. *Proceedings of the 1st International Citrus Symposium*. Riverside, California, 2, 985-989.

Van Gundy, S. D., & Meagher, J. W. (1977). Citrus nematode (*Tylenchulus semipenetrans*) problems worldwide. *1977 International Citrus Congress*, Orlando, FL, 7 pp.

Van Gundy, S. D., & Rackham, R. L. (1961). Studies on the biology and pathogenicity of *Hemicycliophora arenaria. Phytopathology*, 51, 393-397.

Van Gundy, S. D., & Tsao, P. H. (1963). Growth reduction of citrus seedlings by *Fusarium solani* as influenced by the citrus nematode and other soil factors. *Phytopathology*, 53, 488-489.

Van Gundy, S. D., Thomason, I. J., & Rackham, R. L. (1959). The reaction of three *Citrus* spp. to three *Meloidogyne* spp. *Plant Disease Reporter*, 43, 970-971.

Van Gundy, S. D., Stolzy, L. H., Szuszkiewicz, T. E., & Rackham, R. L. (1962). Influence of oxygen supply on survival of plant parasite nematodes in soil. *Phytopathology*, 52, 628-632.

Van Gundy, S. D., Martin, J. P., & Taso, P. H. (1964). Some soil factors influencing reproduction of the citrus nematode and growth reduction of sweet orange seedlings. *Phytopathology*, 54, 294-299.

Van Gundy, S. D., Bird, A. F., & Wallace, H. R. (1967). Aging and starvation in larvae of *Meloidogyne javanica* and *Tylenchulus semipenetrans. Phytopathology*, 57, 599-571.

Verdejo-Lucas, S., & Kaplan, D. T. (2002). The citrus nematode: *Tylenchulus semipenetrans*. In: J. L. Starr, R. Cook & J. Bridge, (eds.). *Plant Resistance to Parasitic Nematodes*. CABI, New York: 207-219.

Verdejo-Lucas, S., Sorribas, F. J., Pons, J., Forner, J. B., & Alcaide, A. (1997). The mediterranean biotypes of *Tylenchulus semipenetrans* in Spanish citrus orchards. *Fundamental and Applied Nematology*, 20, 399-404.

Verdejo-Lucas, S., Sorribas, F. J., Forner, J.B., Alcaide, A., Ornat, C., & Galeano, M. (2000). Evaluating resistance to *Tylenchulus semipenetrans* Cobb in hybrid citrus rootstocks in Spain. *Proceedings of the International Society of Citriculture, IX Congress 2000*. Orlando, FL: 1, 818-822.

Verdejo-Lucas, S., Galeano, M., Sorribas, F. J., Forner, J. B., & Alcaide, A. (2003). Effect on resistance to *Tylenchulus semipenetrans* of hybrid citrus rootstocks subjected to continuous exposure to high population densities of the nematode. *European Journal of Plant Pathology*, 109, 427-433.

Vilardebo, A. (1964). Etude sur *Tylenchulus semipenetrans* Cobb au Maroc II. *AL Awamia, Rabat*, 11, 31-49.

Vilardebo, A., Sqalli, A., & Devaux, R. (1975). Utilisation du DBCP, du phenamiphos et du prophos contre *Tylenchulus semipenetrans* dans les vergers du Maroc. *Fruits*, 30, 313-317.

Walter, D. E. & Kaplan, D. T. (1990). Antagonists of plant-parasitic nematodes in Florida citrus. *Journal of Nematology*, 22, 567-573.

Walker, G. E. & Morey, B. G. (1999) Effects of chemicals and microbial antagonists on nematodes and fungal pathogens of citrus roots. *Australian Journal of Experimental Agriculture*, 39, 629-637.

Wheaton, T. A., Childers, C. C., Timmer, L. W., Duncan, L. W., & Nikdel, S. (1985). Effects of aldicarb on yield, fruit quality, and tree condition on Florida citrus. *Proceedings of the Florida State Horticultural Society*, 98, 6-10.

Whitehead, A. G. (1968). Taxonomy of *Meloidogyne* (Nematodea: Heteroderidae) with descriptions of four new species. *Transactions of the Zoological Society of London*, 31, 263-401.

Willers, P. (1979). Influence of citrus nematode, *Tylenchulus semipenetrans*, on Navel yield in Sundays River Valley orchard. *Citrus and Subtropical Fruit Research Institute*, Nelspruit, SA: 9-10.

Willers, P., & Holmden, E. (1980). The influence of citrus nematode, *Tylenchulus semipenetrans*, on the performance of trees growing under saline conditions. *Information Bulletin, Citrus and Subtropical Fruit Research Institute*, 99, 13-16.

Yassin, A. M. (1974). A note on *Longidorus* and *Xiphinema* species from the Sudan. *Nematologia Mediterranea*, 2, 141-147.

Yokoo, T. (1964). Studies on the citrus nematode (*Tylenchulus semipenetrans* Cobb, 1913) in Japan. *Agricultural Bulletin, Saga University*, 20, 71-109.

Yokoo, T., & Ikegemi, Y. (1966). Some observations on growth of the new host plant, snapdragon (*Antirrhinum majus* L.) attacked by root lesion nematode, *Pratylenchus coffeae* and control effect of some nematicides. *Agricultural Bulletin, Saga University*, 22, 83-92.

Young, T. W. (1954). An incubation method for collecting migratory endoparasitic nematodes. *Plant Disease Reporter*, 38, 794-795.

Section 2

Temperate Fruit Crops and Forest Nematodes

7

SOLEDAD VERDEJO-LUCAS AND M. TALAVERA

INTEGRATED MANAGEMENT OF NEMATODES PARASITIC ON *PRUNUS* SPP.

IRTA, Ctra de Cabrils, Km 2
E- 08348 Cabrils, Barcelona, Spain

Abstract. Parasitic nematodes affecting *Prunus* spp. are reviewed, focusing on root-knot, root lesion, ring and dagger nematode species. Management and control methods include prevention and quarantine, pre-planting measures, as well as methods to lower population densities prior to establish an orchard, like fallow, crop rotation, soil solarization, biofumigation or steam applications. Other methods reviewed include soil fumigation or chemical control procedures with non-fumigant nematicides at pre-planting and seedling treatments. Post-planting measures consider rootstocks resistance, chemical or biological control and other cultural methods. Integrated pest management is then discussed, analysing the sequences of actions, from sample collection to nematodes extraction, identification and quantification, followed by damage estimation and management decision.

1. INTRODUCTION

Prunus is a genus comprising trees and bushes, which includes *Prunus dulcis* (almonds), *P. armeniaca* (apricots), *P. avium* and *P. cerasus* (cherries), *P. persica* (peaches), and *P. domestica*, *P. salicina* hybrids and *P. cerasifera* hybrids (plums and prunes), all of which have cultivars selected for commercial fruit production. The fruit is a drupe with a relatively large "stone", which gives the name of "stone fruits". In most species, the commercial value is the drupe but in almonds, it is given by the seed inside the stone. There are also a number of species and cultivars grown as ornamental plants, due to their profusion of flowers, they are called "flowering cherries".

All *Prunus* species belong to the family Rosaceae, though some authors place them in their own family, called Prunaceae. There are about 430 species coming from one Asian ancestor (Bortiri et al., 2001). They extend mainly throughout the temperate regions in both hemispheres. Because of their considerable value as food and ornamental plants, many *Prunus* species were introduced to regions in which they were not native but soil, moisture and climatic conditions were suitable for their cultivation. The main producers of peaches, cherries and plums in the world are the

A. Ciancio & K. G. Mukerji (eds.), *Integrated Management of Fruit Crops and Forest Nematodes*, 177-193.

United States and some Mediterranean countries such as Italy, Spain, Greece, and Turkey, which are also the largest producers of apricots (Food and Agriculture Organization [FAO], 2004). Production of cherries is also important in Germany, and plums and prunes in Argentina, Chile, China, France, and Russia (FAO, 2004).

Numerous species of plant-parasitic nematodes have been associated with the rhizosphere of *Prunus* spp. and some species of *Meloidogyne*, *Pratylenchus*, *Xiphinema*, *Criconemoides*, *Helicotylenchus*, *Hoplolaimus*, and *Paratylenchus* are parasites of stone fruit trees. *Meloidogyne incognita*, *M. javanica*, *Pratylenchus vulnus*, *P. penetrans*, and *Criconemoides xenoplax* are considered major nematode pests because they cause significant yield losses in many regions of the world (McKenry, 2004). Mature trees can tolerate large populations of nematodes before exhibiting symptoms of damage but young trees grow poorly if replanted in nematode infested sites. Symptoms include plant stunting, loss of vigor, leaf yellowing, early defoliation, and early death of trees (McKenry, 1999).

To alleviate nematode problems, the implementation of procedures of integrated pest management (IPM) based on principles of prevention, reduction of the initial population and host resistance or tolerance should be considered. IPM aims at maintaining nematode populations at acceptable levels, below the damage thresholds, resulting in favorable long-term socio-economic and environmental consequences (Bird, 1981). IPM for nematode control, however, has received little attention, mainly due to the availability of reliable broad-spectrum soil fumigants, which had provided good nematode control in most cases. The ban or restrictions in the use of some fumigants as methyl bromide or 1,3-dichloropropene as a result of environmental and health concerns, has driven research efforts to find other non-chemical alternatives for nematode control. These include the use of host resistance, soil solarization, and cultural practices that will have to be integrated in IPM practices, because no single control method has proven effective by itself alone.

2. PARASITIC NEMATODES AFFECTING *PRUNUS* SPP.

2.1. Root-Knot Nematodes

The four major *Meloidogyne* species, *M. incognita*, *M. javanica*, *M. arenaria*, and *M. hapla* can cause economic damage to stone fruits in different parts of the world (Marull, Pinochet, Felipe, & Cenis, 1994; Simeone & Di Vito, 1992). Three other *Meloidogyne* species of limited distribution have been reported parasitizing peaches, *M. hispanica* in Spain (Hirschmann, 1986), *M. morocciensis* in Morocco (Rammah & Hirschmann, 1990) and *M. floridensis* in Florida, USA (Handoo et al., 2004). Root-knot nematodes cause a 15% loss in vigor and yield of *Prunus* crops (nurseries and orchards) on a worldwide basis. The economic importance of *Meloidogyne* spp. to stone fruits was shown with the use of nematicides and rootstocks with nematode resistance. Sharpe, Pusey, Nyczepir, and Florkowski, (1993) showed that pre-plant fumigation with methyl bromide to control *Meloidogyne* sp. in peaches increased the cumulative yield over three seasons by 2535 Kg per ha.

The species of *Meloidogyne* are sedentary endoparasites. The infective second-stage juveniles hatch from the eggs in moist soil, move freely in the soil and penetrate into the roots just behind the root tip. Once inside the cortical tissue, the juveniles establish feeding sites, where several root cells around nematode's head enlarge to form giant cells that constitute a nutrient sink from which nematodes take nutrients. Juveniles enlarge and swell as they develop to adult females. Most plants react to *Meloidogyne* feeding by rapid cell division and expansion in the cortical area surrounding nematode that results in characteristic knots or galls. Within the galls, pear shaped females lay eggs into a gelatinous matrix, known as egg masses. The life cycle is completed in 4 to 6 weeks depending on soil temperatures. Nematodes are active in warm and moist soils that support growth of the host plants and they can complete several generations in one season, increasing their population densities by several folds.

The aboveground symptoms of root-knot nematode damage include stunting, poor growth, yellowing, early senescence, wilting and reduced foliage and yield as in some other root diseases and nutrient deficiencies. Uneven plant growth is an early symptom of nematode attack caused by the simultaneous root invasion of many juveniles, which produce retardation in plant growth due to great injury in root tissue. Nematode damaged plants are usually located in patches or along the planting row, reflecting an aggregation pattern (Nyczepir & Becker, 1998). Belowground symptoms are the typical galls caused by the establishment of the nematode at the feeding sites. Galls caused by root-knot nematodes on *Prunus* appear as outgrowths of root tissue, distributed along the secondary roots. Number and size of the galls depend on the susceptibility of the host, initial population densities, and numbers of females inside of them and thus, on the severity of the nematode attack. These symptoms results from the damage caused by the nematode to the root system that reduces the ability of the plant to absorb water and nutrients necessary for normal growth. Root-knot nematode damage is more serious to trees growing in sandy soils and can be greater in periods of drought.

2.2. Root Lesion Nematodes

At least ten species of root-lesion nematodes have been found in association with stone fruits throughout the world, *Pratylenchus brachyurus*, *P. convallariae*, *P. crenatus*, *P. hexincisus*, *P. neglectus*, *P. penetrans*, *P. pratensis*, *P. thornei*, *P. vulnus*, and *P. zeae* (Lownsbery, Moody, & Braun, 1974; Nyczepir & Becker, 1998). Of these, only *P. penetrans* and *P. vulnus* are major nematode pests of *Prunus* of economic importance. Usually, *P. penetrans* is associated with cooler temperatures and higher elevations, whereas *P. vulnus* is typically more often associated with warmer temperature regions. Yield losses caused by these nematodes are variable depending on initial population densities, host tolerance and agro-environmental conditions. Thus, *P. vulnus* can reduce marketable fruit of plum and peach by a 16% (McKenry, 1989).

The species of *Pratylenchus* are migratory endoparasites that enter and move within the roots while feeding on cortical cells where they can cause extensive damage. Active juveniles and adults stages have the capability of entering and leaving the roots. Adult females usually deposit eggs singly within individual root cells or necrotic tissue. The presence of root-lesion nematodes in stone fruit orchards can affect tree establishment and fruit production. Aboveground symptoms are tree decline with general characteristics of nutrient deficiency, reduced shoot growth and tree vigor, and reductions in fruit size. Belowground symptoms are reduction in number of feeder roots, darkening, and necrotic lesions.

2.3. Ring Nematodes

The most important species of ring nematodes involved in *Prunus* diseases are *Criconemoides xenoplax* and *C. curvatus*. Both species have been found in North America and Europe (Nicotina, 1990; Nyczepir, Bertrand, Miller, & Motsinger, 1985). Damage caused by ring nematodes to stone fruits is double, a direct damage on the roots and secondly but more important, damage caused by interactions with other pathogens.

Nematodes belonging to this genus have an ectoparasitic feeding habit. The life cycle of *C. xenoplax* is completed in 25-34 days at 22-26ºC. Feeding occurs along roots and at the root tips, and egg deposition takes place close to the root surface. Root systems parasitized by *C. xenoplax* show a lack of feeder roots (Lownsbery, English, Moody, & Shick, 1973). In southeastern United States, *C. xenoplax* in association with *Pseudomonas syringae* predisposes peach trees to a disease complex called "Peach Tree Short Life" (PTSL) (Nyczepir, 1990). Aboveground symptoms of PTSL are shown when trees are 3 to 6 years old, and include chlorosis, wilting, and sudden death of trees after bloom (Ritchie & Zehr, 1995). Although tree loss due to PTSL varies from year to year, a total loss of $ 6 million per year was estimated in South Carolina, United States (Miller, 1994). Nematicide treatments are recommended when population densities exceed 40 *C. xenoplax* per 100 cm^3 of soil in North Carolina (Ritchie & Zehr, 1995). In California, *C. xenoplax* in association with *Pseudomonas syringae* cause in peach the disease complex called Bacterial Canker Complex (BCC) (McKenry, 1989). Bacterial Canker of peach is generally associated with delay of bloom, and in severe cases collapse of trees. A distinction between BCC and PTSL is that cold injury is not associated with BCC. In addition, bacterial spot damage caused by *Xanthomonas campestris* pv. *pruni* becomes more severe on peach trees if the soil is infested with *C. xenoplax* (Shepard, Zehr, & Bridges, 1999).

2.4. Dagger Nematodes

At least seven species of *Xiphinema* have been associated with stone fruit diseases. *X. americanum*, *X. basiri*, *X. brevicollum*, *X. californicum*, *X. diversicaudatum*, *X. rivesi*, and *X. vuittenezi* (Nyczepir & Becker, 1998). These nematodes can reduce the root system and tree vigor, but they are more importantly associated with vectoring nepoviruses that cause serious fruit diseases (Brown, Halbrendt, Robbins, & Vrain, 1993). Dagger nematodes acquire the virus while feeding on virus-infected plants, which can be weed reservoirs, and transmit the virus when feeding on stone fruit trees. These nematodes have an ectoparasitic feeding habit. Eggs are deposited in the rhizosphere, and field data indicate a long life cycle and low reproduction rate, but they are able to persist in soil for several years (Jaffee, Harrison, Shaffer, & Strang, 1987).

Nepoviruses that cause diseases in *Prunus* spp. include Cherry rasp leaf virus (CRLV) that causes the disease known as cherry rasp leaf. Leaves of infected trees are distorted and fruit production is reduced. Tomato ring spot virus (TmRSV) causes Prunus stem pitting (PSP), Prune brown line (PBL), Stanley constriction and decline (SCAD) and yellow bud mosaic (YBMV) diseases. Trees affected by PSP show trunks enlarged and the bark is thick and spongy. Peach trees show frequently chlorosis, and may die if they are young. Peach rosette mosaic virus (PRMV) gives affected trees a dark green color and the internodes are shortened giving rise to rosettes of leaves (Nyczepir & Becker, 1998).

3. MANAGEMENT OF NEMATODES AND CONTROL METHODS

Most plant-parasitic nematodes attacking stone fruits have a wide geographical distribution and a wide host range in addition to the species of the genus *Prunus*. These characteristics have to be considered for their management. Management of nematodes implies the use of various tactics over an extended period. Control implies some actions within a limited period leading to a marked reduction in either the pest population or the damage caused by the pathogen (Thomason & Caswell, 1987). Every single control method has its own limitations and there is no one that can achieve more than 90% control, prolonged in time. Therefore, to achieve satisfactory rates of nematode control, nematode populations should be "managed" by means of integration of several tactics. This integration must take into account the scientific knowledge on the plant-parasitic nematode, its relationship with the host plant and its behavior in local agro-environmental conditions.

Management of nematode problems starts with prevention measures to avoid that uninfested areas where plant-parasitic nematodes are not present become infested. Soil or plant analyses for the occurrence of nematodes should be done with adequate methods and tools to determine whether and when to apply direct control measures. When a specific phytoparasitic nematode is already present in soil, efforts should be addressed first to diminish the initial population densities in soil (pre-planting measures) and then, to moderate population increases and plant damage (post-planting measures).

3.1. Prevention and Quarantine

Crop losses caused by nematodes can be avoided by preventing the introduction of specific nematodes into areas where they have not existed before. Therefore, those areas having a history of nematode problems or replant problems should be avoided in selecting sites for stone fruit production.

Prevention and quarantine methods use tactics that restrict movement of plant and soil from infested areas, and are implemented by regional or national agencies. Occasionally, eradication procedures are taken when the presence of the nematode can be delimited.

The increased national and international exchanges of woody plant materials from nurseries is nowadays the most common mean of dissemination of parasitic nematodes. They are also spread with movement of soil, farm implements, animals, wind and irrigation or runoff water. Planting certified nematode-free rootstock, when available, is extremely important as a management practice to prevent problems with orchard establishment. For container-grown stocks, nematode-free soil and planting media should be used to avoid future problems.

3.2. Pre-planting Measures

Once nematodes are detected in a field, the most effective approach is to reduce initial population densities prior to establish an orchard.

3.2.1. Fallow

Fallowing is a simple method of reducing nematode population densities by starvation. This tactic has proved useful before the establishment of a new orchard when clean fallowing is practiced for at least one or preferably, two years. Weed control is important when fallowing because both *Meloidogyne* and *Pratylenchus* species have a wide range of host plants, and weeds can act as reservoirs of infection that maintain or even build up nematode populations.

3.2.2. Crop Rotation

Rotations with cereals or grasses have been used to suppress nematode populations before establishing stone fruit orchards. Thus, coastal Bermuda grass (*Cynodon dactylon*) has been grown in sites infested with *Meloidogyne* before establishing peach orchards (Bertrand & Nyczepir, 1989). Pre-planting bahia grass (*Paspalum notatum*) or wheat (*Triticum aestivum* 'stacy') increased peach tree growth and survival from PTSL in soil infested with *C. xenoplax*, (Nyczepir & Bertrand, 2000). However, finding viable crops to be introduced in rotation programs is difficult for orchards due to the polyphagous nature of the major parasitic nematodes attacking stone fruits. Furthermore, the value of each crop in the rotation should provide a minimum profit to the farmer.

3.2.3. Site Preparation

Physical disturbance and soil manipulation can accelerate the mortality rate of nematodes due to desiccation or direct exposure to sunlight. When replanting an orchard, removal of the roots that remain in soil from the previous crop is essential, particularly in replant situations because nematodes may survive within the remnant roots or deeper in the subsoil where there is less fluctuation in temperature and moisture levels after removal of the aboveground portions.

Nematode problems can be aggravated by soil physical characteristics such as salts, chemical residues or irrigation problems, all of which limit root development. In addition, some cyanogenic compounds from the old roots are released into the soil and can have a toxic effect on the young trees (Sotomayor, González, & Castro, 2006; Tagliavini & Marangoni, 1991). Correcting physical problems prior to planting can ensure development of a good root system. Deep sub-soiling may be necessary to fracture deep surface soil layers that may restrict root penetration and will reduce nematode population densities at deeper layers. Fallowing combined with a thoroughly soil preparation can greatly reduce nematode population densities without need of additional measures, and will allow a satisfactory tree establishment and growth before nematode populations reach high levels.

3.2.4. Soil Solarization

Soil solarization by covering moistened soil with a clear plastic sheet is an attractive control tactic for warm areas. Its major advantages are the simultaneous control of pests, soil-borne pathogens, weeds and nematodes, and the improvement of soil physical properties (Stapleton, 2000). The main limitations are its dependence on climate, the long duration of the treatment (40 to 60 days), and the fact that sufficient temperature to kill pathogens rarely reach deeper than the first 10 cm of soil depth. Soil solarization used to suppress populations of *C. xenoplax* in peach orchards was effective for up to 19 months, and increased peach tree growth (Nyczepir & Kluepfel, 2007). A solarization technique, which obtains higher soil temperatures than solarization in open fields, is used as a nematicidal treatment for container nurseries (Stapleton, Prather, Mallek, Ruiz, & Elmore, 2002).

3.2.5. Biofumigation

Organic soil amendments play an important role in limiting populations of plant-parasitic nematodes. They may enhance the activity of natural enemies and improve soil fertility and structure. A grain sorghum that suppressed ring nematode under greenhouse and field conditions was tested as a pre-plant biofumigant green manure under orchard conditions in the southeastern United States, and was effective in suppressing population densities for up to 12 months (Nyczepir & Rodríguez-Kabana, 2006).

3.2.6. Steam

Heating the soil to temperatures around 70 ºC by means of aerated steam can be used for soil sterilization in facilities that have heating systems, as those used for heating greenhouses during the cold season. Steam can be useful in nursery facilities for treating potting media and propagating beds. However, the use of steam in open fields requires expensive infrastructures and careful soil preparation to allow steam penetration into soil, in addition to the high cost of fuel and water.

3.2.7. Soil Fumigation

Soil fumigation with chemical products is the most effective approach to control soil-borne pests and pathogens, including nematodes. Pre-plant fumigation must be considered when replanting stone fruit orchards since substantial damage to young trees can occur if nematodes or other soil-borne pathogens are present (McKenry, 1987; McKenry, 1999). Fumigants are volatile chemicals that are delivered into soil by injection, emulsified in liquid formulations via irrigation system or as granules, where they generate gases that have a lethal effect on many pathogens. They are applied before planting because their phytotoxicity, and application rates varies from 50 to 2500 kg/ha.

The success of fumigation depends on the application method, dosage and timing relative to temperature and moisture content of soil. McKenry (1978) identified the movement pattern of several fumigants under different soil conditions and developed a guide to decide application rates and to improve efficacy of fumigation. In addition, fumigation also kills remnant roots. Killing or neutralizing remnant roots of a previous crop prevents obligate parasites such as pathogenic nematodes from using them as a support system for their survival.

After the ban of methyl bromide in developed countries in 2005, the remaining fumigants are other halogenated hydrocarbons (1,3-D and chloropicrin), methyl isothiocyanate liberators (metam sodium, metam potassium, dazomet) iodomethane, and sodium azide. Their use, however, is also restricted or even banned in many regions. For instance, the use of 1,3-D will be banned in the European Union for most crops in 2009 (European Commission, 2007).

1,3-Dichloropropene. It was the alternative to methyl bromide where sandy to sandy loam soils were involved. Finer-textured soils must be deeply dried or the benefits of 1,3-dichloropropene will be reduced (McKenry, 1999). Combinations of 1,3-dichloropropene with chloropicrin and the choice of strip or broadcast applications should be based on proper diagnosis of pest and disease presence.

Chloropicrin. This product if applied to soils that are deeply dried can move as far as 1,3-dichloropropene but has a much faster degradation rate. It is not a great nematicide but is quite stimulatory of tree growth. This product is effective in stone fruit orchards but must be applied deeply and with care to avoid off gassing. Usually, it is applied in mixtures with 1-3 dichloropropene to broad the spectrum of pathogens to be controlled.

Metam sodium or Metam potassium. Both generate methyl isothiocyanate (MIT) within the soil. Because of their low volatility, they move poorly in soil by themselves, and they are not true fumigants, but when uniformly mixed in water can reach 99.9% of nematode control in sandy soils. Time between application and planting should be at least 12 months, otherwise trees will not grow well (McKenry, 1987). Besides, there have been failures in many occasions and thus, the inconsistency of their performance is an important drawback for their use (McKenry, 1999).

Dazomet. It is also a methyl isothiocyanate (MIT) liberator but formulated and applied as granules. Granules are spread onto the soil surface and incorporated to soil. Water is then applied to dissolve granules and liberate the MIT within the soil. Its nematicide effectiveness is often lower than other MIT liberators.

Methyl iodide. The performance of this fumigant in soil shows similar results to that of methyl bromide (Eayre, Sims, Ohr, & Mackey, 2000). If a sandy loam soil is well dried, methyl iodide have provided one year of nematode relief, the same as a strip application with other fumigants. Phytotoxicity, likely a result of residual iodide, has been a problem in plum crops.

Sodium Azide. It can be drenched almost odor-free, but its weakness is its inability to penetrate old roots if pests are within them, because any soil pests not killed can very quickly refill the biological vacuum created.

3.2.8. Chemical Control with Non-fumigants Nematicides at Pre-planting

Many non-fumigant nematicides (see below) have been tested for pre-planting control of nematodes, but in general, the protection provided has been short even at high application rates.

3.2.9. Seedling Treatments

The use of hot water dips to eliminate nematodes from plant material is effective only when the thermal tolerance of the nematode is less than that of the plant material. Immersion of *Meloidogyne* infected dormant seedlings of *P. mahaleb* and 'Lovell' peach in a hot water bath at 48 °C during 30 minutes, 49 °C for 20 minutes or 50°C from 5 to 10 minutes killed effectively nematodes inside the roots (Nyland, 1955).

The use of mycorrhizal seedlings has also been proposed as a potential measure to reduce nematode damage. *Glomus* spp. suppressed *Meloidogyne* reproduction on the peach almond hybrid 'GF-677' but it did not affect growth parameters (Calvet, Pinochet, Hernández-Dorrego, Estaun & Camprubí, 2001).

3.2.10. Resistance

The use of resistant rootstocks is the most effective, economical and environmental sound control method for parasitic nematodes in stone fruits. A rootstock is considered as resistant if greatly inhibits nematode infection and/or reproduction relative to a known susceptible standard. *Prunus* seedlings have been used traditionally as rootstocks because they are well adapted to local environmental and edaphic conditions prevalent in one region. Thus, sweet and sour cherries are generally grafted onto seedlings of *P. mahaleb* or *P. avium* (Iezzoni et al., 1991) and peaches, plums and apricots are grown on *P. persica*, *P. cerasifera*, *P. domestica*, *P. insititia*, or *P. salicina* or onto their interspecific hybrids (McKenry, 1989; Mehlenbacher, Cociu, & Hough, 1991; Scorza & Okie, 1991).

In peach, resistance to *Meloidogyne* spp. is available in *P. persica* and has been introduced into commercial rootstock since the resistance trait is easily transmitted by conventional hybridization (Cook & Evans, 1987; Claverie et al., 2004). The peach-almond hybrid rootstocks 'Nemaguard', 'Nemared', 'Flordaguard' 'Okinawa' 'Cadaman' and, 'Guardian' among others are resistant to *M. incognita* and *M. javanica* (Fernández, Pinochet, Esmenjaud, Salesses, & Felipe, 1994; Marull et al., 1994; McKenry, 1999; Simeone & Di Vito, 1992) although *Prunus* species express different ranges and levels of nematode resistance depending on the source of resistance (Esmenjaud et al., 1997).

In California, 85% of the almond, peach, plum, and nectarine orchards are replanted on 'Nemaguard' (*P. persica* × *P. davidiana*) since it provides field resistance to all races of root-knot nematodes (McKenry, 1987), although, as with every other rootstock, it has limitations such as its susceptibility to *P. penetrans*, *P. vulnus* or *C. xenoplax*, and to bacterial canker and peach tree short life (Zehr, Miller, & Smith, 1976), it does not grow well in alkaline soils or in fine-textured soils where plum rootstocks are better adapted. In plums, there is root-knot nematode resistance in 'Marianna' (*P. cerasifera* × *P. munsoniana*) and 'Myrobalan' (*P. cerasifera*). The so called Ma genes in some 'Myrobalan' clones completely suppress root-knot nematode multiplication and confer a complete spectrum, high level and stable resistance to *M. incognita*, *M. arenaria* and *M. javanica* (Esmenjaud, Minot, Voisin, Bonnet, & Salesses, 1996; Lecouls et al., 1999).

The resistance genes on peach and plum are independent each other and thus, they can be identified, marked, and pyramided into new interspecific hybrid rootstocks based on these species (Claverie et al., 2004). Resistance to root-knot nematodes has been identified in the ornamental peach 'Jeseitou' (Yamamoto & Hayashi, 2002), and apparently the resistance gene of 'Jeseitou' is similar to that of 'Nemared' (Claverie et al., 2004). Simple and complex interspecific hybrids involving one or more sources of *Prunus* material bearing agronomic adaptation are now available. In addition, root-knot nematode resistant rootstocks can provide protection against other pathogens since wounds caused by the nematode may enhance root penetration by other pathogenic agents. This is the case of some resistant myrobalan clons that provide protection against *Agrobacterium tumefaciens* (Rubio-Cabetas, Minot, Voisin, & Esmenjaud, 2001). The rootstock 'Guardian', resistant to *M. incognita* and *M. javanica*, also provides protection against PTSL (Nyczepir, Beckman, & Reighard, 2006).

Resistance and tolerance to *P. penetrans* has been found on rootstocks 'Bailey', 'BY520-8', 'Guardian', 'Rutgers red' and others (Layne, 1987; McFaden-Smith,

Miles, Potter, & Monet, 1998; Potter et al., 1984) but the search for resistance to *P. vulnus* has only detected potential sources of resistance on a few wild plums hybrids (Pinochet, Anglés, Dalmau, Fernández, & Felipe, 1996, Pinochet, Fernández, Calvet, Hernández-Dorrego, & Felipe, 2000). *Prunus* rootstocks from around the world have been screened to determine resistance to both major endoparasites, *M. incognita* and *P. vulnus*, but so far only the plum hybrid 'Bruce' has shown resistance to both nematode species (Dirlewanger et al., 2002; Pinochet et al., 1996). Many of these rootstocks are still being screened against *C. xenoplax* (McKenry, Kaku, & Buzo, 2006). The rootstocks 'Lovell' and 'Guardian' (susceptible and resistant to root-knot nematodes, respectively) have been recommended for their tolerance to *C. xenoplax* and longer survival in PTSL sites (Okie et al., 1994; Wilkins et al., 2002).

Numerous rootstocks are available for *Prunus* but the number of candidates resistant or tolerant to dominant pathogens or pests in one region is limited to a few. Significant efforts have been done in the last two decades to incorporate nematode resistance into new *Prunus* rootstocks. Yet, there is still a demand for new resistant rootstock, and they should combine nematode resistance with desirable agronomic traits including scion compatibility, reduced tree-size, and tolerance to abiotic stresses such as drought or water logging.

In Europe, the *Prunus* rootstock programme, involving three countries (France, Spain, and Italy) has tried to characterize interspecific hybrids with high level of root-knot nematode resistance, easiness for rooting (from 'Myrobalan' plum), adaptation to chlorosis and drought (from almond), and good grafting compatibility for peach (from peach) (Dirlewanger et al., 2002).

It is necessary to point out that resistance frequently affects only to nematode reproduction and not to nematode penetration, therefore, young resistant trees can even suffer damage during the first year after planting. In addition, high soil temperatures modified the resistance response to *M. incognita* on *Prunus* spp. resulting in increased root galling and nematode reproduction (Fernández, Pinochet, & Felipe, 1993).

3.3. Post Planting Measures

Once trees are infected, no full curative methods of nematode control are available. Therefore, post planting measures aims at reducing plant damage or populations increases.

3.3.1. Chemical Control

Two main groups of non-fumigant nematicides, carbamates (oxamyl, carbofuran) and organophosphates (fenamiphos, etoprophos, cadusaphos) are available for use against nematodes after orchard establishment. These nematicides, were firstly available as granular formulations, however, currently emulsionable concentrates are also available for application through irrigation systems. These products are lethal to nematodes and their primary action results of direct contact, but once the product reaches more than 8 cm into soil, their action is mostly due to sublethal effects in nematode behavior, as stimulation of egg hatching, or disorientation. However, repeated applications are necessary to maintain reduced population densities and consistent yield increases. Multiple treatments via drip irrigation can reduce nematode populations as much as 90% but they will not eradicate them (Ferris, McKenry, Jaffee, Anderson, & Jurma, 2004; McKenry, Buzo, & Kaku, 1998).

The efficacy of these nematicides is influenced by several factors as product solubility, soil texture, the amount of organic matter in the soil and microbial degradation that can occur in soils when microbial populations capable of metabolizing the nematicides increase with the repeated use. In addition, the use of some nematicides is restricted in some regions because contamination of ground water or they are not registered for use in stone fruit orchards.

3.3.2. Biological Control

Microbial antagonists can regulate nematode populations through direct parasitism or predation or indirectly via release of toxic metabolites. The nematode antagonists most widely studied include the bacterial obligate parasite *Pasteuria penetrans*, the nematode trapping fungi *Arthrobotrys oligospora*, *Dactylellina dactyloides* and *D. ellipsospora* (syn. *Monacrosporium ellipsosporum)*, *Hirsutella rhossiliensis*, and the fungal egg parasites *Pochonia chlamydosporia* and *Paecilomyces lilacinus*. These antagonists have shown potential as biological control agents under certain conditions but currently, only *P. lilacinus* is commercially available in several countries such as Australia, Colombia, Germany or South Africa. Information on fungal antagonists in stone fruit orchards is limited to few reports (Jaffee & Zehr, 1982; Stirling, McKenry, & Mankau, 1979).

One strain of *Pseudomonas* 'BG33R' isolated from suppressive PTSL soils showed capability of reducing *C. xenoplax* multiplication in vivo and egg hatch in vitro. In field experiments, *Pseudomonas* 'BG33R' maintained *C. xenoplax* populations below the economic threshold for nematicide treatment for up to 18 months (Kluepfel, Nyczepir, Lawrence, Wechter, & Leverentz, 2002).

Biological control may be helpful in conditions where chemical control is not available or affordable, and in orchards under organic farming. It may be more successful at moderate rather than at high pest pressure due to the inverse relationship between nematode densities and level of control achieved with the biological control agent (Bourne & Kerry, 1999).

However, there are still several problems for the commercial use of biological control, particularly in perennial crops. Plant parasitic-nematodes associated with perennial crops live deep in soil, whereas microbial antagonists tend to inhabit the shallowest 15 cm where biological activity is greatest and many of these agents do not have yet a way of mass production, and thus they are not commercially available. Biological control remains at experimental stage, and additional data are needed to increase the knowledge and overcome difficulties for use at present.

3.3.3. Cultural Methods

Many properly established orchards generate high yields in presence of nematodes, but often conditions exerting stress on the plant result in suboptimal production, which in turn may eventually produce losses.

Good agronomic and cultural practices may help to compensate damage caused by nematodes and consequently maintain production at acceptable levels. For instance, controlling weeds will reduce competition for water and nutrients. Mulching can help to reduce water loss as it reduces evaporation and moderates extreme daily soil temperatures. Consequently, the crop environment is modified, promoting tree vigor and increased yield. However, nematode densities may increase on mulched trees due to more favorable conditions for root growth and nematode reproduction. Several ground covers appear promising either as pre-plant or post-plant management strategies. Thus, nimblewill (*Muhlenbergia schreberi*) planted around peach trees suppressed populations of *C. xenoplax*, but did reproduce *M. javanica* or *M. incognita* (Meyer, Zehr, Meager, & Salvo, 1992).

4. INTEGRATED MANAGEMENT

Continued removal and/or restrictions of chemical nematicides from the market is leaving growers with fewer nematode control options and thus, stone fruit cropping is moving towards a more integrated approach for nematode management, with all available control options being used in a compatible manner to reduce nematode populations to levels under damage thresholds.

With the development of the concept of IPM, monitoring for nematode infestations has become important as ever in modern agriculture. Samples should be collected to determine nematode infestation levels in soil and their distribution within the field. Then, this information is used to determine whether action needs to be taken against nematodes. Besides, nematode management requires a thorough understanding of the growth of the host plant, biology, ecology and epidemiology of the nematode, and the influence of the environment on the nematode-plant interaction in a given region (Verdejo-Lucas, 1999).

Briefly, the procedure for a nematode IPM program should be as follows:

1. *Sample collection.* Soil or root samples collected from the field.
2. *Nematode extraction.* Nematodes extracted from soil or roots using proper methods for the suspected nematodes and type of sample.
3. *Nematode identification and quantification.* Nematodes are identified to species level with aid of microscopes, molecular methods and pertinent literature, and they are quantified in aqueous suspensions to estimate population densities.
4. *Estimation of nematode damage.* Nematode densities are compared with damage threshold experimentally determined for the region when available or those in the literature for a particular nematode and crop.

5. *Management decision.* Considering the probability to cause damage according to the population densities on a given crop, a decision is made on whether nematodes should be controlled, and in case, which methods of control should be adopted.

There are very few examples on successful nematode management with IPM in stone fruit orchards, primarily because the above points are difficult to attain with soil as the media. Overall, the focus has been on "pest avoidance" through quarantines, assurances of clean nursery stock, pre-plant soil fumigation, resistant rootstocks, and more recently, suppressive soils. In summary, the approach has been prevention rather than therapy.

IPM strategies for replanting *Prunus* orchards have been proposed by Ritchie and Zehr (1995) and McKenry, Buzo, and Kaku (2006) as follow:

1. Diagnosis for nematode presence or absence is one of the first considerations when deciding the steps for replanting a specific orchard. Determine if any major nematode pests (i.e. *Meloidogyne* sp., *P. vulnus* and/or *C. xenoplax*) are present in the orchard site.
2. Remove trunks and kill root systems with an herbicide.
3. Wait 18 months before replanting. During the wait, correct soil physical, chemical and biological problems, using control methods to reduce nematode populations.
4. Replant on a rootstock with a parentage unrelated to the previous one and if possible resistant or tolerant to nematodes present in the orchard soil.
5. Fertilize at planting with addition of macro and micronutrients.

5. FUTURE PROSPECTS

In the near future, growers will have to adopt IPM systems due to restrictions in the use of broad-spectrum soil fumigants. Resistant rootstocks reduce nematode reproduction considerably, but combination of two or more strategies will be necessary to alleviate nematode problems, since there is seldom a single effective method.

Additional research is needed to find means for interrupting nematode's life cycles, enhancing microbial activity in the rhizosphere to promote plant growth or increase its tolerance to the nematode. Any chemical, microbial, cultural or management approach that is developed must be within the capability of the grower and should meet the necessary environmental and economic requirements. The grower will benefit if these treatments are reliable, practicable and economically justified.

REFERENCES

Bertrand, P. F., & Nyczepir, A. P. (1989). Nematodes. In: S. C. Meyers (Ed.), *Peach production handbook* (pp.146-151). Athens: University of Georgia Cooperative Extension Service.

Bird, G. W. (1981). Integrated nematode management for plant protection. In B. M. Zuckerman et al. (Eds.), *Plant parasitic nematodes* Vol. III (pp. 355-375). New York: Academic Press.

Bortiri, E., Oh, S., Jiang, J., Baggett, S., Granger, A., Weeks, C., Buckingham, M., Potter, D., & Parfitt, D. E. (2001). Phylogeny and systematics of *Prunus* (Rosaceae) as determined by sequence analysis of ITS and the chloroplast trnL-trnF spacer DNA. *Systematic Botany*, 26, 797-807.

Bourne, J. M., & Herry, B. R. (1999). Effect of the host plant on the efficacy of *Verticillium chlamydosporium* as a biological control agent of root-knot nematodes at different nematode densities and fungal application rates. *Soil Biology and Biochemistry*, 31, 75-84.

Brown, D. J. F., Halbrendt, R. T., Robbins, R. T., & Vrain, T. C. (1993). Transmission of nepoviruses by *Xiphinema americanum* group nematodes. *Journal of Nematology*, 25, 349-354.

Calvet, C., Pinochet, J., Hernández-Dorrego, A., Estaun, V., & Camprubí, A. (2001). Field microplot performance of the peach-almond hybrid GF-677 after inoculation with arbuscular mycorrhizal fungi in a replant soil infested with root-knot nematodes. *Mycorrhiza*, 10, 295-300.

Claverie, M., Bosselut, N., Lecouls, A. C., Voisin, R., Lafargue, B., Poizat, C., Kleinhentz, M., Laigret, F., Dirlewanger, E., & Esmenjaud, D. (2004). Location of independent root-knot nematode resistance genes in plum and peach. *Theoretical and Applied Genetics*, 108, 765-773.

Cook, R., & Evans, K. (1987). Resistance and tolerance. In R. H. Brown & B. Kerry (Eds.), *Principles and practice of nematode control* (pp. 179-231). Australia: Academic Press.

Dirlewanger, E., Salesses, G., Bonnet, A., Kleinhentz, M., Arùs, P., Esmenjaud, D., et al. (2002). Breeding for *Prunus* rootstocks cumulating resistance to root-knot nematodes and favorable traits under Meditarrenean environments: a European project. *Acta Horticulturae*, 592, 61-67.

Eayre, C. G., Sims, J. J., Ohr, H. D., & Mackey, B. (2000). Evaluation of methyl iodide for control of peach replant disorder. *Plant Disease*, 84, 1177-1179.

Esmenjaud, D., Minot, J. C., Voisin, R., Bonnet, A., & Salesses, G. (1996). Inheritance of resistance to the root-knot nematode *Meloidogyne arenaria* in Myrobalan plum. *Theoretical and Applied Genetics*, 92, 873-879.

Esmenjaud, D., Minot, J. C., Voisin, R., Pinochet, J., Simard, M.H. & Salesses, G. (1997). Differential response to root-knot nematodes in *Prunus* species and correlative genetic implications. *Journal of Nematology*, 29, 370-380.

European Commission. (2007). Commission Decision of 20 September 2007 concerning the non-inclusion of 1,3-dichloropropene in Annex I to Council Directive 91/414/EEC and the withdrawal of authorisations for plant protection products containing that substance (notified under document number C(2007) 4281). *Official Journal of the European Union*, L249 (50), 11-13.

Fernández, C., Pinochet, J., Esmenjaud, D., Salesses, G., & Felipe, A. (1994). Resistance among new *Prunus* rootstocks and selections to root-knot nematodes in Spain and France. *HortScience*, 29, 1064-1067.

Fernández, C., Pinochet, J., & Felipe, A. (1993). Influence of temperature on the expression of resistance in six *Prunus* rootstocks infected with *Meloidogyne incognita*. *Nematropica*, 23, 195-202.

Ferris, H., McKenry, M. V., Jaffee, B. A., Anderson, C. E., & Jurma, A. (2004). Population characteristics and dosage trajectory analysis for *Mesocriconema xenoplax* in California *Prunus* orchards. *Journal of Nematology*, 36, 505-516.

Food and Agriculture Organization of the United Nations, FAO. (2004). *FAO statistical yearbook*. Rome, Italy: FAO.

Handoo, Z. A., Nyczepir, A. P., Esmenjaud, D., Van der Beek, J. G., Castagnone-Sereno, P., Carta, L. K., Skantar, A. M., & Higgins, J. A. (2004). Morphological, molecular, and differential-host characterization of *Meloidogyne floridensis* n. sp. (Nematoda: Meloidogynidae), a root-knot nematode parasitizing peach in Florida. *Journal of Nematology*, 36, 20–35.

Hirschmann, H. (1986). *Meloidogyne hispanica* n. sp. (Nematoda: Meloidogynidae), the 'Seville' root-knot nematode. *Journal of Nematology*, 18, 520-532.

Iezzoni, A., Schmidt, H., & Albertini, A. (1991). Cherries (*Prunus*). In J. N. Moore & J. R. Ballington (Eds.), *Genetic resources of temperate fruit and nut crops* (pp. 109-174). Wagenigen, The Netherlands: International Society of Horticultural Sciences.

Jaffee, B. A., Harrison, M. B., Shaffer, R. L., & Strang, M. B. (1987). Seasonal population fluctuation of *Xiphinema americanum* and *X. rivesi* in New York and Pennsylvania orchards. *Journal of Nematology*, 19, 369-378.

Jaffee, B. A., & Zehr, E. I. (1982). Parasitism of the nematode *Criconemella xenoplax* by the fungus *Hirsutella rhossiliensis*. *Phytopathology*, 72, 1378-1381.

Kluepfel, D. A., Nyczepir, A. P., Lawrence, J. E., Wechter, W. P., & Leverentz, B. (2002). Biological control of the phytoparasitic nematode *Mesocriconema xenoplax* on peach trees. *Journal of Nematology*, 34, 120-123.

Layne, R. E. (1987). Peach rootstock. In R. C. Rom & R. F. Carlson (Eds.), *Rootstocks of fruit crops* (pp. 385-416). New York, USA: John Wiley & Sons.

Lecouls, A. C., Rubio-Cabetas, M. J., Minot, J. C., Voisin, R., Bonnet, A., Salesses, G., Dirlewanger, E. & Esmenjaud, D. (1999). RAPD and SCAR markers linked to the Ma1 root-knot nematode resistance gene in Myrobalan plum (*Prunus cerasifera* Her.). *Theoretical and Applied Genetics*, 99, 328-335.

Lownsbery, B. F., English, H., Moody, E. H., & Shick, F. J. (1973). *Criconemoides xenoplax* experimentally associated with a disease of peach. *Phytopathology*, 63, 994-997.

Lownsbery, B. F., Moody, E. H., & Braun, A. L. (1974). Plant parasitic nematodes in California prune orchards. *Plant Disease Reporter*, 58, 633-636.

Marull, J., Pinochet, J., Felipe, A., & Cenis J. L. (1994). Resistance verification in *Prunus* selections to a mixture of thirteen *Meloidogyne* isolates and resistance mechanisms of a peach-almond hybrid to *M. javanica*. *Fundamental & Applied Nematology*, 17, 85-92.

McFaden-Smith, W., Miles, N. W., Potter, J. W., & Monet, R. (1998). Greenhouse evaluation of *Prunus* rootstocks for resistance and tolerance to the root lesion nematode (*Pratylenchus penetrans*). *Acta Horticulturae*, 465, 723-729.

McKenry, M. V. (1978). Selection of preplant fumigation. *California Agriculture*, 32, 15-16.

McKenry, M. V. (1987). Control strategies in high value crops. In R. H. Brown & B. R. Kerry (Eds.), *Principles and practice of nematode control in crops* (pp. 329-349). Australia: Academic Press.

McKenry, M. V. (1989). Nematodes. In J. H. LaRue & R. S. Johnson (Eds.), *Peaches, plums and nectarines. Growing and handling for the fresh market. Publication 3331* (pp. 139-147). Oakland, CA: University of California. Division of Agriculture and Natural Resources.

McKenry, M. V. (1999). The replant problem and its management. Fresno, CA, USA. Catalina Publishing.

McKenry, M. V. (2004). Three nematode genera and the damage they cause for plum producers. *Journal of Nematology*, 36, 333. (Abstract).

McKenry, M. V., Buzo, T., & Kaku, S. (1998). Efficacy of five post-plant nematicides via drip irrigation to first-year *Prunus* spp. *Journal of Nematology*, 30, 505. (Abstract).

McKenry, M. V., Buzo, T., & Kaku, S. (2006). Replanting stone fruit orchards without soil fumigation. *International Conference on Methyl Bromide Alternatives and Emission Reductions*. 6, 2006. Orlando, FL. 28.1-28.2.

McKenry, M. V., Kaku, S., & Buzo T. (2006). Development of a nematode-rootstock profile for *Prunus* spp. *Journal of Nematology*, 38, 282. (Abstract).

Mehlenbacher, S. A., Cociu, V., & Hough, L. F. (1991). Apricot (*Prunus*). In J. N. Moore & J. R. Ballington (Eds.), *Genetic resources of temperate fruit and nut crops* (pp. 63-107). Wagenigen, The Netherlands: International Society of Horticultural Sciences.

Meyer, J. R., Zehr, E. I., Meager, R. L. Jr., & Salvo, S. K. (1992). Survival and growth of peach trees and pest populations in orchard plots managed with experimental ground covers. *Agriculture, Ecosystems and Environment*, 41, 353-363.

Miller, R. W. (1994). Estimated peach tree losses 1980 to 1992 in South Carolina – causes and economic impact. In: A. P. Nyczepir et al. (Eds.), *Proceedings of the 6th Stone Fruit Decline Workshop* (pp. 121-127). Fort Valley, GA: USDA-ARS.

Nicotina, M. (1990). Nematode living in the rhizosphere of peaches in decline in Lazio. *Informatore Agrario*, 46, 63-64.

Nyczepir, A. P. (1990). Influence of *Criconemella xenoplax* and pruning time on short life of peach trees. *Journal of Nematology*, 22, 97-100.

Nyczepir, A. P., & Becker, J. O. (1998) Fruit and Citrus Trees. In K. R. Barker et. al. (Eds.), *Plant Nematode interactions, Agronomy Monograph 36* (pp. 637-684). Madison, WI: America Society of Agronomy - Crop Science Society of America - Soil Science Society of America.

Nyczepir, A. P., Beckman, T. G., & Reighard, G. L. (2006). Field evaluation of 'Guardian'™ peach rootstock to different root knot nematode species. *Acta Horticulturae*, 713, 303-309.

Nyczepir, A. P., & Bertrand, P. F. (2000). Preplanting bahia grass or wheat compared for controlling *Mesocriconema xenoplax* and short life in a young peach orchard. *Plant Disease*, 84, 789-793.

Nyczepir, A. P., Bertrand, P. F., Miller, R. W., & Motsinger, R. W. (1985). Incidence of *Criconemella* spp. and peach orchard history in short-life and non-short-life sites in Georgia and South Carolina. *Plant Disease*, 69, 874-877.

Nyczepir, A. P., & Kluepfel, D. A. (2007). Impact of solarization, rootstock and *Pseudomonas synxantha* on *Criconemoides xenoplax* populations and tree growth in peach tree short life site. *Journal of Nematology*, 39, 75. (Abstract).

Nyczepir, A., & Rodriguez-Kabana, R. (2006). Effectiveness of biofumigation for ring nematode control in a young peach orchard. In *Proceedings of the 1st International Symposium on Biofumigation: A possible alternative to methyl bromide*, 1, 64-65.

Nyland, G. (1955). Killing root knot nematodes in some stone fruit tree rootstocks. *Plant Disease Reporter*, 39, 573-575.

Okie, W. R. Reighard, G. L., Beckman, T. G., Nyczepir, A. P., Reilly, C. C., Zehr, E. I., Newall, W. C., & Cain, D. W. (1994). Field-screening *Prunus* for longevity in the southeastern United States. *HortScience*, 29, 673-677.

Pinochet, J., Anglés, M., Dalmau, E., Fernández, C., & Felipe, A. (1996). *Prunus* rootstock evaluation to root-knot and lesion nematodes in Spain. *Journal of Nematology*, 28(S), 616-623.

Pinochet, J., Fernández, C., Calvet, C., Hernández-Dorrego, A., & Felipe, A. (2000). Selection against *Pratylenchus vulnus* populations attacking rootstocks. *HortScience*, 35, 1333-1337.

Potter, J. W., Dirks, V. A., Jonhson, P. W., Olthof, T. H. A., Layne, R. E. C., & McDonell, M. M. (1984). Response of peach seedlings to infection by the root-lesion nematode *Pratylenchus penetrans* under controlled conditions. *Journal of Nematology*, 16, 317-322.

Rammah, A., & Hirschmann, H. (1990). *Meloidogyne morocciensis* n. sp. (Meloidogyninae), a root-knot nematode from Morocco. *Journal of Nematology*, 22, 279-291.

Ritchie, D. F., & Zehr, E. I. (1995). Peach tree short life. In J. M. Ogawa et al. (Eds.), *Compendium of stone fruit diseases* (pp. 45-46). St. Paul, MN: APS Press.

Rubio-Cabetas, M. J., Minot J. C., Voisin, R., & Esmenjaud, D. (2001). Interaction of root-knot nematodes (RKN) and the bacterium *Agrobacterium tumefaciens* in roots of *Prunus cerasifera*: evidence of the protective effect of the Ma RKN resistance genes against expression of crown gall symptoms. *European Journal of Plant Pathology*, 107, 433-441.

Scorza, R., & Okie, W. R. (1991). Peaches (*Prunus*). In J. N. Moore & J. R. Ballington (Eds.), *Genetic resources of temperate fruit and nut crops* (pp. 175-231). Wagenigen, The Netherlands: International Society of Horticultural Sciences.

Sharpe, R. P., Pusey, P. L., Nyczepir, A. P., & Florkowski, W. J. (1993). Yield and economics of intervention with peach tree short life disease. *Journal of Production Agriculture*, 6, 241-244.

Shepard, D. P., Zehr, E. I. & Bridges, W. C. (1999). Increased susceptibility to bacterial spot of peach trees growing in soil infested with *Criconemella xenoplax*. *Plant Disease*, 83, 961-963.

Simeone, A. M., & Di Vito, M. (1992). Reactions to nematodes of selections of peach rootstocks. *Acta Horticulturae*, 315, 197-202.

Sotomayor, C., Gonzalez, E. & Castro, J. (2006). Effect of amygdalin on growth of Nemaguard peach seedlings. *Acta Horticulturae*, 721, 111-116.

Stapleton, J. J. (2000). Soil solarization in various agricultural production systems. *Crop protection*, 19, 837-841.

Stapleton, J. J, Prather, T. S,, Mallek, S. B,, Ruiz, T. S, & Elmore, C. L. (2002). High temperature solarization for production of weed-free container soils and potting mixes. *HortTechnology*, 12, 697-700.

Stirling, G. R., McKenry, M. V., & Mankau, R. (1979). Biological control of root-knot nematode (*Meloidogyne* spp.) on peach. *Phytopathology*, 69, 806-809.

Tagliavini, M., & Marangoni, B. (1991). Growth of peach as affected by decomposition of own root residues in soil. *Plant and Soil*, 145, 253-260.

Thomason, I. J., & Caswell, E. P. (1987). Principles of nematode control. In R. H. Brown & B. R. Kerry (Eds.), *Principles and practice of nematode control in crops* (pp. 87-130). Australia: Academic Press.

Verdejo-Lucas, S. (1999). Nematodes. In R. Albajes et al. (Eds.), *Integrated pest and disease management in greenhouse crops* (pp. 61-68). The Netherlands: Kluwer Academic Publishers.

Wilkins, B. S., Ebel, R. C., Dozier, W. A., Pitts, J., Eakes, D. J., Himelrick, D. G., Beckman, T., & Nyczepir, A. P. (2002). Field performance of Guardian™ peach rootstock selections. *HortScience*, 37, 1049-1052.

Yamamoto, T., & Hayashi, T. (2002). New root-knot nematode resistance genes and their STS markers in peach. *Scientia Horticulturae*, 96, 81-90.

Zehr, E. I., Miller, R. W., & Smith, F. H. (1976). Soil fumigation and peach rootstocks for protection against peach tree short life. *Phytopathology*, 66, 689-694.

DANIEL ESMENJAUD[1] AND A. BOUQUET [2]

SELECTION AND APPLICATION OF RESISTANT GERMPLASM FOR GRAPEVINE NEMATODES MANAGEMENT

[1]INRA, UMR IPMSV, Equipe de nématologie, 06903 Sophia-Antipolis Cedex, France

[2]INRA, UMR-DIAPC, Equipe Vigne, 34060 Montpellier Cedex 1, France

Abstract. The status of selection studies on resistant grapevine rootstocks for management of root-knot nematodes *Meloidogyne* spp. and of *Xiphinema index*, vector of *Grapevine fanleaf virus*, is reviewed. The biology, ecology, symptomatology and control of root-knot nematodes are revised, for application in the selection and breeding of resistant rootstocks. Data on resistant *Vitis* and *Muscadinia* material, as well as on the genetics, mechanisms and durability of resistance are also provided. The pathogenicity of *X. index* and of other grapevine nematode vectors is then summarized, as concerns the biology, transmission and classical nematode control. Breeding efforts for selection of grape resistant rootstocks are then reported, focusing on resistance features of *M. rotundifolia* and *V. vinifera* × *M. rotundifolia* F_1 hybrids, obtained in California and France, and new prospects are foreseen. Resistance to other nematodes and rootstock applications for control of multiple nematode pests are also discussed.

1. INTRODUCTION

Grapevine is mainly grown under temperate and Mediterranean climates. This crop has been imported and cultivated into diverse countries throughout the world and is thus parasitized by both native and introduced nematodes (Brown et al. 1993; Nicol et al. 1999). At the world scale the major nematode pests of table and wine grapes are the root-knot nematodes (RKN) of the genus *Meloidogyne* (endoparasitic and sedentary) and the ectoparasitic virus vector species of the genus *Xiphinema* (dagger nematodes).

RKN can reduce grapevine yields by as much as 20 % particularly in the USA (Anwar & McKenry, 2000) whereas the dagger nematode *X. index* is a real concern in old grape-growing areas as the vector of *Grapevine fanleaf virus* (GFLV), the first world virus disease of grapevine (Martelli & Savino, 1991; Andret-Link et al. 2004a). Other grape damaging species are the root-lesion nematode *Pratylenchus vulnus* (endoparasitic, migratory), the ring nematode *Criconemoides xenoplax* (ectoparasitic) and the citrus nematode *Tylenchulus semipenetrans* (McKenry et al.

A. Ciancio & K. G. Mukerji (eds.), *Integrated Management of Fruit Crops and Forest Nematodes*, 195-214.

2001a; 2001b). All these species were traditionally controlled with chemicals but the ban of nematicides highlights the need for a more accurate development and use of cultural (i.e. rotation), physical (i.e. solarisation) and genetic (i.e. plant resistance) alternatives.

The deployment of resistant plant material, although it appears as one of the most promising alternatives, is rendered more difficult by the perennial status of the crop. Indeed, breeding resistant grapes has to face two major constraints: the long generation intervals that economically impose the limitation of the breeding strategies to a few plant cycles and the need for such strategies to reliably guarantee the resistance durability. Conversely perennial crops are generally grafted and resistance breeding can be done on rootstocks material which may be genetically independent from the scion whether a high grafting compatibility is warranted.

In this chapter we will illustrate the current use of the resistance strategy to control the diverse nematodes affecting grapes and the work in progress for a wider use of those possibilities through resistance breeding in the future. Our illustration will be mainly focused on the *Meloidogyne* spp. and the dagger nematode *X. index*. We will conclude our chapter by data that take into account altogether the diverse nematodes.

2. THE ROOT-KNOT NEMATODES *MELOIDOGYNE* SPP.

2.1. Biology, Ecology, Symptoms and Control

Root-knot nematodes are extremely polyphagous pests developing on most crops and weeds and are mainly localized in sandy soils. Among the over 50 species described, only four can be considered as affecting vines throughout the world: *M. arenaria*, *M. incognita*, *M. javanica* and *M. hapla*. The first three species reproduce by mitotic parthenogenesis and are mainly located under Mediterranean and hot climates whereas *M. hapla* (meiotic parthenogenesis or amphimixis) has a more temperate distribution.

Meloidogyne spp. are sedentary endoparasitic nematodes with the second-stage juvenile as the sole motile stage. These juveniles hatch from the egg mass grouping up-to 1500 eggs and migrate through the soil to find a host plant root. They penetrate the root-tip and move intercellularly to the vascular cylinder to induce a feeding site composed of specialized cells designated as 'giant cells'. The juveniles develop into third- and fourth-stage juveniles, and to female adults, all being swollen fixed stages imbedded into a characteristic gall. Filiform males leaving the root and moving freely into the soil generally occur under unfavourable developmental conditions. Under optimal climatic conditions, the complete cycle is only 4-5 weeks long, what produces several generations per season.

RKN infection results in a reduction of vigour and a loss of yield and also causes an increased susceptibility to other biotic (phylloxera or crown gall disease) or abiotic (drought) stresses. RKN identification, a problem of major importance (Dalmasso, 1973), had been classically based on perineal patterns and on the differential host range test (Hartman & Sasser 1985; Jepson, 1987). Because such information have appeared clearly imperfect and can have lead to mis-identification (Stanton & O'Donnel, 1998), they have been progressively replaced by enzymatic (Janati et al., 1982) and DNA (Zijlstra et al. 1997; 2000) markers. In the data reported below for the behaviour of *Vitis* plant material to RKN, this putative mis-identification has to be considered.

2.2. Selection and Breeding of Resistant Rootstocks

2.2.1. Evidence of Resistant Vitis *and* Muscadinia *Material*

Evaluation of the response of *Vitis* selections and *Vitis* interspecific hybrids has been performed by Snyder in California as early as in 1936. These tests included phylloxera resistant rootstocks and identified mainly the American *Vitis* species *V. champinii*, *V. longii* (syn. *V. solonis*), *V. doaniana* and *V. cinerea* as potential sources of resistance to RKN. Those sources were confirmed by Lider (1954) with root-knot nematode populations identified as *M. incognita* var. *acrita* Chitwood. Lider also observed the complete resistance of *Muscadinia rotundifolia*. In North Carolina, Nesbitt (1974) noted the resistance of *M. rotundifolia* to the three species *M. arenaria*, *M. incognita* and *M. javanica* and this was confirmed by Bloodworth et al. (1980).

In France, Boubals (1954) and Bouquet and Dalmasso (1976) respectively evaluated resistance to *M. arenaria* and a population of the complex *M. incognita-arenaria* in diverse rootstock accessions of American *Vitis* cultivars previously selected in France for resistance to phylloxera and observed a similar relative ranking of those stocks although the second population was more aggressive than the first one. Resistance to *M. incognita* was then reported by Walker et al. (1994c) in selections of *V. aestivalis*, *V. champinii*, *V. cinerea*, *V. rufotomentosa* and *V. rupestris* and in all *M. rotundifolia* sources.

Attacks by the meiotic species *M. hapla*, thus expected to have a higher genetic variability than the mitotic species *M. arenaria*, *M. incognita* and *M. javanica*, have been reported in France. Using nine populations of this RKN, Dalmasso and Cuany (1976) have shown poor galling in *V. riparia* cv. Gloire de Montpellier, and 5 BB (*V. riparia* × *V. berlandieri*) and found 41 B (*V. vinifera* × *V. berlandieri*) extremely susceptible. However, Bouquet et al. (1982) have shown great differences of galling between the two rootstocks 41 B and Fercal, two limestone-resistant rootstocks with *V. vinifera* and *V. berlandieri* in their parentage.

2.2.2. Breeding for Resistance

In the USA, severe damage and nematicide removal have driven plant breeders in close collaboration with nematologists to focus their efforts on the development of new rootstocks with a wide-spectrum and durable resistance to RKN nematodes together with desirable agronomical traits. The first generation of resistant rootstocks specially bred for RKN resistance relied on *V. champinii* as this species had been recognized as carrying a high level of RKN resistance. The rootstocks

Ramsey (= Salt Creek) and Dog Ridge, considered as belonging to this latter *Vitis* species, were selected and expressed a wide resistance to RKN (Kasimatis and Lider 1967) and particularly to *M. incognita* (Loubser and Meyer 1987).

Rootstocks Freedom and Harmony, both hybrids of *V. champinii* with 1613C (a complex hybrid with *V. longii* and *V. vinifera* in its parentage) also resulted resistant in many RKN infected locations. Nevertheless in California, cases of severe local attacks by *M. incognita* were reported as early as in 1954 on *V. champinii* and *V. longii*, by Lider (1954, 1959, 1960) and then by Cain et al. (1984) on Freedom and Harmony, by McKenry (1992) and McKenry and Kretsch (1995) on Ramsey, Harmony, Freedom and 1613C. The latter authors also noticed severe damage by a virulent population of *M. arenaria*. This *M. arenaria* population and other resistance-breaking biotypes have then focused resistance breeding efforts of several teams: a number of new Californian selections were or are currently released, of which some are the RS hybrids (RS-2, RS-3 and RS-9) of Ramsey and Schwarzmann (*V. riparia* x *V. rupestris*) (Anwar et al. 2002) and the accessions 6-19B, 10-17A and 10-23B that are complex hybrids involving *V. candicans*, *V. riparia* and *V. rupestris* (Anwar and McKenry 2000; Anwar and McKenry 2002a and b). Because of its resistance to the dagger nematode *X. index*, the *M. rotundifolia* source has been used to create the *X. index*-resistant hybrid rootstock VR 039-16 (*V. vinifera* x *M. rotundifolia*) (Walker et al. 1991) but surprisingly this cultivar resulted susceptible to most tested RKN (McKenry et al. 2001b; McKenry and Anwar 2006).

Thus within the genus *Vitis*, the accessions from a same *Vitis* species may express different resistance responses to a given RKN population. This illustrates the two levels of intraspecific variability of the interaction, at the levels of the *Vitis* species and of the *Meloidogyne* species, respectively. By contrast the species *M. rotundifolia* is probably resistant to all RKN species.

Experience from rootstock selection and breeding reported previously confirmed that screening of primary resistance sources is an essential step of breeding programs. Evaluation of plant material, even conducted rigorously under greenhouse conditions based on a wide variety of nematode species and populations, may not always allow to predict the response to a particular field nematode isolate. Nevertheless, at the vineyard scale, resistance-breaking populations may remain localized if appropriate viticultural practices are applied to prevent spreading, considering that nematodes are soil dwelling organisms with a limited dissemination *per se*.

2.2.3. Genetics of Resistance

As certain *Vitis* species were shown to carry a high degree of resistance, a study of the genetics of RKN resistance was of major interest for breeding programs. A single dominant allele for resistance to *M. incognita* was found, either as homozygous or as heterozygous *in V. champinii*, *V. mustangensis* and 1613C by Lider (1954). Bloodworth et al. (1980) observed that the resistance of *Muscadinia rotundifolia* to the three species *M. incognita*, *M. arenaria* and *M. javanica* was predominantly dominant in *Euvitis* × *Muscadinia* hybrids. Using 807 offsprings of 46 families representing *Euvitis* × *Muscadinia* hybrids backcrossed to *vinifera*, Firoozabady and Olmo (1982) estimated the heritability of the resistance to be $h^2=0.39$.

More recently, Cousins and Walker (2002) confirmed the monogenic hypothesis of resistance using a non-virulent *M. incognita* population with the sources Ramsey, Dog Ridge, Harmony, Freedom, 1613C and 1616C. Cousins et al. (2003) studied the resistance to *Meloidogyne* populations (including the *M. arenaria* virulent isolate and the above-mentioned non-virulent *M. incognita* population) with a source close to *V. champinii* identified as belonging to the species *V. mustangensis* Buckley that had been firstly found resistant by Lider (1954) to *M. acrita*. According to these authors, a single dominant allele would confer the resistance to both the virulent *Meloidogyne* spp. and the avirulent *M. incognita* nematodes. These results on genetics of resistance confirm those concerning other perennials, such as plum (Esmenjaud et al., 1996; Lecouls et al., 1997) and peach (Claverie et al., 2004), or annuals such as tomato (Williamson, 1998) and pepper (Djian et al., 2001) in which major dominant genes were shown to control more or less broad resistance spectra and to confer a more or less high level of resistance.

2.2.4. Resistance Mechanisms and Durability

It is important to characterize the nematode behaviour together with the tissular and cellular mechanisms of plant resistance. Recording the evolution of nematode populations exposed to the selection pressure of resistant material and of putative corresponding resistance genes may allow to predict the resistance durability. Very little is known on the resistance mechanisms of grapes to RKN. However studies were conducted in California using diverse virulent populations of *M. arenaria*, *M. incognita* and other mixed RKN populations. The two resistance-breaking populations (so called pathotypes) of *M. arenaria*, pathotypes 'Freedom' and 'Harmony' (Cain et al. 1984) are capable of overcoming resistance in the currently used rootstocks Freedom and Harmony (Anwar et al., 2000).

The pathotype Freedom is highly virulent and all nematode stages occur earlier and in greater numbers than in the pathotype Harmony which can be considered as moderately virulent. For this latter population, resistance mechanisms have been studied by Anwar and McKenry (2000) on the two new resistant selections, RS3 which belongs to the RS series and 10-23B, one of the complex hybrids previously mentioned. Both exhibited hypersensitive resistance reactions (HR) in the epidermis, cortical cells and along the differentiating vascular bundle and halted or delayed migration of juveniles.

In 10-23B, this early HR was completed by a late resistance expression characterized by the development of undersized adult females and lack of reproduction. Two other recently selected resistant rootstocks (10-17A and 6-19B, belonging to the complex hybrids mentioned above) inoculated with the same *M. arenaria* 'Harmony' population expressed a similar early hypersensitive reaction with a complete absence of swollen juveniles or adult stages (Anwar & McKenry,

2002b). By contrast, this *M. arenaria* 'Harmony' population produced cell necrosis and underdeveloped giant cells in the resistant rootstocks RS-9 and Teleki 5C (*V. riparia* × *V. berlandieri*) with a delayed development of adults and limited egg production (Anwar & McKenry, 2002a). Thus, the diverse accessions express a more or less early (HR reaction stopping the juveniles during the migration) or late (formation of imperfect giant cells by undersized females) resistance response. The authors hypothesize that virulent populations have appeared and developed gradually along the years in the presence of the supposedly weak resistance mechanisms of Freedom and Harmony rootstocks. Concomitantly, resistance mechanisms to other RKN species have remained intact (Anwar & McKenry, 2000, 2002a; 2002b; Anwar et al., 2002).

3. GFLV VECTOR NEMATODE *XIPHINEMA INDEX*

3.1. Xiphinema index *and the Other Virus Vector Nematodes on Grape*

Xiphinema index was first described in 1950 by Thorne and Allen as a pest of fig trees in California and reported in 1958 by Raski and Radewald as parasitizing grape. The nematode, also associated with woody perennial crops such as rose, mulberry and pistachio (Weiner & Raski, 1966), has a world-wide distribution closely related to that of grapevine (Barbercheck et al., 1985; Cohn, 1969; Dalmasso & Caubel, 1966; Feldman & Pontis, 1964). In addition to being a potentially serious root pest of grapes, with extreme root injury under artificial inoculations (Raski & Radewald, 1958), this nematode was reported by Hewitt et al. (1958) to be the vector of GFLV that causes one of the most destructive diseases of the grapevine, the infectious degeneration, so-called 'court-noué'.

Another species, *X. italiae*, widely distributed in the Mediterranean Basin, was claimed to be a vector of GFLV in Israel (Cohn et al., 1970) but these results were not corroborated by Italian data obtained from nine geographic populations of this nematode (Catalano et al. 1992). Other nematodes of grapevine belonging to the genera *Xiphinema* and *Longidorus* are known to be the vectors of various grapevine virus diseases (Frazier et al., 1970; Martelli, 1978).

Xiphinema diversicaudatum was proven to be the vector of the *Arabis mosaic nepovirus* (ArMV) (Jha & Posnette, 1959; Harrison & Cadman, 1959) and the *Strawberry latent ringspot sadwavirus* (SLRV). GFLV and ArMV constitute the viral complex of the infectious degeneration.

Xiphinema americanum sensu lato (Lamberti & Bleve-Zacheo, 1979) is considered to be the vector of *Peach rosette mosaic nepovirus*, *Tobacco ringspot nepovirus* (Fulton, 1962) and *Tomato ringspot nepovirus* (Teliz et al., 1966). *Xiphinema bricolensis* and *X. pacificum* were found widely distributed in Canadian vineyards (Graham et al., 1988) and their role in transmission of these grapevine viruses is highly probable.

Longidorus attenuatus and *L. elongatus* are considered to be the vectors of *Tomato black ring nepovirus* (TBRV). As for RKN, the problem of identification of the different species of *Xiphinema* is of major importance and can be solved by the use of PCR with specific primers (Wang et al., 2002; Huebschen et al., 2004).

3.2. *Biology, Vection and Classical Control of* X. index

3.2.1. *Biology and Vection*

The feeding of the nematode occurs at actively growing root tips and induces the formation of galls which contain enlarged multinucleate cells with dense cytoplasm (Brown et al. 1995; Wyss 2000). *Xiphinema index* reproduces asexually by meiotic parthenogenesis (Dalmasso and Younes 1969) and males are very scarce, what differentiates this species from *X. diversicaudatum* which is amphimictic. In a given field, the horizontal dispersion of nematodes from the initial introduction point depends mainly on the number of grapevine generations grown in the field since the nematode introduction (Dalmasso 1970).

Vertical distribution in the soil is also a factor to be considered and preferential localization and survival of the nematode in the deep soil layers are major constraints for chemical control (Esmenjaud et al. 1992).

The site of retention of the virus is located in the alimentary tract of the nematode (Taylor and Robertson 1970). The transmission process is characterized by a high degree of specificity between GFLV and *X. index* (Belin et al. 2001; Andret-Link et al. 2004b; Andret-Link and Fuchs 2005). Recent studies have shown that even in absence of roots, the nematode can survive in the soil at least during four years (Demangeat et al. 2005) and the virus can still be detected in the surviving nematodes by RT-PCR (Demangeat et al. 2004), a method more sensitive and reliable than ELISA (Bouquet 1983a; Esmenjaud et al. 1993, 1994).

3.2.2. *Classical Control*

Side-dressing treatments of Californian vineyards with 1,2-dibromo-3-chloropropane (DBCP) were reported to be effective against *X. index* and widely used also against root-knot nematodes (Raski & Schmitt, 1964) but DBCP was withdrawn in 1976 and the only chemical alternatives available were preplant soil treatments with methyl bromide, 1,3-dichloropropene or aldicarb (Lear et al., 1981; Esmenjaud et al., 1988). But chemical fumigation of vineyard soil before replanting only temporarily reduces numbers of *X. index*. Moreover, chemical treatments are very costly and have a reduced efficiency in deep, wet or clay soils, even when the placement conditions are optimal. Furthermore because of their high environmental impact, nematicides are submitted to increasing regulatory restrictions.

As the nematodes may persist for up to ten years on root fragments in the soil (Raski et al., 1965), a long-term fallow or non-host crop cultivation of at least 7 years is recommended for their eradication (Brown et al., 1993). Such a long interruption of grapevine cultivation, confirmed by the data of Demangeat et al. (2005), showing that the virus can still be detected in the nematodes kept starving for four years, is not an attractive prospect for the grower. It could be shortened by killing the roots of the diseased vines before their removal, with systemic herbicides such as glyphosate. But the efficiency of the technique is not guaranteed and depends on the vine age and the soil depth. Moreover, the harmlessness of glyphosate for health and environment has been recently questioned and its use is also coming under increasing regulatory restrictions. There is, therefore, a need for a better and integrated control method for *X. index* in replant vineyards that had already been recommended by Raski et al. in 1983.

3.3. Selection and Breeding of Resistant Rootstocks

Alternatives to nematicides have to consider the two main limiting factors of GFLV control which are the location of surviving nematodes in deep soil layers (Esmenjaud et al., 1992) and the long virus survival in the nematodes between two successive grapevine crops (Demangeat et al., 2005). Nematode resistant rootstocks and correlative delayed GFLV spread appear as the most efficient alternative to face these constraints. Resistant rootstocks would provide protection from nematode damage for the life of the crop. As for RKN resistance breeding, this is a challenging task since the long generation time of grapevine increases the risk for resistance breaking.

3.3.1. Vitis *Breeding*

Resistance ratings of some *Vitis* species and cultivars have been reported according to both visible symptoms and changes in numbers of *X. index* on the roots of tested plants in greenhouse over an inoculation period of eight months (Kunde et al., 1968). The highest resistance was found in *V. candicans*, *V. solonis*, *V. arizonica*, *V. rufotomentosa* and *V. smalliana*. Moderate resistance was observed in *V. riparia*, *V. rubra* and *V. slavinii*. Among commercial rootstocks, 1613C showed moderate resistance and all others, including Salt Creek (= Ramsey) and Dog Ridge, were quite susceptible. Based upon this information, 40 crosses were made among a number of resistant and susceptible *Vitis* species and cultivated varieties, from which 33 gave progenies with enough seedlings for adequate testing that was based on root damage ratings rather than nematode counts (Meredith et al., 1982). For this reason, in conformance with the terminology widely accepted in the plant pathology literature (Buddenhagen, 1981), and the interest of accuracy, the term tolerance was used for describing the results.

On the basis of segregation patterns observed in the progenies, two genetic models for *X. index* tolerance were proposed. The first and simplest model involves one gene with tolerance being dominant. In the second model, two genes, one dominant and one recessive, condition tolerance (Meredith et al., 1982). Tolerance/susceptibility ratios observed in several F_2 generations (sibling crosses tolerant × susceptible) of *V. rupestris* × *V. arizonica* seedling populations (Walker & Jin, 1998; 2000) suggest that tolerance is inherited as a single major gene, heterozygous in *V. arizonica*, as proposed by Meredith et al. (1982). Male parents used in the crosses with *V. rupestris* were thought firstly to be *Muscadinia* cultivars, but proved later to be mainly wild plants of *V. arizonica* (Riaz et al., 2007). A genetic linkage map was established, using 116 progeny plants from the cross of two half-sib genotypes *V. rupestris* and *V. arizonica* (Doucleff et al., 2004) and the locus for tolerance to *Xiphinema index* was placed on the linkage group 19 (Xu et al., 2008).

The results of Kunde et al. (1968) were partially confirmed by Harris (1983) who screened 38 rootstock cultivars and *Vitis* hybrids, using resistance ratings based on visible root symptoms changes in the nematode populations over 16 months. Among those rootstocks, he found 1613C moderately resistant, and observed that this resistance was genetically transmitted to the two rootstocks Freedom and Harmony, issued from crosses between 1613C and Dog Ridge. Similarly, he observed that the resistance of *V. arizonica*, *V. candicans*, *V. rufotomentosa* and *V. solonis* was genetically inherited but he could not conclude for the resistance level of the cultivars Dog Ridge and Ramsey (= Salt Creek). These rootstocks were screened for field performance over 10 years. Two hybrids between *V. rufotomentosa* and *V. vinifera*, 171-13 and 171-52, obtained in California, look the most promising, but the indexing results against GFLV were inconclusive with regard to the resistance to the virus transmission through nematode feeding (Harris, 1988). However, another hybrid from the same series, 171-6, was tested during eight years in a Californian vineyard site infected with GFLV and showed foliar symptoms of the virus and poor fruit yield, but strong vegetative growth and high pruning weight (Lider & Goheen, 1986; Walker et al., 1994b).

Using the same resistance ratings as Kunde et al. (1968) and Harris (1983), Malan and Meyer (1993) tested the resistance of 31 rootstock cultivars with a South African population of *X. index* and confirmed the resistance of 1613C, Freedom and Harmony and the susceptibility of Ramsey and Dog Ridge. Using four different populations of *X. index* from Italy, California, Israel and France, Coiro et al. (1990) found a high level of resistance to reproduction in 1613C, but also in Dog Ridge, which had been evaluated as quite susceptible by Kunde et al. (1968). These authors found also a high level of resistance to the Californian population on Ramsey.

However, using ELISA tests, Malan and Meyer (1993) detected the presence of GFLV in the roots of all cultivars tested after four months and the systemic spread of the virus in the top leaves during two growing seasons, from six to 18 months after inoculation. There was no evidence of resistance to the transmission of GFLV through feeding of *X. index* in any of the rootstocks studied, even though some had a low reproduction potential for the nematode and no root damage was observed. Boubals and Pistre (1978) concluded similarly after a large survey of fifty commercial rootstocks and *Vitis* species. Five years after inoculation, all the plants tested showed foliar symptoms of GFLV. The ratings of root damage and nematode populations after 10 months, confirmed the susceptibility of Dog Ridge and Salt Creek, but not the resistance of 1613C, and showed a moderate level of resistance in *V. riparia* 'Gloire de Montpellier' and the rootstock 3309C, that was also observed later by Harris (1983).

The results of Boubals and Pistre are interesting in that they showed that great differences of susceptibility/tolerance could exist among varieties of the same *Vitis* species. For example, *V. riparia* 'Grand glabre' and *V. riparia* 'Gloire' are moderately resistant, but *V. riparia* 'Messner n. 9' is very susceptible. The evidence of poor correlation between the resistance to transmission of GFLV and the resistance to nematode feeding in *Euvitis* species was also observed by Staudt and Kassemeyer (1990): *V. riparia* 'Gloire' and *V. rupestris* 'du Lot' showed similar rates of GFLV infection (92-95%), but in the same conditions *V. arizonica* showed a rate of 64%.

Although the susceptibility of *V. cinerea* to *X. index* was reported by Kunde et al. (1968) and Boubals and Pistre (1978), a high resistance of the rootstock cultivar 'Börner' (*V. riparia* × *V. cinerea*) to nematode feeding and GFLV infection was claimed (Becker 1989; Becker & Sopp, 1990) but is still controversial (Sopp et al., 1998; Ipach et al., 2000). In a field trial conducted in France in heavily infected soil, the rates of GFLV infection of scions of Cabernet-Sauvignon grafted on Börner and SO4 are equivalent six years after planting (Bouquet, unpublished data). Moreover, Börner proved to be very susceptible to lime-induced chlorosis, contrary to the moderately tolerant SO4.

3.3.2. Vitis × Muscadinia *Breeding*

3.3.2.1. Resistance Features of Muscadinia rotundifolia

The most interesting findings of Boubals and Pistre (1978) were the high resistance of some species of *Ampelopsis* and *Parthenocissus*, specially *A. aconitifolia* and *P. quinquefolia*, but above all the resistance of *Vitis* (*Muscadinia*) *rotundifolia* that did not show any foliar symptoms of GFLV five years after inoculation. The resistance of this species to GFLV by nematode feeding was confirmed in further studies (Bouquet, 1981; Bouquet & Danglot, 1983; Staudt & Weischer, 1992; Sopp et al., 1998).

According to the classification of *Ampelidaceae* (Planchon 1887), *Vitis rotundifolia* (the muscadine grape) belongs to the section *Muscadinia* of the genus *Vitis* and is distinct from the section *Euvitis* (the true grapes or bunch grapes). Considering their morphological, anatomical and caryological characteristics (Bouquet 1980b), these two sections are so distantly related that the section *Muscadinia* can be raised to generic rank as proposed by Small (1913), reserving the genus *Vitis* for the bunch grapes.

The muscadine grape was the first American grape species to be cultivated in Southeastern United States. Though its acreage is limited to less than 2000 ha, this fruit has a long history in commercial and backyard culture (Olien, 1990). But despite its high resistance to Phylloxera (Boubals, 1966; Pouget, 1975) and root-knot nematodes (Lider, 1954; see also previous paragraph on RKN resistance), *M. rotundifolia* is not suitable as rootstock because of its graft-incompatibility with *V. vinifera* (Bouquet & Hevin, 1978; Bouquet, 1980a) and its poor rooting ability (Goode et al., 1982).

3.3.2.2. V. vinifera × M. rotundifolia *F_1 Hybrids Obtained in California*

In California, some F_1 hybrids *V. vinifera* × *M. rotundifolia* produced in 1948 (Patel & Olmo, 1955) were screened for use as phylloxera resistant rootstocks

(Davidis & Olmo, 1964) and tested in field trials. Among this material included in 1979 in a field screening for resistance to fanleaf degeneration, the hybrids VR 039-16 and VR 043-43 excelled (Lider & Goheen 1986; Walker et al. 1989) and were patented and released as resistant to fanleaf degeneration (Walker et al. 1991).

However, in short-time greenhouse tests made in Germany, VR 039-16 was found resistant but VR 043-43 susceptible (Staudt & Kassemeyer, 1990). Since that time, both rootstocks have shown that they allow movement of GFLV into scions grafted on them, but prevent virus'disruptive effect on fruit set (Walker et al., 1994a; 1994b). In laboratory tests and field trials for resistance to type B phylloxera, VR 039-16 resulted resistant but VR 043-43 susceptible (Granett et al., 1987). In addition, VR 043-43 expressed a high susceptibility to lime-induced chlorosis (Bavaresco et al., 2005). VR 039-16 is currently the sole rootstock recommended for fanleaf vineyard sites in California.

Nevertheless, there is a number of problems associated with this rootstock. To date no evidence of Phylloxera susceptibility has been found in vineyards planted on VR 039-16, but the collapse of VR 043-43 questions the durability of the resistance of VR 039-16, and the long term reliability of this rootstock. VR 039-16 is also susceptible to RKN (see previous paragraph) which are found with *X. index* in several parts of the Californian vineyard. Because of its *Muscadinia* parentage, this rootstock is also difficult to propagate, induces high vigour in its scions and tends to have a very long cycle of growth leading to problems with wood maturity on mother-vines and scions. All these constraints led to renew breeding efforts on a *V. rupestris* × *M. rotundifolia* basis (Walker & Jin, 2000; Doucleff et al., 2003; 2004).

3.3.2.3. V. vinifera × M. rotundifolia *F_1 and Backcrosses Obtained in France*

Hybridization between *V. vinifera* and *M. rotundifolia* was performed since 1974 in France. Despite the genetic barriers between the two species, numerous F_1 hybrids were obtained (Bouquet, 1980b; 1983c) and screened for phylloxera resistance (Bouquet, 1983b) and GFLV resistance through nematode feeding (Bouquet, 1983c). But these F_1 hybrids could not be used as rootstocks as they inherited the cultural drawbacks of their *Muscadinia* parents, particularly their poor rooting ability and their extreme susceptibility to lime chlorosis. Despite their high sterility, a few F_1 hybrids have been successfully backcrossed and the resistance to *X. index* was introduced in *Vitis* genotypes potentially usable as rootstocks resistant to virus spread (Bouquet et al., 2000).

One of these genotypes (Mtp 3146-1-87) was obtained from the cross of a highly nematode-resistant and partially fertile F_1 hybrid with the rootstock 140 Ruggeri (*V. berlandieri* × *V. rupestris*), that is highly tolerant to chlorosis and drought and widely used in the Mediterranean vineyards. Grafted under Cabernet-Sauvignon scions and tested in field trial since 1999, Mtp 3146-1-87 shows a considerable delay in contamination by GFLV in a highly-infested soil, comparatively to the rootstock SO4. Its tolerance to lime-induced chlorosis is intermediate between SO4 and 140 Ruggeri. In a healthy soil, the vigour conferred to the scions and the fruit yield are lower than those of SO4 and 140 Ruggeri, but they are much higher in a GFLV-infested soil (Bouquet et al. 2003a).

3.3.3 Future Prospects

The improvement of some cultural aptitudes of Mtp 3146-1-87, particularly the rooting ability of its hardwood cuttings, requires its crossing with other rootstock cultivars, but might reduce its field resistance to virus spread. A valuable strategy could be to pyramidize genes for feeding resistance from both *M. rotundifolia* and *Vitis* species. In this purpose, work is in progress in France to detect genes for at least partial feeding resistance to *X. index* among anciently- and newly- introduced *Vitis* germplasm (N. Ollat and D. Esmenjaud, unpublished data). Other studies in France deal with the specific genetic diversity in *X. index* in the objective of selecting representative populations to challenge the durability of the resistance from Mtp 3146-1-87 and related *Muscadinia* interspecific material.

Another strategy could be to combine the feeding resistance genes of *M. rotundifolia* with a biotechnologically engineered resistance, such as the resistance induced by the GNA (*Galanthus nivalis*) gene encoding a lectine. Transformation of the rootstocks Freedom, 101-14 and Teleki 5C with this gene has been reported (Viss & Driver, 1996). A long term strategy would also consist in combining the feeding resistance to *X. index* with resistance to GFLV. But attempts to find genes for viral resistance in the *Vitis* germplasm were unsuccessful (Lahogue & Boulard, 1996), despite results firstly encouraging (Walker et al., 1985; Walker & Meredith, 1990) but not corroborated.

Since two decades, biotechnologically-engineered resistance of grapevine to GFLV is under way in numerous laboratories worldwide. The coat protein gene of GFLV was introduced successfully in several rootstock cultivars (Krastanova et al., 1995; Mauro et al., 1995; Golles et al., 2000; Krastanova et al., 2000; Mauro et al., 2000; Valat et al., 2006) and some transgenic plants have shown field resistance in infested soils (Vigne et al., 2004). Until now, attempts to transform directly the *Xiphinema* resistant variety Mtp 3146-1-87 were unsuccessful, due to problems of somaclonal variation in the embryogenic cultures (Bouquet, unpublished results).

A promising strategy would be to transfer by hybridization the coat protein gene from transgenic rootstocks to *X. index* resistant rootstocks (Bouquet et al. 2003b). In this strategy, one may hypothesize that resistance to nematode would act as a filter limiting the virus load introduced by *X. index* during feeding attempts and thus increasing the subsequent efficiency of a biotechnologically-engineered resistance to the virus multiplication in the plant.

4. RESISTANCE TO OTHER NEMATODES AND ROOTSTOCK CONTROL OF MULTIPLE NEMATODE PESTS

Besides the root-knot and dagger nematodes responsible for most damage on grapes, there are some other nematodes that can be a concern for grape growers (Nicol et al., 1999). These are mainly the root-lesion nematode *Pratylenchus vulnus*, the ring nematode *Criconemoides xenoplax* and the citrus nematode *Tylenchulus semipenetrans*.

Host suitability of diverse rootstocks to *P. vulnus* has been studied in the USA (McKenry, 1992; McKenry & Kretsch, 1995; McKenry et al., 2001b; McKenry and Anwar, 2006), in Australia (Sauer, 1977) and in Spain (Pinochet & Raski, 1977; Pinochet et al., 1992) and has shown a high variability with most plant material being susceptible. Indeed, Sauer (1977) could only find a satisfactory level of resistance in the rootstocks 3306C and 3309C (*V. riparia* × *V. rupestris*) while McKenry et al. (2001b) only evidenced resistance in accessions of *V. champinii* (Ramsey) or derived from it (K51-32; *V. champinii* × *V. riparia*).

No resistance source has been found against the ring nematode *C. xenoplax* (Sauer, 1977; Walker, 1994; McKenry et al., 2001a; McKenry & Anwar, 2006). A good level of resistance to the citrus nematode *T. semipenetrans* has only been observed in Ramsey (McKenry et al., 2001b; McKenry & Anwar, 2006), K51-32 (McKenry et al., 2001b) and SO4 (McKenry & Anwar, 2006). As resistance is generally studied separately for each nematode species, a global synthesis would be needed but there are very few reports of such synthetic or at least comparative studies across nematode species. Ramsdell et al. (1996), comparing *C. xenoplax*, *P. penetrans*, *X. americanum* and *M. hapla*, found this latter RKN as the most virulent in microplots experiments conducted in Michigan (USA).

In a perspective of selection and application of resistant germplasm for grapevine nematode management, resistance to certain major nematode pests has to be associated with a satisfactory behaviour of resistance or tolerance to less predominant nematodes. McKenry et al. (2001a and b) and McKenry and Anwar (2006) have reported a global approach in which all main endo and ectoparasitic nematodes affecting directly grapes have been considered. Among a set of rootstocks commonly used in USA, those exhibiting the most wide resistance to endoparasitic nematodes are Ramsey and Dog Ridge, and a few derived stocks (Harmony, Freedom, 1613C) together with K51-32.

For *X. index*, VR039-16, Schwartzmann and Freedom result as the sole American stocks exhibiting resistance. In an attempt to evidence the nematode and grape rootstock interaction in a field situation where occurrence of more than one species is common, they have evaluated resistance (final/initial numbers) and tolerance (vine growth with/without nematode inoculum) using separate populations of RKN (resistant-breaking and not breaking populations), *P. vulnus*, *C. xenoplax*, *X. index* and also mixed RKN populations and mixed populations of *Meloidogyne* plus *X. index* or *P. vulnus*. They could record both the nematode dynamics and the plant reactions. RKN, *X. index* and *C. xenoplax* all developed faster and caused greater damage than other nematodes. RKN sampled from fields with a history of feeding on grape showed the highest development. Vines appeared to tolerate slow developing or less pathogenic nematode populations. RKN resistant rootstocks were often stimulated by nematode attacks but were affected by resistance-breaking populations.

REFERENCES

Andret-Link, P., & Fuchs, M. (2005). Transmission specificity of plants viruses by vectors. *Journal of Plant Pathology*, 87, 153-165.

Andret-Link, P., Laporte, C., Valat, L., Ritzenthaler, C., Demangeat, G., Vigne, E., et al. (2004a). Grapevine fanleaf virus: still a major threat to the grapevine industry. *Journal of Plant Pathology*, 86, 183-195.

Andret-Link, P., Schmitt-Keichinger, C., Demangeat, G., Komar, V., & Fuchs, M. (2004b). The specific transmission of grapevine fanleaf virus by its nematode vector *Xiphinema index* is solely determined by the viral coat protein. *Virology*, 320, 12-22.

Anwar, S. A., & McKenry, M. V. (2000). Penetration, development and reproduction of *Meloidogyne arenaria* on two new resistant *Vitis* spp. *Nematropica*, 30, 9-17.

Anwar, S. A., & McKenry, M. V. (2002a). Developmental response of a resistance breaking population of *Meloidogyne arenaria* on *Vitis* spp. *Journal of Nematology*, 34, 28-33.

Anwar, S. A., & McKenry, M. V. (2002b). Penetration and development of *Meloidogyne arenaria* on two new grape rootstocks. *Journal of Nematology*, 34, 143-145.

Anwar, S. A., McKenry, M. V. & Faddoul, J. (2000). Reproductive variability of field populations of *Meloidogyne* spp. on grape rootstocks. *Journal of Nematology*, 32, 265-270.

Anwar, S. A., McKenry, M. V., & Ramming, D. (2002). A search for more durable grape rootstock resistance to root-knot nematode. *American Journal of Enology and Viticulture*, 53, 19-23.

Barbercheck, M., Smith, P. C., & Heyns, J. (1985). Occurrence and distribution of *Xiphinema* in vineyards on the Bree River valley. *Phytophylactica*, 17, 27-30.

Bavaresco, l., Presutto, P., & Civardi, S. (2005). VR 043-43 : A lime-susceptible rootstock. *American Journal of Enology and Viticulture*, 56, 192-195.

Becker, H. (1989). The new rootstock Börner (in German). *Der Deutsche Weinbau*, 44, 960-962.

Becker, H., & Sopp, E. (1990). *Rootstocks with immunity to phylloxera and nematode resistance.* Paper presented at the 5th Intern Symp Grape Breeding, St-Martin/Pfalz, Germany, September 1989. *Vitis* (special issue), 294.

Belin, C., Schmitt, C., Demangeat, G., Komar, V., Pinck, L., & Fuchs, M. (2001). Involvement of RNA2-encoded proteins in the specific transmission of grapevine fanleaf virus by its nematode vector *Xiphinema index*. *Virology*, 290, 161-171.

Bloodworth, P. J., Nesbitt, W. B., & Barker, K. R. (1980). *Resistance to root-knot nematodes in Euvitis x Muscadinia grape hybrids.* Paper presented at the 3rd Intern Symp Grape Breeding, Davis, California, 1980. UCAL ed., pp. 275-292.

Boubals, D. (1954). Parasitic nematodes of grapevine (in French). *Progrès Agricole et Viticole*, 141, 173-182 and 204-208.

Boubals, D. (1966). A study of the distribution and causes of resistance to radicicolous *Phylloxera* in the *Vitaceae* (in French). *Annales d'Amélioration des Plantes*, 16, 145-184.

Boubals, D., & Pistre, R. (1978). Resistance of some *Vitaceae* and usual rootstocks to *Xiphinema index* and contamination by fanleaf virus (in French). In R. Pouget & J. P. Doazan (Eds.), *Grapevine Genetics and Breeding* (pp. 199-207). Paris: INRA.

Bouquet, A. (1980a). Differences observed in the graft compatibility between some cultivars of Muscadine grape (*Vitis rotundifolia* Michx) and European grape (*Vitis vinifera* L. cv. Cabernet-Sauvignon). *Vitis*, 19, 99-104.

Bouquet, A. (1980b). *Vitis x Muscadinia hybridization: a new way in grape breeding for disease resistance in France.* Paper presented at the 3rd International Symposium on Grape Breeding, Davis, California, June 1980. UCAL ed., pp. 42-61.

Bouquet, A. (1981). Resistance to grape fanleaf virus in Muscadine grape inoculated with *Xiphinema index*. *Plant Disease*, 65, 791-793.

Bouquet, A. (1983a). Enzyme-Linked Immunosorbent Assay to detect grape fanleaf virus in its vector *Xiphinema index* Thorne & Allen (in French). *Compte-rendu de l'Académie des Sciences de Paris*, Série III, 271-273.

Bouquet, A. (1983b). Phylloxera resistance of *Vitis vinifera* × *Muscadinia rotundifolia* hybrids (in French). *Vitis*, 22, 311-323.

Bouquet, A. (1983c). *Contribution to the study of the species Muscadinia rotundifolia (Michx) Small and its hybrids with Vitis vinifera L. Applications in breeding* (in French). University of Bordeaux, France.

Bouquet, A., & Dalmasso, A. (1976).. Resistance and susceptibility of grape rootstocks to a nematode population (*Meloidogyne* sp.) originating from the Southwest of France (in French). *Connaissance de la Vigne et du Vin*, 10, 161-174.

Bouquet, A., & Danglot, Y. (1983). Search for grape rootstocks resistant to the transmission of grape fanleaf virus (GFLV) by the nematode *Xiphinema index* Thorne & Allen (in French). *Agronomie*, 3, 957-963.

Bouquet, A., & Hevin, M. (1978). Green-grafting between Muscadine grape (*Vitis rotundifolia* Michx) and bunch grapes (*Euvitis* sp.) as a tool for physiological and pathological investigations. *Vitis*, 17, 134-138.

Bouquet, A., Dalmasso, A., & Bongiovanni, M. (1982). Susceptibility of the rootstocks Fercal and 41B to the nematode *Meloidogyne hapla* (in French). *Progrès Agricole et Viticole*, 99 (23), 576-577.

Bouquet, A., Danglot, Y., Torregrosa, L., Bongiovanni, M., Castagnone-Sereno, P., Esmenjaud, D., et al. (2000). *Breeding rootstocks resistant to grape fanleaf virus spread, using Vitis × Muscadinia hybridisation*. Paper presented at the 7th International Symposium on Grape Genetics and Breeding, Montpellier, France, July 1998. *Acta Horticulturae*, 528, 517-526.

Bouquet, A., Torregrosa, L., & Chatelet, P. (2003a). Combined use of conventional and biotechnological methods in the selection of rootstocks with durable resistance to the transmission of grape fanleaf virus (in French). *Progrès Agricole et Viticole*, 120 (22-23), 507-512 and 528-532.

Bouquet, A., Marck , G., Pistagna, D., & Torregrosa, L. (2003b). *Transfer of grape fanleaf virus coat protein gene through hybridization with Xiphinema index resistant genotypes to obtain rootstocks resistant to virus spread.* Paper presented at the 8th International Conference on Grape Genetics and Breeding, Kecskemet, Hungary, August 2002. *Acta Horticulturae*, 603, 325-334.

Brown, D. J. F., Dalmasso, A., & Trudgill, D. L. (1993). Nematode pests of soft fruits and vines. In K. Evans, D. L. Trudgill & J. M. Webster (Eds.), *Plant parasitic nematodes in temperate agriculture*. Wallingford, UK: CAB International, 427-461.

Brown, D. J. F., Robertson, W. M., & Trudgill, D. L. (1995). Transmission of viruses by plant nematodes. *Annual Review of Phytopathology*, 33, 223-249.

Buddenhagen, I. W. (1981). Conceptual and practical considerations when breeding for tolerance or resistance. In R.C. Staples & G. H. Toenniessen (Eds.), *Plant Disease Control: Resistance and Susceptibility*. New York: John Wiley and Sons.

Cain, D. W., McKenry, M. V., & Tarailo, R. E. (1984). A new pathotype of root-knot nematode on grape rootstocks. *Journal of Nematology*, 16, 207-208.

Catalano, L., Savino, V., & Lamberti, F. (1992). Presence of grape fanleaf nepovirus in populations of Longidorid nematodes and their vectoring capacity. *Nematologia Mediterranea*, 20, 67-70.

Claverie, M., Bosselut, N., Lecouls, A. C., Voisin, R., Poizat, C., Dirlewanger, E., et al. (2004). Location of independent root-knot nematode resistance genes in plum and peach. *Theoretical & Applied Genetics*, 108, 765-773.

Cohn, E. (1969). The occurrence and distribution of species of *Xiphinema* and *Longidorus* in Israel. *Nematologica*, 15, 179-192.

Cohn, E., Tanne, E., & Nitzany, F. E. (1970). *Xiphinema italiae*, a new vector of grapevine fanleaf virus. *Phytopathology*, 60, 181-182.

Coiro, M. I., Taylor, C. E., Borgo, M., & Lamberti, F. (1990). Resistance of grapevine rootstocks to *Xiphinema index*. *Nematologia Mediterranea*, 18, 119-121.

Cousins, P., & Walker, M. A. (2002). Genetics of resistance to *Meloidogyne incognita* in crosses of grape rootstocks. *Theoretical & Applied Genetics*, 105, 802-807.

Cousins, P., Lauver, M., & Boyden, L. (2003). *Genetic analysis of root-knot nematode resistance derived from Vitis mustangensis.* Paper presented at the 8th International Conference on Grape Genetics and Breeding, Kecskemet, Hungary, August 2002. *Acta Horticulturae*, 603, 149-154.

Dalmasso, A. (1970). Direct influence of some ecological factors on the biological activity and the distribution of the French species of the *Longidoridae* family (Nematoda: Dorylaimida) (in French). *Annales de Zoologie et Ecologie Animale*, 2, 163-200.

Dalmasso, A. (1973). Reproduction pattern of species belonging to the genus *Meloidogyne*. *OEPP/EPPO Bulletin*, 3, 67-73.

Dalmasso, A., & Caubel, G. (1966). Distribution of species of the genera *Xiphinema* and *Longidorus* found in France (in French). *Compte-rendu des séances de l'Académie d'Agriculture de France*, 440-446.

Dalmasso, A., & Cuany, A. (1976). Resistance of grapevine rootstocks to different populations of the nematode *Meloidogyne hapla* (in French). *Progrès Agricole et Viticole*, 25, 800-807.

Dalmasso, A., & Younes, T. (1969). Ovogenesis and embryogenesis in *Xiphinema index* (Nematoda: Dorylaimida). *Annales de Zoologie et Ecologie Animale*, 1 (3), 265-279.

Davidis, U. X., & Olmo, H. P. (1964). The *Vitis vinifera* × *V. rotundifolia* hybrids as phylloxera resistant rootstocks. *Vitis*, 4, 129-143.

Demangeat, G., Komar, V., Cornuet, P., Esmenjaud, D., & Fuchs, M. (2004). Sensitive and reliable detection of Grapevine fanleaf virus in a single *Xiphinema index* nematode vector. *Journal of Virological Methods*, 112, 79-86.

Demangeat, G., Voisin, R., Minot, J. C., Bosselut, N., Fuchs, M., & Esmenjaud, D. (2005). Survival of *Xiphinema index* in vineyard soil and retention of grapevine fanleaf virus over extended time in the absence of host plants. *Phytopathology*, 95, 1151-1156.

Djian-Caporalino, C., Pijarowski, L., Fazari, A., Samson, M., Gaveau, L., O'Byrne, C., et al. (2001). High-resolution genetic mapping of the pepper (*Capsicum annuum* L.) resistance loci *Me3* and *Me4* conferring heat-stable resistance to root-knot nematodes (*Meloidogyne* spp.). *Theoretical & Applied Genetics*, 103, 592-600.

Doucleff, M., Jin, Y., & Walker, M. A. (2003). *Mapping Xiphinema index resistance in V. rupestris x M. rotundifolia.* Paper presented at the 8th International Conference on Grape Genetics and Breeding, Kecskemet, Hungary, August 2002. *Acta Horticulturae*, 603, 79-81.

Doucleff, M., Jin, Y., Gao, F., Riaz, S., Krivanek, A. F., & Walker, M. A. (2004). A genetic linkage map of grape, utilizing *Vitis rupestris* and *Vitis arizonica. Theoretical & Applied Genetics*, 109, 1178-1187.

Esmenjaud, D., Pistre, R., & Bongiovanni, M. (1988). Nematicide activity of aldicarb in deep and clayey soils against *Xiphinema index* Thorne & Allen, 1950 (Nematoda: Longidoridae) vector of grapevine fanleaf virus (in French). *Mededelingen Van De Faculteit Landbouwwetenschappen Rijksuniversiteit Gent*, 53/2b, 885-891.

Esmenjaud, D., Walter, B., Valentin, G., Guo, Z. T., & Cluzeau, D. (1992). Vertical distribution and infectious potential of *Xiphinema index* (Thorne & Allen, 1950) (Nematoda: Longidoridae) in fields affected by grapevine fanleaf virus in vineyards in the Champagne region of France. *Agronomie*, 12, 395-399.

Esmenjaud, D., Walter, B., Minot, J. C., Voisin, R., & Cornuet, P. (1993). Biotin-avidin ELISA detection of grapevine fanleaf virus in the vector nematode *Xiphinema index*. *Journal of Nematology*, 25, 401-405.

Esmenjaud, D., Pinck, L., Walter, B., & Abad, P. (1994). Detection of a region of the coat protein gene of grapevine fanleaf virus by RT-PCR in the nematode vector *Xiphinema index*. *Plant Disease*, 78, 1087-1090.

Esmenjaud, D., Minot, J. C., Voisin, R., Bonnet, A., & Salesses, G. (1996). Inheritance of resistance to the root-knot nematode *Meloidogyne arenaria* in Myrobalan plum. *Theoretical & Applied Genetics*, 92, 873-879.

Feldman, J. M., & Pontis, R. E. (1964). *Xiphinema index* in vineyard soils in Mendoza, Argentina. *Plant Disease Reporter*, 48, 373.

Firoozabady, E., & Olmo, H. P. (1982). The heritability of resistance to root-knot nematode (*Meloidogyne incognita acrita* Chit.) in *Vitis vinifera* × *V. rotundifolia* hybrid derivatives. *Vitis*, 21, 136-144.

Frazier, N., Fulton, J. P., Thresh, J. M., Converse, R. H., Varney, E. H., & W.B., H. (1970). Virus and virus-like diseases of the grapevine. In *Virus diseases of small fruits and grapevine (a handbook)*. Berkeley, USA: University of California: Section 5, 195-271.

Fulton, J. P. (1962). Transmission of *Tobacco ringspot virus* by *Xiphinema americanum*. *Phytopathology*, 52, 375.

Golles, R., da Camara-Machado, A., Minafra, A., Savino, A., Saldarelli, P., Martelli, G. P., et al. (2000). *Transgenic grapevines expressing coat protein gene sequences of Grapevine fanleaf virus, Arabis mosaic virus, Grapevine virus A and Grapevine virus B.* Paper presented at the 7th Intern Symposium on Grapevine Genetics and Breeding, Montpellier, France, July 2006. *Acta Horticulturae*, 528, 305-311.

Goode, D. K., Krewer, G. W., Lane, R. P., & Daniell, J. W. (1982). Rooting studies of dormant muscadine grape cuttings. *HortScience*, 17, 644-645.

Graham, M. B., Ebsary, B. A., Vrain, T. C., & Webster, J. M. (1988). Distribution of *Xiphinema bricolensis* and *X. pacificum* in vineyards of the Okanagan and Similkameen valleys, British Columbia. *Canadian Journal of Plant Pathology*, 10, 259-262.

Granett, J., Goheen, A. C., Lider, L. A., & White, J. J. (1987). Evaluation of grape rootstocks for resistance to type A and type B grape phylloxera. *American Journal of Enology and Viticulture*, 38, 298-300.

Harris, A. R. (1983). Resistance of some *Vitis* rootstocks to *Xiphinema index*. *Journal of Nematology*, 15, 405-409.

Harris, A. R. (1988). *Xiphinema index*-resistant *Vitis* rootstocks screened for comparative field performance in a Chasselas vineyard replant site. *Vitis*, 27, 243-251.

Harrison, B. D., & Cadman, C. H. (1959). Role of a dagger nematode (*Xiphinema* sp.) in outbreaks of plant disease caused by *Arabis mosaic virus*. *Nature*, 184, 1624-1626.

Hartman, K., & Sasser, J. N. (1985). Identification of *Meloidogyne* on the basis of differential host test and perineal pattern morphology. In J. N. Sasser & C. C. Carter (Eds.), Vol. 2, *An advanced treatise on Meloidogyne*. North Carolina State University: Raleigh, NC USA: 69-77.

Hewitt, W. B., Raski, D. J., & Goheen, A. C. (1958). Nematode vector of soil-borne fanleaf virus of grapevines. *Phytopathology*, 48, 586-595.

Huebschen, J., Kling, L., Ipach, U., Zinkernagel, V., Bosselut, N., Esmenjaud, D., et al. (2004). Validation of the specificity and sensitivity of species-specific primers that provide a reliable molecular diagnostic for *Xiphinema diversicaudatum*, *X. index* and *X. vuittenezi*. *European Journal of Plant Pathology*, 110, 779-788.

Ipach, U., Kling, L., & Rudel, M. (2000). *Transmission of grapevine fanleaf virus by Xiphinema index to different newly breeded rootstocks in greenhouse and field trials*. Paper presented at the 13th ICVG Conference, Adelaide, Australia, March 2000.

Janati, A., Bergé, J. B., Triantaphyllou, A. C., & Dalmasso, A. (1982). New data on the use of isoesterases for identiciation of *Meloidogyne* spp (in French). *Revue de Nématologie*, 5, 147-154.

Jepson, S. B. (1987). *Identification of root-knot nematodes Meloidogyne species*. CAB International, Wallingford, Oxon, UK: 265 pp.

Jha, A., & Posnette, A. F. (1959). Transmission of a virus to strawberry plants by a nematode (*Xiphinema* sp.). *Nature*, 184, 962-963.

Kasimatis, A. N., & Lider, L. A. (1967). Grape rootstock varieties. *California Agriculture Extension Service*, 47.

Krastanova, S., Perrin, M., Barbier, P., Demangeat, G., Cornuet, P., Bardonnet, N., et al. (1995). Transformation of grapevine rootstocks with the coat protein gene of grapevine fanleaf nepovirus. *Plant Cell Reports*, 14, 550-554.

Krastanova, S., Lig, K. S., Zhu, H. Y., Xue, B., Burr, T. J., & Gonsalves, D. (2000). *Development of transgenic grapevine rootstocks with genes from grapevine fanleaf virus and grapevine leafroll associated closterovirus 2 and 3*. Paper presented at the 7th International Symposium on Grapevine Genetics and Breeding, Montpellier, France, July 1998. *Acta Horticulturae*, 528, 367-372.

Kunde, R. M., Lider, L. A., & Schmitt, R. V. (1968). A test of *Vitis* resistance to *Xiphinema index*. *American Journal of Enology and Viticulture*, 19, 30-36.

Lahogue, F., & Boulard, G. (1996). Investigations on natural resistance genes for two grapevine virus: The fanleaf degeneration and the leafroll disease (in French). *Vitis*, 35, 43-48.

Lamberti, F., & Bleve-Zacheo, T. (1979). Studies on *Xiphinema americanum* sensu lato with descriptions of fifteen new species (Nematoda, Longidoridae). *Nematologia Mediterranea*, 7, 51-106.

Lear, B., Goheen, A. C., & D.J., R. (1981). Effectiveness of soil fumigation for control of fanleaf-nematode complex in grapevine. *American Journal of Enology and Viticulture*, 31, 208-211.

Lecouls, A. C., Salesses, G., Minot, J. C., Voisin, R., Bonnet, A., & Esmenjaud, D. (1997). Spectrum of the *Ma* genes for resistance to *Meloidogyne* spp. in Myrobalan plum. *Theoretical & Applied Genetics*, 95, 1325-1334.

Lider, L. A. (1954). Inheritance of resistance to a root-knot nematode (*Meloidogyne incognita* var. *acrita* Chitwood) in *Vitis* spp. *Proceedings of the Helminthological Society of Washington*, 21, 53-60.

Lider, L. A. (1959). Nematode resistant rootstocks for California vineyards. *California Agricultural Experiment Station leaflet*, No. 114.

Lider, L. A. (1960). Vineyard trials in California with nematode resistant grape rootstock. *Hilgardia*, 30, 123-152.

Lider, L. A., & Goheen, A. C. (1986). *Field resistance to the grapevine fanleaf virus-Xiphinema index complex in interspécific hybrids in Vitis*. Paper presented at the 4th Intern Symp Grapevine Breeding, Verona, Italy, April 1985. *VigneVini*, 13 (suppl.12), 166-169.

Loubser, J. T., & Meyer, A. J. (1987). Resistance of grapevine rootstocks to *Meloidogyne incognita* under field conditions. *South African Journal of Enology and Viticulture*, 8, 70-74.

Malan, A. P., & Meyer, A. J. (1993). Interaction between a South African population of *Xiphinema index* and different grapevine rootstocks. *South African Journal of Enology and Viticulture*, 14, 11-14.

Martelli, G. P. (1978). Nematode-borne viruses of grapevine, their epidemiology and control. *Nematologia Mediterranea*, 6, 1-27.

Martelli, G. P., & Savino, V. (1991). Fanleaf degeneration. In: Pearson, R. C., & Goheen, A. C. (Eds). *Compendium of Grape Diseases*. American Phytopathological Society, St Paul MN: 49-49.

Mauro, M., Toutain, S., Walter, B., Pinck, L., Otten, L., Coutos-Thevenot, P., et al. (1995). High efficiency regeneration of grapevine plants transformed with the GFLV coat protein gene. *Plant Science*, 112, 97-106.

Mauro, M., Coutos-Thevenot, P., Boulay, M., Barbier, P., Walter, B., Valat, L., et al. (2000). *Analysis of 41B (Vitis vinifera x V. berlandieri) grapevine rootstocks for grapevine fanleaf virus resistance.* Paper presented at the 7th Intern Symp Grapevine Genetics and Breeding, Montpellier, France, July 1998. *Acta Horticulturae*, 528, 313-319.

McKenry, M. V. (1992). Nematodes. In D. L. Flaherty, L. P. Christensen, W. T. Lanini, J. J. Marois, P. A. Phillips & L. T. Wilson (Eds.), *Grape pest management*. Oakland: University of California.

McKenry, M. V., & Kretsch, J. O. (1995). *It is a long road from the finding of a new rootstock to the replacement of a soil fumigant.* Annual International Research Conference on methyl bromide alternatives and emission reductions.

McKenry, M. V., Kretsch, J. O., & Anwar, S. A. (2001a). Interactions of selected *Vitis* cultivars with ectoparasitic nematodes. *American Journal of Enology and Viticulture*, 52, 304-308.

McKenry, M. V., Kretsch, J. O., & Anwar, S. A. (2001b). Interactions of selected *Vitis* cultivars with endoparasitic nematodes. *American Journal of Enology and Viticulture*, 52, 310-316.

McKenry, M. V., & Anwar, S. A. (2006). Nematode and grape rootstock interactions including an improved understanding of tolerance. *Journal of Nematology*, 38, 312-318.

Meredith, C. P., Lider, L. A., Raski, D. J., & Ferrari, N. L. (1982). Inheritance of tolerance to *Xiphinema index* in *Vitis* species. *American Journal of Enology and Viticulture*, 33, 154-157.

Nesbitt, W. B. (1974). Breeding resistant grape rootstocks. *HortScience*, 9, 359-361.

Nicol, J. M., Stirling, G. R., Rose, B. J., May, P., & Van Heeswijck, R. (2000). Impact of nematodes on grapevine growth and productivity: current knowledge and future directions, with special reference to Australian viticulture. *Australian Journal of Grape and Wine Research*, 2, 109-127.

Olien, W. C. (1990). The muscadine grape: botany, viticulture, history and current industry. *HortScience*, 25, 732-739.

Patel, G. I., & Olmo, H. P. (1955). Cytogenetics of *Vitis* : I. The hybrid *V. vinifera* x *V. rotundifolia*. *American Journal of Botany*, 42, 141-159.

Pinochet, J., & Raski, D. J. (1977). Observations on the host-parasite relationship of *Pratylenchus vulnus* on grapevine, *Vitis vinifera*. *Journal of Nematology*, 9, 87-88.

Pinochet, J., Verdejo, S., Soler, A., & Canals, J. (1992). Host range of a population of *Pratylenchus vulnus* in commercial fruit, nut, citrus and grape rootstocks in Spain. *Journal of Nematology*, 24, 693-698.

Planchon, J. E. (1887). Monography of true *Ampelideae* in *Monographia Phanerogamarum* A et C de Candolle (in French): 5, 305-364.

Pouget, R. (1975). Method of contamination of grapevine roots *in vitro* by radicicolous *Phylloxera*: Application to the search for resistant rootstocks (in French). *Connaissance de la Vigne et du Vin*, 3, 165-176.

Ramsdell, D. C., Bird, G. W., Warner, F. W., Davenport, J. F., Diamond, C. J., & Gillet, J. M. (1996). Field pathogenicity studies of four species of plant-pathogenic nematodes on French-American hybrid grapevine cultivars in Michigan. *Plant Disease*, 80, 334-338.

Raski, D. J., & Radewald, J. D. (1958). Reproduction and symptomatology of certain ectoparasitic nematodes on roots of Thompson seedless grape. *Plant Disease Reporter*, 42, 941-943.

Raski, D. J., & Schmitt, R. V. (1964). Grapevine responses to chemical control of nematodes. *American Journal of Enology and Viticulture*, 15, 199-203.

Raski, D. J., Hewitt, W. B., Goheen, A. C., & Taylor, R. H. (1965). Survival of *Xiphinema index* and reservoirs of fanleaf virus in fallowed vineyard soil. *Nematologica*, 11, 349-352.

Raski, D. J., Goheen, A. C., Lider, L. A., & Meredith, C. P. (1983). Strategies against grapevine fanleaf virus and its nematode vector. *Plant Disease*, 67, 335-339.

Riaz, S., Vezzulli, S., Harbertson, E. S., & Walker, A. M. (2007). Use of molecular markers to correct grape breeding errors and determine the identity of novel sources of resistance to *Xiphinema index* and Pierce's disease. *American Journal of Enology and Viticulture*, 58, 494-498.

Sauer, M. R. (1977). Nematode resistant grape rootstocks. *Australian Dried Fruit News*, 5, 10-14.

Schindler, A. F. (1957). Parasitism and pathogenicity of *Xiphinema diversicaudatum*, an ectoparasitic nematode. *Nematologica*, *2*, 25-31.

Small, J. K. (1913). *Flora of the Southeastern United States*. New York. 2nd ed., New York, 1394 pp.

Snyder, E. (1936). Susceptibility of grape rootstocks to root-knot nematodes. *US Dept Agric Circular*, 405, 1-15.

Sopp, E., Rühl, E. H., & Holst, H. (1998). Resistance of rootstocks to the virus transmitting nematode *Xiphinema index*. *Wein-Wissenchaft* (Wiesbaden), 53, 3-6.

Stanton, J. M., & O'Donnell, W. E. (1998). Assessment of the North Carolina differential host test for identification of Australian populations of root-knot nematodes (*Meloidogyne* spp.). *Australian Plant Pathology*, 27, 104-111.

Staudt, G., & Kassemeyer, H. H. (1990). *Resistance to transmission of grapevine fanleaf virus by Xiphinema index in some Vitis species and hybrids*. Paper presented at the 5th International Symposium on Grape Breeding, St-Martin/Pfalz, Germany, September 1989. *Vitis* (special issue) 223-227.

Staudt, G., & Weisher, B. (1992). Resistance to transmission of grapevine fanleaf virus by *Xiphinema index* in *Vitis rotundifolia* and *Vitis munsoniana*. *Wein-Wissenchaft* (Wiesbaden), 47, 56-61.

Taylor, C. E., & Robertson, W. M. (1970). Sites of virus retention in the alimentary tract of the nematode vectors, *Xiphinema diversicaudatum* (Micol.) and *X. index* (Thorne and Allen). *Annals of Applied Biology*, 66, 375-380.

Teliz, D., Grogan, R. G., & Lownsberry, B. F. (1966). Transmission of tomato ringspot, peach yellow bud mosaic and grape yellow vein viruses by *Xiphinema americanum*. *Phytopathology*, 58, 658-663.

Thorne, G., & Allen, M. (1950). *Paratylenchus hamatus* n. sp. and *Xiphinema index* n. sp., two nematodes associated with fig roots. *Proceedings of the Helminthological Society of Washington*, 17, 27-35.

Valat, L., Fuchs, M., & Burrus, M. (2006). Transgenic grapevine rootstock clones expressing the coat protein or movement protein genes of grapevine fanleaf virus: Characterization and reaction to virus infection upon protoplast electroporation. *Plant Science*, 170, 739-747.

Vigne, E., Komar, V., & Fuchs, M. (2004). Field safety assessment of recombination in transgenic grapevines expressing the coat protein gene of grapevine fanleaf virus. *Transgenic Research*, 13, 165-179.

Viss, W. J., & Driver, D. J. (1996). *Key grapevine pests and diseases in North America and resistance through genetic engineering*. Paper presented at the Brighton Crop protection Conference, Pest and Diseases. 3B2: 125-130.

Walker, G. E. (1995). Nematodes associated with grapevine foundation planting at Loxton. *Australian Grapegrower & Winemaker*, 381, 34-40.

Walker, M. A., & Meredith, C. P. (1990). *The genetics of resistance to grape fanleaf virus in Vitis vinifera*. Paper presented at the 5th International Symposium on Grape Breeding, St-Martin-Pflaz, Germany, September 1989. *Vitis* (special issue) 228-238.

Walker, M. A., & Jin, Y. (1998). Development of resistant rootstocks to control *Xiphinema index* and fanleaf degeneration. *Acta Horticulturae*, 473, 113-120.

Walker, M. A., & Jin, Y. (2000). *Breeding Vitis rupestris × Muscadinia rotundifolia rootstocks to control Xiphinema index and fanleaf degeneration*. Paper presented at the 7th International Symposium on Grapevine Genetics and Breeding, Montpellier, France, July 1998. *Acta Horticulturae*, 528, 511-515.

Walker, M. A., Meredith, C. P., & Goheen, A. C. (1985). Sources of resistance to grapevine fanleaf virus (GFV) in *Vitis* species. *Vitis*, 24, 218-228.

Walker, M. A., Wolpert, J. A., Vilas, E. P., Goheen, A. C., & Lider, L. A. (1989). Resistant rootstocks may control fanleaf degeneration of grapevine. *California Agriculture*, 43, 13-14.

Walker, M. A., Lider, L. A., Goheen, A. C., & Olmo, H. P. (1991). VR 039-16. *HortScience*, 26, 1224-1225.

Walker, M. A., Wolpert, J. A., & Weber, E. (1994a). Viticultural characteristics of VR hybrid rootstocks in a vineyard site infected with grapevine fanleaf virus. *Vitis*, 33, 19-23.

Walker, M. A., Wolpert, J. A., & Weber, E. (1994b). Field screening of grape rootstock selections for resistance to fanleaf degeneration. *Plant Disease*, 78, 134-136.

Walker, M. A., Ferris, H., & Eyre, M. (1994c). Resistance in *Vitis* and *Muscadinia* species to *Meloidogyne incognita. Plant Disease*, 78, 1055-1058.

Wang, X., Bosselut, N., Castagnone, C., Voisin, R., Abad, P., & Esmenjaud, D. (2002). PCR multiplex identification of single individuals of the Longidorid nematodes, *Xiphinema index*, *X. diversicaudatum*, *X. vuittenezi* and *X. italiae* using specific primers from ribosomal genes. *Phytopathology*, 93, 160-166.

Weiner, A., & Raski, D. J. (1966). New host records for *Xiphinema index* Thorne & Allen. *Plant Disease Reporter*, *30*, 27-28.

Williamson, V. M. (1998). Root-knot nematode resistance genes in tomato and their potèntial for future use. *Annual Review of Phytopathology*, 36, 277-293.

Wyss, U. (2000). *Xiphinema index*, maintenance and feeding in monoxenic culture. In K. Maramorosch & F. Mahmood (Eds.), *Maintenance of human, animal and plant pathogen vectors*. Science Research Associates. Chicago: 251-281.

Xu, K., Riaz, S., Roncoroni, N. C., Jin, Y., Hu, R., Zhou, R., & Walker, M. A. (2008). Genetic and QTL analysis of resistance to *Xiphinema index* in a grapevine cross. *Theoretical and Applied Genetics*, 116, 305-311.

Zijlstra, C. (1997). A fast PCR assay to identify *Meloidogyne hapla*, *M. chitwoodi*, and *M. fallax* and to sensitively differentiate them from each other and from *M. incognita* mixtures. *Fundamental and Applied Nematology*, 20, 505-511.

Zijlstra, C., Donker-Venne, D. H. T. M., & Fargette, M. (2000). Identification of *Meloidogyne incognita*, *M. javanica* and *M. arenaria* using sequence characterised amplified region based PCR assays. *Nematology*, 2, 847-853.

TATYANA BILEVA[1], BORYANA CHOLEVA[2], SUE HOCKLAND[3], AURELIO CIANCIO[4]

MANAGEMENT OF VIRUS-TRANSMITTING NEMATODES WITH SPECIAL EMPHASIS ON SOUTH-EAST EUROPE

[1] *Agricultural University, Faculty of Plant Protection and Agroecology, 12 Mendeleev str., 4000 Plovdiv, Bulgaria*

[2] *Sofia University, Faculty of Biology, 8 Dr.Tzankov Blv., 1164 Sofia, Bulgaria*

[3] *Plant Health Group, Central Science Laboratory, Sand Hutton, York YO41 1LZ, UK*

[4] *Istituto per la Protezione delle Piante, CNR, Via Amendola 122/D 70126 Bari, Italy*

Abstract. Available strategies for the management of nematode vectors of plant viruses are reviewed, focusing on the nematode vector species, their associated viruses, as well as their geographic distribution and spread. Diagnostic procedures including morphological identification of virus vectors, plant tests and transmission assays as well as the application of molecular detection tools are reviewed, in the light of preventive and phytosanitary procedures. Management of GFLV on grapevine requires production of healthy plants for certification and marketing schemes, to be used in soils found free of its vector, *Xiphinema index*. In fields already infested, some integrated management options may be applied including, in order of importance, agronomic practices with long rotations (5-7 years) intercropping with poor or antagonistic hosts, chemical control, application of organic amendments and natural products, biofumigation, nematicidal plants and biological control agents. Given the risk that due to some nematodes' parthenogenetic reproduction epidemics may arise even from a single individual vector, emphasis must be given to preventive and continuous field monitoring procedures.

1. INTRODUCTION

Two families of polyphagous ectoparasitic nematodes, Longidoridae (Dorylaimida) and Trichodoridae (Triplonchida), include species capable of transmitting plant

A. Ciancio & K. G. Mukerji (eds.), *Integrated Management of Fruit Crops and Forest Nematodes*, 215-242.

viruses. In particular, twenty-four nematode species within Longidoridae transmit twelve viruses of the genus *Nepovirus* and one of *Sadwavirus*, whereas all three members of the genus *Tobravirus* are vectored by thirteen nematode species belonging to the genera *Paratrichodorus* and *Trichodorus* (Trichodoridae) (Taylor & Brown, 1997; Decraemer & Robbins, 2007).

Among nematode-transmitted viruses, *Grapevine fanleaf nepovirus* (GFLV) is the causal agent of an economically important decline of grapevine, causing stem and leaf deformations and extensive leaf yellowing. GFLV is widespread in many regions of the world where grapevines are grown, and the yield losses can be severe, with progressive epidemic foci developing in infested vineyards within a few years, following initial infestation (Martelli & Savino, 1991; Martelli, 2002).

The vector of GFLV is *Xiphinema index*. Other important nematode vectors transmitting economically important viruses include *Xiphinema diversicaudatum*, which transmits *Strawberry latent ringspot sadwavirus* (SLRSV) and *Arabis mosaic nepovirus* (ArMV), *X. rivesi*, vector of *Cherry raspberry leaf nepovirus* (Cheravirus, CRLV), *Tobacco ringspot nepovirus* (TRSV) and *Tomato ringspot nepovirus* (ToRSV), and *X. americanum sensu stricto*, vector of TRSV, TomRSV, CRLV. *Peach rosette mosaic nepovirus* (PRMV) is transmitted by *X. americanum sensu lato*.

Xiphinema index, as many other Longidorid nematodes, is a long-lived species whose life-cycle may exceed one year. It may remain viruliferous for long periods of time and is capable of acquiring GFLV particles even when feeding on root debris scattered in soil after plants removal (Raski et al., 1965). Long-term retention of virus particles has also been observed experimentally for other nematode-virus associations, as demonstrated for individuals of *X. rivesi* stored for up to three years at 1-3°C in soil, which remained viable and capable of transmitting TomRSV for at least two years (Bitterlin & Gonsalves, 1987).

Given the specific association between the virus and its vector, GFLV particles acquired during feeding remain adsorbed on the inner lining of the stylet and oesophagus, where they may be retained for long periods of time. They are released through salivation when the nematodes feed on new roots, but are lost at moult (Taylor & Robertson, 1970; Wang et al., 2002).

Nematodes belonging to *X. americanum sensu lato* in North America have been shown to be vectors of four economically important nepoviruses that are listed in the EPPO lists of pests recommended for regulation, namely CRLV, PRMV, TRSV and ToRSV. Thus in Europe, non-European populations of *Xiphinema* species belonging to *X. americanum sensu lato* are listed in Annex IAI of the EU Plant Health Directive 2000/29/EC, being considered as harmful organisms, not known to occur in any part of the community and relevant for the entire community. Brown et al. (1994) reported that the species *Xiphinema americanum sensu stricto*, *X. californicum* and *X. rivesi* transmitted CRLV, TRSV and ToRSV and noted the broad spectrum of virus transmission capabilities of these North American populations, compared to the relatively narrow specificity of transmission that exists between indigenous European nepoviruses and their vector species. *Xiphinema bricolensis* (now known as *X. bricolense*) transmitted only the two serologically distinguishable strains of ToRSV but was more efficient vector of the peach stem pitting (PSP) strain than of the prune brown line (PBL) strain of this virus. However, the identity of other species that may transmit viruses, coupled with the difficulty of distinguishing many species in this group from one another, has resulted in phytosanitary legislation which continues to refer to all species morphologically similar to *X. americanum sensu stricto* as *X. americanum sensu lato.*

The number of species in *Xiphinema americanum sensu lato* has risen to about 51, but despite several morphological and molecular studies there remains considerable taxonomic debate about the number of species in the group (Coomans et al., 2001). For this reason phytosanitary legislation in Europe continues to list *X. americanum sensu lato* (non-European populations).

Until recently, no European populations of *X. americanum sensu lato* have been shown to transmit the quarantine-listed viruses, but Širca et al. (2007a) reported transmission of TRSV and ToRSV to bait plants by a Slovenian population of *X. rivesi* with no known links to imported consignments. This knowledge reinforces the current requirement to prevent spread of the listed nepoviruses in Europe.

Among the other nematode vector genera, *Longidorus apulus* is the vector of *Artichoke italian latent nepovirus* (AILV) in Southern Italy (Lamberti & Roca, 1987). Other isolates of the same virus are transmitted by *L. fasciatus* in Greece (Roca et al., 1982; Rana & Kyriakopoulou, 1982; Kyriakopoulou, 1996). *Longidorus elongatus* is the vector of *Raspberry ringspot nepovirus* (RRSV) and *Tomato black ring nepovirus* (TBRV). The causal agent of a cherry rosette disease, *Cherry rosette nepovirus* (CRV), is transmitted by *L. arthensis*, whereas *L. macrosoma* is vector of RRSV (Klinger et al., 1985; Brown et al., 1995).

Nematodes of the didelphic genera *Paratrichodorus* and *Trichodorus* are vectors of the three viruses of the genus *Tobravirus*: *Tobacco rattle tobravirus* (TRV), *Pea early-browning tobravirus* (PEBV) and *Pepper ringspot tobravirus* (PepRSV). These viruses cause economically important diseases, especially in potato and ornamental bulbous crops (Harrison & Robinson, 1986). *Paratrichodorus pachydermus* and *Trichodorus primitivus* are important nematode vector vectors of TRV, one of the causal agents of the spraing disease in potato. Detailed lists of nematode vectors and viruses are already available in the literature (Lamberti & Roca, 1987; Brown et al., 1995; MacFarlane, 2003).

1.1. Geographic Distribution and Spread

Nematode distribution maps are useful tools in the study of vector epidemiology and spatial spread, and surveys should be used at regular intervals in order to update the information. This requires, however, the availability of trained taxonomists and specialized identification personnel. Distribution of Longidoridae in Europe has been the subject of an intense investigation in the last few decades and Taylor & Brown (1997) produced a review. Since then, research has continued to record many new species, including *Longidorus artemisiae* (Rubtsova et al., 1999), *L. balticus* (Brzeski et al., 2000), *L. carpathicus*, *L. juglandicola* and *L. piceicola* (Liskova et al., 1997), *L. cretensis* (Tzortzakakis et al., 2001), *L. cylindricapitatus* (Krnjaic et al., 2005), *L. dalmassoi* (Peneva et al., 1999), *L. danuvii* (Barsi et al., 2007), *L. fagi* (Peneva et al., 1997), *L. seinhorsti* (Peneva et al., 1998), *L. sturhani* (Rubtsova et al.,

2001), *Paralongidorus iberis* and *P. monegrensis* (Escuer & Arias, 1997), *P. litoralis* (Palomares et al., 2008), *X. pirinense* (Mincheva et al., 2008) and *X. silvesi* (Roca & Bravo, 1998), as well as expand knowledge on distribution (Agostinelli et al., 2008; Kumari & Decraemer, 2007; 2008; Lamberti et al., 2000; Liskova & Brown, 2003; Širca et al., 2007b). Such work emphasises the continuing need to identify longidorid nematodes to species before commencing any control programme, but comprehensive testing of potential virus vectors has not been done. In addition, the identification and hence virus vector status of members of *Xiphinema americanum sensu lato* remains controversial. Members of this group exhibit a lack of specificity in the transmission of North American nepoviruses which contrasts with the specific associations between European nepoviruses and their vector nematode species. Recent studies of European populations of *X. rivesi* suggest they may be able to transmit non-European nepoviruses (Širca et al., 2007a), thus supporting current legislation that prohibits movement of material so infected.

Xiphinema index is one of the best-studied nematode vectors worldwide. This species is common in the Mediterranean basin where grapevine is cultivated (Lamberti, 1981a; 1981b; Hanna et al., 2008), and has been reported from grapevine cultivated areas in Europe (Albania, Austria, Bulgaria, Croatia, Czech Republic, France, Germany, Greece, Crete, Hungary, Italy, Malta, Moldova, Poland, Portugal, Romania, Russia, Slovakia, Slovenia, Spain and Balearic Islands, Canary Islands, Switzerland, Ukraine, Federal Republic of Yugoslavia), Asia (Armenia, Azerbaijan, Cyprus, Republic of Georgia, India, Haryana, Uttar Pradesh, West Bengal, Iran, Iraq, Israel, Pakistan, Tajikistan, Turkey, Turkmenistan, Uzbekistan), Africa (Algeria, South Africa, Tunisia), North America (USA), South America (Argentina, Brazil, Chile, Peru), Oceania, Australia and New Zealand (Anon, 2000).

Trichodorid nematodes are distributed worldwide and have been frequently reported from Europe and North America on several host plants, including weeds, which may also act as natural reservoirs of transmitted viruses. Decraemer and Robbins, (2007) indicated that didelphic *Trichodorus* spp. originated in the Northern hemisphere, whereas the most probable origin of *Paratrichodorus* spp. is the tropical and subtropical regions. The distribution of these species appears also related to their dispersal with non-indigenous plants or associated soil, introduced as a consequence of agriculture or other human activities. Monodelphic genera (*Allotrichodorus*, *Ecuadorus* and *Monotrichodorus*) for which no virus transmission has yet been reported, originated from Neotropical regions and are considered as endemic in Central and South America (Decraemer, 1995; Decraemer & Robbins, 2007).

Although virus-vector associations generally show a high degree of specificity in Europe, knowledge about the local occurrence of nematode populations and/or virus isolates is important, since differences in the effectiveness of transmission may arise among vector populations and/or within different genomes of the same virus, whose diversity may account for variable epidemic outcomes. For example, tests showed that *Paratrichodorus allius*, the only known vector of TRV in the Columbia River basin of Washington and Oregon (Mojtahedi & Santo, 1999), transmitted some, but not all, isolates of TRV (Crosslin et al., 2003). *Longidorus apulus* is vector of AILV in Southern Italy (Lamberti & Roca, 1987), but a serologically distinguishable strain, isolated from artichoke plants showing symptoms of a yellowing disease described as "artichoke patchy chlorotic stunting" (APCS), is transmitted with high frequencies in Greece by populations of a close species, *L. fasciatus*, occurring in the Northeast Peloponnesus region (Roca et al., 1982; Kyriakopoulou, 1985; Kyriakopoulou, 1996).

Data on field distribution are also important in evaluating the environmental spread of viruses and associated vectors, and their persistence over time. A study of the distribution changes of two similar populations of *X. diversicaudatum* in an uncultivated woodland habitat in England and a cultivated soil in Scotland after 30 and 24 years respectively showed no variation in nematode distribution, but lower densities were found in the undisturbed habitat. In the cultivated soil, nematodes showed aggregation in discrete populations associated with cropping in the intervening years, with a reduced horizontal spreading. Although SLRSV and AMV were found in 1966, only the latter virus was observed in 1991 (Taylor et al., 1994).

Cropping may play a significant role in the dispersal of virus and vectors. In vineyards naturally infested by *X. index* in France, Esmenjaud et al. (2008) observed that soil passing through machinery was a key factor in the dispersal of the nematode and GFLV between fields. In a population dynamics study of *X. americanum sensu stricto* and ToRSV transmission, data showed that the virus spread in a raspberry field at a rate of 70 cm per year, suggesting that virus dispersal, in the absence of nematode-infested soil movement in the field, was limited by its systemic diffusion in plants (Pinkerton et al., 2008; Bitterlin & Gonsalves, 1987).

Due to their importance in international trade and plant protection, the identification and distribution of nematode vectors at the local or regional scale should often be monitored in routine surveys, or before propagating material is used or shipped. Vineyards are also routinely sampled by farmers and extension officers, to detect the presence before transplanting, or to monitor the spread of, *X. index* and/or listed viruses.

In Southern Italy, *X. index* often occurs in association with GFLV and more than 90% of populations sampled were viruliferous or capable to transmit isolates of the virus (Catalano et al., 1992). In recent years in Apulia, this nematode has been found in 28% of samples, taken from fields before transplanting or from vineyards already cultivated.

Knowledge about the distribution and occurrence of nematode vectors can also shed light on the reasons for the sudden appearance of a viral disease. Recent surveys in Italy showed that GFLV is still present in some propagation material. Since *X. index* was not found in the rootstock production fields, the spread was assumed to have occurred by means of infected plants (Bica et al., 2002).

In a recent survey in Bulgaria, several vector nematodes were recorded, including *X. index*, which is frequently associated with GFLV and widely distributed in this country, even in vineyards with heavy (clay) soils in the Southern regions. (Fig.1A). This species predominantly occurs in vineyards and, as elsewhere, also on fig roots, on which it actively multiplies but does not transmit viruses.

Xiphinema italiae is often monitored where grapevine is cultivated. Its capacity as a GFLV vector is controversial, since the transmission reported for a Middle East population was not experimentally reproduced with any other population of this species (Cohn et al., 1970; Lamberti & Roca, 1987; Catalano, 1992). This species is widespread in Bulgaria, mainly on sandy soils along river valleys of the rivers Danube, Maritza, Struma as well as in the Black Sea region (Fig. 1B). Apart from grapevine, it has been found in association with apple, peach, apricot, almond, mulberry, citrus and chestnut. *X. diversicaudatum* is also widespread in South and South-West Bulgaria (Fig. 1C). It is frequently found in association with raspberry and rose as well as blackcurrant, grapevine, pepper, pine tree and grasses, in moist soils ranging from sandy loam to heavy clays.

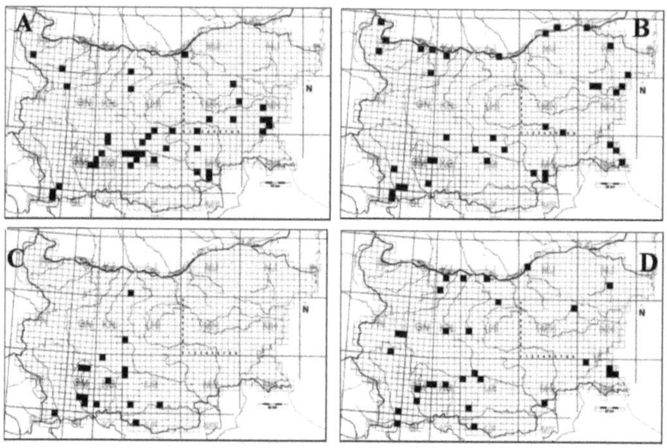

Figure 1. Distribution of Xiphinema index *(A),* X. italiae *(B),* X. diversicaudatum *(C) and* Longidorus elongatus *(D) in Bulgaria.*

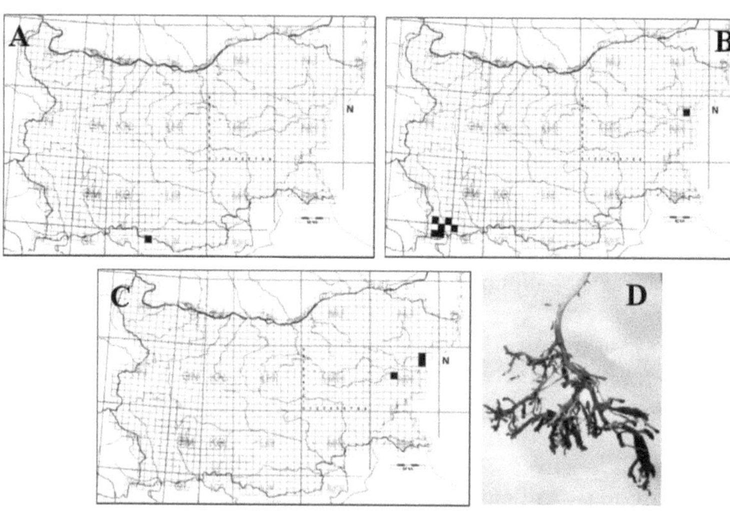

Fig. 2. Distribution of Longidrus macrosoma *(A),* L. pisi *(B) and* Paralongidorus maximus *(C) in Bulgaria and galling on ivy* (Hedera sp.) *root tips induced feeding of latter species (D).*

Longidorus elongatus is also widespread in Bulgaria on a wide host range including many crops of economic importance like grapevine, fruit trees, berries, vegetables and grasses (Fig. 1D). Other species of *Longidorus* include *L. macrosoma*, reported only in South Bulgaria at Arda, in association with raspberry, and *L. pisi*, reported in this country in sandy to sandy-loam soils, in association with tobacco (preferred host), grapevine, tomato, pepper, peach, corn, peanut and chestnut (Fig. 2A, B).

Finally, *Paralongidorus maximus*, a large nematode about 12 mm long, was reported from Norten-East Bulgaria (Galata, Varna) in association with vineyards and ivy (Fig. 2C). This species is vector of AMV, SLRSV, CLRV (Jones et al., 1981) and RRSV (Brown & Trudgill, 1998). Feeding by *P. maximus* causes typical galls on parasitized roots (Fig. 2D).

2. VECTOR AND VIRUS DIAGNOSIS

Nematodes belonging to *X. americanum sensu lato* have a very wide host range of both herbaceous and woody plants in agriculture, horticulture and forestry. As free-living ectoparasites they are also found in soil residues or growing media and some species can overcome dry periods and survive for years in soil even in the complete absence of host plants. In the absence of virus infection, the aerial parts of plants show no symptoms unless there is a heavy infestation, when roots show swellings close to the root tips. With high populations, typical symptoms of root damage such as reduction in vigour or signs of drought, may be seen. These nematodes, like most ectoparasitic plant-parasitic nematodes, can only be detected by extraction from soil or growing media. The most usual method is by the Flegg modified Cobb technique but other elutriation methods can also be used.

2.1. Morphological Identification of Virus Vectors

The identification of virus vector nematodes has, until recently, depended upon the use of morphological keys, utilising the morphological characters of each group and species. This requires the essential skills of trained specialists and taxonomists, and still forms an integral part of the classification of species. Increasingly, molecular tools are supporting such work, but care should be taken if these are used as the sole tool of identification, to ensure that protocols use accredited populations as controls and have also taken into account the full range of species that might be present, in order to reduce the risk of false positives. This is especially important in quarantine laboratories where new, invasive species might be expected.

The diagnosis of the genus *Xiphinema* has been recently described by Coomans et al., (2001), but the identification key for *Xiphinema* spp. (apart from the *X. americanum* group) of Loof and Luc (1990) is still widely used, with supplements (Loof & Luc, 1993; Loof et al., 1996). Investigations into the identity of '*X. americanum*' started in 1979 when Lamberti & Bleve-Zacheo studied populations from disparate geographical areas and concluded that they were dealing with 25 different species, 15 regarded as new. The species were difficult to distinguish and thus the term *X. americanum sensu lato* is used to describe species that are morphologically similar to *X. americanum*, with the latter species being more correctly know as *X. americanum sensu stricto.* The current taxonomic situation of *X. americanum sensu lato* is not clear, due to the high number of species defined in it (about 51 at present), the weak differences reported between many species, the lack of data on intraspecific variability for the majority of the species as well as insufficient illustrations for many of them. Whilst Lamberti et al. have published keys to this group (2000; 2004), a thorough revision of the group, combined with a molecular study, is still required to clarify the number of putative species. Currently international protocols are being developed and progress can be followed on the EPPO web-site (www.eppo.org).

The identification key for species of the genus *Longidorus* produced by Chen et al., (1997) is widely used but it is in need of revision due to the number of new species that have been described since its publication. The most recent key to species in the family Trichodoridae was provided by Decraemer & Baujard (1998).

2.2. Transmission Assays

Traditional bioassays on host plants are carried out in controlled trials and depend on long-term feeding of the virus-vector nematodes, to reveal their transmission capability. Tests with host plants require the subsequent collection of data on acquisition and/or retention time, virus transmission efficiency and detection of induced symptoms (i.e. ringspots). Several weeks are needed for detection with this procedure, with difficulties arising due to the survival of the nematodes or plants, evidence and clarity of symptoms and/or plant sensitivity to the viral disease (Trudgill & Brown, 1972; Jones et al., 1981; Brown et al., 1995). Although time consuming, this approach is universally recognised and is still indispensable to prove the status of a species as a vector.

2.3. Molecular Detection

A reliable diagnostic procedure should be available for successful management of virus-vector nematode populations, to implement preventive certification schemes or to monitor nematodes and viruses spreading in fields over time. A wide range of detection procedures have thus been developed and applied in epidemiology and transmission studies, to test the vectoring capacity (capability?) of several nematode species (Roberts & Brown, 1980). A wide range of antibodies and molecular probes allow fast and accurate assays, and have significantly improved the resolution power of early diagnostic procedures, which were based on immunosorbent or transmission electron microscopy tests capability (Esmenjaud et al., 1993; 1994; Raski et al., 1965; Taylor & Raski, 1964; Taylor & Robertson, 1970).

GFLV detection in plant tissues is routinely performed with CP (coat protein) specific antibodies in ELISA (Esmenjaud et al., 1993; 1994) or by means of PCR assays (Rowhani et al., 1993). The latter technique may also be used to check the presence of the GFLV particles in nematodes (Esmenjaud et al., 1994). For the quantification of the viral load in the vectors, however, different methods may be chosen, depending on the detection goals and the target organism, including ELISA or quantitative PCR (Boonham et al., 2002; 2003; Esmenjaud et al., 1993; 1994; Taylor & Robertson, 1970).

Detection of TRSV and TomRSV by means of fluorescent antibodies was performed in single specimens *X. americanum* and this approach proved to be useful in localizing the virus particles in different regions of the nematode food canal (Wang et al., 2002).

A routine detection procedure should be simple enough to allow detection among a batch of soil-extracted specimens and sufficiently cheap to reduce the cost of consummables. In Germany, *Longidorus attenuatus*, *L. elongatus*, *L. macrosoma* and *Paralongidorus maximus* are economically important grapevine pests since they vector RRSV and TBRV). These species occur in vineyard soil with other non-vectoring but morphologically similar longidorid species, namely *L. helveticus*, *L. profundorum* and *L. sturhani*. Species-specific primers were developed from ribosomal DNA for all seven nematode species to facilitate taxonomic identification and reliably discriminate from closely related longidorid species and soil nematode communities. Primers were assessed for reliability by screening different populations of each species and allowed multiplex assays to detect the three target nematode species in the same PCR reaction (Hübschen et al., 2004a).

Species-specific primers were also developed for *X. diversicaudatum*, *X. index*, *X. italiae* and *X. vuittenezi* for multiplex tests which were sufficiently sensitive to detect all developmental forms of these nematodes (Wang et al., 2003). The procedure appeared valid to distinguish the target nematodes also from other longidorid species occurring in vineyard soils. Only a minor sensitivity was observed for the primers designed for *X. index*, which occasionally yielded an amplification product also from the DNA of *L. elongatus* (Hübschen et al., 2004b).

For viruses like GFLV, amounts of RNA-2 molecules were used to get a first estimate of the viral load in the nematodes. However, for counting the number of viral particles in the vector, antibody-based techniques represent a valid tool, since GFLV B particles may contain two RNA-2 or, in alternative, a single RNA-1 molecule (Belin et al., 2001). However, quantitative RT-PCR was used to estimate the amount of GFLV RNA-2 in *X. index*, with self-hybridization fluorescent probes designed on the CP gene sequence of an Italian isolate. Details on how these probes function, including their thermodynamics, have been published elsewhere (Tyagi et al., 1998; Thelwell et al., 2000; Whitcombe et al., 1999).

For detection of the virus isolates, recent advances in diagnostics with DNA-based probes offer a specificity and reliability higher than serological techniques (Tyagi et al., 1998; Thelwell et al., 2000; Whitcombe et al., 1999). In fact, recognition through antibodies may not detect changes occurring at the nucleic acid level, i.e. when nucleotide polymorphisms do not alter the aminoacid composition of the coding proteins. Furthermore, DNA-based technologies offer a higher flexibility in the choice of the target nucleic acid sequence to amplify, as well as a variety of detection protocols.

GFLV detection in specimens of *X. index* focused on RT-PCR and other techniques (Esmenjaud et al., 1994; Wetzel et al., 2001; 2002). Techniques based on probe recognition of specific DNA fragments (i.e. TaqMan, molecular beacons) increased the speed and resolution of diagnostics, although they require RNA-to-DNA reverse transcription. These techniques are useful when routine samplings must be performed, avoiding the separation of the amplified DNA from the reaction mix. With these type of probes, fluorescence arises only from a successful hybridization, eliminating the need for subsequent electrophoretic analyses and DNA sequencing. Compared to traditional methods, probes like Scorpions® and molecular beacons offer a detection sensitive at the single-nucleotide level of the target sequence, saving time and sequencing costs (Wetzel et al., 2002). Direct analysis of specific DNA regions with molecular beacons showed a high performance in Real-Time detection assays from single specimens, and the detection time was in the order of a few hours after nematodes were isolated from the soil extract. When these assays are coupled to a fragment specifically amplified from the vector, the parallel identification of the virus and the nematode is also possible (Finetti-Sialer & Ciancio, 2005).

A benefit of this molecular approach is the possibility of low cost multiplex detection with fluorescence reading devices and of field epidemiology studies, which may be performed by processing samples in batches. For higher sensitivity and reliability, however, several nematodes should be used, since the efficiency of template preparation and processing also affect the assay. Other factors, i.e. virus prevalence in the vector population and its spread in the field, also influence the detection efficiency.

In theory, the lower resolution limit of molecular probes correspond to a single transcript. Considering that not all nematodes may be carrying virus, several samples should be processed in batch, to increase the assay reliability. In replicated tests, an average 10% of a *X. index* population showed no RT-PCR amplification (Finetti-Sialer & Ciancio, 2005). Using a 'housekeeping' gene may also help in preventing false negative results, possibly provided with the same level of expression of the gene of interest.

Finally, reliable detection of field isolates is information-dependent, since it requires knowledge of the virus isolates present, in order to identify conserved and isolate-specific regions. The application of specific probes (i.e. Scorpions, molecular beacons) to detect single nucleotide polymorphisms increases resolution, allowing isolate level detection, since these probes only recognize their own complementary targets. This property is useful when a discrimination among isolates of the same virus is required, or when low amounts of template are available, as is the case of single nematodes.

Molecular probes may also allow identification of single nucleotide changes, as a reliable alternative to DNA sequencing (Tyagi et al., 1998; Whitcombe et al., 1999). This property is useful especially when the nucleotide substitutions do not affect the coded aminoacid sequence, and no separation is possible if antibodies were applied. Further advantages derive by the single nucleotide mismatch sensitivity of this class of molecules, which allow a wider choice in probe design and selection of the sequence target. These properties reflect the information stored in the genome to be detected and have practical implications when virus identification is required in quarantine or diagnostic applications.

3. INTEGRATED MANAGEMENT

Whitehead (1998) gave a useful summary of control strategies for virus vector species. Several procedures must be applied, depending on the scale of the problem, from the regional to the field scale. Factors involved in the selection of the most appropriate management schemes include the crop and the dispersal of the transmissible viruses, the resources locally available, including the sources of resistance genes, and the organic or agronomic practices, including intercropping, as well as the permitted use of chemicals.

3.1. Prevention and Quarantine Procedures

Xiphinema index and the *X. americanum* group are the most commonly listed virus vector nematodes in quarantine legislation and the latter group (non-European populations) is listed in IAI of the EU Plant Health Directive 2000/29/EC, as harmful organisms whose introduction into, and spread within, all member states shall be banned. Other virus vectors are considered to pose a minor risk unless they are intercepted on imported consignments as alien species. This may occur despite the prohibition of the import of untreated soil by virtually every country, as soil often adheres to tubers and plant roots, as well as machinery. Pest risk analyses may determine that such nematodes then become regulated non-quarantine pests; with the withdrawal of many agrochemicals, there is seldom any alternative but to destroy such consignments.

For native species of virus vectors, phytosanitary measures can, in many cases, be used as a model for both preventative and sustainable control programmes. Certification and marketing schemes for the production of plant material for propagation include some aspect of freedom from plant-parasitic nematodes. This is usually achieved by growing plants in approved growing media according to officially agreed protocols, or transporting plants with bare roots. They also include hygiene and cultural methods to minimise the possibility of infestations and infections. More details of phytosanitary procedures can be found in Hockland et al. (2006).

Data on the spread and epidemiology of nematode vectors and associated viruses play a fundamental role in the development of quarantine and prevention schemes, at different scales. Quarantine procedures target both the vector and the virus, since both the nematodes and the transmitted viruses require monitoring and detection efforts. For example, since GFLV can also be mechanically transmitted, the most common procedure aimed at preventing its dispersal is the use of virus-free certified plants. This represents a fundamental strategy to avoid diffusion of GFLV and, in general, of other nematode-transmitted viruses. Thus propagation material certified as virus-free is produced in monitored, pre-multiplicaton fields. This preventive procedure aims at halting the virus dispersal, especially among rootstocks and nurseries, at a scale wider than the single field, by selecting and planting healthy vines. Recent progress in diagnostic procedures (see above) allow the identification of the virus in the host plants in batch, permitting the exclusion of infected propagation material at an acceptable cost (Esmenjaud et al., 1993; 1994; Rowhani et al., 1993).

3.2. Agronomic Practices

Agronomic practices may help in lowering the density of most vectors, but due to the prevalence of polyphagous habits they cannot result in complete exclusion. For this reason, the feeding habits and preferential hosts of the vector should be known. In some cases, such as the disease soybean severe stunt (SSS), caused by *Soybean severe stunt virus* (SSSV), first described in Delaware, USA, a significant reduction occurred when two-year rotations were applied to lower the densities of its putative vector, *X. americanum sensu stricto*. In experimental trials, two years of continuous corn or grain sorghum, wheat followed by 'HT-5203' soybean, or 2-year fallow, reduced both density of *Xiphinema* species in the soil and SSS severity. These observations were supported by greenhouse studies with corn, wheat, marigold, castor and fallow, whereas rotations with soybean tolerant HT-5203 as a single crop for 2 years increased nematode density and SSS severity (Evans et al., 2007).

However, when vector nematodes and/or their viruses are widespread in the agroecosystems or in nearby natural environments, preventive and/or agronomic practices need to be usefully integrated by replicated field monitoring and inspections, since nematodes are commonly dispersed by several means, including irrigation water (Roccuzzo & Ciancio, 1991), wind or soil movement by machinery. Infested fields normally require a long-term quarantine period (4-7 years) before re-planting, and replicated samplings and tests are needed to determine the nematodes densities and their perceived threat as vectors.

3.3. Chemical Control

When a vector nematode species is found in the field, the possibilities for halting its spread in the infested soil or lowering its density by means of chemical treatments such as suitable fumigants or nematicides depend on several factors, including crop rotation (perennial or annual), soil type, roots depth, vector endemism in the surrounding areas and availability of resistant plant germplasm.

Due to the polyphagy of several species, the virus persistance and nematodes occurrence in field margins and/or other uncropped areas of the field, including deepest soil layers in the cropping zone, a complete nematode eradication is often difficult to achieve. A rotation with non-host crops lasting for several years is always necessary when vines are removed from a field infested with *X. index*. This canonical management strategy may be usefully integrated with treatments such as 1,3-dichloropropene (1,3D) (Lamberti, 1991). A significant reduction may be obtained through injected fumigants, applied at various rates, if available (Chapman, 1983). Depending on the legislation applied, fumigants use is, however, limited by the recent ban of most widely used products, i.e. methyl bromide, by the ecological effects on the soil environment or water, as well as by the risk to operator safety that some formulations may cause. 1,3D is, for example, phytotoxic and its persistance in soil is dependent on temperature and water content, as well as on soil physical (texture) and chemical (pH, organic matter) profiles (Lamberti & Basile, 1982). Whether fumigants are used or not after the quarantine period, replicated samplings from the field at specified times (especially spring) are always recommended, to verify the density and status of the vector population, in particular for its transmission capability.

In an eight year study in Northern Italy, application of a granular nematicide, fenamiphos (at rates up to 60 kg a.i./ha) did not appear capable of controlling the density and population dynamics of a *X. diversicaudatum* population attacking peach. This population, however, showed a low efficacy in virus transmission, and shorter periods of intercropping appeared sufficient to minimize the risk of SLRSV spreading, if healthy plants were used at transplant (Lamberti et al., 1993).

In Scotland, dichloropropane-dichloropropene (D-D) applied before planting at 224 or 448 kg/ha showed the best control of potato spraing disease caused by TRV and treatments were more effective in autumn. Methomyl at 9 kg/ha and dazomet at 168 kg/ha greatly decreased the TRV spread in the first year after treatment (Cooper & Thomas, 1971). Dazomet and D-D also showed good control of *L. elongatus* and virus transmission (Taylor & Gordon, 1970).

Quintozene, a fungicide with nematicidal activity, and methomyl both decreased numbers of *L. elongatus* only slowly, but quintozene was shown to prevent virus transmission for up to four years (Murant & Taylor, 1965; Taylor & Murant, 1968). In raspberry plantations, fumigant nematicides like dazomet or D-D, applied prior to planting at 336 kg a.i./ha, controlled *L. elongatus*, reducing transmission of RRSV, whereas two carbamate nematicides (aldicarb or oxamyl) showed only moderate nematode and virus transmission control. Quintozene (89-6 kg a.i./ha) decreased numbers of *L. elongatus* slowly and controlled virus transmission in the field. The acquisition and transmission of TBRV by *L. elongatus* in strawberries was inhibited by oxamyl for less than six months (Trudgill & Alphey, 1976).

3.4. Nematode Resistance in Plants

There have been limited efforts to identify and develop resistance in crops to virus vector nematodes, apart from a number of studies on resistance of grapevine rootstocks to *X. index*, carried out in controlled conditions, which have shown the occurrence of resistant germplasm suitable for exploitation in breeding programs. Hybrids O39-16 (*Muscadinia rotundifolia* × Almería) and Harmony (Dog Ridge × C 1613) for example, allowed low rates of reproduction for *X. index* (Aballay et al., 1998). For an update on resistance to *X. index* and/or GFLV in grapevine rootstocks see Chapter 8 in this Volume. However, such work has yielded sufficient success to justify greater efforts (Starr & Bendezu, 2002). Very little of the available germplasm resources of most crops has been examined for resistance or tolerance to virus vectors, probably because their main economic importance lies in the rapid transmission of the viruses they transmit. However, a plant resistance strategy might reduce the spread of virus diseases. Work continues developing transgenic tobacco plants resistant to TRV (Vassilakos et al., 2008).

4. ORGANIC MANAGEMENT

4.1. Organic and Natural Products

One of the basic priorities of organic and sustainable agriculture is the exploitation of non-synthetic chemicals and cultural methods to control pathogens and parasites, and to increase soil fertility. In this way, sustainable methods of control may be achieved. Interest and investigation of alternative methods and means for integrated control of nematodes has increased, including the possible exploitation of raw industrial materials and organic amendments. In low-input agriculture, some long-term experiments using different substrata obtained by recycling plant residues and other organo-biological products for control of virus vector nematodes are providing possible solutions.

An example of novel methods being investigated concerns the effect of the monocellular green alga *Chlorella vulgaris* (Golden apple, Bulgaria), used to control nematodes (Choleva et al., 2007). An assay on grapevine plantlets cv. Sira infested by *X. index* (2 specimens per test-glass and plant) showed an activity of the algal product (Choleva et al., 2005). After 60 days a lower growth in the infested control was observed, with the typical initial root necrosis caused by *X. index* feeding, as well as a reduced development of leaves and stems. Strong stimulating effects of *C. vulgaris* (0.01g) on plant growth and development were observed in plants infested by *X. index*, in comparison with plants infested with *X. index* and non-infested, untreated controls (Table 1).

Table 1. Plants biometric data from an in vitro *assay with* Chlorella vulgaris

Treatments *	Mean height (cm)	Mean root length (cm)	Plant weight (mg)	
			Fresh	Dry
1 - Control with *X. index*	14.2	3.1	200	39.2
2 - Infested with *X. index* + 0.01 g *Chlorella*	18.4	7.3	150	25.4
3 - Infested with *X. index* + 0.025 g *Chlorella*	10.2	2.1	120	19.8
4 - Control + 0.01 g *Chlorella*	20.3	8.2	160	32.9
5 - Control + 0.025 g *Chlorella*	8.1	1.5	96	16.4
6 - Non-infested, untreated control	16.2	4.1	200	48.9

* Three replications used.

The lowest concentration (0.01 g, treatments 4 and 6) showed suppressive effects against *X. index* and stimulated growth, in comparison with untreated plants, whether or not infested with *X. index* (Fig. 3).

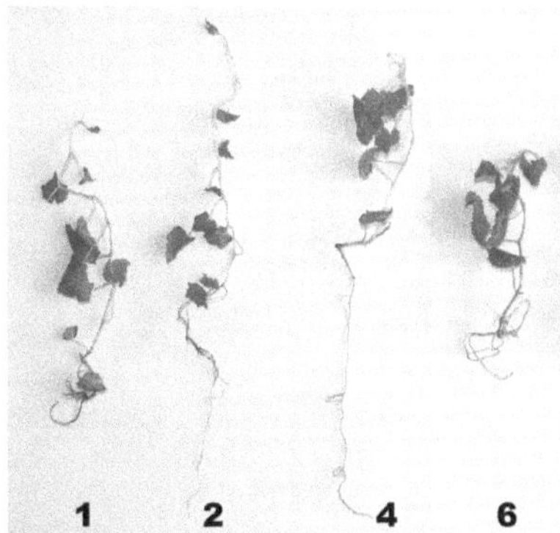

Figure 3. Comparative effect of Chlorella vulgaris *on plants development. For treatments see Table 1.*

In pot experiments with grapevine seedlings cv. Cabernet Sauvignon-157 infested with *X. index* and treated with *C. vulgaris*, carried out during 4 months, phenological data (biomass, height, stem knobs and dry weight), were analyzed. The treatments are shown in Table 1. Density of *X. index* in the infested treatments was 20 specimens / 100 cm^3 soil.

Table 2. Pot assays with grapevine seedlings.

Treatments	*Xiphinema index*	*Chlorella vulgaris* (g)
1	+	0.5
2	+	1.0
3	+	2.0
4	+	-
5	-	-
6	-	0.5
7	-	1.0
8	-	2.0

The effect of *C. vulgaris* on grape seedlings showed promising results, confirmed by plant growth. Three-fold differences occurred for biomass and height in treated infested plants (treatments 2 and 7) in comparison with infested and non-infested controls (treatments 4 and 5) (Fig. 4).

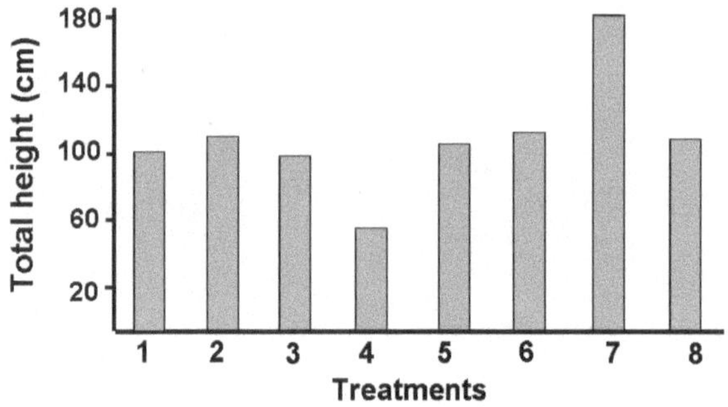

Figure 4. Influence of Chlorella vulgaris *on height of grape seedlings infested by* Xiphinema index.

Stimulating effects of *C. vulgaris* on the root system of grape seedling are shown in Fig. 5.

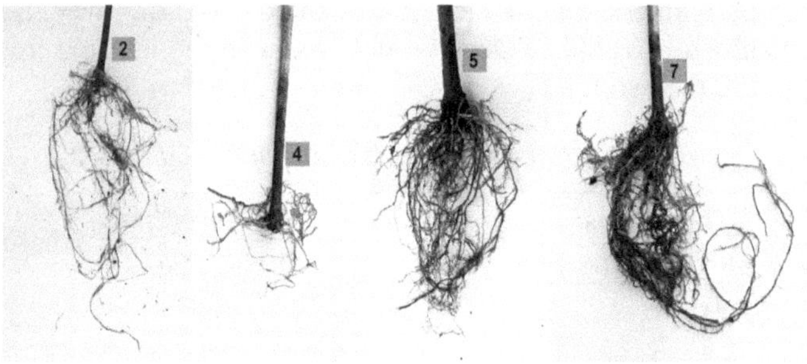

Figure 5. Comparative effect of Chlorella vulgaris *on grape seedlings roots development, with or without* Xiphinema index. V_2 = X. index + *1 g* Chlorella*;* V_4 = X. index *only;* V_5 *= untreated control without nematodes;* V_7 *= not-infested +1 g* Chlorella.

A pot test with grapevine seedlings cv. Cabernet Sauvignon and rootstock $S0_4$ was carried out for six months to study the effect of water nutrient solutions (Biovet ® BG substratum) undiluted or at 1:2, 1:3 on *X. index* (100 specimens per pot with 400 cm^3 sterilized soil) in five treatments and replications. Treatments were: V_1 (pots with *X. index* watered weekly with Biovet solution 1:2); V_2 (pots with *X. index* watered every week with Biovet solution 1:3); V_3 (infested control, pure water); V_4 (non-infested control watered with H_2O); V_5 (non-infested control watered with Biovet solution 1:2). The composition of the Biovet compositing substratum was: pH 6.8-7.7; total nitrogen: 50-75 mg/100 g; phosphorous: (P_20_5) 60-100 mg/100 g; potassium (K_20): 150-225 mg/100 g; calcium (CaO): 280-330 mg/100 g; magnesium (MgO): 40-60 mg/100 g; organic matter: 43-45%; salts: 0.25- 0.46 mmos/100 g. After seven months results showed a significant reduction of *X. index* numbers with Biovet solution 1:2 for the grapevine seedlings of both varieties, in comparison with initial populations. In the treatment with Biovet solution 1:3, a slower increase of the *X. index* population was found. Simultaneously, the population of *X. index* increased five to six-fold in the untreated soil of either Cabernet and rootstock $S0_4$; values about two times higher than in treatments with solution 1:3 and six times higher for solution 1:2 being recorded. More than a three-fold stimulation effect was noted for plant size and biomass if the plants are watered with the nutrient solution 1:2, in comparison with plants in sterilized soil, where the seedlings showed clear nematode damage (Fig. 6).

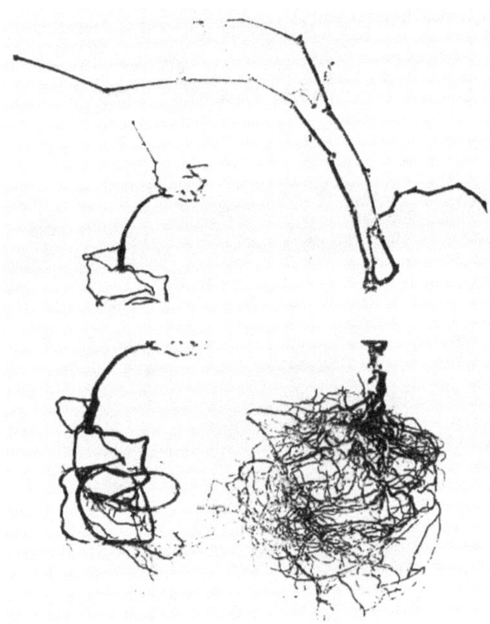

Figure 6. Effect of the Biovet solution 1:2 on test plant compared to infested one (left).

In a one-year field experiment to study the influence of the Biovet substratum in a vineyard cv. Cabernet Sauvignon, infested with *X. index*, three treatments were used, each one on an average area of 0.3 ha: V_1: 10 t of Biovet substratum incorporated in soil with *X. index* (initial density, Pi = 41 nematodes/100 cm^3 soil); V_2: 15 t of Biovet incorporated in soil with *X. index* (Pi = 50 nematodes/100 cm^3 soil); V_3: control with natural manure and *X. index* (Pi = 39 nematodes/100 cm^3 soil). Data showed increased plants growth and reduction of the *X. index* densities in treatments V_1 and V_2 (10 and 15 t of substrate) in comparison to the infested control soil.

In a second field experiment to study the influence of the composting substratum on black currant strongly infested with *Longidorus distinctus* and *Xiphinema diversicaudatum*, the treatments used were: infested soil with nematodes (control) and soil infested and treated with Biovet and potassium humate (Humustim). At planting, 50 cm^3 of the organic product was mulched on the black currant plants. Simultaneous Humustim was applied as a fertilizer trough a leaf spray (80 cm^3 / 20 l) and then at 10-day intervals. In the experiments with Biovet, the substratum from composting microbial waste product was used.

After six months the black currant trial showed significant reduction of *L. distinctus* (only 1 specimen found) and no *X. diversicaudatum*, in the treatment with Biovet (50 cm^3) in comparison with initial populations desnities (*L. distinctus* Pi = 49; *X. diversicaudatum* Pi =10) (Fig. 7A).

The population of *X. diversicaudatum* multiplied by five-six fold, whereas *L. distinctus* increased three-fold in the control. A more than three fold increase was noted for the size and biomass of plants in soil infested and treated with Biovet, in comparison with control soil in which the plants showed clear nematode damage (Fig. 7B).

Figure 7. Black currant field infested by Longidorus distinctus *(A) and comparison (B) between infested black currant plants treated (arrow) and untreated with Biovet.*

4.2. Biofumigation and Nematicidal Plants

Biofumigation is a term originally used to describe the use of bioactive brassicaceous plant products for pest, disease and weed control in agriculture and horticulture. The technique has gained interest in recent years due to the phase-out of synthetic soil fumigants and a general interest in more environmentally sensitive plant production systems worldwide. The concept is based on capturing benefits from the bioactive products of the glucosinolate-myrosinase system in plants, which originally evolved as part of their own defence system.

There is a plethora of terms describing crops that are being developed to achieve sustainable benefits for growers; catch crops to attract pathogens which are then disposed, cover crops to prevent weed growth, green manure crops to improve soil condition and structure, and biofumigants to produce natural substances that were the precursors of chemical fumigants. However, with increasing pressure for growers to adopt sustainable farming practices and be economically viable in the short term, there is a demand for crops that can combine functions. Hence breeders may enhance the properties of particular lines so that future crops may be multi-functional, e.g. a variety that shows particular benefits against plant-parasitic nematodes can be improved by increasing its biomass, so that it also serves to return more organic material to the soil and hence improve soil structure, water holding capacity and levels of beneficial micro-organisms. However, growers naturally consider yields and profits as measures of the benefits of a particular farming system so new crops or strategies must also be beneficial from an economical point of view.

Choice of crop is also important: ideally they should be poor hosts, but some, like certain brassicas, can be good hosts for some groups such as trichodorids (Sue Hockland, personal communication). To date there has been relatively little work done on the effectiveness of such crops on virus vector species. Thus it will continue to be important for growers to assess the susceptibility of particular fields so that the process of growing cover crops or biofumigants does not encourage pathogens. As an example, Hairy vetch (*Vicia villosa*) is used as a cover crop for weed control but it will increase numbers of some root-lesion nematodes such as *Pratylenchus penetrans*.

A number of plant products have been tested to control nematodes and thus interfere with the virus transmission process (Birch et al. 1993). This management approach may have a practical exploitation in organic agriculture or when a reduction in the nematode population in a field is required, but variable results have been achieved and accumulating evidence suggests a long-term management plan is required for best results. Rather, such methods should be seen as part of an integrated programme utilising many sustainable methods combined for a more significant overall effect. Biological control, by its very nature, will rarely be as effective as agrochemicals, and, given the nature of the organisms involved, a single vector nematode specimen may be sufficient to transmit the disease. Low vector densities do not always result in the loss of virus epidemics in the field, as each nematode has the potential to transmit virus.

Nevertheless, several plants are known for their nematicidal effects in vitro or in vivo. Tests showed that root sap of *Asparagus officinalis* was toxic to *Paratrichodorus minor*, while intercropping of *A. officinalis* with tomato suppressed nematode populations (Rohde & Jenkins, 1958). One nematicidal compound was identified as asparagusic acid (Takagusi et al., 1975). Also *Crotalaria* spp., *Sesamun indicum*, *Sinapis alba*, *Tagetes* spp. and *Vigna unguiculata*, appeared capable of reducing populations of *P. minor* in soil (McSorley & Dickson, 1995).

Other in-vitro studies also showed suppressive effects of plant extracts towards *Xiphinema index* and *X. americanum sensu lato*, including aerial parts of *Chamomilla recutita*, wormseed (*Chenopodium ambrosioides*), *Cosmos bipinnatus Oxalis rosea*, *Vestia lycioides* and roots of *Zinnia elegans* (Insunza et al., 1998).

Amendment with green parts of *C. bipinnatus* incorporated in soil at 20 cm depth after flowering showed significant reduction in numbers of *X. americanum sensu lato*, when compared with untreated control (Aballay et al., 2001). On grapevine plants infested by *X. index*, the incorporation of *Brassica juncea* cv.

Nemfix reduced the nematode density by almost 65% compared to the more than 80% reduction with fenamiphos which, however, lowered the total plants fresh weight, possibly due to phytotoxic effects. At incorporation rates of 2% (w/v) also brown mustard (*Brassica juncea*), thyme (*Thymus vulgaris*), wormseed and rue (*Ruta graveolens*) showed potentials as green manure crops for control of *X. index* (Aballay et al., 2004).

4.3. Biological Control Agents

Several nematode antagonists have been reported in association with virus-vector species, and include aquatic fungi, hyphomycetes and bacteria. High levels of parasitism of *X. rivesi* and *X. americanum sensu stricto* by zoosporic fungi were observed during storage in funnels for extraction or during storage in wet soil (Jaffee, 1986). When working with longidorid nematodes, parasitism by aquatic fungi may easily be observed, and a simple method to infect nematodes is to keep them in the water soil extract, at room temperature for a few days. Aquatic fungal species, such as *Catenaria anguillulae*, are common soil inhabitants and have a worldwide occurrence. In Italy, apart from *C. anguillulae*, further unknown species of *Lagenidium*, *Phytophthora* or *Pythium* were often observed on populations of *X. diversicaudatum* or *X. index* (Fig. 9; 10A). Some of these species appeared virulent and were observed to produce sporangia even before the nematode was completely killed, as shown by the struggling movements of the parasitised victim (Fig. 9A, B).

Although the efficacy of aquatic fungi in regulating the nematodes population in the field appears limited by the water amounts in soil, being effective only when the saturated soil capacity is reached, their role as members of a wider community of antagonists deserves further investigation, in view of the current interest in exploitation of biological suppression or organic management. Other endoparasitic fungi include *Hirsutella rhossiliensis*, a hyphomycete parasitic on *X. diversicaudatum* (Ciancio et al., 1986).

Among specialized bacteria, several *Pasteuria* spp. have been observed in *Xiphinema* (Fig. 10B) and *Longidorus* spp., and were reported from different areas of the world (Ciancio, 1995a). However, studies mainly concerned the bacterium biodiversity rather than its effectiveness in controlling the population of the vector nematodes. In a two-year field study on the population dynamics of *X. diversicaudatum* attacking peach and parasitized by a specific *Pasteuria* sp. (Fig. 10B), the nematode population showed fluctuation around a mean density of 78 individuals/100 cc soil, a density high enough to ensure an high risk of virus transmission. Modelling of the host prevalence revealed a potential of the bacterium in regulating the nematode population at an equilibrium density of 120 specimens/100 cc soil, with prevalence around 7-10%, and an extinction risk attainable if a 20% prevalence or higher levels were achieved (Ciancio, 1995b). As a consequence, only inundative treatments with the bacterial endospores (not yet available due to the difficulties for culturing of several *Pasteuria* spp.) should be able to develop local epidemics, leading to the extinction of the nematode population in treated microcosms.

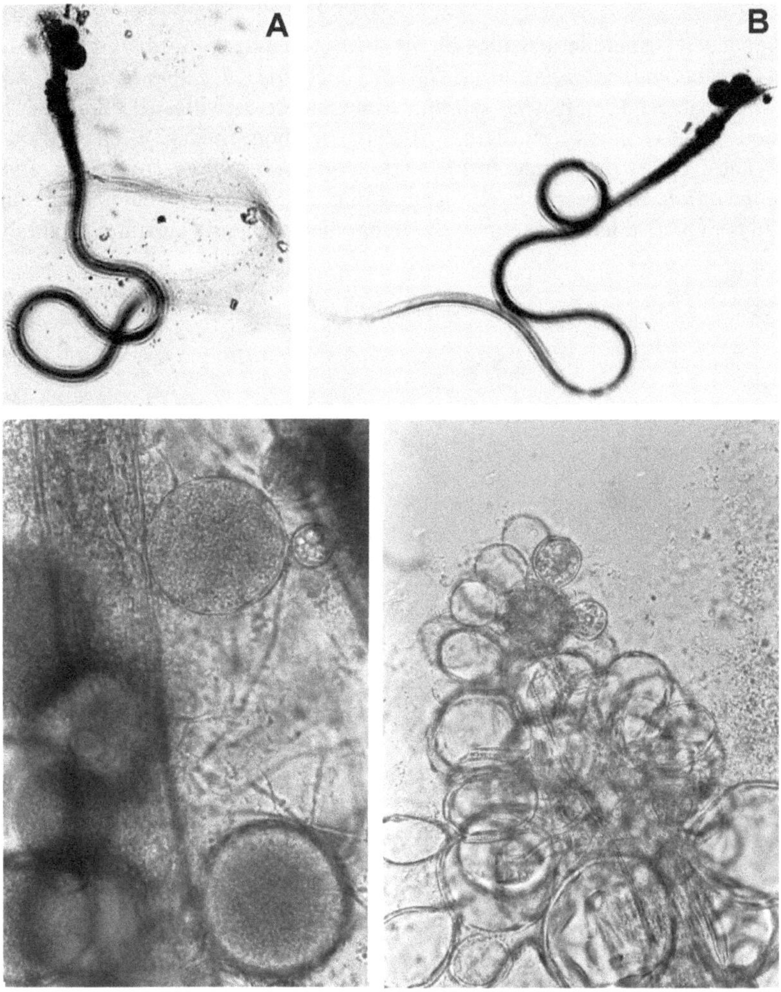

Figure 9. Parasitism of Xiphinema diversicaudatum *by a* Pythium *sp. forming zoosporangia during the early stages of infection, as shown by the active movements of the host nematode (A,B). Zoosporangia at later infection stages cover the whole nematode body before (C) and after (D) zoospores discharge.*

Given the large number of soil bacterial species still unknown and their biodiversity range, which may exceed 2000 species per g of soil (Torvisk et al., 1990), it is possible that several new bacteria will be discovered in the near future, with higher biocontrol capabilities, suitable for biological management of vector nematodes. Bacterial species isolated from roots of nematicidal plants or potato and characterized by an antifungal activity against the fungus *Rhizoctonia solani* were tested in greenhouse on potato (cv. Saturna) for their nematicidal activity against

P. pachydermus and *T. primitivus*, vectors of TRV. In naturally-infested soil, some isolates reduced nematode densities by 50–100%. In a trichodorid and TRV-infested soil, *Stenotrophomonas maltophilia*, *Bacillus mycoides*, *Pseudomonas* sp., and one unidentified bacterium consistently reduced nematode densities (by 56.7–74.4%) on inoculated potato tubers (cv. King Edward) without negative effects on plant growth. The isolates originated from potato, *Plantago major*, *Thymus vulgaris* and *Asparagus officinalis*, respectively, suggesting that plants producing nematicidal compounds may also harbour nematode-antagonistic bacteria (Insunza et al., 2002).

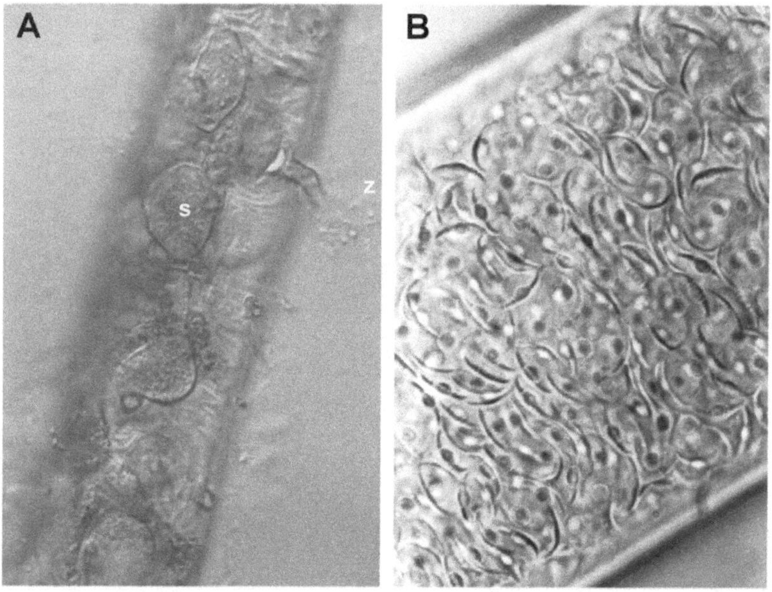

Figure 10. Parasitism of Xiphinema index *by* Catenaria anguillulae *(A) showing zoospores (z) emerging through a zoosporangium germ tube (s). Endopores of* Pasteuria *sp. within* X. diversicaudatum *(B).*

Following some first observations on ovaries of *X. americanum sensu lato*, Coomans et al. (2000) later reported the occurrence of the transovarial transmission of obligate intracellular bacteria also in *X. brevicollum*. Subsequent research revealed the occurrence of "*Candidatus* Xiphinematobacter species" in *X. brevicollum*, *X. americanum*, and *X. rivesi* (Vandekerckhove et al., 2000). These fastidious or unculturable bacteria are endosymbiotic and vertically transmitted. Although their role in the host biology is unknown, it is probable that they evolved from ancestor endoparasitic species, and may affect the nematodes fecundity or even speciation, as occurs with other bacteria found in insects. Further research work is, however, is needed to clarify aspects of their biology and life cycle.

CONCLUSIONS

The wide range of distribution and the risks associated with even low densities of vector nematodes necessitates the adoption of several preventive actions to be deployed before planting, including the use of certified virus-free material. Given the difficulties in managing a vector infestation during the crop production, prevention must be seen as the first barrier against this threat. When no plant-resistant material is available and/or if the crop is in the ground, farmers have to deal with the risk of epidemics reducing profits by using any possible available management strategy.

In poor agricultural conditions, when no further preventive technologies are available, data has shown that the addition of organic amendments, such as composting flax residues and waste products from the microbial industry and green algae, may help farmers reduce the problem whilst also avoiding environmental contamination due to pesticides, increase yields and lower nematode densities. The dry extracts of *C. vulgaris* also showed direct positive effects on nutrition of grape seedlings infested by *X. index*, encouraging its practical application for organic agricultural production. However, organic or biological management of virus vector nematodes alone cannot be considered sufficient, since in theory only one nematode specimen (several species are also parthenogenetic, i.e. *X. index*) may be needed to increase a population or transmit the disease. Although microbial antagonists are a component of the soil microflora with potentials in regulating nematodes densities in soil, biological control of virus vector nematodes will require the testing and application of antagonists capable of maintaining the host population at very low levels. This goal is very difficult to achieve in biological management, since any antagonist needs sufficient host numbers to survive in soil.

Given the present constraints in the use of fumigants or chemicals, including the ban of methyl bromide and the paucity of resistant germplasm available for all crops exposed to nematode vectors, the wise use of two or more of the cited techniques combined with a careful preventive approach based on soil management and monitoring, the application of long rotations lasting for several years, and the use of certified material for planting, appears to be the most likely strategy to lower the risk of epidemics due to nematode-trasmitted viruses.

REFERENCES

Aballay, E., Benavides, F. D., & Vieira, A. (1998). Resistencia de algunos portainjertos a una población chilena de *Xiphinema index*. *Nematologia Mediterranea*, 26, 185-188.

Aballay, E., Flores, P., & Insunza, V. (2001). Efecto nematicida de ocho especies vegetales sobre *Xiphinema americanum* sensu lato, en *Vitis vinifera* L. var. Cabernet Sauvignon en Chile. *Nematropica*, 31, 95-102.

Aballay, E., Sepúlveda, R., & Insunza, V. (2004). Evaluation of five nematode-antagonistic plants used as green manure to control *Xiphinema index* Thorne et Allen on *Vitis vinifera* L. *Nematropica*, 34, 45-51.

Agostinelli, A., Nagy, P., Coiro, M. I., Hecker, K. & Lamberti, F. (2008). Distribution and morphological characterization of *X. pachtaicum*, *X. simile* and *X. brevicollum* from Hungary. *Helminthologia*, 45, 2, 96-102.

Anonymous (2000). *Xiphinema index*. [Distribution map]. *CAB International*. Distribution Maps of Plant Diseases (Edition 1), Map 819.

Barsi, L., Lamberti, F. & De Luca, F. (2007). Morphological and molecular characterisation of *Longidorus danuvii* sp.n. and *L. silvae* Roca, 1993 (Nematoda: Dorylaimida) from Serbia. *Nematology*, 9, 4, 585-598.

Belin, C., Schmitt, C., Demangeat, G., Komar, V., Pinch, L. & Fuchs, M. (2001). Involvement of RNA2-encoded proteins in the specific transmission of Grapevine fanleaf virus by its nematode vector *Xiphinema index*. *Virology*, 291,161-171.

Bica, D., Nicolosi, E., Costa, A., Colombo, A. & Buonocore, E. (2002). Indagine sulla presenza dei principali virus e nematodi della vite in Sicilia. *Informatore Fitopatologico*, 52, 64-67.

Birch, N. E., Robertson, W. M., & Fellows, L. E. (1993). Plant products to control plant parasitic nematodes. *Pesticide Science*, 39, 141-145.

Bitterlin, M. W. & Gonsalves, D. (1987) Spatial distribution of *Xiphinema rivesi* and persistence of tomato ringspot virus and its vector in soil. *Plant Disease*, 71, 408-411.

Boonham, N., Smith, P., Walsh, K., Tame, J., Morris, J., Spence, N., et al. (2002). The detection of Tomato spotted wilt virus (TSWV) in individual thrips using real time fluorescent RT-PCR (TaqMan). *Journal of Virological Methods, 101, 37-48.*

Boonham, N., Walsh, K., Smith, P., Madagan, K., Graham, I., & Barker, I. (2003). Detection of potato viruses using microarray technology: towards a generic method for plant viral disease diagnosis. *Jorunal of Virological Methods*, 108, 181-187.

Brown, D. J. F., Robertson, W. M., & Trudgill, D. L. (1995). Transmission of viruses by plant nematodes. *Annual Review of Phytopathology*, 33, 223-249.

Brown, D. J. F., & Trudgill, D. L.(1998).Nematode transmission of plant viruses - a 30 year perspective. Available at *http://www.scri.ac.uk/scri/file/individualreports/1998/22NEMATO.PDF.*

Brzeski, M. W., Peneva, V., & Brown, D. J. F. (2000). *Longidorus balticus* sp.nov. (Nematoda: Longidoridae) from coastal sand dunes in northeast Poland. *Annales Zoologici,* 50, 3, 321-325.

Catalano, L. (1992). I nematodi vettori di virus. *Vignevini*, 19(5), 39-47.

Catalano, L., Savino, V., & Lamberti, F. (1992). Presence of Grapevine fanleaf virus in populations of Longidorid nematodes and their vectoring capacity. *Nematologia mediterranea*, 20, 67-70.

Chapman, P. J. (1983). The effectiveness of 1,3-dichloropropene for controlling virus vector nematodes with reference to the MAFF Certification Scheme for strawberry nursery stock. *Plant Pathology*, 32, 273-279.

Chen, Q.-W., Hooper, D. J., Loof, P. A. A., & Xu, J. (1997). A revised polytomous key for the identification of species of the genus *Longidorus* Micoletzky, 1922 (Nematoda: Longidoridae). *Fundamental and Applied Nematology*, 20, 1, 15-28.

Choleva, B., Bileva, T., Tzvetkov, J., & Barakov, P. (2005). Preliminary study of the green algae Chlorella (*Chlorella vulgaris*) for control on the root-knot nematode (*Meloidogyne arenaria*) in tomato plants and ectoparasite *Xiphinema index* in grape seedlings. Communications in Agricultural and Applied Biological Sciences, Ghent University, 70, 915-926.

Choleva, B., T. Bileva, Tzvetkov, J. (2007). Organo-biological means and methods for control of plant parasitic nematodes as alternative of agrochemicals. Ecology and Future, Vol. VI (4), p. 43-49; Bulgarian Journal of Ecological Science, Sofia.

Cohn, E., Tanne, E., & Nitzany, F. E. (1970). *Xiphinema italiae* a new vector of grapevine fanleaf virus. *Phytopathology*, 60, 181-182.

Ciancio A. (1995a). Phenotypic adaptations in *Pasteuria* spp. nematode parasites. *Journal of Nematology*, 27, 328-338.

Ciancio, A. (1995b). Density dependent parasitism of *Xiphinema diversicaudatum* by *Pasteuria penetrans* in naturally infested soil. *Phytopathology*, 85, 144-149.

Ciancio, A. Logrieco, A., & Lamberti, F. (1986). Parasitism of *Xiphinema diversicaudatum* by the fungus *Hirsutella rhossiliensis*. *Nematologia Mediterranea*, 14, 187-192.

Cooper, J. I., & Thomas, P. R. (1971). Chemical treatment of soil to prevent transmission of tobacco rattle virus to potatoes by *Trichodorus* spp. *Annals of Applied Biology*, 69, 23-34.

Coomans, A., Huys, R., Heyns, J. & Luc, M. (2001). Character analysis, phylogeny and biogeography of the genus *Xiphinema* Cobb, 1913 (Nematoda: Longidoridae). Royal Museum for Central Africa, Leuvensesteenweg 13, B-3080 Tervuren, Belgium.

Coomans, A., Vandekerckhove, T. T. M., & Claeys, M. (2000). Transovarial transmission of symbionts in *Xiphinema brevicollum* (Nematoda: Longidoridae). *Nematology,* 2, 443–449.

Crosslin, J. M., Thomas, P. E., & Hammond, R.W. (2003).Genetic variability of genomic RNA 2 of four tobacco rattle tobravirus isolates from potato fields in the Northwestern United States. *Virus Research,* 96, 99-105.

Decraemer, W. (1995). The family Trichodoridae: stubby rot and virus vector nematodes. Kluwer, Dordrecht, NL: 360 pp.

Decraemer, W., & Baujard, P. (1998). A polytomous key for the identification of species of the family Trichodoridae Thorne, 1935 (Nematoda: Triplonchida). *Fundamental and Applied Nematology,* 21, 1, 37-62.

Decraemer, W., & Robbins, R. T. (2007). The who, what and where of Longidoridae and Trichodoridae. *Journal of Nematology*, 39, 295-297.

Escuer, M., & Arias, M. (1997). *Paralongidorus iberis* sp.n. and *P. monegrensis* sp.n. from Spain with a polytomous key to the species of the genus *Paralongidorus* Siddiqi, Hooper & Khan, 1963 (Nematoda: Longidoridae). *Fundamental and Applied Nematology,* 20, 2, 135-148.

Esmenjaud, D., Walter, B., Minot, J. C., Voisin, R. & Cornuet, P. (1993). Biotin-avidin ELISA detection of Grapevine fanleaf virus in the vector nematode *Xiphinema index*. *Journal of Nematology*, 25, 401-405.

Esmenjaud, D., Abad, P., Pinck, L., & Walter, B. (1994). Detection of a region of the coat protein gene of Grapevine fanleaf virus by RT-PCR in the nematode vector *Xiphinema index*. *Plant Disease*, 78,1087-1090.

Evans, T. A., Miller, L. C., Vasilas, B. L., Taylor, R. W., & Mulrooney, R. P. (2007). Management of *Xiphinema americanum* and Soybean Severe Stunt in soybean using crop rotation. *Plant Disease*, 91, 216-219.

Finetti-Sialer, M. M., & Ciancio, A. (2005). Isolate-specific detection of *Grapevine fanleaf virus* from *Xiphinema index* through DNA-based molecular probes. *Phytopathology*, 95, 262-268.

Hanna, E., Digiaro, M., Elbeaino,T., Choueiri, E., Jawhar, J., & Martelli, G. P. (2008). Incidence of viruses and nematode vectors in Lebanese vineyards.*Journal of Phytopathology*, 156, 304–310.

Harrison, B. D., & Robinson, D. J. (1986). Tobraviruses. In: *The Plant Viruses*, (Van Regenmortel M.H.V., and Fraenkel-Conrat H., eds.). Plenum Press: New York: 339-369.

Hockland, S., Inserra, R. N., Millar, L., & Lehman, P. S. (2006). International Plant Health – Putting Legislation into Practice. In: Plant Nematology. Perry, R. N. and M. Moens (Eds.). CABI Publishing. Wallingford, UK : 327-345.

Hübschen, J., Kling, L., Ipach, U., Zinkernagel, V., Brown, D. J. F., & Neilson, R. (2004a). Development and validation of species-specific primers that provide a molecular diagnostic for virus-vector longidorid nematodes and related species in German viticulture. *European Journal of Plant Pathology*, 110, 883–891.

Hübschen, J., Kling, L., Ipach, U., Zinkernagel, V., Bosselu, N., Esmenjau, D., et al. (2004b). Validation of the specificity and sensitivity of species-specific primers that provide a reliable molecular diagnostic for *Xiphinema diversicaudatum*, *X. index* and *X. vuittenezi*. *European Journal of Plant Pathology*, 110, 779–788.

Insunza, V., Aballay, E., & Macaya, J. (1998). Acci6n nematicida de extractos acuosos de 30 plantas en poblaciones chilenas de *Xiphinema index* y *X. americanum sensu lato*. *Nematropica* 28, 134–135.

Insunza, V., Alström, S., & Eriksson, K. B. (2002). Root bacteria from nematicidal plants and their biocontrol potential against trichodorid nematodes in potato. *Plant and Soil*, 241, 271–278.

Jaffee, B. A. (1986). Parasitism of *Xiphinema rivesi* and *X. americanum* by zoosporic fungi. *Journal of Nematology*, 18, 87-93.

Jones, A. T., McElroy, F. D., & Brown, D. J. F. (1981). Tests for transmissin of cherry leaf roll virus using *Longidorus*, *Paralongidorus* and *Xiphinema* nematodes. *Annals of Applied Biology*, 99, 143-150.

Klinger, J., Kunz, P., & Hauri, H. P. (1985). Die Verbreitung des Vektornematoden der Pfeffingerkrankheit in Kerschenanbaugebier der Nordwestschweiz. *Schweizerische Zeitschrift für Obst- und Weinbau*, 121, 782-789.

Krnjaic, D., Roca, F., Krnjaic, S. & Agostinelli, A. (2005). *Longidorus cylindricapitatus* sp.n. (Nematoda: Longidoridae) from Serbia. *Nematology,* 7, 6, 803-808.

Kumari, S., & Decraemer, W. (2008). First report of the dagger nematode *Xiphinema dentatum* (Nematoda: Longidoridae) in a deciduous forest in the Czech Republic. *Plant Disease,* 92, 9, 1370.

Kumari, S., & Decraemer, W. (2007). The genus *Longidorus* (Nematoda: Longidoridae) from Bohemia and South Moravia in the rhizosphere of fruit orchards and vineyards. *Helminthologia,* 44, 4, 193-203.

Kyriakopoulou, T. P. (1985). Viruses observed on artichoke in Greece. 3rd National Phytopathological Conference, Summaries of invited and research papers, Volos, Greece, 16–18 October 1985, p. 28.

Kyriakopoulou, P. E. (1996). Artichoke Italian latent virus causes artichoke patchy chlorotic stunting disease. *Annals of Applied Biology*, 127, 489–497.

Lamberti, F. (1981a). Combating nematode vectors of plant viruses. *Plant Disease*, 65, 113-117.

Lamberti, F. (1981b). Plant nematode problems in the Mediterranean region. *Helminthological Abstracts*, 50, 145-166.

Lamberti, F. (1991). Nematodi parassiti della vite e relativa lotta. *Vignevini*, 18, (11), 43-48.

Lamberti, F., & Basile, M., (1982). Chemical control of nematode vectors. In: F. K. Harris and K. Maramorosch (Eds). *Pathogens, vectors and plant diseases*. Academic Press, NY, USA: 57-69.

Lamberti, F., Hockland, S., Agostinelli, A., Moens, M., & Brown, D. J. F. (2004). The *Xiphinema americanum* group. III. Keys to species identification. *Nematologica mediterranea* 32, 53-56.

Lamberti, F., Molinari, S., Moens, M., & Brown, D. J. F. (2000). The *Xiphinema americanum* group. I. Putative species, their geographical occurrence and distribution, and regional polytomous identification keys for the group. *Russian Journal of Nematology*, 8, 1, 65-84.

Lamberti, F., Landriscina, S., Ciancio, A., & Catalano, L. (1993). Lotta contro *Xiphinema diversicaudatum* nematode vettore del SLRV su pesco in Piemonte. *Informatore Fitopatologico, (5)* 43, 57-59.

Lamberti, F., & Roca, F. (1987). Present status of nematodes as vectors of plant viruses. In: J. A. Veech and D. W. Dickson (Ed.). Vistas on Nematology. Hyattsville, Maryland, USA: Society of Nematologists Inc.: 321-328.

Liskova, M., & Brown, D. J. F. (2003). Longidoridae (Nematoda: Dorylaimida) in the Slovak Republic. *Helminthologia*, 40, 3, 165-172.

Liskova, M., Robbins, R. T., & Brown, D. J. F. (1997). Descriptions of three new *Longidorus* species from Slovakia (Nemata: Longidoridae). *Journal of Nematology*, 29, 3, 336-348.

Loof, P. A. A. & Luc, M. (1990). A revised polytomous key for the identification of species of the genus *Xiphinema* Cobb, 1913 (Nematoda: Longidoridae) with exclusion of the *X. americanum*-group. *Systematic Parasitology* 16, 35-66.

Loof, P. A. A., & Luc, M. (1993). A revised polytomous key for the identification of species of the genus *Xiphinema* Cobb, 1913 (Nematoda: Longidoridae) with exclusion of the *X. americanum*-group: Supplement 1. *Systematic Parasitology*, 24, 185-189.

Loof, P. A. A., Luc, M., & Baujard, P. (1996). A revised polytomous key for the identification of species of the genus *Xiphinema* Cobb, 1913 (Nematoda: Longidoridae) with exclusion of the *X. americanum*-group: Supplement 2. *Systematic Parasitology*, 33, 23-29.

MacFarlane, S. A. (2003). Molecular determinants of the transmission of plant viruses by nematodes. *Molecular PlantPathology*, 4, 211-215.

Martelli, G. P. (2002). Le principali virosi della vite oggi. *Informatore Fitopatologico*, 52,18-27.

Martelli, G. P., & Savino, V. (1991). Fanleaf degeneration. In: Pearson, R. C., Goheen, A. (Eds.), Compendium of Grape Diseases. APS Press, St. Paul, MN: 48– 49.

Mincheva, Y., Lazarova, S., & Peneva, V. (2008). *Xiphinema pirinense* n.sp (Nematoda: Dorylaimida: Longidoridae), a new species from Bulgaria with a digitate tail. *Systematic Parasitology*, 70, 3, 215-222.

Mojtahedi, H., & Santo, G. S. (1999). Ecology of *Paratrichodorus allius* and its relationship to the corky ring-spot disease of potato in the Pacific northwest. *American Journal of Potato Research*, 76, 273-280.

Murant, A. F., & Taylor, C. E. (1965). Treatment of soil with chemicals to prevent transmission of tomato blackring and raspberry ringspot viruses by *Longidorus elongatus* (de (Man). *Annals of Applied Biology*, 55, 227-237.

Palomares-Rius, J. E., Subbotin, S.A., Landa, B.B. Vovlas, N., & Castillo, P. (2008). Description and molecular characterisation of *Paralongidorus litoralis* sp.n. and *P. paramaximus* Heyns, 1965 (Nematoda: Longidoridae) from Spain. *Nematology*, 10, 87-101.

Peneva, V., Choleva, B., & Nedelchev, S. (1997). Description of *Longidorus fagi* n.sp. (Nematoda, Dorylaimida) with an identification key to the species of the same genus occurring in Bulgaria. *Systematic Parasitology*, 36, 2, 115-122.

Peneva, V., Loof, P.A.A. & Brown, D. J. F. (1998). *Longidorus seinhorsti* sp.n. (Nematoda: Dorylaimoidea) from The Netherlands. *Fundamental and Applied Nematology*, 21, 5, 605-609.

Peneva, V., Loof, P. A. A. & Brown, D. J. F. (1999). *Longidorus dalmassoi* sp.n. (Nematoda: Dorylaimoidea) from France. *Nematology*, 1, 5, 557-560.

Pinkerton, J. N., Kraus, J., Martin, R. R., & Schreiner, R. P. (2008). Epidemiology of *Xiphinema americanum* and *Tomato ringspot virus* on red raspberry, *Rubus idaeus*. *Plant Disease*, 92, 364-371.

Rana, G. L., & Kyriakopoulou, T. P. (1982). Artichoke Italian latent and artichoke mottle crinkle viruses in artichoke in Greece. *Phytopathologia Mediterranea*, 21,101–104.

Raski, D. J., Hewitt, W. B., Goheen, A. C., Taylor, C. E., & Taylor, R. H. (1965). Survival of *Xiphinema index* and reservoirs of fanleaf virus in fallowed vineyard soil. *Nematologica*, 11, 349-352.

Roberts, I. M., & Brown, D. J. F. (1980). Detection of six nepoviruses in their nematode vectors by immunosorbent electron microscopy. *Annals of Applied Biology*, 96, 187-192.

Roca, F., & Bravo, M. A. (1998). Description of *Xiphinema silvesi* n.sp. from Portugal (Nematoda: Longidoridae). *Fundamental and Applied Nematology*, 21, 4, 389-392.

Roca, F., Rana, G. L., & Kyriakopoulou, P. E. (1982). *Longidorus fasciatus* Roca et Lamberti vector of a serologically distinct strain of artichoke Italian latent virus in Greece. *Nematologia Mediterranea*, 10, 65-69.

Roccuzzo, G., & Ciancio, A. (1991). Note on nematodes found in irrigation water in southern Italy. *Nematologia Mediterranea*, 19, 105-108.

Rowhani, A., Chay, C., Golino, D. A., & Falk, B. W. (1993). Development of a polymerase chain reaction technique for the detection of Grapevine fanleaf virus in grapevine tissues. *Phytopathology*, 83, 749-753.

Rubtsova, T. V., Subbotin, S. A., Brown, D. J. F. & Moens, M. (2001). Description of *Longidorus sturhani* sp.n. (Nematoda: Longidoridae) and molecular characterisation of several longidorid species from Western Europe. *Russian Journal of Nematology*, 9, 2, 127-136.

Rubtsova, T. V., Chizhov, V. N., & Subbotin, S. A. (1999). *Longidorus artemisiae* sp.n. (Nematoda: Longidoridae) from roots of *Artemisia sp.* Rostov region, Russia. *Russian Journal of Nematology*, 7, 1, 33-38.

Širca, S., Stare, B. G., Plešco, I. M., Marn, M. V., & Urek, G. (2007a). *Xiphinema rivesi* from Slovenia transmit *Tobacco ringspot virus* and *Tomato ringspot virus* to Cucumber Bait Plants. *Plant Disease*, 91, 6, 770.

Širca, S., Stare, B. G., Plesko, I. M., Marn, M.V., & Urek, G. (2007b). First record of *Longidorus juvenilis* and *L. leptocephalus* (Nematoda: Dorylaimida) in Slovenia and their morphometrical and ribosomal DNA sequence analysis. *Russian Journal of Nematology*, 15, 1, 1-8.

Starr, J. L., & Bendezu, I. F. (2002). Ectoparasitic Nematodes. In: Plant Resistance to Parasitic Nematodes. CABI Publishing. Wallingford, UK: 229-239.

Taylor, C. E., & Brown, D. J. F. (1997). Nematode vectors of plant viruses. CABI Publishing, Wallingford, UK: 72-85.

Taylor, C. E., & Raski, D. J. (1964). On the transmission of grape fanleaf by *Xiphinema index*. *Nematologica*, 10, 489-495.

Taylor, C. E., & Robertson, W. M. (1970). Sites of virus retention in the alimentary tract of the nematode vectors, *Xiphinema diversicaudatum* (Micol.) and *X. index* (Thorne and Allen). *Annals of Applied Biology*, 66, 375-380.

Taylor, C. E., & Gordon, S. C. (1970). A comparison of four nematicides for the control of *Longidorus elongatus* and *Xiphinema diversieaudatum* and the viruses they transmit. *Horticultural Research*, 10, 133-141.

Taylor, C. E., & Murant, A. F. (1968). Chemical control of raspberry ringspot and tomato black ring viruses in strawberry. *Plant Pathology*, 17, 171-178..

Taylor, C. E., Brown, D. J. F., Neilson, R., & Jones, A. T. (1994). The persistence and spread of *Xiphinema diversicaudatum* in cultivated and uncultivated biotopes. *Annals of Applied Biology*, 124, 469-477.

Thelwell, N., Millington, S., Solinas, A., Booth, J., & Brown, T. (2000). Mode of action and application of Scorpion primers to mutation detection. *Nucleic Acid Research*, 28, 3752-3761.

Torsvik, V., Goksøyr, J., & Daae, F. L. (1990). High diversity in DNA of soil bacteria. *Applied and Environmental Microbiology*, 56, 782-787.

Trudgill, D. L., & Alphey, T. J. W. (1976). Chemical control of the Virus-vector nematode *Longidorus elongatus* and of *Pratylenchus crenatus* in raspberry plantations. *Plant Pathology*, 25, 15-20.

Trudgill, D. L., & Brown, D. J. F. (1972). Ingestion, retention and transmission of two strains of raspberry ringsport virus by *Longidorus macrosoma*. *Journal of Nematology*, 10, 85-89.

Tyagi, S., Bratu, D. P., & Kramer, F. R. (1998). Multicolor Molecular Beacons for allele discrimination. *Nature Biotechnology*, 16, 49-53.

Tzortzakakis, E. A., Peneva, V., Terzakis, M., Neilson, R., & Brown, D. J. F. (2001). *Longidorus cretensis* n.sp. (Nematoda: Longidoridae) from a vineyard infected with a foliar 'yellow mosaic' on Crete, Greece. *Systematic Parasitology*, 48, 2, 131-139.

Vandekerckhove, T. T. M., Willems, A., Gillis, M., & Coomans, A. (2000). Occurrence of novel verrucomicrobial species, endosymbiotic in *Xiphinema americanum*-group species (Nematoda,

Longidoridae) and associated with parthenogenesis. *International Journal of Systematic and Evolutionary Microbiology*, 50, 2197–2205.

Vassilakos, N., Bem, F., Tzima, A., Barker, H., Reavy, B., Karanastasi, E. & Robinson, D. J. (2008). Resistance of transgenic tobacco plants incorporating the putative 57-kDa polymerase read-through gene of Tobacco rattle virus against rub-inoculated and nematode transmitted virus. *Transgenic Research,* 17, 5, 929-941.

Wang, S., Gergerich, R. C., Wickizer, S. L., & Kim, K. S. (2002). Localization of transmissible and nontransmissible viruses in the vector nematode *Xiphinema americanum*. *Phytopathology*, 92, 646-653.

Wang, X., Bosselut, N., Castagnone, C., Voisin, R., Abad, P., & Esmenjaud, D. (2003). Multiplex polymerase chain reaction identification of single individuals of the Longidorid nematodes *Xiphinema index, X. diversicaudatum, X. vuittenezi*, and *X. italiae* using specific primers from ribosomal genes. *Phytopathology*, 93, 160–166.

Wetzel, T., Meunier L., Jaeger, U., Reustle, G. M., & Krezal, G. (2001). Complete nucleotide sequences of the RNA 2 of German isolates of Grapevine fanleaf and Arabis mosaic. *Virus Research*, 75,139-145.

Wetzel, T., Jardak, R., Meunier, L., Ghorbel, A., Reustle, G. M., & Krczal, G. (2002). Simultaneous RT/PCR detection and differentiation of arabis mosaic and grapevine fanleaf nepoviruses in grapevines with a single pair of primers. *Journal of Virological Methods*, 101, 63-69.

Whitcombe, D., Theaker, J., Guy, S. P., Brown, T., & Little, S. (1999). Detection of PCR products using self-probing amplicons and fluorescence. *Nature Biotechnology*, 17, 804-807.

Whitehead, A.G. (1998). Ectoparasitic nematodes of roots (*Belonolaimus, Criconemella, Hoplolaimus, Longidorus, Trichodorus, Paratrichodorus, Tylenchorhynchus* and *Xiphinema*. In: Plant nematode Control. CAB International, Wallingford, UK, 53-89.

SADDIGHEH FATEMY

INTEGRATED MANAGEMENT OF PISTACHIO NEMATODES

Nematology Department
Plant Protection Research Institute
P.O. Box 1454 Tehran 19395
Theran IRAN

Abstract. The production of pistachio is reviewed, with data on main parasitic nematode, including root-knot and lesion nematodes. The distribution of pistachio nematodes and the management options available are listed, including agronomic management, use of resistant rootstocks, biological control with applications of the bacterium *Pasteuria penetrans* or the fungus *Pochonia chlamydosporia*, as well as soil solarization.

1. INTRODUCTION

The word pistachio is a Persian loanword, coming into English through Italian and is a cognate to the modern Persian word *Pesteh*. Pistachio originated from West and Central Asia where still large areas of natural populations exist (Zohary, 1952). Ancient trees of *Pistacia vera* nearly 1800 years old are believed to be living today in Syria (Hadj-Hassan, 1988). Pistachio reached the Mediterranean by way of central Iran not before medieval times. This crop is cultivated mainly in the Middle East, North Africa, Mediterranean regions of Europe and California.

2. PISTACHIO PRODUCTION AND PROPERTIES

2.1. Pistachio Production

Total world production of pistachio in 2006 was estimated as 568855 metric tons., of which 41% was produced in Iran, followed by USA (22%), Turkey (19%), Syria (9.5%) and China (6%) (Table 1) (FAO, 2006). World production is increasing and fewer quantities are also produced in countries like Greece, Italy, Tunisia, Pakistan, Uzbekistan, Madagascar, Mexico and Mauritius.

Pistachio is a tree reaching up to 10m in height, dioecious and deciduous with fruits produced in clusters. The fruit is a drupe containing an elongated seed (a nut in the culinary sense, but not a true botanical nut) with a hard, whitish shell and a

A. Ciancio & K. G. Mukerji (eds.), *Integrated Management of Fruit Crops and Forest Nematodes*, 243-252.

striking kernel which has a mauve skin and light green flesh, with a distinctive flavour. When the fruit ripens, the hull changes from green to an autumnal yellow/red and the shells split partially opens (Tous & Ferguson, 1996).

The pistachio nut *Pistacia vera*, (Anacardiaceae) is the only edible species within Pistacia. The genus comprises 11 species: *P. atlantica*, *P. vera*, *P. cabulica*, *P. chinensis*, *P. falcata*, *P. integerrima*, *P. kurdica*, *P. mutica*, *P. palaestina*, *P. terebinthus* and *P. khinjuk* (Zahary, 1952).

Table 1. Main pistachio producing countries, total cultivated area and yields, for 2006 (source: FAO).

Countries	Area (ha)	Production (metric tons)
Azerbaijan	3	15
China	16000	36000
Cyprus	150	18
Greece	5035	9365
Iran	440025	229657
Italy	3635	2719
Kyrgyzstan	100	100
Madagascar	514	152
Pakistan	201	632
Syria	22000	60000
Tunisia	19562	1206
Turkey	40000	110000
USA	42525	122470
Uzbekistan	1893	203

Some of the other species like *P. lentiscus* are grown as an ornamental and for their aromatic resin and oil extraction (Joley, 1960). *Pistacia mutica* (Mastic tree) and *P. khinjuk* (Khonjok tree) form forest stands in their natural habitat (Abrishami, 1995), and are widely distributed in Egypt, Iraq, Iran and eastwards to Kashmir (Kawashty et al., 2000) and south eastern Turkey. Also, *P. atlantica* grows in the Mediterranean and Aegean regions of Turkey and *P. terebinthus* (terebinth tree) is widely distributed in the Middle East and Southern Europe (Aydin & Ozcan, 2002).

2.2. Nutritional Value

Pistachio nuts are eaten fresh, roasted or salted. They are used in dishes as well as in sweets like baklava, cakes and ice creams. The kernels are nutritious (Table 2). Their fat contains 69% oleic acid and 19.8% linoleic acid and their oils are used in the cosmetic and pharmaceuticals industries (Tous & Ferguson, 1996).

There is scientific evidence indicating consumption of 1.5 ounces (42.5g) per day of most nuts, such as pistachios, as part of a diet low in saturated fat and cholesterol, may reduce the risk of heart disease (FDA, 2003). Nuts also reduce levels of LDL, the bad cholesterol and may calm acute stress reaction (http://pistachiohealth.com/nutrition.html).

In Iran, pistachio is reported to have three naturally growing species: *P. vera*, *P. mutica* and *P. khinjuk* (Sabeti, 1966). *Pistacia vera* is native to Eastern Province of Khorasan, the variety is called Sarakhs, and some of the trees are believed to be more than 900 years old (Abrishami, 1995). Pistachio production areas are located between 700 and 3000 meters above sea level. The mature trees tolerate a wide range of temperature (from -20°C to 45°C). Long and warm summers with low humidity (RH <35%) are suitable for production. Water supply for irrigation is mainly provided from deep wells which are mostly high in salinity (Javanshah et al., 2000). Kerman Province in the South, with an average annual rainfall of 130 mm, is the largest commercial pistachio growing area in Iran. Pistachio seeds having high genetic diversity were brought to Kerman from Khorasan Province, and there are more than 70 pistachio cultivars grown in Kerman alone.

Table 2. Nutritional value per 100 g of pistachio nuts, dry roasted

Energy	570 kcal (2390 kj)	
Carbohydrates	27.65 g	
- Sugars	7.81 g	
- Dietary fiber	10.3 g	
Fat	45.97 g	
Protein	21.35 g	
Thiamin (Vit. B1)	0.84 mg	65% [1]
Riboflavin (Vit. B2)	0.158 mg	11%
Niacin (Vit. B3)	1.425 mg	10%
Pantothenic acid (B5)	0.513 mg	10%
Vitamin B6	1.274 mg	98%
Folate (Vit. B9)	50 μg	13%
Vitamin C	2.3 mg	4%
Calcium	110 mg	11%
Iron	4.2 mg	34%
Magnesium	120 mg	32%
Phosphorus	485 mg	69%
Potassium	1042 mg	22%
Zinc	2.3 mg	23%
Manganese	1.275 mg	

[1] Percentages relative to USA recommendations for adults (source: USDA nutrient database).

Due to its economic importance, the second largest exported commodity after oil, there is renewed interest in Pistachio production within Iran. There is need for resistance to many unfavourable conditions like soil and water salinity, water deficiency and soil-borne pathogens. Commercial pistachio production is expanding rapidly to Semnan, Yazd, Esfahan, Khorasan, Tehran, Zanjan, Sistan and Baluchistan and Fars provinces where climate is less suitable for production of other crops. The major pistachio varieties grown in Iran are Ohady and Kaleghochi (Esmail-pour, 1998).

3. NEMATODES MANAGEMENT

3.1. Pistachio Nematodes

Fifteen genera of plant parasitic nematodes have been associated with pistachio crops, in most cases their pathogenicity and economic importance have not been indicated (Table 3). An additional complication is that in many parts of the world pistachio is grown with companion crops that could be supporting prevailing nematode species.

Several nematode species have been reported from orchards of pistachio in different parts of the country (Table 3). Although their status and economic importance have not been evaluated thoroughly, yet in most surveys, symptoms of tree yellowing, malnutrition, and decline have been attributed to root knot nematodes. *Meloidogyne javanica* and *M. incognita* are frequently isolated from the largest production region of Kerman, and also Yazd, Esfahan and Semnan (Madani, et. al., 1995a; Farivar Mehin, 1984; Askarian et. al., 2006; Barooti & Hoseininejad, 2004). In Rafsanjan, the main pistachio producer city of Kerman province, *M. javanica* produces more than 5 generations per year, nearly 18 galls and 50 females per g root; and in both Kerman and Semnan regions, population of 7-10 second-stage juveniles / g soil of *P. vera* have been detected (Banihashemi & Kheiri, 1995; Farivar Mehin, 1986a).

At present, most of the commercial pistachios in Iran are on rootstocks susceptible to nematodes. The reported priority of pistachio growers, is adaptability and tolerance of trees to adverse environmental conditions including salinity, drought and freezing (Banihashemi, University of Shiraz, Iran, personal communication).

In California, *Meloidogyne* spp., *Pratylenchus neglectus*, *Xiphinema americanum* and *Paratylenchus hamatus* have been reported from pistachio orchards which up to now have not been of any serious threat to this crop (Table 3) (McKenry & Kretsch, 1984; Kodira & Westerdahl, 1995). During the early years of California's pistachio industry, own-rooted *P. vera* cv Kerman was recognized as susceptible to nematodes and *Phytophthora*. *Pistacia atlantica* and *P. terebinthus* quickly became the preferred rootstocks. However, these two rootstocks were not resistant to Verticillium wilt. At present, *P. integerrima* PG 1 provides tolerance to Verticillium Wilt and UCB 1, a hybrid of *P. atlantica* × *P. integerrima*, provides increased vigor, earlier production and tolerance to Verticillium Wilt, cold injury and salinity. With few exceptions these improved rootstocks, as well as *P. atlantica* and *P. terebinthus*, have provided resistance to *Meloidogyne* spp. and *Pratylenchus vulnus*. From the 1950s to the present each of these has at one time or another become the preferred rootstock of the California pistachio industry. However, growers should be suspicious that nematode resistance may not be present in all cultivars of each *Pistacia* spp. (Westerdahl & McKenry, 2002).

Table 3. Nematodes reported in association with pistachio trees, worldwide.

Criconema mutabile	*Pratylenchus* sp .
Criconema spp.	
Filenchus spp.	*P. neglectus*
Geocenamus app.	*P. penetrans*
G. rugosus	*P. vulnus*
G. brevidens	*P. thornei*
Helicotylenchus digonicus	*Rotylenchulus macrodoratus*
Heterodera sp.	*R. macrosomus*
H. mediterranea	*Rotylenchus* sp.
H. marioni	*Scutylenchus quettensis*
Longidorus africanus	*Trichodorus* sp.
Meloidogyne sp.	*Trophurus* spp.
M. javanica	*Tylenchus* spp.
M. incognita	*Tylenchorhynchus* spp.
Merlinius	*X. index*
Paralongidorus litoralis	*X. vuittenezi*
Paratylenchus hamatus	*X. americanum*
P. projectus	*X. pachtaicum*
Paratylenchus spp.	*Xiphinema* spp.
Pratylenchoides spp.	*Zygotylenchus guevarai*

In Turkey, the only records of nematodes associated with pistachio are from Sanliurfa Province, located in the Southeast of Anatolian region, where piatachio is one of the main income source of non-irrigated crops (Figure 1). The rootstocks are mainly *P. vera* grown in plastic tubes, transferred to main land between 1-2 years and grafted within 6-7 years (Yildiz, 2007). Sanliurfa area has a heavy, clay soil type which is classified as vertisols (Aydemir, 2001). There is no report of nematode problem in Turkey so far, the only work is from a recent survey in which various densities of mainly ectoparasitic nematodes including *Rotylenchulus macrosomus*, *Criconema* spp. *Paratylenchus* spp., *Geocenamus* spp., *Trophurus* spp., *Trichodorus* sp. *Tylenchorhynchus* spp. and *Pratylenchoides* spp. (Table 3) have been found in association with pistachio in Sanliurfa Province (Yildiz, 2007). The importance and economic status of these species are yet to be determined (S. Yildiz, Harran University, Sanliurfa, Turkey, pers. comm.).

In Spain, a population of lesion nematode *Pratylenchus vulnus* reproduced well on rootstocks of *Pistacia*, and reproduction factors (Pf/Pi) of 8.3, 6.3 and 5.6 were measured on *P. terebinthus*, *P. vera* and *P. atlantica*, respectively. The report mentions the occurrence of more than $4 \cdot 10^{3}$ nematodes · g root $^{-1}$ of *P. terebinthus*, without any comment on damage or yield reduction induced by the nematode feeding (Pinochet et al., 1992). Also, a new longidorid species, *Paralongidorus litoralis*, has been described from *P. lentiscus* (Palomares-Rius et al., 2008).

Figure 1. A traditonal pistachio orchard in Sanliurfa, Turkey (courtesy of A. Ikinci, *Harran University, Sanliura, Turkey).*

In Italy, *P. lentiscus* and *P. vera* were susceptible host to *Heterodera mediterranea* (Vovlas & Inserra, 1983) and Vovlas (1983) has reported heavy infection and galled roots of *P. vera* by *Rotylenchulus macrodoratus*. Furthermore, *Heterodera marioni*, *Meloidogyne javanica*, *Pratylenchus neglectus* and *Xiphinema* spp. have been reported from pistachio in Sicily (Greco & Nucifora, 1999).

In Pakistan, *Scutylenchus quettensis*, *Helicotylenchus digonicus* and *Pratylenchus penetrans* were reported with high numbers on pistachio in Baluchistan region with populations fluctuating with changes in soil temperature (Qasim & Hashim, 1988).

3.2. Management

In regions where various biotic and abiotic stresses are damaging crops, a package of sustainable strategies is needed and this includes assurances that root systems be maintained healthy. Nematode resistant/ tolerant rootstocks, use of pest free soil in nursery production, recognition and control of other diseases which might interact with nematodes, biological control, soil solarization, soil improvement and amendments, better orchard management, and crop health maintenance are examples of environmentally safe measures that should be packaged into an overall strategy of nematode management.

There is within *Pistacia* spp. a wide range of genetic diversity. Examples have been provided herein to indicate there is high probability for finding resistance to nematodes and other soil-borne pests. With a proper breeding and selection program focused on the full spectrum of root depleting agents, including nematodes, improved rootstocks may be achieved.

Recently in California root knot and root lesion nematodes were detected among a few cultivars of *P. integerrima* and *P. atlantica.* It follows that understanding and careful measures are now needed as there is a re-examination of the association of these nematodes to present rootstocks (McKenry, personal communication). This incident and other sporadic incidents perhaps formed the basis for the report by Koenning and colleagues (1999) wherein they quote USA losses of 1-5 % , due to *Meloidogyne* spp., *Pratylenchus* spp. and *Xiphinema* spp. in California pistachio production (M. V. McKenry, University of California, USA, pers. Comm.).

In Iran, several experiments have been carried out, in order to examine the reaction of different wild species and commercial cultivars of pistachio to root knot nematodes. All commercial cultivars of *P. vera* Akbari, Kaleh Ghoochi, Owhadi, Ahmadaghaii, Shahpasand, Aliabadi, Abbasali and Khanjari were susceptible to *M. javanica* (Mohammadi Moghadam et al., 2006). Furthermore, work of Farivar Mehin (1984) suggests that *P. palaestina*, and *P. atlantica* were host to *M. incognita*, and *P. vera* was a susceptible host to *M. javanica*, *M. arenaria* and *M. incognita*. Results from different trials are all in agreement that, *P. vera* is the most susceptible and *P. mutica* the most resistant species to root knot nematodes; and *P. palaestina*, *P. atlantica* and *P. khinjuk* being moderately resistant/tolerant of different populations of nematodes tested (Farivar Mehin, 1986b; Madani et al., 1995b).

Natural enemies are an important means for control of nematodes. The bacterium *Pasteuria penetrans* is an obligate parasite of root-knot nematodes (Sayre & Starr, 1988) which has caused natural suppression of root-knot nematodes in West Africa (Mankau, 1980), and on vines in South Australia (Stirling & White, 1982). In an experiment by Karimipourfard & Damadzadeh (2006) an isolate of *P. penetrans* controlled final population of *Meloidogyne* spp. on *P. vera* cultivar Badami-reez by more than 70% under greenhouse conditions.

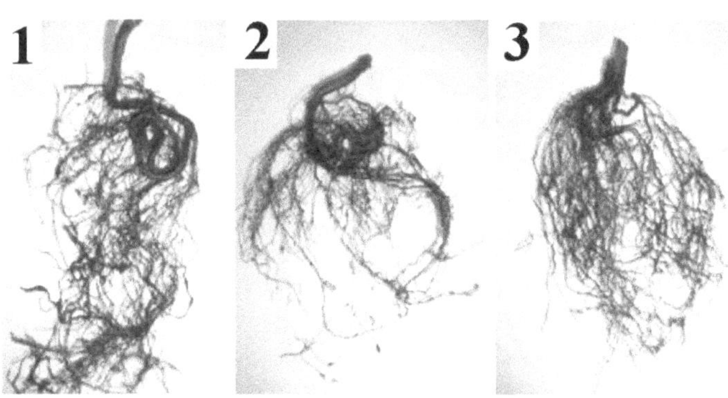

Figure 2. Comparison of pistachio roots parasitised by Meloidogyne *sp. and treated with* Pochonia chlamydosporia *var.*chlamydosporia: *1) roots with nematode and fungus, 2) with nematode only and 3) control without nematode and fungus.*

Pochonia chlamydosporia var. *chlamydosporia* (Goddard) Zare & Gams (synonym: *Verticillium chlamydosporium* Goddard) is a facultative parasite of cyst and root-knot nematodes (Willcox & Tribe, 1974; Kerry, 1975; Godoy et al., 1983; Fatemy et al., 1999). Different isolates have shown promising potential as biological control agents on sedentary nematodes (Ayatollahy et al., 2008; Sorribas et al., 2003; Atkins et al., 2003). One isolate of this fungus parasitized eggs of *M. javanica* by more than 70% on agar, whereas only 5% of the second stage juveniles were killed in culture filtrate in vitro condition (Ebadi et al., 2008). In a recent experiments strains of this fungus were able to control final density of *Meloidogyne* spp. on *P. vera*, by nearly 60% in natural soil, and nematode reproduction was reduced to below 1 (Figure 2) (Fatemy, unpublished).

Soil solarization is one of the methods which has been effective against Verticilium wilt in pistachio (Stapleton et al., 1993), olive trees (Tjamos, 1991; Katan and DeVay, 1991), almonds, apricots (Stapleton et al., 1993); and also against certain nematodes (Abu-Gharbieh et al., 1991). It can be used in tropical climates or during warm seasons in temperate regions, by trapping radiation and heat, the temperature raises to levels which can suppress soil-born pathogens and pests including weeds (Katan & DeVay, 1991). It also induces complex changes in the biological, physical and chemical properties of the soil that improve plant development, growth, quality and yield for several years (Stapleton, 1994; Katan & DeVay, 1991). Furthermore, in arid regions with drought and shortage of water supply, it may help preserve water in established orchards, reduce period of irrigation and promote soil microbial population. In pre-plant orchards, clear polyethylene films, and in the established plantation use of black polyethylene films to prevent excess heat from damaging trees, have been recommended (Stapleton et al., 1993; Duncan et al., 1992).

REFERENCES

Abrishami M. H. (1995). Persian Pistachio. A comprehensive history. Iran University Press: 669 pp.

Abu-Gharbieh, W., Saleh, H. and Abu-Blan, H. (1991). Use of black plastic for soil solarizaton and post-plant mulching. In: DeVay, J. E., Stapleton, J. J. Elmore, C. E. (eds). Soil Solarization. Plant Production and Protection Paper, 109. Food and Agriculture Organization of the United Nations, Rome, pp. 229-242.

Askarian, H., Sharifnabi, B., Olia, M. & Mehdikhani Moghaddam, E. (2006). Identification of *Meloidogyne javanica* on pistachio using Polymerase Chain Reaction. 17th Iranian Plant Protection Cong., 2-5 Sept., Karaj, Iran. 311.

Atkins, S. D., Hidalgo-Diaz, L, Kalisz, H., Mauchline, T. H., Hirsch, P. R. & Kerry, B. R. (2003). Development of a new management strategy for the control of root-knot nematodes (*Meloidogyne* spp.) in organic vegetable production. *Pest Management Science*, 59: 183-189.

Ayatollahy, E., Fatemy, S., & Etebarian, H. R. (2008). Potential for biocontrol of *Heterodera schachtii* by *Pochonia chlamydosporia* var *chlamydosporia* on sugar beet. *Biocontrol Science and Technology*, 18, 2: 157-167.

Aydin, C., & Ozcan, M. (2002). Some physic-mechanic properties of terebinth (*Pistacia terebinthus* L.) fruits. *Journal of Food Engineering*, 53, 97-101.

Aydemir, S. (2001). Palygorskite-influenced Vertisols and Vertic like soils in the Harran Plain in the Southeastern Turkey. PhD. Thesis, Texas A&M University, Soil and Crop Sciences Department, College Station, TX 77843, USA.

Banihashemi, Z. & Kheiri, A. (1995). The occurrence of root knot nematode (*Meloidogyne javanica*) on pistachio in Damghan *Iranian Plant Pathology*, 31, 37-38.

Barooti, S. & Hoseininejad, S. A. (2004). Identification of plant parasitic nematodes in some pistachio orchards in Kerman. 16th Iranian Plant Protection Cong., 28 Aug- 1 Sept., Tabriz, Iran. 377.

Duncan, R. A., Stapleton, J. J. & McKenry, M. V. (1992). Establishment of orchards with black polyethylene film mulching: Effect on nematode and fungal pathogens, water conservation and tree growth. *Journal of Nematology, Supplement*, 24 (4S), 681-687.

Ebadi, M., Fatemy, S. & Riahi, H. (2008). Antagonistic activity of an isolate of *Pochonia chlamydosporia* var. *chlamydosporia* on root-knot nematode in vitro. 5th International Congress of Nematology, 13-18 July, Brisbane, Queensland, Australia, 293 (abstract).

Esmail-Pour, A. (1998). Distribution, use and conservation of pistachio in Iran. In Toward a comprehensive documentation and use of Pistachio genetic diversity in Central and West Asia, North Africa and Europe. Report of the IPGRI Workshop, 14-17 Dec., Irbid, Jordan: 16-18.

FAO. (2006). Food and Agricultural Organization of the United Nations, FAOSTAT.

Farivar Mehin, H. (1984). Study of the root knot nematodes on pistachio in Iran. 1st International Congress of Nematology, Guelph, Ontario, Canada, 5-10 Aug., P 26-27.

Farivar Mehin, H. (1986a). Study on root knot nematodes of pistachio. Final report of Plant Pests and Diseases Research Laboratory, Rafsanjan. 15 pp.

Farivar Mehin, H. (1986b). Root knot nematodes on pistachio in Kerman. 8th Iran. Plant Protection Congress, 2-5 September: 136.

Fatemy, S., AhmadianYazdi, A., Ahmadi, A., Parvizy, R., Pakniat, M., Barooti, S., et al. (1999). Fungal parasites of *Heterodera schachtii* in Iran. *Pakistan Journal of Nematology*, 7(1), 61-66.

FDA. (2003). Office of Nutritional Products, Labeling and Dietary Supplements, Center for Food Safety and Applied Nutrition. Qualified Health Claims: letter of enforcement discretion-Nuts and Coronary Heart Disease (Document No. 02P-0505).

Godoy, G., Rodriguez-Kabana, R. & Morgan-Jones, G. (1983). Fungal parasites of *Meloidogyne arenaria* eggs in an Alabama soil. A mycological survey and greenhouse studies. *Nematropica*, 13, 201-203.

Greco, F., & Nucifora, S. (1999). I fitofagi del pistachio. *Informatore Agrario*, 55(26), 65-71.

Hadj-Hassan, A. (1988). Characters of most important Syrian pistachio female varieties widely cultivated in Aleppo. ACSAD, Damascus, Syria: 25 pp.

Javanshah, A., Avanzato, D., & Hokmabadi, M. H. (2000). Iran's pistachio industry. *Rivista di Frutticoltura e di Ortofloricoltura*, 62 (10), 90-92.

Joley, L. E. (1960). Experiment with propagation of the genus Pistachia. *Proceedings of the International Propagation Society*, 10, 287-292.

Karimipourfard, H., & Damadzadeh, M. (2006). The effect of *Pasteuria penetrans* to control of *Meloidogyne* spp. on pistachio. 17th Iranian Plant Protection Congress, 2-5 September, Esfahan: 375 (abstract).

Katan, J., & DeVay, J. E. (1991). Soil Solarization. CRC Press, Boca Raton, Ann Arbor. Boston, London.

Kawashty, S. A., Mosharrafa, S. A. M., El-Gibali, M., & Saleh, N. A. M. (2000). The flavonoids of four *Pistacia* species in Egypt. *Biochemical Systematics and Ecology*, 28, 915-917.

Kerry, B. R. (1975). Fungi and the decrease of cereal cyst nematode populations in cereal monoculture. *EPPO Bulletin*, 5, 361-553.

Kodira, U. C., & Westerdahl, B. B. (1995). Pistachio pest management guidelines. UC IPM, IPM Education and Publications, University of California, Davis: 12-13.

Koenning, S. R., Overstreet, C., Noling, J. W., Donald, P. A., Becker, J. O. & Fortnum, B. A. (1999). Survey of crop losses in response to phytoparasitic nematodes in the United States for 1994. *Journal of Nematology, Supplement* 31, (4S), 587-618.

Madani,M. Akhiani, A., & Damadzadeh, M. (1995a). Species and races of root knot nematodes (*Meloidogyne* spp.) on pistachio. 12th Iran. Plant Protection Congress, 2-5 September, Karaj, Iran: 248.

Madani, M. Kheiri, A., & Akhiani, A. (1995b). Evaluation of greenhouse reaction of Pistachia vera cultivars and wild masses to *Meloidogyne incognita*-R2. 12th Iran. Plant Protection Congress, 2-5 September, Karaj, Iran: 247.

Mankau, R. (1980). Biological control of *Meloidogyne* populations by *Bacillus penetrans* in West Africa. J. Nematology, 12: 230.

Mckenry, M. V., & Kretsch, J. O. (1984). Nematodes in pistachio orchards. *California Agriculture*, 38, 21.

Mohammadi Moghadam, M., Mortazavi, A. M., & Tanha Maafi, Z. (2006). Reaction of pistachio cultivars to root knot nematodes in field condition. 17th Iranian Plant Protection Cong., 2-5 September, Karaj, Iran: 370.

Palomares-Rius, J., Subbotin, S. A, Landa, B. B., Vovlas, N., & Castillo, P. (2008). Description and molecular characterisation of *Paralongidorus litoralis* sp. n. and *P. paramaximus* Heyns, 1965 (Nematoda: Longidoridae) from Spain. *Nematology*, 10, 87-101.

Pinochet, J., Verdejo, S., Soler, A., & Canals, J. (1992). Host range of a population of *Pratylenchus vulnus* in commercial fruit, nut, citrus and grape rootstocks in Spain. *Journal of Nematology, Supplement*, 24 (4S), 693-698.

Qasim, H., & Hashmi, S. (1988). Seasonal population fluctuation of nematodes on pistachio in Baluchistan. *International Nematology Newsletter*, 5, (3), 50-53.

Sabeti, H. (1966). Native and exotic trees and shrubs of Iran. Tehran University Press, Publication No. 1937, 430 pp.

Sayre, R. M., & Starr, M. P. (1988). Bacterial diseases and antagonists of nematodes. In: Poinar, G. A. and Jansson, H. B. (ed.). Diseases of nematodes, Vol. 1. CRC Press. Boca Raton, Fl.: 69-101.

Sorribas, F. J., Ornat, C., Galeano, M., & Verdejo-Lucas, S. (2003). Evaluation of a native and introduced isolate of *Pochonia chlamydosporia* against *Meloidogyne javanica. Biocontrol Science and Technology*, 13, 707-714.

Stapleton, J. J. (1994). Solarization as a framework for alternative soil disinfestations strategies in the interior valleys of California. Proceedings of the 1994 Annual International Research Conference on Methyl Bromide Alternatives and emissions reductions. Kissimmee, F1.

Stapleton, J. J., Paplomatas, E. J., Wakeman, R. J., & DeVay, J. E. (1993). Establishment of apricot and almond trees using soil mulching with transparent (solarization) and black polyethylene film: effects on Verticillium wilt and tree health. *Plant Pathology*, 42, 333-338.

Stirling, G. R., & White, A. M. (1982). Distribution of a parasite of root-knot nematodes in South Australian vine yards. *Plant Disease*, 66, 52-53.

Tjamos, E. C. (1991). Recovery of olive trees with Verticillium wilt after individual application of soil solarization in established olive orchards. *Plant Disease*, 75, 557-562.

Tous, J., & Ferfuson, L. (1996). Mediterranean fruits. In: Progress in new crops. Janick, J (ed.). ASHA Press, Arlington, VA. 416-430.

Vovlas, N. (1983). Gall formation on *Pistacia vera* by *Rotylenchulus macrodoratus. Journal of Nematology*, 15, 148-150.

Vovlas, N., & Inserra, R. N. (1983). Biology of *Heterodera mediterranea. Journal of Nematology*, 15, 571-576.

Westerdahl, B. B., & McKenry, M. V. (2002). Diseases caused by nematodes. In: Compendium of nut crop diseases in temperate zones. Teviotdale, B. L., Michailides, T. J. & Pscheidt, J. W. (eds). The American Phytopathological Society: 11-14.

Willcox, J., & Tribe, H. T. (1974). Fungal parasitism in cysts of *Heterodera*. I. Preliminary investigations. *Transaction of the British Mycological Society*, 62, 505-594.

Yildiz, S. (2007). Studies on the nematode fauna and biodiveristy of Sanliurfa. PhD. Thesis. School of Natural and Applied Sciences, Çukurova University, Adana, Turkey.

Zohary, M. A. (1952). Monographical study of the genus *Pistacia. Palestinian Journal of Botany* (Jerusalem), 5,187-228.

MANUEL M. MOTA[1], KAZUYOSHI FUTAI [2] AND PAULO VIEIRA[1]

PINE WILT DISEASE AND THE PINEWOOD NEMATODE, *BURSAPHELENCHUS XYLOPHILUS*

[1]NemaLab-ICAM, Departamento de Biologia, Universidade de Évora, 7002-554 Évora, PORTUGAL

[2]Laboratory of Environmental Mycoscience, Graduate School of Agriculture, Kyoto University, Sakyo-ku, Kyoto, 606-8502, JAPAN

Abstract. Pine wilt disease (PWD) is one of the most damaging events affecting conifer forests (in particular *Pinus* spp.), in the Far East (Japan, China and Korea), North America (USA and Canada) and, more recently, in the European Union (Portugal). In Japan it became catastrophic, damaging native pine species (*Pinus thunbergii* and *P. densiflora*), and becoming the main forest problem, forcing some areas to be totally replaced by other tree species. The pine wood nematode (PWN) *Bursaphelenchus xylophilus*, endemic, with minor damage, to North America, was introduced in Japan in the early XX century and then spread to Asia (China and Korea) in the 1980s. In 1999 it was detected for the first time in Portugal, where, due to timely detection and immediate government action, it was initially (1999-2008) contained to a small area 30 km SE of Lisbon. In 2008, the PWN spread again to central Portugal, the entire country now being classified as "affected area". Being an A1 quarantine pest, the EU acted to avoid further PWN spreading and to eradicate it, by actions including financial support for surveyes and eradication, annual inspections and research programs. Experience from control actions in Japan included aerial spraying of insecticides to control the insect vector (the Cerambycid beetle *Monochamus alternatus*), injection of nematicides to the trunk of infected trees, slashing and burning of large areas out of control, beetle traps, biological control and tree breeding programs. These actions allowed some positive results, but also unsuccessful cases due to the PWN spread and virulence. Other Asian countries also followed similar strategies, but the nematode is still spreading in many regions. In Portugal, despite lower damage than Asia, PWD is still significant with high losses to the forestry industry. New ways of containing PWD include preventing movement of contaminated wood, cutting symptomatic trees and monitoring. Despite a national and EU legislative body, no successful strategy to control and eventually eradicate the nematode and the disease will prevail without sound scientific studies regarding the nematode and vector(s) bioecology and genetics, the ecology and ecophysiology of the pine tree species, *P. pinaster* and *P. pinea* , as well as the genomics and proteomics of pathogenicity (resistance/ susceptibility).

1. INTRODUCTION

For millions of years the distribution of the world's biota has been constrained by natural barriers. However, with increasing globalization and the breaking down of

A. Ciancio & K. G. Mukerji (eds.), *Integrated Management of Fruit Crops and Forest Nematodes*, 253-274.

geographical boundaries, new biological invasions by non-indigenous species have become a global environmental issue, often causing severe outbreaks with economic and ecological disruption in various ecosystems (Liebhold et al., 1995; Sakai et al., 2001).

In forest ecosystems the pinewood nematode (PWN), *Bursaphelenchus xylophilus* (Steiner & Buhrer, 1934) Nickle, 1970, is considered one of the most important pests and pathogens in the world. The general fear of establishment of the PWN, the causal agent of the pine wilt disease (PWD), into countries where conifer forests assume great importance, stems from the devastating damage caused by this nematode to pine forests (Mamiya, 2004; Mota & Vieira, 2008; Shin & Han, 2006). The introduction of the PWN into non-native areas (outside of North America) is primarily associated with trade and the global flow of forest products (Bergdahl, 1999; Webster, 2004).

Unmanufactured wood, especially in raw log form, has been identified as one of the most high-risk pathways of movement of forest insects and pathogens into new environments, between continents (Evans et al., 1996; Tkacz, 2002). Many of the *Bursaphelenchus* species, including the PWN, have been routinely intercepted in packaging and wood products in several countries, e.g. Austria (Tomiczek et al., 2003), China (Gu et al., 2006), Finland (Tomminen, 1991) and Germany (Braasch et al., 2001). Furthermore, the recent detections of the PWN in packaging wood imported from countries considered free of this pest, due to the repeated use and circulation of this type of wood material, e.g. Brazil, Belgium, Italy and Spain, (Gu et al., 2006), undoubtedly stresses the importance of trade globalization for the potential entry/establishment of this pathogen into endemic forests worldwide.

The damage by this invasive species is clearly demonstrated by the devastation caused in non-native regions where the disease became established, e.g. Japan and China (Yang, 2004; Shimazu, 2006). The introduction of this nematode into non-native areas has resulted in huge annual losses due to the effects on increased mortality and growth loss of the pine forest (26 million m^3 of timber lost since 1945 in Japan), and by the increased costs in management procedures and disease control (Mamiya, 2004; Mota & Vieira, 2008; Shimazu, 2006). In addition, the introduction of this pest has resulted in vast and irreversible changes to the native forest ecosystems including tree species conversions, wildlife habitat destruction, soil and water conservation and loss of biodiversity (Kiyohara & Bolla, 1990; Suzuki, 2002).

The PWN is already established for more then 100 years in Japan (Yano, 1913), and in the past two decades the new reports of pine wilt disease came mainly from East Asia (Cheng et al., 1983; Yi et al., 1989). However, in 1999 the PWN was reported for the first time in Portugal and in Europe (Mota et al., 1999). Following this finding, there has been considerable activity in both delineating the extent of the infested area and preventing the spread to the remainder of the country and the European Union (EU) (EC, directive 2001/218/EC). The potential threat of the PWN to coniferous forests is real and the most effective way of reducing this threat is to be more restrictive to the importation of wood products, and to carry a rigorous inspection system for wood material (Evans et al., 1996; Bergdahl, 1999; Gu et al., 2006). Therefore, specific measures have been applied in Portugal in order to control the PWN and its insect vector, and in each EU member country, national surveys were performed to determine whether the nematode is present in other territories beside Portugal (directive 2001/218/EC).

The current situation in Portugal assumes great importance not only because of the economic implications, but also through the destruction of the pine forest in the area where the PWN became established (Setúbal Península). On the other hand, pine forests occupy a huge area of the continental territory ($1.25 \cdot 10^6$ ha) representing one of the greatest natural resources of the country, namely in the form of timber (*Pinus pinaster*), wood products and pine nuts (*Pinus pinea*). Consequently, strict requirements have been imposed on all wood movements from the affected area to other regions in Portugal, as well as to other EU member states. These measures have had serious implications for the timber industry within the affected area, creating a significant impact on the national economy and markets of wood industries (Rodrigues, 2008) (Fig. 1). Unfortunately, these measures have not been successful in preventing the spread of the PWN in Portugal.

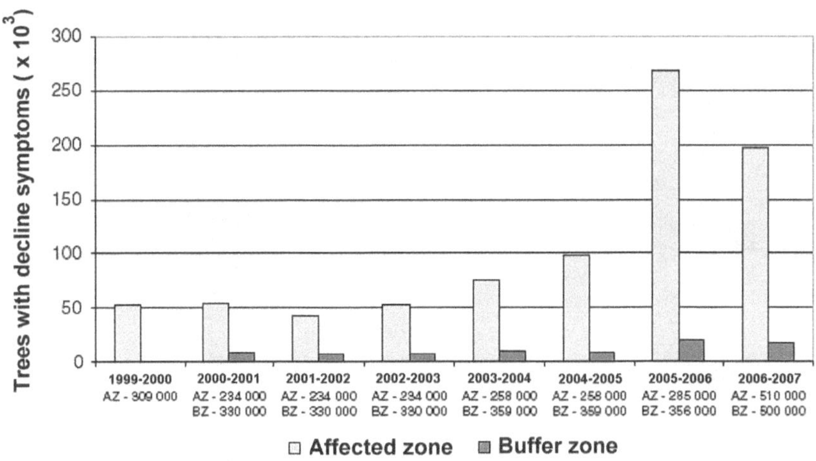

*Figure 1. Evolution of declining maritime pine (*Pinus pinaster*) trees in the 1999-2007 demarcated area in Portugal (Setúbal Peninsula) (from Rodrigues, 2008).*

The occurrence of pine wilt disease in Portugal was initially (1999-2008) limited to a relatively small area (ca. 500 000 ha). Nevertheless, the danger of spread of this disease assumes a high phytosanitary risk because of the wide distribution of both the insect vector (*Monochamus galloprovincialis* Oliv.) and the known susceptible host (*Pinus pinaster* Ait.) in Portugal (Rodrigues, 2008). Until recently, no consensus has emerged on the possible pathway of the PWN introduction in Portugal. This is partly due to a scarceness of studies using different sources of isolates from the affected area in the country.

Several hypotheses have been put forward to explain this introduction, such as from endemic areas where the nematode naturally occurs (North America), or non-endemic areas where the nematode behaves as an exotic pest (Asia) (Iwahori et al.,

2004; Mota et al., 2004). They were recently tested, suggesting a possible double introduction of the PWN in Portugal (Metge & Burgermeister, 2006), both from East Asian countries. Although this study incorporates a large number of different isolates from different regions of the world, concerning Portugal it is restricted to the use of three isolates only, and representative of a small area of the full affected area. Recently, a more complete genetic analysis has been made using 24 isolates from the original demarcated area (Setúbal Peninsula) (Fig. 2) and the results clearly indicate a lack of genetic diversity among isolates as well as a confirmation of the proximity with East Asian populations of the PWN (Vieira et al., 2007).

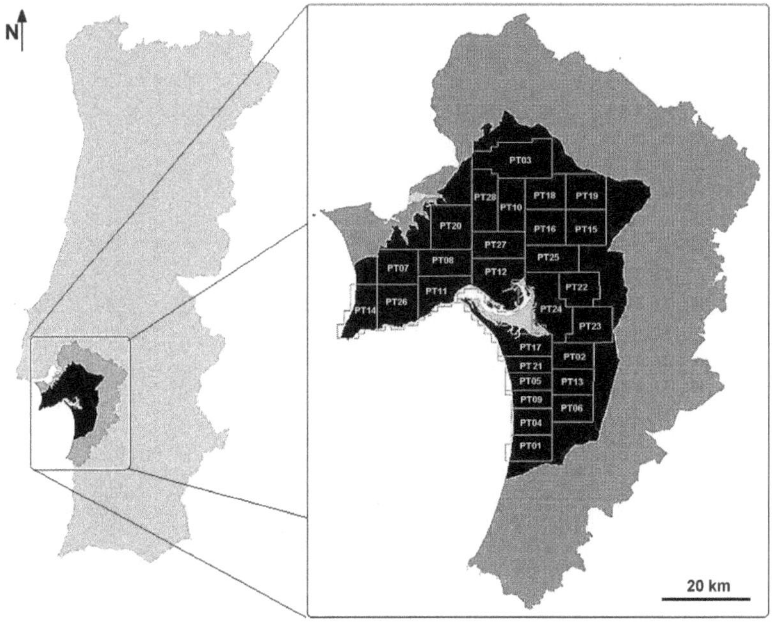

Figure 2. Portugal (continental, left) and location of the 1999-2008 quarantine area. Location of Bursaphelenchus xylophilus *isolates (right) from different blocks within the affected area. Black: the area affected by the PWN; dark grey: the buffer area, established in 1999 for safety reasons (free of PWN) (from Vieira* et al., *2007).*

2. PWN DISTRIBUTION AND DISEASE DISSEMINATION

PWN is considered a native species from North America, where it is distributed throughout Canada and USA (Robbins, 1982; Bowers et al., 1992; Sutherland & Peterson, 1999), and also with a single report from Mexico (Dwinell, 1993). In these regions, the PWN has been associated with several conifer species: blue spruce and white spruce (*Picea* spp.), atlas cedar and deodara cedar (*Cedrus* spp.), eastern larch and european larch (*Larix* spp.), balsam fir (*Abies* spp.) and Douglas fir (*Pseudotsuga* spp.), however, it is mainly found in pine species (*Pinus* spp.) (Robins, 1982; Bowers et al., 1992).

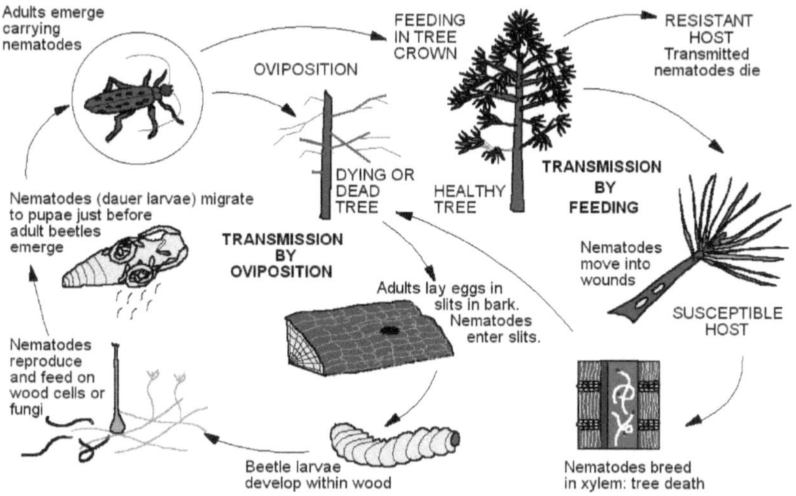

Figure 3. Schematic representation of the inter-relationships between the pinewood nematode, Bursaphelenchus xylophilus, *and its insect vector (*Monochamus *spp.) (adapted from* Evans, 1996*).*

Bursaphelenchus xylophilus has both phytophagous (transmission by feeding) and mycophagous (transmission by oviposition) phases of development (Fig. 3). The nematode is carried by *Monochamus* beetles that feed on twigs in the crowns of healthy trees (known as "maturation feeding"). Later, the female beetles lay their eggs in damaged or dying trees as well as in freshly cut stems with bark. Fourth-stage (J_{IV}) dispersal juveniles ("dauer" larvae,) of *B. xylophilus* are carried under the elytra (wing cases) and in the tracheae (breathing tubes) of the beetles and migrate into the tree through the wounds caused by feeding or ovipositing beetles.

Transmission during maturation feeding is the initiation of the phytophagous phase of the nematode, which has the greatest importance for the potential development of pine wilt disease. In a suitable tree species and under favorable climatic conditions, the nematodes multiply rapidly in susceptible trees, feed on plant tissues and move from the cambium into the xylem. Their generation time is 6 days at 20°C and 3 days at 30°C. The nematodes contribute to plant death by blocking water conductance (cavitation) through the xylem. The damaged trees become available for oviposition by *Monochamus* spp. females; therefore, nematodes also enter the tree through the oviposition slits in the bark. In dead trees, the nematodes feed on fungi, in particular on blue stain fungi (*Ceratocystis, Gliocladium*). *Monochamus* larvae develop initially in the cambium and then burrow into the wood, where the nematodes congregate in the vicinity of the pupal chambers formed by the mature beetle larvae. When the new beetle emerges, the nematodes migrate into the tracheae and to the area beneath the elytra of the beetles. The presence of suitable fungi in the trees encourages nematode reproduction and survival and, consequently, increases the number of nematodes carried by the emerging beetles (Mamiya, 1984; Linit, 1988; Evans et al., 1996).

The introduction and spread of this species into new areas has also been aided by the high phenotypic plasticity of the nematode, including excellent adaptation for resistance in the host tree (i.e. long periods of starvation) and dispersion (ectophoretic insect association) (Mamiya, 1984). In the native host species of North America, the nematode does not cause disease, since both plant and nematode have co-evolved for a very long time and thus the trees have become resistant/tolerant to its presence (Kiyohara & Bolla, 1990), except in some exotic *Pinus* spp. plantations (Evans et al., 1996). On the other hand, this scenario changes drastically when this organism reaches non-native habitats.

It is assumed that the presence of the PWN in Japan is the result of an accidental introduction by means of contaminated wood products from the USA (California) to the southern Japanese island of Kyushu, in the beginning of the 20th century (Yano, 1913). However, only in 1971 was the PWN associated with the high mortality of pine trees and identified as the causal agent of PWD, mainly of Japanese black pine (*P. thunbergii*) and Japanese red pine (*P. densiflora*) (Kiyohara & Tokushige, 1971). In spite of the numerous efforts to control the nematode and the insect vector (*M. alternatus*), the disease spread throughout the entire country, with the exception of the most Northern prefectures of Aomori and Hokkaido, occupying nowadays 28% of the total pine forest area (580 000 ha) (Mamiya, 2004; Shimazu, 2006).

During the eighties, the PWN was reported in other east Asia countries as well. In 1983 it was found for the first time in mainland China, associated with dead and dying Japanese black pine, in Nanjing (Jiangsu Province) (Cheng et al., 1983). The situation in China assumes great importance firstly by the continuous spreading of the disease (up to date affecting 75000 ha, and more then 20 million pine trees destroyed) among different regions of the country (Jiangsu Province, Anhui Province, Guangdong Province, Zhejiang Province, Shandong Province and Hubei Province) mainly due to human factors, and secondly by the potential threat to other areas where all the conditions that determine the establishment of the disease are present, and which are still free of the PWN (Yang, 2004).

In Taiwan the first report of the PWN occurred in 1985, identified from a luchu pine (*P. luchuensis*) stand displaying 50% mortality, in the Taipei prefecture (Tjean & Jan, 1985a). It has also been reported from Japanese black pine in Taoyeun prefecture (Tjean & Jan, 1985b).

In 1989, the PWN was detected in South Korea, in Pusan (the largest harbor city located in the extreme southern part of the country), associated with the Japanese black pine and Japanese red pine (Yi et al., 1989). Although the area of distribution of the disease was controlled until 1997, and limited in relatively small areas in the southern part of the country (La et al., 1999), in the last years a continuous spread of the disease has been observed, and more recently it has been reported simultaneously from new different areas (Mokpo, Sinan, Yeongam, Daegu, Gumi, Andong, Gyeongbuk, Gangneung and Donghae), constituting today the major forest pest in the country (Shin & Han, 2006).

In 1999, the PWN was reported for the first time in Portugal, and in Europe, associated with maritime pine (*P. pinaster*) (Mota et al., 1999), and with a single species as the insect vector (*M. galloprovincialis*) (Sousa et al., 2001). After the initial detection, a national survey was carried out along the pine forests, and a quarantine area was established where the nematode occurred, in the Peninsula of Setúbal (ca. 30 km SE of Lisbon).

Simultaneously, research focusing on the bioecology of the nematode (see following section) were initiated. Regarding the insect vector, *Monochamus galloprovincialis*, and besides a national survey (Sousa et al., 2002), information was obtained on nematode entry and population dynamics inside the vector (Naves et al., 2006a), feeding and oviposition (Naves et al., 2007; Naves et al., 2006b), flight patterns, traps, reproduction (Naves et al., 2006c).

The initial PWN affected area covered 510,000 ha, surrounded by a buffer zone of 500,000 ha more, for safety reasons. Although the initial affected area persisted as almost identical from 1999 ro 2007, in the last survey/eradication campaign the number of declining trees in the demarcated area increased significantly within the affected zone (Rodrigues, 2008), followed by an expansion of the demarcated area, particularly to the south of the country (Sines, corresponding to the south point), and very recently to the central areas of Arganil and Lousã. As a result of this trend, in 2007 prevention measures were established by the EU, i.e., the implementation of a 3 km phytosanitary strip surrounding the initial quarantine area, where all pine trees were cut and removed until the end of 2007 (Rodrigues, 2008). The effectiveness of this strip was questioned at the time and now with the new areas of implantation of the nematode (ca. 200 km North of the initial affected area) has become useless.

3. PINEWOOD NEMATODE TAXONOMY

3.1. Morphological Approaches

The genus *Bursaphelenchus* was established by Fuchs (1937) and includes nematodes that are associated with insects and dead or dying trees, mainly conifers, and which have an ectophoretic stage. Most species are fungal feeders and are either transmitted to dead or dying trees during oviposition by insect vectors, or to healthy trees during maturation feeding of their insect vectors (Hunt, 1993). The genus is mainly distributed in the northern hemisphere, however a few number of species have been reported outside of this geographical range (South Africa), associated with plantations of pine species (for a detailed information see Ryss et al., 2005).

The current concern on the introduction of the PWN into new areas has increased the interest and the knowledge of this genus and the number of species recorded worldwide. Up to date, the genus comprises nearly 100 described species, 10 of which where described in the last two years, mainly from east Asia (Hunt, 2008; Ryss et al., 2005). In Portugal, until the report of the PWN in 1999, no knowledge of this genus was available. At the moment, 10 species have been reported for the country, associated with maritime pine trees (Penas et al., 2004), including the description of a new species to science, *B. antoniae* Penas, Metge, Mota and Valadas, 2006 (Penas et al., 2006).

The economic importance posed by the PWN clearly reinforced the need for an accurate diagnosis of the species, where morphological studies remain the standard method for routine identification. Different criteria may be used to divide the large number of nominal species of the genus *Bursaphelenchus*, into smaller and more convenient species groupings. Tarjan and Baéza-Aragon (1982) were the first to attempt the assembly of morphological identification keys for this genus, providing a detailed classification of the spicule characters and other useful morphological diagnostic data. Braasch (2001), and for the species associated with conifer trees in Europe (28 at that time), proposed the establishment of the species groups based on the number of lateral lines (nine different groups), followed by the distribution of the male papillae, spicule shape, presence and size of the female vulval flap and the shape of female tail.

Yet, an integrated morphological identification system to all the species of the genus has been lacking. Furthermore, the fact that more then 70% of these species occur in pine trees makes the identification even more uncertain. Therefore, Ryss et al. (2005) ellaborated a synopsis of the genus in order to provide an identification system to all the nominal species, where the spicule structure is the main diagnostic character to separate the species into groups. The six species groups (*aberrans*-group, *borealis*-group, *eidmanni*-group, *hunti*-group, *piniperdae*-group and *xylophilus*-group) are merely recognized as identification units in order to facilitate species identification. However, some of these groups could be considered as natural, i.e. phylogenetically related (e.g. the *xylophilus*-group) (Ryss et al., 2005).

Despite the clear separation of the members of the *xylophilus*-group (*B. baujardi*; *B. conicaudatus*; *B. doui*; *B. fraudulentus*; *B. kolymensis*; *B. luxuriosae*; *B. mucronatus*; *B. singaporensis*; *B. xylophilus*) from other groups based solely on the male spicule shape, the variability and overlapping in range of several other taxonomic characters within some species of this group is such that their accurate identification is difficult.

One of the major characters used for distinguishing the PWN from all other members is the shape of the female tail, i.e. rounded, and lacking a distinct mucron. However, specimens of *B. xylophilus* from North America show a wide variation in female tail shape, showing variations from rounded to a mucronated form, similar to the female tail of *B. mucronatus* (Wingfield et al., 1983). In addition to the morphological similarities between *B. xylophilus* and *B. mucronatus*, these two species are capable of genetic exchange, either directly or via intermediate forms (De Guiran & Bruguier, 1989), which clearly compromise the identification at the species level using morphological data only. Furthermore, the presence of males or juvenile stages alone deemed to be an unreliable method in the identification at the species level within the *xylophilus*-group, as well as for the differentiation of geographic isolates.

3.2. Molecular Approaches

Due to the difficult identification and constrains of morphological observations between *Bursaphelenchus* species, alternative molecular tools have become a valuable instrument for species and sub-specific separation. Initially these molecular tools were mainly developed for the differentiation of some species of the *xylophilus*-group, such as *B. xylophilus* and *B. mucronatus*, in order to achieve a better understanding of the relationships, and the clear identification of the *B. xylophilus* isolates.

The first methods used for the *Bursaphelenchus* species identification and isolates separation were based on protein profiles (Hotchkin & Giblin, 1984) and enzyme electrophoresis (De Guiran et al., 1985). However, the value of these methods was limited by differential gene expression during the life cycle of the nematode or by the response to external environmental influences (Harmey and Harmey, 1993). Immunological approaches have also been used for species-specific identification, using polyclonal antibodies that could differentiate specific antigens of certain *B. xylophilus* isolates (Lawler & Harmey, 1993), as well as monoclonal phage antibodies (Fonseca et al., 2006).

With the expansion of DNA-based methodologies, new alternatives, independent of the development stage and phenotypic variation due to external influences (Harmey & Harmey, 1993), have been able to detect genetic variation that can be exploited or adapted for taxonomic and diagnostic purposes. Bolla et al. (1988) differentiated *B. xylophilus* pathotypes using restriction enzyme analyses and hybridization with total genomic DNA. Others have used cloned DNA hybridization probes from *C. elegans* (Abad et al., 1991), or *Bursaphelenchus*, based on ribosomal probes (Webster et al., 1990), DNA probes (Abad et al., 1991; Tàres et al., 1992) and satellite DNA (Tàres et al., 1994), for a more reliable characterization of the species, and for the differentiation of specific and intraspecific groups.

The development of the polymerase chain reaction (PCR) promoted the improvement of some of the previous methods, and the establishment of new methods where only small amounts of DNA are required. The amplification of specific genomic regions is a highly effective methodology to detect inter- and intra-specific variations among taxa. Species-specific DNA fragments have been amplified using primers derived from a cloned repetitive DNA sequence (Harmey & Harmey, 1993). ITS-RFLP has been used mainly for *Bursaphelenchus* species identification (Burgermeister et al., 2005; Metge et al., 2008), while other methods have been carried out for the specific-species detection of *B. xylophilus*, namely PCR-based diagnostics with species-specific primers (Kang et al., 2004; Matsunaga & Togashi, 2004; Li et al., 2004; Leal et al., 2005; Leal et al., 2008), real-time PCR assay (Cao et al., 2005), and PCR amplification using satellite DNA-based primers (Castagnone et al., 2005; Castagnone-Sereno et al., 2008).

Concerning the assessment of the relationships among isolates with different geographical origins the following molecular methods have been applied: sequencing of heat shock protein genes, hsp70 (Beckenbach et al, 1992), sequence of rDNA ITS regions (Iwahori et al., 1998; Beckenbach et al., 1999; Zhang et al., 2001; Kanzaki & Futai, 2002; Megte et al., 2008), sequence of D2 and D3 of the 28S gene (Zheng et al., 2003; Metge et al., 2008). The random amplified polymorphic DNA technique (RAPD) has also been used for the study of intra-specific variation of PWN isolates from China (Zheng et al., 1998; Zhang et al.,

1999), Japan (Kusano et al., 1999), and a mixture of different geographical isolates (Braasch et al., 1995; Irdani et al., 1995a, 1995b; Wang et al., 2001; Zhang et al., 2002). Recently, a more integrated study has been conducted using several isolates each from the native regions (Canada and USA) and non-indigenous areas (China, Japan, Korea and Portugal) (Metge & Burgermeister, 2006).

4. PWN INTRODUCTION IN PORTUGAL AND THE EU

The way of introduction of the PWN to non-endemic areas has been primarily attributed to several hypotheses related with human activities, especially by the movement of infected wood products, between long (among continents and countries) and short (within a country) levels of distance. However, the short distance level of the disease spreading is attributed to the biological development of the insect vector as well. The genetic diversity of an exotic species in a new established area is always dependent on the diversity of the initial colonizers. An understanding of the role played in the Portuguese situation has been hindered by the lack of detailed studies from the isolates distributed in this region (Vieira et al., 2007).

The native forms of an organism are the major source of genetic variation, regularly displaying a higher level of genetic diversity when compared with those populations found in non-native areas and due to its artificial establishment. The effect of human activities on spreading the PWN into new areas is well documented, and variation on the PWN, at different levels, can explain a substantial part of the within-isolate variation observed from different geographical areas. Genetic variation among the PWN isolates is certainly not new. According to previous studies, the isolates collected from the USA and Canada exhibit a high level of diversity, the greatest level of diversity being reached among isolates collected in some areas of Canada (Iwahori et al., 1998). On the other hand, isolates found in the non-endemic areas express a low level of genetic diversity. Indeed, even in some of the non-native areas the genetic variation reaches some heterogeneity among some of the PWN isolates. Nevertheless, the degree of this variation could be limited by several hypotheses, i.e. the origin of the isolate (endemic area vs. non-endemic area), or by the number of introduced isolates. Furthermore, the number of individuals present in the infected wood products that reach the new site of infection could also limit the genetic variation of the initial introduction.

In Portugal, the extension of this genetic variation has not been clear. Recently, the origin of the PWN in Portugal was stated as being from an Asia region, and by a possible double introduction. If the introduction of this pathogen occured at least twice (even from non-native regions), different levels of genetic variability among the affected area in Portugal are to be expected, since a relative degree of variability in the Portuguese isolates was shown (Metge & Burgermeister, 2006). Still, this result might be due to a genetic shift of one of the isolates kept in fungal culture for a long period of time (Chapter II). The fact that the Portuguese *B. xylophilus* isolates show a high genetic similarity, using RAPD-PCR and satellite DNA clearly exclude the idea of a possible double introduction in Portugal (Vieira et al., 2007). Furthermore, the Portuguese isolates display a close genetic similarity with the East Asia isolate, confirming the results previously obtained by other authors (Metge & Burgermeister, 2006).

4.1. Dispersal of the PWN Within the Affected Area in Portugal

According to the data generated from other countries, the detection of the PWN is consistently coincident with port areas, associated with the trade of goods between countries. Initially the main concern came from those countries where the PWN was already naturally or artificially established. However, the report of several detections of PWN in wood products originating from PWN-free countries, increased the unpredictable introduction of this pathogen into new areas. It has been shown (Vieira et al., 2007) that the lack of genetic diversity among the PWN isolates in Portugal reflect a single introduction. Furthermore, the proximity of the international sea harbor in the Setúbal Península could determine the initial point of introduction.

The evolution of a forest disease within a country is guided by a widely studied framework involving two main processes: *i)* transport of contaminated wood by human activities and *ii)* biological development of the insect vector. In addition, the insect vector species occurs throughout the affected area (Sousa et al., 2001; 2002). Such overlapping distribution of the insect vector coupled with human activity in moving wood may have provided the main source of spreading of the pine wilt disease in Portugal.

5. CONTROL MEASURES FOR PWD

Controlling PWD and the PWN is not an easy task. The complex biological system (Fig. 3) involves knowledge concerning many aspects of the bioecology of both the nematode and the insect vector, coupled with knowledge regarding the degree of susceptibility/ resistance of the tree, as well as the environmental factors (climate and soil) that play a pivotal role in the development of the disease.

Most of the knowledge and success regarding the control of PWD has stemmed from the dramatic Japanese experience during the XXth century, followed by the more recent experiments and results obtained from China and Korea. For details on the speific actions taken in these countries, see Mota and Vieira (2008). Europe, and namely Portugal, has limited experience concerning tactics and strategy for an effective and sustainable control of PWD, which is easily understandable due to the relatively recent (1999) detection of the PWN and the need to take immediate actions for prompt containment of the disease (Rodrigues, 2008). However, an urgent coordinated effort between research and forest authorities is badly needed in order to stop the spread of the nematode beyond the borders of Portugal. The European Union (EU) should also contribute to this effort, as a pre-emptive action, in order to avoid the appearance of the nematode in other Southern, or even Central European countries where climatic conditions, the presence of the insect vector and of several highly-susceptible pine species would be catastrophic for EU forestry.

5.1. Control Measures Before the Discovery of PWN as the Causal Agent

When the first outbreak of pine wilt disease occurred at Nagasaki in Kyushu Island in 1905, local people in Japan made considerable efforts to eradicate the epidemic forest disease, though they did not recognize PWD as an epidemic. The dead trees were felled down and debarked completely to stamp out the first incidence of the PWD by 1915. Pine wilt disease, however, recurred at a harbor town in Hyogo prefecture, in the western part of the mainain in 1921, and also at another harbor town, northern part of Kyushu Island in 1925 (Fig. 4). Then, PWD gradually spread surrounding regions year by year. In the 1940s PWD remarkably expanded its distribution not only into surrounding regions but also to remote regions such as Shikoku Island and Kanto districts, eastern part of Mainland.

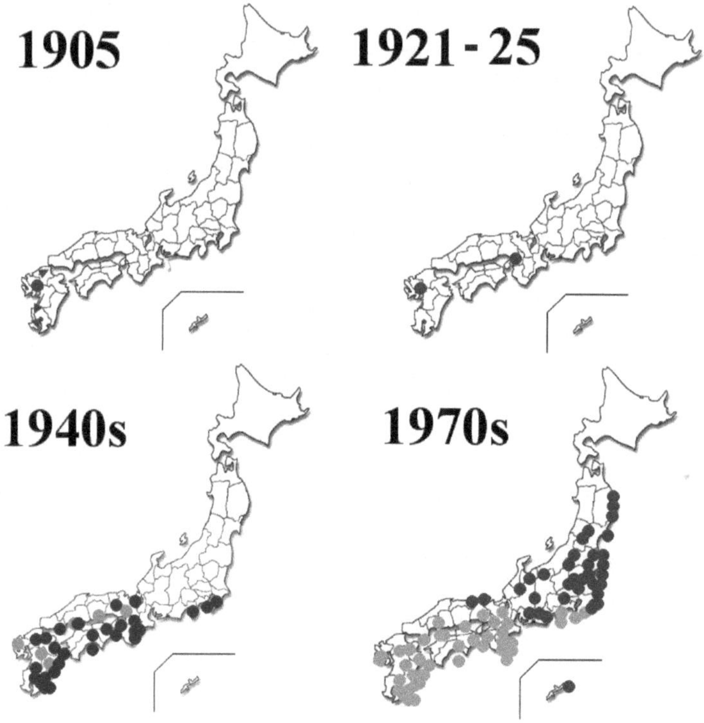

Figure 4. Spread of pine wilt disease (PWD) in Japan, during the XXth century. 1905: the first outbreak of PWD was reported from Nagasaki prefecture in Kyushu island. 1921-1925: in mainland Japan (Honshu island), the first occurence of PWD was reported from Hyogo prefecture. In1925, PWD recurred in a harbor town, 50 km apart from the first recorded place, and spread into the surrounding regions (grey dots). In 1940s: PWD spread over a wide area in southeastern Japan, and then moved to eastern Japan. In 1970s: PWD spread to a wide area of northeastern Japan.

Because of World War II, the Japanese people had to live very harsh times in the 1940s and therefore dead pine trees were apt to be left in stands. Furthermore, it became difficult to eradicate dead Japanese black pine, *Pinus thunbergii*, during wartime because the harbor area where black pines were dominant became restricted area, for military reasons. This background facilitated the vector beetle *Monochamus alternatus* to build up their population, and thereby remarkably increased the damage.

Before the discovery of PWN as the causal agent of PWD, most Japanese scientists had attributed the massive loss of pine trees to pine bark and wood borers. So several measures such as felling and burning, immersion in water, and spraying insecticides were recommended to control PWD. The insecticides used in this period were carbon disulfide and chloropicrin.

After World War II, General Headquarters (GHQ) of the Allied occupation military was seriously concerned about the devastated pine forests, and charged a forest entomologist, Dr. R. L. Furniss, to inspect pine forests damaged by PWD. After intensive field survey and discussion with Japanese experts he submitted two reports indicating seven issues to be revised: *i)* to establish a special organization that would be in charge of controlling forest insect pests, *ii)* as a part of the organization, special survey crews should be involved in evaluating the exact status of the infestation so that control projects could be properly planned, *iii)* of several control measures adopted till then, the best available method under the conditions in Japan was felling, peeling and burning dead pine trees.

Other methods used so far were of no use, but immersing infested logs for several weeks was effective, *iv)* governmental subsidization should be limited to epidemic outbreak, *v)* to carry out the recommended control methods effectively, relevant statute should be modified, *vi)* to keep the population of forest insect pests under control, appropriate silvicultural treatments were needed, *vii)* more experts trained in forest management and protection were especially needed (Furniss 1950; 1951).

The GHQ adopted these recommendations and urged the Japanese government to implement the control measures recommended by Furniss. The extensive control efforts following Furniss's recommendations, together with plentiful labor available then, succeeded in reducing the damage. Thus, the annual loss of pine trees due to PWD was reduced in the 1950s and until early 1960s. The life style of public people in Japan, however, changed remarkably in this period and pine needles and fallen twigs that had been used as fuel and/or fertilizer became abandoned and thus accumulated, which contributed to eutrophication of the forest soil. Soil eutrophication damaged the mycorrhizal relationship of pine trees, and thereby imposed serious stress on pine trees. Annual loss of pine trees increased again in the middle to the later half of the 1960s.

To establish a control method for PWD, a new national project was organized (1968–1971). This project team found that the insect pests that had supposedly been the causal agent of pine death could not lay their eggs on healthy trees, and the trees had reduced resin exudation as an early wilting symptom before the attack of insects (Nitto et al., 1966; 1967). Therefore, the national project had to change the study target from insect pests to other unknown factors such as microorganisms, edaphic factors, meteorological factors, and so on. In 1968, Tokushige, a tree pathologist of the project team found *Bursaphelenchus* nematodes and confirmed its pathogenicity against pine trees by a series of well-designed inoculation tests (Tokushige & Kiyohara, 1969; Kiyohara & Tokushige, 1971).

After a massive search for vector insects, the Japanese pine sawyer, *Monochamus alternatus* was found to be the sole vector of the nematode, one which transfered pathogenic nematodes from dead to healthy pine trees (Mamiya & Enda, 1972; Iwasaki & Morimoto, 1972). When the complete infection cycle of PWD was thus clarified, traditional control measures were abandoned and new ones, which set the vector beetle as a target, were applied.

5.2. Control Measures After the Discovery of the PWN and its Vector, Monochamus alternatus

After the discovery of PWN and its vector beetle, various control efforts were focused mainly on the vector beetle, *Monochamus alternatus*.

5.2.1. Physical Control

Among physical control measures, "felling, debarking and burning" which are rather traditional control methods, are still effective to eradicate vector beetles. This method, however, is laborious, and entails danger of forest fires and may facilitate some thermophilic pathogens such as *Rhizina undulata* (Sato, 1974). To avoid danger of forest fires, dead trees felled down were also burried under soil or submerged in water, though either of these measures was more laborious than burning.

5.2.2. Chemical Control

Before the discovery of the PWN, control measures had been targeted at larvae of bark or wood borers inhabiting in dead pine trees. For control purposes, therefore, the most predominant chemical measure was sanitation spraying with such insecticides as BHC and DDT on the bark of felled pine trees. These, however, were banned for use against forest pests in 1971 because of its residual toxicity to mammals. To control the newly-found vector of PWD, various insecticides were examined, and organophosphate insecticides such as fenitrothion and fenthion seemed to be the most effective and were applied instead of BHC and DDT.

Based on the information of the infection cycle of PWD (Fig. 3), scientists recommended the use of insecticides preventively to living trees; when *Monochamus* beetles emerge from dead pine trees their reproductive organs are not yet matured (Katsuyama et al. 1989), and they therefore move to surrounding healthy pine trees to feed on the bark of young shoots and thereby they become reproductively active ("maturation feeding"). Meanwhile, pathogenic PWNs enter pine trees via feeding wounds made by *Monoachamus* beetles, and the trees ultimately become diseased. Insecticides such as fenitrothion and fenthion may be sprayed over the crown of pine trees. This measure does not kill vector beetles directly but protect living trees from feeding of *Monochamus* beetles, and so has been called "preventive spraying". When this new measure was applied by aerial spraying, however, public people, some scientists and some media opposed the application for fear that these insecticides would harm the environment.

When this preventive spray was applied to forests when and where PWD was rampant, healthy living trees would be protected from PWN infection, while trees that were asymptomatic carriers and those that have been infected beforehand in the season could become diseased and then be killed even after preventive spraying. These actions seemed to fail in controling PWD, and gave people and the media arguments against the government. So the national and local governments became very cautious in applying aerial spraying with insecticides, and carried out these actions just in limited areas and/or in limited periods with as little amount of insecticides as possible. The *Monochamus* beetle, however, could often fly a few kilometers or more. When the insecticide lost its toxicity, the beetles could visit the area from untreated surroundings and kill pine trees that had received insecticide beforehand. Thus cautious application made the measure more ineffective.

To reduce environmental damage by insecticides, fumigation with methyl bromide, EDB, NCS and so on was applied after dead pine trees were felled down, cut into small-sized logs, and piled up. This method is apparently laborious, and time-consuming. Discarding the vinyl sheets used for covering the pile of dead pine logs is another problem after fumigation.

Prophylactic trunk injection of a nematicide is alternative method to control PWD. A company has applied a vermicide (morantel tartarate) to living pine trees, and succeeded in protecting them against PWD infection. Some other chemicals such as levamisol hydrochloride, methyl phenphos, emamectin benzoate, milbemectine have also been used as candidates for trunk injection nematicides. Among them, emamectin benzoate, milbemectine and nemadectin are antagonists of gamma-aminobutyric acid (GABA)-receptor, morantel tartarate and levamisol hydrochloride are muscle activity blockers, and mesulfenfos is an acetylcholine esterase inhibitor. Thus, these chemicals used for trunk injection were not necessary to kill nematodes in pine tissues, but may disturb nematode activity and/or reproduction, thereby facilitating host resistance against PWN. This measure (trunk injection) is very effective to control PWN, but the cost of the chemicals and that for manpower are so expensive that most owners of forests hesitate to use this measure.

5.2.3. Biological Control

To reduce application of insecticide for PWD control, some natural enemies have been examined as biological control agents against the vector beetle (further indicated as M), and PWN (further indicated as N). Among them were woodpeckers (M), predaceous insects (M) such as *Trogossita japonica*, *Dastarcus longulus*, and parasitoid insects (M) such as *Sclerodermus* spp.

Recent research from Portugal has also provided some interesting information on the potential of certain parasitoids for the biocontrol of *Monochamus galloprovincialis* (Naves et al., 2005). This paper includes a good review of parasitoids from East Asia and North America.

Entomoparasitic fungi (M) such as *Beauveria* spp. have been examined their effects in control *Monochamus* beetles (Fig. 5). These fungi seem to be effective, but it is often difficult to apply in the field because of indirect contamination of other useful insects such as silk worm and honey bee. Trapping fungi (N) and entomopathogenic nematodes (M) have also been examined for their ability to control PWN and the vector beetle, respectively. These biological control measures have not yet been practiced because these require more cost and labor than chemical ones. Exception is the case of *Sclerodermus* species in China, which have been reproduced in bulk and applied to pine forest with successful result (Fig. 5).

5.2.4. Breeding of Resistant Hosts

Trees of the genus *Pinus* propagate predominantly by sexual reproduction, so genes are mingled by pollination every year. Thus genetic diversity is very high among progenies. Host resistance against PWD takes advantage of genetic diversity, various among individual pine trees. Host resistance against PWD seems to be determined not by a single gene but by multi genes, though the host resistance mechanism has not yet been elucidated.

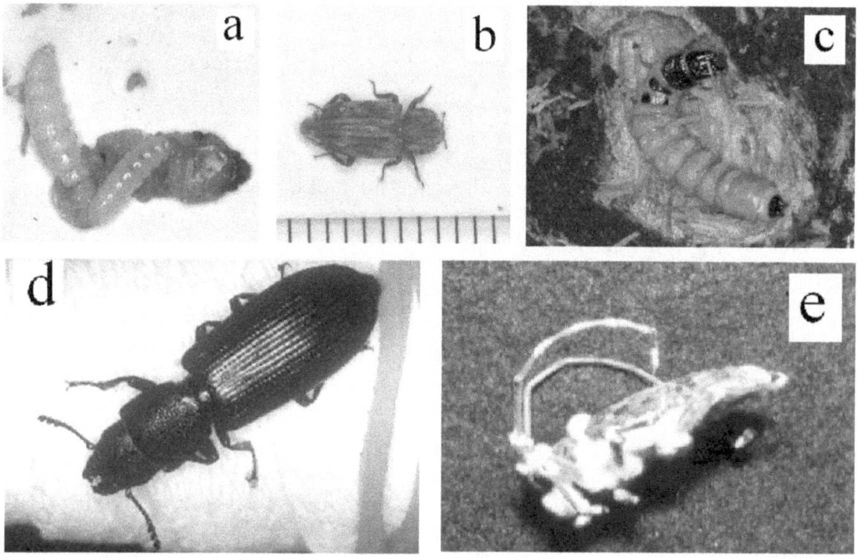

Fig. 5. Natural enemies of the Monochamus *vector:* Dastarcus longulus *(a, b) ;* Trogossita japonica *(c, d);* Monochamus *beetle infected with* Beauveria *sp. (e).*

When PWD rages fiercely through a pine stand, several surviving trees may remain due to a somewhat higher resistance. From such remaining pine trees, scientists have collected scions or seeds to breed resistant clones. When these candidate plants (grafts or seedlings) grow enough to serve for inoculation tests, they are inoculated once with the PWN, then the surviving plants receive another inoculation.

Pine grafts or seedling surviving two inoculation tests are regarded as resistant clones. Since the beginning of this project in 1978, 135 and 41 plants have been selected so far as resistant clones against PWD for *P. densiflora* (Japanese red pine) and *P. thunbergii* (Japanese black pine), respectively. These resistant clones have been propagated by grafting and cutting, and the resulting seedlings are being distributed over various regions of Japan. This tactic seems to be a reasonable way to make Japanese pine forests more resistant, but can not protect pine trees being exposed to PWD at present. As in the case of Dutch elm disease, and plant parasitism by *Meloidogyne* spp., once resistance-breaking individuals develop within the PWN population, resistant clones obtained after long selection procedure may be easily defeated.

6. A BLIND SPOT IN PWD CONTROL STRATEGY: THE ASYMPTOMATIC CARRIER AND ITS SOPHISTICATED DETECTION METHOD

To prevent pine wilt disease (PWD) from spreading over pine forests, elimination of pine trees killed by PWN is desirable, although this method is very laborious and time-consuming. If such dead trees are left in the field, pathogenic nematodes and their vector, *Monochamus* beetles, could spread from tree to tree without any difficulty. In the Kyoto University arboretum, where many precious foreign pine species are planted in the field, all pine trees killed by PWD have been eradicated thoroughly before the next pine wilt season.

Despite intensive efforts in removing dead trees from the stands, new dead trees tend to appear in the vicinity of the stumps of trees killed in the previous year, and wilting recurs in the same pine stand every year. To understand the reason why PWD recurs at the same stand even after thorough eradication of dead pine trees, a long-term survey at a stand of Korean pine, *Pinus koraiensis*, has been undertaken, and thus revealed the important role of asymptomatic carriers in spreading PWD to surrounding pine trees. When PWD-infected pine trees survive asymptomatically, and begin the symptom appearance far later than usual and overlapped with the following season of the beetles' activity, such trees could play a role as strong attractants to the vector beetles, posing a danger to pine stands (Futai, 2003).

To remove asymptomatic carriers from pine forests, a rapid and accurate detection of the PWN is needed. The population of PWN in asymptomatic trees is generally too low to be detected by traditional methods such as the Baerman funnel method. To detect low densities of PWN from living pine trees, a new diagnostic method based on a simple DNA extraction and nested-PCR has been developed (Takeuchi et al., 2005). This new method has been applied to two natural stands (Japanese black pine and a Japanese red pine) and found that many trees of either pine species contained PWN, though some of them displayed no external and/or internal symptoms (Takeuchi & Futai 2007). Thus some trees of Japanese black and red pine survived for one or more years after PWN infection without any symptoms, suggesting that they may have been overlooked during eradication, and may play a role in initiating new PWD occurrences.

7. CONCLUSIONS

Pine wilt disease constitutes a major threat to forest ecosystems worldwide, both from the economical point of view as well as from the environmental (landscape) perspective. In countries, such as Japan, China and Korea, where the disease is present and the pinewood nematode well established, forest authorities have undertaken extensive and very costly efforts to contain the disease, and to prevent further spread. In many cases, these actions have not been successful due to the high susceptibility of the tree species and the agressive virulence of the nematode. In Kyoto, Japan, for example, some large areas of local pine species have simply been replaced by other tree species such as oaks. In other more localized situations, such as religious temples or national scenic sites (e.g., Amanohashidate, Kyoto), PWD control programs using various approaches (resistance varieties, chemical control, etc..) have been successful, albeit at a high economical cost, but defrayed by the high cultural and environmental value.

The relatively recent detection of the nematode in the EU (Mota et al., 1999), poses a serious threat and challenge to European forestry officials and national plant protection authorities. Although the nematode is present, for the time being, in Portugal, the EU must maintain a continuing effort in: *1)* supporting surveying and control measures in Portugal; *2)* increasing the level of inspections at ports of entry, namely sea ports, in order to guarantee a rigorous interception of potential sources of PWN from non-EU countries; *3)* establishing a European network of diagnostic labs; *4)* establishing a EU-level research network involving the major scientific centers, to study the bio-ecology of the nematode and insect vectors, as well as the natural conditions that may enable the establsihment of the PWN in other areas of the EU.

The issue of PWD is one that constitutes a good example of the urgent need for a concerted action, not only at the EU level, but also worldwide due to the important economical sector of wood trade.

REFERENCES

Abad, P., Tàres, S., Bruguier, N., & Guiran, G. (1991). Characterization of the relationships in the pinewood nematode species complex (PWNSC) (*Bursaphelenchus* spp.) using a heterologous *Unc-22* DNA probe from *Caenorhabditis elegans*. *Parasitology* 102: 303-308.

Braasch, H. (2001). *Bursaphelenchus* species in conifers in Europe: distribution and morphological relationships. *EPPO Bulletin* 31: 127-142.

Braasch, H., Tomiczek, C., Metge, K., Hoyer, U., Burgermeister, W., Wulfert, I., & Schönfeld, U. (2001). Records of *Bursaphelenchus* spp. (Nematoda, Parasitaphelenchidae) in coniferous timber imported from the Asian part of Russia. *Forest Pathology* 31: 129-140.

Beckenbach, K., Smith, M., & Webster, J. (1992). Taxonomic affinities and intra- and interspecific variation in *Bursaphelenchus* spp. As determined by polymerase chain reaction. *Journal of Nematology* 24: 140-147.

Beckenbach, K., Blaxter, M., & Webster, J. (1999). Phylogeny of *Bursaphelenchus* species derived from analysis of ribosomal internal transcribed spacer DNA sequences. *Nematology* 1: 539-548.

Bergdahl, D. (1999). Threat of pine wilt disease to coniferous forests around the world. In Sustainability of pine forests in relation to pine wilt and decline. Proceedings of the Symposium, Tokyo, Japan, 26-30 October 1998. Futai K., Togashi K. and Ikeda T. (Eds). Kyoto, Japan, Shokado Shoten: 136-139.

Bolla, R., Weaver, C., & Winter, R. (1988). Genomic differences among pathotypes of *Bursaphelenchus xylophilus*. *Journal of Nematology*, 20, 309-316.

Bowers, W., Hudak, J., Raske, A., Magasi, L., Myren, D., Lachance, D., et al. (1992). Host and vector surveys for the pinewood nematode, *Bursaphelenchus xylophilus* (Steiner and Buhrer) Nickle (Nematoda: Aphelenchoididae) in Canada. Information Report Newfoundland and Labrador Region, Forestry Canada (N-X-285): 55 pp.

Burgermeister, W., Metge, K., Braasch, H., & Buchbach, E. (2005). ITS-RFLP patterns for differentiation of 26 *Bursaphelenchus* species (Nematoda: Parasitaphelenchidae) and observations on their distribution. *Russian Journal of Nematology* 13: 29-42.

Castagnone-Sereno, P., Castagnone, C., François, C., & Abad, P. (2008). Satellite DNA as a versatile genetic marker for *Bursaphelenchus* In Pine wilt disease: a worldwide threat to forest ecosystems. Mota, M. & Vieira, P. (Eds.). Springer, Dordrecht, NL: 187-196.

Cao, Y., Ma, H., Yang, W., Bai, G., Li, H., Hu, J., & Wang, Y. (2005). Development and application of immunomagnetic separation ELISA for identification of *Bursaphelenchus xylophilus*. *Forest Research*, 18, 585-589.

Cheng, H. R., Lin, M., Li, W., & Fang, Z. (1983). The occurrence of a pine wilting disease caused by a nematode found in Nanjing. *Forest Pest and Disease*, 4, 1-5.

Evans, H., McNamara, D., Braasch, H., Chadouef, J., & Magnusson, C. (1996). Pest risk analysis (PRA) for the territories of the European Union (as PRA area) on *Bursaphelenchus xylophilus* and its vectors in the genus *Monochamus*. *EPPO Bulletin*, 26, 199-249.

Fonseca, L., Curtis, R., Halsey K., Santos, M. C., Abrantes, I. M., & Santos, M. S. N. A. (2006). Morphological, molecular and serological characterization of *Bursaphelenchus xylophilus* isolates. Pine wilt disease: a worldwide threat to forest ecosystems. International Symposium, Lisbon, 10-14 July, 2006: 63 (Abstract).

Furniss, R. L. (1951) Forest insect control in Japan. GHQ, SCAP, National Resources Section. Preliminary Study, 45: 23 pp.

Furniss, R. L. (1950) Recommendations for forest insect in Japan. GHQ, SCAP, National Resources Section, 8 pp.

Futai, K. (2003) Role of asymptomatic carrier trees in epidemic spread of pine wilt disease. *Journal of Forestry Research*, 8, 253-260

Gu, J., Braasch, H., Burgermeister, W., & Zhang, J. (2006). Records of *Bursaphelenchus* spp. intercepted in imported packaging wood at Ningbo, China. *Forest Pathology*, 36, 323-333.

De Guiran, G., Lee, M., Dalmasso, A., & Bongiovanni, M. (1985). Preliminary attempt to differentiate pinewood nematodes (*Bursaphelenchus* spp.) by enzyme electrophoresis. *Revue de Nématologie*, 8, 88-90.

De Guiran, G., & Bruguier, N. (1989). Hybridization and phylogeny of the pine wood nematode (*Bursaphelenchus* spp.). *Nematologica*, 35, 321-330.

Hunt, D. J. (2008). A checklist of the Aphelenchoidea (Nematoda: Tylenchina). *Journal of Nematode Morphology and Systematics*, 10, 99-135.

Kang, J., Choi, K., Shin, S., Moon, I., Lee, S., & Lee, S. (2004). Development of an efficient PCR-based diagnosis protocol for the identification of the pinewood nematode, *Bursaphelenchus xylophilus* (Nematoda: Aphelenchoididae). *Nematology*, 6, 279-285.

Kanzaki, N., & Futai, K. (2002). Phylogenetic analysis of the phoretic association between *Bursaphelenchus conicaudatus* (Nematoda: Aphelenchoididae) and *Psacothea hilaris* (Coleoptera: Cerambycidae). *Nematology*, 6, 759-771.

Katsuyama, N., Sakurai, H., Tabata, K., & Takeda, S. (1989). Effect of age of post-feeding twig on the ovarian development of Japanese pine sawyer, *Monochamus alternatus*. *Research Bulletin of the Faculty of Agriculture, Gifu University*, 54, 81-89.

Kishi, Y. (1995). *The pine wood nematode and the Japanese pine sawyer*. Thomas Company, Tokyo: 301 pp.

Kiyohara, T., & Tokushige, Y. (1971). Inoculation experiments of a nematode, *Bursaphelenchus* sp., onto pine trees. *Journal of the Japanese Forestry Society*, 53, 210-218.

Kiyohara, H., & Bolla, R. I. (1990). Pathogenic variability among populations of the pinewood nematode, *Bursaphelenchus xylophilus. Forest Science*, 36, 1061-1076.

Harmey, J., & Harmey, M. (1993). Detection and identification of *Bursaphelenchus* species with DNA fingerprinting and polymerase chain reaction. *Journal of Nematology* 25: 406-415.

Hotchkin, P., & Giblin, R. (1984). Comparison of electrophoregrams from *Bursaphelenchus* spp. (Aphelenchoididae). *Revue de Nematologie*, 7, 319: 320.

Iwahori, H., Tsuda, K., Kanzaki, N., Izui, K., & Futai, K. (1998). PCR-RFLP and sequencing analysis of ribosomal DNA of *Bursaphelenchus* nematodes related to pine wilt disease. *Fundamental and Applied Nematology*, 21, 655-666.

La, Y., Moon, Y., Yeo, W., Shin, S., & Bak, W. (1999). Recent status of pine wilt disease in Korea. In: Futai, K., Togashi, K. and Ikeda, T. (Eds*).* Sustainability of pine forests in relation to pine wilt and decline. Proccedings of the Symposium, Tokyo, Japan, 26-30 October 1998. Kyoto, Japan, Shokado Shoten: 239-241.

Leal, I., Allen, E., Humble, L., Green, M., & Rott, M. (2008). Application of conventional PCR and real-time PCR diagnostic methods for detection of the pinewood nematode, *Bursaphelenchus xylophilus*, in wood samples from lodgepole pine In: Pine wilt disease: a worldwide threat to forest ecosystems. Mota, M. & Vieira, P. (eds.). Springer: 197-210.

Liebhold, A., MacDonald, W., Bergdahl, D., & Mastro, V. (1995). Invasion by exotic forest pests: A threat to forest ecosystems. *Forest Science Monographs*, 30, 1-49.

Linit, M. (1988). Nematode-vector relationships in the pine wilt disease system. *Journal of Nematology* 20: 227-235.

Mamiya, Y. (1984). The pine wood nematode. In: Nickle, W.R. (Ed.). *Plant and insect nematodes.* New York and Basel. Marcel Dekker: 589-627.

Mamiya, Y. (2004). Pine wilt disease in Japan. In: Mota, M. and Vieira, P. (Eds). *The pinewood nematode*, Bursaphelenchus xylophilus. *Nematology Monographs and Perspectives*, *1*, 9-20.

Mamiya, Y., & Enda, N. (1972) Transmission of *Bursaphelenchus lignicolus* (Nematoda: Aphelencchioididae) by *Monochamus alternatus* (Coleoptera: Cerambycidae). *Nematologica* 18: 159-162.

Matsunaga, K., & Togashi, K. (2004). A simple method for discriminating *Bursaphelenchus xylophilus* and *B. mucronatus* by species-specific polymerase chain reaction primer pairs. *Nematology*, 6, 273-277.

Metge, K., Braasch, H., Gu, J., & Burgermeister, W. (2008). Variation in ITS and 28S rDNA of *Bursaphelenchus* species (Nematoda: Parasitaphelenchidae). In Pine wilt disease: a worldwide threat to forest ecosystems. Mota, M. & Vieira, P. (Eds.). Springer, NL: 151-154.

Metge, K., & Burgermeister, W. (2006). Intraspecific variation in provenances of *Bursaphelenchus xylophilus* (Nematoda: Aphelenchoididae) revealed by ISSR and RAPD fingerprints. *Journal of Plant Diseases and Protection*, 113, 1-8.

Metge, K., Braasch, H., Gu, J., & Burgermeister, W. (2006). Intraspecific variation in provenances of *Bursaphelenchus xylophilus* (Nematoda: Aphelenchoididae) revealed by ISSR and RAPD fingerprints. *Russian Journal of Nematology*, 14, 147-158.

Morimoto, K., & Iwasaki, A. (1972). Role of *Monochamus alternatus* (Coleoptera: Cerambycidae) as a vector of *Bursaphelenchus lignicolus* (Nematoda: Aphelencchioididae). *Journal of the Japanese Forestry Society*, 54, 177-183.

Mota, M., & Vieira, P. (2008). *Pine wilt disease: a worldwide threat to forest ecosystems*. Springer, NL: 428 pp.

Mota, M., Braasch, H., Bravo, M. A., Penas, A. C., Burgermeister, W., Metge, K., & Sousa, E. (1999). First report of *Bursaphelenchus xylophilus* in Portugal and in Europe. *Nematology*, 1, 727-734.

Mota, M., Bonifácio, L., Bravo, M., Naves, P., Penas, C., Pires, J., Sousa, E. & Vieira, P. (2004). Discovery of pine wood nematode in Portugal and in Europe. In M. Mota and P. Vieira, (Eds). The pinewood nematode, *Bursaphelenchus xylophilus*. *Nematology Monographs and Perspectives 1*, 1-5.

Naves, P., Kenis, M., & Sousa, E. (2005). Parasitoids associated with *Monochamus galloprovincialis* (Oliv.) (Coleoptera; Cerambycidae) within the pine wilt nematode-affected zone in Portugal. *Journal of Pesticide Science*, 78, 57-62.

Naves, P. M., Camacho, S., Sousa, E. M., & Quartau, J. A. (2006a). Entrance and distribution of the pine wood nematode *Bursaphelenchus xylophilus* on the body of its vector of *Monochamus galloprovincialis* (Coleoptera: Cerambycidae). *Entomologia Generalis*, 29, 071-080.

Naves, P., Sousa, E., & Quartau, J. A. (2006b). Feeding and oviposition preferences of *Monochamus galloprovincialis* for certain conifers under laboratory conditions. *Entomologia Experimentalis et Applicata*, 120, 99-104.

Naves, P., Sousa, E., & Quartau, J. A. (2006c). Reproductive traits of *Monochamus galloprovincialis* (Coleoptera: Cerambycidae) under laboratory conditions. *Bulletin of Entomological Research*, 96, 289-294.

Naves, P. M., Camacho, S., Sousa, E. M., & Quartau, J. A. (2007). Transmission of the pine wood nematode *Bursaphelenchus xylophilus* through feeding activity of *Monochamus galloprovincialis* (Col., Cerambycidae). *Journal of Applied Entomology* 131, 21-25.

Nitto, M., Oda, K., Kato, Y., Yamane, A., & Enda, N. (1966). Studies on pine wood borers: On host trees that receive beetle's oviposition. *Proceeding of Japanese Forest Society*, 77, 376-379 (in Japanese).

Nitto, M., Oda, K., & Kato, Y. (1967). Studies on pine wood borers: On host pine trees that receive beetle's oviposition. *Proceeding of Japanese Forest Society*, 78,193-195 (in Japanese)

Penas, C., Correia, P., Bravo, M., Mota, M., & Tenreiro, R. (2004). Species of *Bursaphelenchus* Fuchs, 1937 (Nematoda: Parasitaphelenchidae) associated with maritime pine in Portugal. *Nematology*, 6, 437-453.

Penas, C., Metge, K., Mota, M., & Valadas, V. (2006). *Bursaphelenchus antoniae* sp. n. (Nematoda: Parasitaphelenchidae) associated with *Hylobius* sp. from *Pinus pinaster* in Portugal. *Nematology*, 8, 659-669.

Robbins, K. (1982). Distribution of the pinewood nematode in the United States. In: Appleby, J.E. and Malek, R.B. (Eds) *Proceedings of the national pine wilt disease workshop. Illinois National History Survey*. Champaign, IL: 3-6.

Rodrigues, J. (2008). National erradication programme for the pinewood nematode in Portugal. In: *Pine wilt disease: a worldwide threat to forest ecosystems.* Mota, M. & Vieira, P. (eds.). Springer, Dordrecht, NL: 5-14.

Ryss, A., Vieira, P., Mota, M., & Kulinich, O. (2005). A synopsis of the genus *Bursaphelenchus* Fuchs, 1937 (Aphelenchida: Parasitaphelenchidae) with keys to species. *Nematology*, 7, 393-458.

Sakai, A., Allendorf, F., Holt, J., Lodge, D., Molofsky, J., With, K., et al. (2001). The population biology of invasive species. *Annual Review of Ecology and Systematics*, 32, 305-332.

Sato, K. (1974). Studies on *Rhizina* root rot causing group dying of pine trees. *Bulletin of the Forest and Forestry Production Research Institute*, 268, 13-48.

Shimazu, M. (2006). Current status on research and management of pine wilt disease in Japan. Current status on research and management of pine wilt disease, International Symposium, October 20. Korea Forest Research Institute, Seoul, Korea: 1-18.

Shin, S., & Han, H. (2006). Current status on research and management of pine wilt disease in Korea. Current status on research and management of pine wilt disease, International Symposium, October 20. Korea Forest Research Institute, Seoul, Korea: 31-44.

Sousa, E., Bravo, M. A., Pires, J., Naves, P., Penas, A. C., Bonifácio, L., & Mota, M. (2001). *Bursaphelenchus xylophilus* (Nematoda: Aphelenchoididae) associated with *Monochamus galloprovincialis* (Coleoptera: Cerambycidae) in Portugal. *Nematology*, 3,89-91.

Sousa, E., et al. (2002). Preliminary survey for insects associated with *Bursaphelenchus xylophilus* in Portugal. *Bulletin OEPP/EPPO Bulletin*, 32, 499-502.

Sutherland, J. & Peterson, M. (1999). The pinewood nematode in Canada: history, distribution, hosts, potential vectors and research. In: Futai, K., Togashi, K. and Ikeda, T. (Eds*). Sustainability of pine forests in relation to pine wilt and decline. Proccedings of the Symposium, Tokyo, Japan, 26-30 October 1998.* Kyoto, Japan, Shokado Shoten, pp. 247-253.

Suzuki, K. (2002). Pine wilt disease – a threat to pine forest in Europe. *Dendrobiology*, 48, 71-74.

Takeuchi, Y., Kanzaki , N., & Futai, K. (2005). A nested PCR-based method for detecting the pinewood nematode, *Bursaphelenchus xylophilus*, from pine wood. *Nematology*, 7, 775-782.

Takeuchi, Y. & Futai, K. (2007). Asymptomatic carrier trees in pine stands naturally infected with *Bursaphelenchus xylophilus*. *Nematology*, 9, 243-250.

Tàres, S., Abad, P., Bruguier, N. & Guiran, G. (1992). Identification and evidence for relationships among geographical isolates of *Bursaphelenchus* spp. using homologous DNA probes. *Heredity*, 68, 157-164.

Tàres, S., Lemontey, J., Guiran, G. & Abad, P. (1994). Use of species-specific satellite DNA from *Bursaphelenchus xylophilus* as a diagnostic probe. *Phytopathology*, 84,294-298.

Tarjan, A. and Baéza-Aragon, C. (1982). An analysis of the genus *Bursaphelenchus* Fuchs, 1937. *Nematropica*, 12, 121-135.

Tkacz, B. (2002). Pest risks associated with importing wood to the United States. *Canadian Journal of Plant Pathology*, 24, 111-116.

Tokushige, Y., & Kiyohara, T. (1969). *Bursaphelenchus* sp. in the wood of dead pine trees. *J. Japn. For. Soc.* 51, 193-195.

Tomiczek, C., Braasch, H., Burgermeister, W., Metge, K., Hoyer, U., & Brandstetter, M. (2003). Identification of *Bursaphelenchus* spp. isolated from Chinese packaging wood imported to Austria. *Nematology*, 5, 573-581.

Tomminen, J. (1991). Pinewood nematode, *Bursaphelenchus xylophilus*, found in packing case wood. *Silva Fennica*, 25, 109-111.

Tzean, S. & Jan, S. (1985a). The occurrence of pinewood nematode, *Bursaphelenchus xylophilus*, in Taiwan. *Proceedings 6th ROC Symposium of Electron Microscopy*, 38-39.

Tzean, S. & Jan, S. (1985b). Pine wilt disease caused by pinewood nematode (*Bursaphelenchus xylophilus*) and its occurrence in Taiwan. *Phytopathologist and Entomologist, NTU* 12, 1-19.

Vieira, P., W. Burgermeister, M. Mota, K. Metge & G. Silva. 2007. Lack of genetic variation of *Bursaphelenchus xylophilus* in Portugal revealed by RAPD-PCR analyses. *Journal of Nematology*, 39, 118-126.

Webster, J., Anderson, R., Baillie, D., Beckenbach, K., Curran, J. & Rutherford, T. (1990). DNA probes for differentiating isolates of the pinewood nematode species complex. *Revue de Nematologie*, 13, 255-263.

Webster, J. (2004). The pine wood nematode: implications of factors past and present for pine wilt disease. In: Mota, M. and Vieira, P. (Eds). *The pinewood nematode*, Bursaphelenchus xylophilus. *Nematology Monographs and Perspectives*, 1, 55-64.

Wingfield, M., Blanchette, A. & Kondo, E. (1983). Comparison of the pine wood nematode, *Bursaphelenchus xylophilus* from pine and balsam fir. *European Journal of Forest Pathology*, 13, 360-373.

Yang, B. (2004). The history, dispersal and potential threat of pine wood nematode in China. In: Mota, M. and Vieira, P. (Eds). *The pinewood nematode*, Bursaphelenchus xylophilus. *Nematology Monographs and Perspectives*, 1, 21-24.

Yano, M. (1913). [Investigation on the cause of pine mortality in Nagasaki Prefecture]. *Sanrinkoho* 4: 1-14.

Yi, C., Byun, B., Park, J., Yang, S. & Chang, K. (1989). First finding of the pine wood nematode, *Bursaphelenchus xylophilus* (Steiner et Buhrer) Nickle and its insect vector in Korea. *Research Reports of the Forestry Research Institute Seoul*, 38, 141-149.

Zhang, K., Lin, M., Wen, L. & Xu, W. (1999). Genetic variation of *Bursaphelenchus xylophilus* and *B. mucronatus* geographical isolates of China as shown by RAPD's. Pp. 65-69 *in* K. Futai, K. Togashi and T. Ikeda, eds. Sustainability of pine forests in relation to pine wilt and decline. Proceedings of International Symposium, Tokyo, 27-28 Oct., 1998. Tokyo: Nakanishi Printing.

Zhang, L., Kong, F. & Yang, B. (2002). Intra and interspecific variation in *Bursaphelenchus xylophilus* and *B. mucronatus* revealed by mtDNA polymorphism. *Forest Research*, 15, 7-12.

NICOLA SASANELLI

OLIVE NEMATODES AND THEIR CONTROL

Istituto per la Protezione delle Piante, CNR,
Via G. Amendola 122/D, 70126 Bari, Italy

Absctract. The pathogenicity, geographic distribution and damage of plant parasitic nematodes associated with olive are revised, for main species of the genera *Helicotylenchus*, *Meloidogyne*, *Pratylenchus*, *Tylenchulus*, *Gracilacus*, *Rotylenchulus*, *Ogma*, *Heterodera* and *Xiphinema*. Research data on olive nematodes are discussed, focusing on the effects of parasitism and plant growth, cultivars and rootstocks susceptibility, nematodes interactions with the soil-borne pathogen *Verticillium dahliae*, replant problems and control strategies.

1. INTRODUCTION

Olive (*Olea europaea* L. subsp. *europaea*) is one of the most ancient fruit crops in traditional agriculture around the Mediterranean basin. The native area of olive should be identified in the Caucasian region (Armenia, Pamir and Turkestan – Western Asia). This culture was introduced in the Mediterranean region by Fenician and Greek peoples, as reported in ancient texts by Marcus Tullius Cicero, Pliny the Elder and Aristophanes.

Olive today is intensively and extensively grown in the subtropical regions of Australia, North and South America (mainly California, Argentina, Chile, Mexico, Peru) southern Africa, western Asia (mainly Jordan, Iraq, Iran) and expecially in Europe, in Mediterranean areas where almost 95% of the world olive surface is concentrated. Leading producers are Italy (33%), Spain (23%) and Greece (18%) (Fig. 1).

Several plant parasitic nematodes have been found in association with olive, wherever this crop is grown. The first nematode record, concerning a root-knot nematode *Meloidogyne* sp., proceeded from the USA (Buhrer et al., 1933). Over 100 species belonging to 47 genera of plant parasitic nematodes have been reported in association with olive, but the nature of this relationship was evaluated in detail for some of them only (Peña-Santiago, 1990; Lamberti & Vovlas, 1993; Nico et al., 2002). A list of plant parasitic nematodes reported in the rhizosphere and/or on olive roots is given in Table 1.

Only a few genera and species are considered capable of affecting growth of olive trees, including *Meloidogyne* spp., the root-lesion nematodes *Pratylenchus penetrans* and *P. vulnus*, the spiral nematodes *Helicotylenchus* spp., the citrus

A. Ciancio & K. G. Mukerji (eds.), *Integrated Management of Fruit Crops and Forest Nematodes*, 275-315.

nematode *Tylenchulus semipenetrans*, *Gracilacus peratica*, *Rotylenchulus macrodoratus* and the longidorids *Xiphinema index* and *X. elongatum* (Graniti, 1955; Diab & El-Eraki, 1968; Lamberti & Baines, 1969; 1970; Abrantes et al., 1992; Lamberti & Vovlas, 1993; Nyczepir & Halbrendt, 1993; Sasanelli et al., 1997; 1999; Sasanelli & D'Addabbo, 2002).

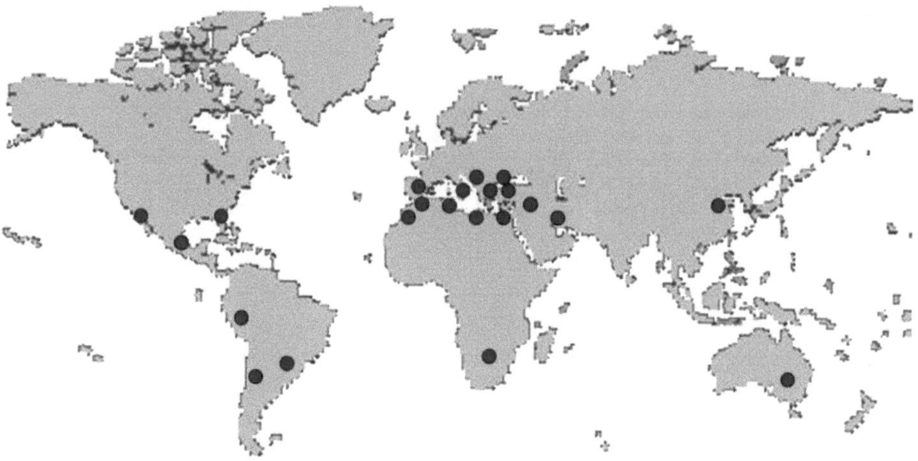

Figure 1. Main olive growing areas in the world.

Several sedentary species also attack olive: *Trophotylenchulus saltensis* was described from olive roots in Jordan and the cyst nematode *Heterodera mediterranea* was observed to feed and multiply on olive roots.

All these species are important rootstock pests and may reduce plants yield and vigour. They damage plant by directly attacking roots and subsequently predisposing them to secondary infections by bacteria and fungi, causing replant and olive preplant problems, worldwide. Direct damages are due to the trophic activity of second stage juveniles on roots, through mechanical action of their stylet. The eventual alterations induced in the root tissues limit plants developement and hence the uptake of nutrient solutions. The stylet mechanical action opens new ways to penetration by different soil pests, as demonstrated during the last two decades by the occurrence of severe wilt symptoms caused by *Verticillium dahliae*. This fungus has frequently been found in association with *Meloidogyne* spp. in olive growing areas (Ciccarese, 1998; Franc & Wheller, 1993; Lamberti et al., 2001b).

Research on nematodes associated with olive focused on various aspects, including their occurrence and geographical distribution, the effects on trees and rootstock susceptibility, the interaction between phytoparasitic nematodes and soil borne fungi, replant problems and control strategies.

Table 1. Plant-parasitic nematodes found in the rhizosphere and/or on roots of olive in the world.

Genus and species	Country	References
Aglenchus agricola	Spain	Peña-Santiago (1990)
Amplimerlinius amplus	Portugal	Siddiqi (1976)
A. macrurus	Jordan	Hashim (1979)
A. paraglobigerus	Spain	Nico et al. (2002)
Amplimerlinius sp.	Spain	Peña-Santiago (1990)
Aphelenchoides spp.	Italy	Scognamiglio et al. (1968; 1971); Fiume (1978)
Aphelenchus avenae	Greece	Hirschmann et al. (1966)
	Spain	Peña-Santiago (1990)
Coslenchus cancellatus	Spain	Peña-Santiago (1990)
C. costatus	Greece	Hirschmann et al. (1966)
C. lateralis	Spain	Peña-Santiago (1990)
Criconemoides informis	Jordan	Hashim (1979)
	Spain	Nico et al. (2002)
Criconema spp.	Greece	Vlachopoulus (1991)
Criconemoides spp.	Greece	Vlachopoulus (1991)
Criconemoides siculum	Italy	Vovlas (1982)
C. xenoplax	Spain	Nico et al. (2002)
Ditylenchus anchiliposomus	Spain	Peña-Santiago (1990)
Ditylenchus virtudeasae	Spain	Tobar-Jimenez (1964)
Ditylenchus spp.	Italy	Scognamiglio et al. (1968)
	Spain	Peña-Santiago (1990)
Dolichodorus heterocephalus	Italy	D'Errico et al. (1977)
Filenchus filiformis	Greece	Hirschmann et al. (1966)
F. sandneri	Spain	Peña-Santiago (1990)
F. thornei	Spain	Peña-Santiago (1990)
Gracilacus peratica	Italy	Scognamiglio et al. (1968); Inserra & Vovlas (1977)
	Portugal	Abrantes et al. (1987)
G. teres	Spain	Santiago & Geraert (1991)
Gracilacus sp.	Spain	Peña-Santiago (1990)
Helicotylenchus digonicus	Cyprus	Philis & Siddiqi (1976)
	Jordan	Hashim (1983)
	Spain	Peña-Santiago (1990)
H. dihystera	Cyprus	Philis & Siddiqi (1976)
	Egypt	Tarjan (1964)
	Italy	Vovlas & Inserra, (1981)
	Jordan	Bridge (1978)
	Spain	Romero & Arias (1969)
	Zimbabwe	Sher (1966)
H. erythrinae	Italy	Graniti (1955)
H. neopaxilli	Italy	Inserra et al. (1979a)
H. oleae	Italy	Inserra et al. (1979a)
H. pseudorobustus	Greece	Hirschmann et al. (1966)
	Spain	Nico et al. (2002)

Table 1 continued

H. tunisiensis	Israel	Sher (1966)
	Jordan	Hashim (1979)
	Portugal	Abrantes et al. (1987)
H. vulgaris	Spain	Nico et al. (2002)
Helicotylenchus spp.	Algeria	Lamberti et al. (1975b)
	Chile	Gallo & Jimenez (1976)
	Greece	Vlachopoulus (1991)
Hemicycliophora sp.	Chile	Gallo & Jimenez (1976)
	Spain	Nico et al. (2002)
Heterodera mediterranea	Italy	Vovlas & Inserra (1983)
	Spain	Castilo et al. (1999)
Hoplolaimus aorolaimoides	Portugal	Abrantes et al. (1987)
Hoplolaimus spp.	Egypt	Diab & El-Eraki (1968)
Longidorus africanus	Egypt	Tarjan (1964)
L. closelongatus	Greece	Lamberti et al. (1996)
L. cretensis	Greece	Tzortzakakis et al. (2001)
L. macrosoma	Spain	Peña-Santiago (1990)
L. siddiqii	Jordan	Hashim (1979)
Meloidogyne acrita	China	Yong & Zhong (1980)
M. arenaria	Chile	Jimenez (1982)
	China	Yong & Zhong (1980)
M. baetica	Spain	Castillo et al. (2003)
M. hapla	Chile	Jimenez (1982)
	Israel	Minz (1961)
	Portugal	Santos (1982)
M. incognita	Chile	Jimenez (1982)
	China	Yong & Zhong (1980)
	India	Sethi et al. (1988)
	Israel	Minz (1961)
	Italy	Lamberti & Di Vito (1972), Inserra & Vovlas (1981)
	Jordan	Abu-Gharbieh et al. (1978), Hashim (1979)
	Lebanon	Saad & Nienhaus (1969)
	Libya	Edongali (1989)
	Portugal	Abrantes (1980)
	Spain	Nico et al. (2002)
M. javanica	Chile	Jimenez (1982)
	China	Yong & Zhong (1980)
	Egypt	Diab & El-Eraki (1968)
	Greece	Hirschmann et al. (1966)
	Israel	Tarjan (1953)
	Italy	Lamberti & Di Vito (1972), Inserra & Vovlas (1981)
	Jordan	Hashim (1979)
	Libya	Edongali (1989)
	Spain	Nico et al. (2002)
	USA	Lamberti & Lownsbery (1968)

Table 1 continued

M. lusitanica	Portugal	Abrantes & Santos (1991)
	Spain	Nico et al. (2002)
	Portugal	Santos & Abrantes (1980)
Meloidogyne spp.	Portugal	Macara (1971)
	USA	Buhrer et al. (1933)
Merlinius brevidens	Cyprus	Philis & Siddiqi (1976)
	Greece	Hirschmann et al. (1966)
	Jordan	Hashim (1979)
	Spain	Peña-Santiago (1990)
Neolobocriconema olearum	Jordan	Hashim (1984a)
Neopsilenchus magnidens	Spain	Peña-Santiago (1990)
Nothocriconema princeps	Portugal	Abrantes et al. (1987)
Ogma rhombosquamatum	Italy	Vovlas & Inserra (1981)
	Portugal	Abrantes et al. (1987)
O. civellae	Zimbabwe	Metha & Raski (1971)
Paraphelenchus pseudoparietinus	Spain	Peña-Santiago (1990)
Paratrichodorus minor	Spain	Nico et al. (2002)
P. teres	Spain	Nico et al. (2002)
Paratrophurus loofi	Spain	Peña-Santiago (1990)
Paratylenchus arculatus	Spain	Nico et al. (2002)
P. baldacii	Spain	Peña-Santiago (1990)
P. microdorus	Spain	Peña-Santiago (1990)
P. vandenbrandei	Italy	Inserra et al. (1976)
Paratylenchus sp.	Italy	Scognamiglio et al. (1968)
Pratylenchus coffeae	Australia	Colbran (1964)
P. crenatus	Italy	Inserra et al. (1976)
P. fallax	Spain	Nico et al. (2002)
P. microdorus	Spain	Nico et al. (2002)
P. neglectus	Greece	Hirschmann et al. (1966)
	Italy	Inserra et al. (1976)
	Spain	Peña-Santiago (1990)
P. penetrans	Australia	McLeod et al. (1994)
	Italy	Inserra et al. (1976)
	Spain	Nico et al. (2002)
P. thornei	Jordan	Hashim (1979)
P. vulnus	Australia	McLeod et al. (1994)
	Algeria	Lamberti et al. (1975b)
	Spain	Nico et al. (2002)
	Italy	Lamberti (1969a)
	USA	Serr & Day (1949); Condit & Horne (1938)
Pratylenchoides ritteri	Spain	Nico et al. (2002)
Psilenchus sp.	Greece	Vlachopoulus (1991)
Radopholus sp.	Greece	Vlachopoulus (1991)
Rotylenchulus macrodoratus	Greece	Koliopanos & Vovlas (1977)
	Italy	Vovlas & Lamberti (1974)

Table 1 continued

R. macrosomus	Israel	Dasgupta et al. (1968)
	Spain	Castillo et al. (2003)
R. reniformis	Greece	Hirschmann et al. (1966)
Rotylenchus robustus	Portugal	Abrantes et al. (1987)
R. cypriensis	Jordan	Hashim (1984b)
Rotylenchus sp.	Greece	Vlachopoulus (1991)
	Spain	Nico et al. (2002)
Trichodorus aequalis	Spain	Peña-Santiago (1990)
T. giennensis	Spain	Nico et al. (2002)
T. primitivus	Portugal	Almeida et al. (1989)
T. taylori	Italy	Waele et al. (1982)
Trophotylenchulus saltensis	Jordan	Hashim (1983)
Tylenchorhynchus aduncus	Spain	Nico et al. (2002)
T. clarus	Jordan	Hashim (1979)
T. dubius	Greece	Hirschmann et al. (1966)
	Spain	Nico et al. (2002); Peña-Santiago (1990)
T. goffarti	Jordan	Hashim (1979)
T. huesingi	Spain	Nico et al. (2002)
T. mamillatus	Spain	Nico et al. (2002)
T. striatus	Greece	Hirschmann et al. (1966)
T. tenuis	Jordan	Hashim (1984b)
Tylenchulus semipenetrans	Australia	Colbran (1955; 1964)
	Greece	Vlachopoulus (1991)
	Italy	Inserra & Vovlas (1978; 1981)
	USA	Baines (1951)
Tylenchus arcuatus	Spain	Peña-Santiago (1990)
Xiphinema aequum	Italy	Roca & Lamberti (1988)
X. barense	Italy	Lamberti et al. (1986)
X. californicum	USA	Lamberti (1969b)
X. elongatum	Egypt	Diab & El-Eraki (1968)
X. index	Greece	Vlachopoulus (1991)
X. ingens	Italy	Lamberti et al. (1975a)
	Jordan	Hashim (1979)
X. italiae	Italy	Martelli et al. (1966)
X. macroacanthum	Italy	Lamberti et al. (1989)
X. pachtaicum	Jordan	Hashim (1979)
	Spain	Nico et al. (2002)
X. sahelense	Spain	Arias (1975)
X. turcicum	Spain	Peña-Santiago (1990)
X. vuittenezi	Spain	Arias (1975)
Zygotylenchus guevarai	Spain	Peña-Santiago (1990)

2. SYMPTOMS AND PATHOGENICITY OF NEMATODES ON OLIVE

The nature of the association with olive has not yet been evaluated for all the nematode species reported, but damage has been studied for main species within the genera *Gracilacus*, *Helicotylenchus*, *Heterodera*, *Meloidogyne*, *Ogma*, *Pratylenchus*, *Rotylenchulus*, *Tylenchulus* and *Xiphinema*.

Helicotylenchus erythrinae, *H. oleae* and *H. dihystera* cause necrosis and/or brown lesions in the feeder olive roots and consequently a growth delay, with reduction of the root systems and a gradual chlorosis of leaves (Graniti, 1955; Diab & El-Eraki, 1968; Inserra et al., 1979a).

Lamberti & Baines (1969) and Sasanelli et al. (1997) have demonstrated by glasshouse trials the reaction of different olive cultivars and, in particular, of rootstock DA12I to *M. incognita* and *M. javanica*, together with the induced reduction of plants growth (Table 2). *Meloidogyne incognita* and *M. javanica* reproduce on olive roots differently according to the host (Tables 3 and 4).

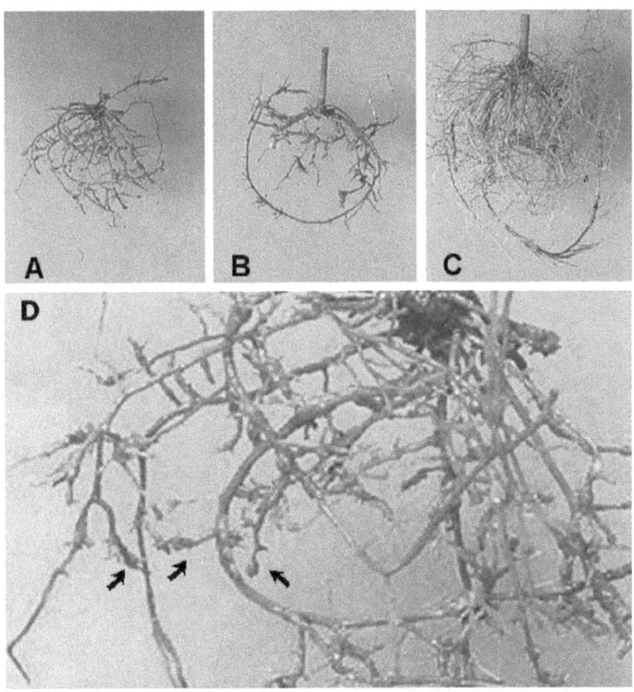

Figure 2. Roots of olive cv. *Yusti attacked by* Meloidogyne incognita *(A) and* M. javanica *(B) in comparison to uninfested control (C). Details of induced galls (D).*

Olive roots infected by root-knot nematodes are deformed by spheroidal or elongate galls located at the root apex and along its axis (Fig. 2). Olive galls are particularly evident when the density of root-knot nematode populations is higher than 1-2 eggs and juveniles/ml of soil. Sections of root galls showed that nematode feeding stimulates the formation of several (3-5) large giant cells around the cephalic region, located in the root vascular cylinder (Fig. 3). Their granulated cytoplasm is dense and homogenous and contains numerous hypertrophied nuclei.

Abnormal and interrupted xylem elements as well as direct injury of xylem and parenchyma can be observed in many sections (Sasanelli et al., 2000).

Abrantes & Santos (1991) described *M. lusitanica* as capable to induce serious growing problems to olive trees. Recently, a new *Meloidogyne* species, *M. baetica* has been found to feed and reproduce on wild olive trees. Studies of host-parasite relationships showed a typical susceptible reaction in naturally infected plants as well as in olive planting stocks (cvs. Arbequina and Picual) (Castillo et al., 2003).

Table 2. Effect of **Meloidogyne incognita** *and* **M. javanica** *on the growth of eight olive cultivars.*

Cultivar	*Meloidogyne* species	Top fresh weight [a] (g)	Top dry weight (g)	Root fresh weight (g)	% increase from initial values			
					Main shoot length (mm)	Shoot diam (mm)	Stem diam (mm)	Numbers of nodes
Cellina di Nardò	*M.incognita*	4.0 **	1.9 **	6.8 **	47 **	45.9	10.1	56 *
	Control (not inoc.)	5.4	2.5	5.3	181	68.5	14.1	118
	M.javanica	5.7	2.7	6.5 *	131	64.5	10.5	119
Cima di Bitonto	*M.incognita*	3.8 **	2.2 **	4.9	77	59.5	9.6	194
	Control	6.1	2.9	6.0	150	60.2	10.8	140
	M.javanica	6.5	3.0	6.6	170	77.6	8.7	132
Coratina	*M.incognita*	3.0 **	1.5 **	4.0	156	78.3	8.5	129
	Control	4.5	2.3	4.4	214	88.7	24.3	157
	M.javanica	5.7	2.7	6.2 **	303	129.3	18.2	272
DA 12 I (Rootstock)	*M.incognita*	3.6 **	1.7 **	5.2 **	462	170.9	9.5 **	218
	Control	5.7	2.6	7.1	470	141.5	37.8	236
	M.javanica	3.9 **	1.7 **	5.0 **	587	143.8	9.5 **	215
Frantoio	*M.incognita*	4.3	2.2 *	5.8	54 *	54.9	6.3	54 *
	Control	6.2	3.0	5.7	142	82.3	13.1	112
	M.javanica	5.8	2.7	5.7	97	58.7	15.2	71
FS 17	*M.incognita*	3.2	1.5	4.2	40	35.5	2.3	63
	Control	3.6	1.7	4.3	63	19.3	9.0	112
	M.javanica	5.7	2.5	6.5 **	88	46.4	14.7	99
Leccino	*M.incognita*	3.8 **	1.9 *	5.8 **	140	85.8	14.3	133
	Control	5.5	2.5	7.0	164	72.6	23.4	196
	M.javanica	4.7	2.0	5.5 **	240	83.6	18.4	216
Yusti	*M.incognita*	1.7 **	0.8 *	2.4 **	99	45.6 *	4.7 **	97 *
	Control	4.2	2.0	3.0	203	87.8	16.1	206
	M.javanica	2.6 **	1.2 *	2.3 **	194	38.5 *	11.9	170

[a] Statistically different from control according to Student's *t* test. * for P=0.05; ** for P=0.01.

According to Seinhorst's model $y = m + (1-m)\,z^{(P-T)}$ (Seinhorst, 1965; 1979) the pathogenicity of root-knot nematodes was demonstrated in pot trials (Sasanelli et al., 2002a). The low tolerance limits (T) found for the tested olive germplasm (cultivar FS17 and rootstock DA12I) indicated that growth of young olive plants can be strongly suppressed by these nematodes.

Olive plants severely attacked by *Pratylenchus vulnus* have been found in Italy (Lamberti, 1969a), USA (Condit & Horne, 1938) and Algeria (Lamberti et al.,

1975b). Parasitized plants show severe defoliation, chlorotic leaves, short internodes and longitudinal lesions with root necroses. Many italian cultivars have been reported to be susceptible to *P. vulnus* (Table 5) (Lamberti & Baines, 1969; Sasanelli & D'Addabbo, 2002). Among them, cv FS 17 appears particularly susceptible allowing the root lesion nematode to reproduce at high levels (Fig. 4). Only cv Verdalion appeared as a relatively poor host (Lownsbery & Serr, 1963). Histopathology of lesions and severe damage on cortical root tissues were illustrated (Inserra et al., 1979b; 1981).

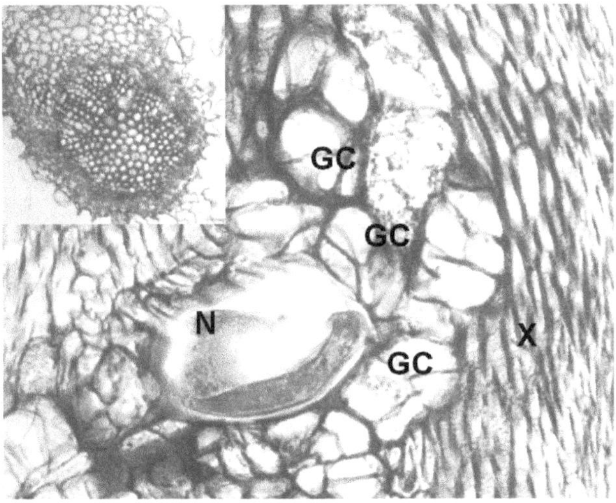

Figure 3. Gall cross section a showing changes induced by Meloidogyne incognita *on roots of olive* cv. *Ascolana. Expanded giant cells (GC) surround the nematode (N) and adjacent xylem elements (X). Insert picture shows the cross section of an healthy root.*

Tylenchulus semipenetrans, a specialized parasite of *Citrus*, may infect olive trees in USA (California) and Italy. Although its population densities on olive are lower than those observed on citrus, high numbers of nematodes may inhibit olive growth (Lamberti et al., 1976). Inserra et al. (1980) suggested that the known populations could be subdivided into four biotypes and that only the "citrus biotype" may infect olive trees. A population of *T. semipenetrans* found in California on olive was more infective and reproduced more rapidly on two olive cultivars than on *Citrus sinensis* (Lamberti & Baines 1970). Histological changes induced on roots were similar to those caused on susceptible *Citrus* spp.

Table 3. Reproduction of M. incognita *on eight olive cultivars.*

Cultivar	Root gall index	♀♀ per g of root	Eggs and juveniles per g of root	Total population (soil and roots) per cm³ of soil	$r = Pf/Pi$	Resistance rating
Cellina di Nardò	4.1 AB*	19 A	3531 AB	72 A	6.0 A	Susceptible
Cima di Bitonto	4.5 A	13 A	1808 C	34 B	2.9 B	Moderately Susceptible
Coratina	3.2 CD	25 AB	678 D	8 C	0.7 C	Resistant
DA 12 I (Rootstock)	3.9 ABC	25 AB	3360 ABC	47 B	3.9 B	Moderately Susceptible
Frantoio	3.5 BCD	21 A	2113 BCD	33 B	2.7 B	Moderately Susceptible
FS 17	2.9 D	28 AB	1861 BCD	25 BC	2.1 BC	Moderately Susceptible
Leccino	3.7 ABC	23 A	1899 BCD	26 BC	2.1 BC	Moderately Susceptible
Yusti	3.9 ABC	43 B	4345 A	31 B	2.6 B	Moderately Susceptible

* Data flanked in any column by the same letters are not statistically different according to Duncan's multiple range test (P=0.01).

Table 4. Reproduction of M. javanica *on eight olive cultivars.*

Cultivar	Root gall index	♀♀ per g of root	Eggs and juveniles per g of root	Total population (soil and roots) per cm³ soil	Pf/Pi	Resistance rating
Cellina di Nardò	3.9 AB*	27 AB	5516 AB	92 B	7.7 B	Susceptible
Cima di Bitonto	4.5 A	28 AB	8071 A	159 A	13.2 A	Highly Susceptible
Coratina	2.5 CD	2 C	87 E	7 D	0.6 D	Resistant
DA 12 I (Rootstock)	3.4 B	34 AB	3722 BC	60 BC	5.0 BC	Moderately Susceptible
Frantoio	3.2 BC	44 A	3834 BC	51 BC	4.2 C	Moderately Susceptible
FS 17	2.2 D	19 BC	1827 CDE	34 CD	2.8 BC	Moderately Susceptible
Leccino	2.1 D	6 C	368 DE	5 D	0.4 D	Resistant
Yusti	3.4 B	32 AB	3188 BCD	24 CD	2.0 CD	Moderately Resistant

* Data flanked in any column by the same letters are not statistically different according to Duncan's multiple range test (P = 0.01).

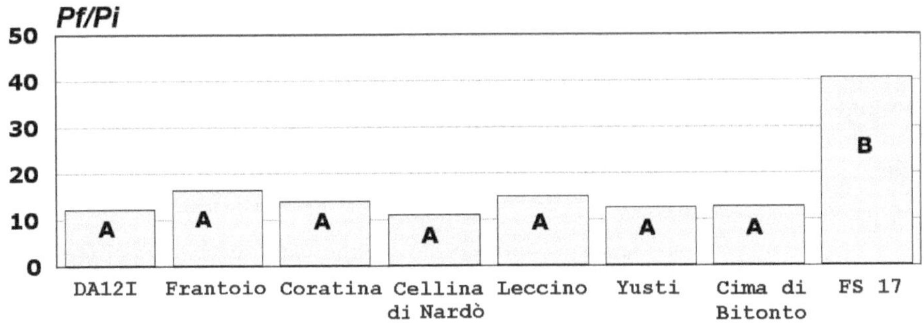

Figure 4. Reproduction rate (Pf /Pi) of Pratylenchus vulnus *on olive varieties.*

Gracilacus peratica feeds ectoparasitically on olive roots by inserting its long stylet into a cortical cell (Inserra & Vovlas, 1977). The cell on which the nematode fed usually shows thickened and lignified walls near the stylet penetration sites. Also *G. teres* has been found in association with olive roots in the province of Jean in Spain (Santiago & Geraert, 1991).

Table 5. Effect of Pratylenchus vulnus *on growth of eight olive varieties.*

Cultivar		Top fresh weight (g)[a]	Root fresh weight (g)	Increase from initial values (%)			
				Main shoot length	Shoot diam	Stem diam	Number of nodes
Cima di Bitonto	Control	20.8 **	10.4 **	136.3*	136.3*	49.0 *	166.8
	Inoculated	7.7	4.3	77.1	77.1	22.1	118.3
Cellina di Nardò	Control	7.3**	8.2 *	61.7	61.7	27.5	127.7 *
	Inoculated	3.5	3.5	40.7	40.7	22.4	47.4
Coratina	Control	14.1	6.1	210.8 **	210.8 **	40.6	171.4
	Inoculated	12.3	6.6	98.0	98.0	24.3	125.1
FS 17	Control	10.1	11.7	81.3	81.3	42.4	174.1
	Inoculated	7.5	7.8	89.3	89.3	41.3	164.9
DA12 I (rootstock)	Control	18.9	14.2	202.9	202.9	47.3	194.6
	Inoculated	12.7	12.5	133.4	133.4	36.1	154.6
Leccino	Control	27.3	20.4	262.0	262.0	130.8	378.4
	Inoculated	27.5	14.4	262.4	262.4	133.4	325.6
Yusti	Control	10.0	7.2 *	174.6 *	174.6 *	40.8	291.7 *
	Inoculated	5.1	2.8	62.1	62.1	23.5	119.3
Frantoio	Control	15.9	10.6 *	148.3*	148.3*	50.5	216.1**
	Inoculated	10.4	6.8	87.7	87.7	25.6	104.2

[a] Statistically different from inoculated plants according to Student's *t* test for P=0.05 (*) and P=0.01 (**).

Rotylenchulus macrodoratus is the most common nematode species on olive in Italy, where it occurs in about 20 % of soil samples (Inserra & Vovlas, 1981). Also *R. reniformis* has been studied in detail on olive (Hirschmann et al., 1966). Histological studies on infected roots showed that the parasites induce the formation of an enlarged mononucleate "nurse cell", with dense cytoplasm and hypertrophic nucleus, from which it feeds (Vovlas & Inserra, 1976).

Table 6. Effect of Xiphinema index *on growth of olive cultivar FS 17 and rootstock DA 12I.*

Varieties	Population density (Females/pot)	Weight (g)		Increase from initial values (%)			
		Top fresh	Root	Shoot diam	Shoot lenght	Stem diam	Number of nodes
DA 12I (rootstock)	0	8.9[(1)] a[(2)]	9.4 a	75.5 a	149.6 a	19.5 a	135 a
	10	8.3 a	7.5 b	68.5 a	123.0 a	16.8 a	109 ab
	20	7.6 ab	7.1 b	46.5 b	108.2 ab	14.4 ab	90 bc
	40	7.9 ab	7.3 b	46.0 b	64.6 b	14.6 ab	72 bc
	80	6.3 b	5.9 c	39.8 b	58.2 b	8.5 b	62 c
FS 17	0	6.2 a	6.1 a	41.7 a	91.9 a	22.0 a	88 a
	10	6.0 a	5.9 a	35.6 ab	98.9 a	15.4 ab	94 a
	20	5.4 a	5.1 a	42.7 a	67.2 ab	16.0 ab	84 a
	40	5.0 a	4.9 a	38.9 a	77.5 ab	16.1 ab	86 a
	80	4.7 a	4.8 a	25.8 b	46.2 b	9.4 b	70 a

(1) Each value is average of ten replicates.
(2) For each cv. data flanked in any column by the same letters are not statistically different according to Duncan's multiple range test (P = 0.05).

Xiphinema index, a species most commonly associated with fig and grapevine, has been also recovered from the rhizophere of olive in Italy (Coiro, pers. comm.) and Greece (Vlachopoulos, 1991), and was found to affect growth of cv. FS 17 and rootstock DA 12I (Sasanelli et al., 1999) (Table 6). Although both cultivars cannot be considered as good hosts for *X. index*, the nematode is able to reproduce on them, as shown by the presence of juveniles in soil (Table 7). High levels of root and leaf phenols were associated with lower gall numbers and lower nematode reproduction rate (Table 8). Root phenols are involved in the mechanisms of plant reaction to root-knot nematodes (Ridolfi et al., 1998) and the same mechanism probably applies in the relationship between olive and *X. index* (Ridolfi et al., 2001). Final population densities of *X. elongatum* around 500 specimens per pot were reported in Egypt to reduce growth of olive seedlings by 65% within 6 months (Diab & El-Eraki, 1968).

Table 7. Reproduction of Xiphinema index *in the rhizosphere of olive cv. FS17 and rootstock DA12I.*

Population density (♀♀ /pot)	Root gall index			Females / l soil			Juveniles / l soil			Reproduction rate ($r = Pf/Pi$)		
	DA 12I	FS 17	t [3]	DA 12I	FS 17	t	DA 12I	FS 17	t	DA 12I	FS 17	t
0	0[1] a[2]	0 a	-	0 a	0 a	-	0 a	0 a	-	0 a	0 a	-
10	0.7 b	0.6 b	-	5.0 ab	7.0 ab	-	3.6 ab	15.0 ab	*	0.4 d	1.2 b	-
20	0.6 b	1.0 bc	-	7.4 b	10.8 b	-	4.8 ab	32.6 bc	**	0.3 cd	1.1 b	*
40	0.8 b	0.6 b	-	11.2 b	8.0 ab	-	6.2 b	33.0 bc	**	0.2 bc	0.5 a	-
80	0.8 b	1.4 c	**	7.4 b	25.0 c	*	4.8 ab	39.0 c	**	0.1 ab	0.4 a	**

(1) Each value is an average of ten replicates;
(2) Data flanked in any column by the same letters are not statistically different according to Duncan's multiple range test (P = 0.05).
(3) Student's t test significant at P = 0.05 (*) and P = .001 (**).

Table 8. Effect of population density of Xiphinema index *on phenols content, PAL and POD in roots and leaves of olive cvs DA 12I and FS 17.*

Population density (♀♀/pot)	Phenols content (mg/g f.w.)[1]			PAL (U/g f.w.)			POD (U/g f.w.)		
	DA 12 I	FS 17	t [3]	DA 12 I	FS 17	t	DA 12 I	FS 17	t
Roots									
0	9.4 b A[2]	7.2 a A	**	344 a A	232 a A	**	36 a A	323 a A	**
10	9.2 ab A	7.2 a A	-	226 b B	201 b B	*	55 ab A	309 a A	*
20	8.6 a A	7.2 a A	-	210 bc BC	192 b B	*	171 bc A	175 b B	**
40	8.9 ab A	8.3 b B	**	173 cd BC	164 c C	-	194 c A	293 a A	**
80	11.3 c B	7.9 b AB	-	164 d C	160 c C	**	2833 d B	302 a A	**
Leaves									
0	24.3 a A	21.9 a A	-	2575 a AB	2032 a A	-	34 a A	1986 a A	**
10	27.8 b B	22.0 a A	-	2331 a A	2077 ab A	-	440 b B	4830 d D	**
20	27.0 b AB	22.9 a AB	-	3223 b B	3433 c B	**	634 c C	5433 e E	**
40	26.9 b AB	27.9 b B	-	3033 b B	2295 b A	-	497 b B	3861 c C	**
80	33.8 c C	27.8 b B	-	4877 c C	1889 a A	**	49 a A	3116 b B	-

(1) Each value is average of three replications.
(2) Data flanked, for roots or leaves, in any column by the same letters are not statistically different according to Duncan's test (small letters for P=0.05; capital letters for P=0.01);
(3) Significantly different according to Student's t test for P=0.05 (*) and P=0.01 (**). Comparison between varieties were on absolute control values (population density = 0) and on % variation from controls for the other densities.

Xiphinema diversicaudatum is the vector of ArMV (*Arabis mosaic nepovirus*) and SLRSV (*Strawberry latent ringspot sadwavirus*). No evidence of ArMV sintomatology occurs on olive trees, and the virus lives in a latent form. However, occasional symptoms of SLRSV with malformations of drupes were found in orchards. In every case the nematode attack may cause necrosis on the olive roots (Saponari et al., 2001) with damages observed particularly in nurseries.

Among sedentary ectoparasitic nematodes, *Ogma rhombosquamatum* was also observed to feed on olive roots (Vovlas & Inserra, 1981). The nematode was found at densities of 280-360 specimens per g of fresh root, and its feeding activity induced thickening of the cell walls, with hypertrophy of nuclei and nucleoli.

Heterodera mediterranea feeds and reproduces on olive roots forming syncytia and inducing a disorder in the stellar structures (Vovlas & Inserra, 1983; Castillo et al., 1999).

Finally, *Trophotylenchulus saltensis* was found to parasitize olive roots in Jordan but its parasitic habit and the eventual damage caused were not reported (Hashim, 1983).

3. INTERACTIONS BETWEEN NEMATODES AND SOIL FUNGI

The interactions between *M. incognita*, *P. vulnus* and the fungus *Verticillium dahliae* were studied on one year old rooted cuttings of olive cvs Leccino and Pendolino. For each cultivar, some plants were inoculated only with *M. incognita* (4000 eggs/pot), *P. vulnus* (200 nematodes/pot) or *V. dahliae* (50 ml/pot of a conidial suspension at concentration of 4 x 10^7 conidia/ml), whereas other treatments included plants inoculated with either nematode species plus the fungus, plants inoculated with all the three pathogens and uninoculated controls (Lamberti et al., 2001a; 2001b). External symptoms were assessed 6 and 18 months after fungus inoculation to evaluate the progress of verticillium-wilt. Eighteen months after fungus inoculation, when the test was discontinued, plant growth parameters were determined together with vascular discoloration, root gall index and nematode reproduction rate.

All plants inoculated with *V. dahliae*, alone or in combination with nematodes showed, after six months, consistent symptoms of wilting (Table 9). For the cv. Pendolino, wilting was related to the simultaneous presence of *M. incognita* and *P. vulnus*. However, such different behaviour of the cv. Pendolino ceased 18 months after inoculation and all plants inoculated with the fungus appeared wilted, independently of the presence of either or both nematodes (Table 9). The symptoms of vascular discoloration (Fig. 5) at the end of the experiment followed the same pattern. However, significant positive correlations between final root gall index, wilting and vascular discoloration were observed.

Table 9. Effect of nematode and fungus inoculations on wilting and stem discoloration of olive cuttings (cvs Leccino and Pendolino).

Pathogen	Wilting				Vascular discoloration	
	6 months after inoculation		18 months after inoculation			
	Leccino [(1)]	Pendolino	Leccino	Pendolino	Leccino	Pendolino
Control	0.0 a A	0.0 a A	0.0 a A	0.0 a A	0.0 a A	0.0 a A
V. dahliae	0.9 b B	1.2 b B	2.6 b B	3.0 b B	2.4 b B	2.8 bc B
M. incognita + *V. dahliae*	1.4 b B	1.7 bc B	3.1 b B	3.1 b B	3.0 b B	3.4 bc B
P. vulnus + *V. dahliae*	1.2 b B	1.0 b B	2.9 b B	3.0 b B	2.6 b B	2.4 b B
M. incognita + *P. vulnus* + *V. dahliae*	1.4 b B	2.1 c B	3.2 b B	3.4 b B	3.1 b B	3.7 c B

(1) Data flanked by the same letters in any column are not statistically different according to Duncan's Multiple Range Test (small letters for P=0.05; capital letters for P=0.01).

Inoculation of single nematodes species or combinations of two species did not affect stem diameter of the cv. Leccino as did *V. dahliae* alone or in combination with the two nematode species. This effect was not observed for the combination or with either nematode species alone. The stem growth of cv. Pendolino was also not affected by either nematode species when introduced alone, but it appeared affected by the combination of *M. incognita* and *P. vulnus* and, in all cases, when *V. dahliae* was also introduced (Table 10). Similarly, the main shoot diameter of cv. Leccino was not affected by the presence of *P. vulnus* alone or in combination with *M. incognita*, but it was suppressed when *V. dahliae* was present (Table 10). For main shoots, all nematode and/or fungus inoculations suppressed the growth of cv. Pendolino. The main shoot growth was suppressed by all nematode and fungus inoculations. Suppression was much more severe for cv. Leccino, when all the three pathogens were present (Table 10). Also the node numbers of cv. Pendolino were affected by any pathogen, alone or in combination. However cv. Leccino showed major damage only when *P. vulnus* or *V. dahliae* were in single inoculation or when the two nematode species were simultaneously present (Table 10).

Single inoculation of either nematode species did not affect the root fresh weight of both olive cultivars (Tables 11-12), but combinations of nematodes and *V. dahliae* or the fungus alone significantly reduced root growth, compared to uninoculated control (Fig. 6).

Figure 5. Vascular discoloration induced by Verticillium dahliae *in the stem of a young olive tree.*

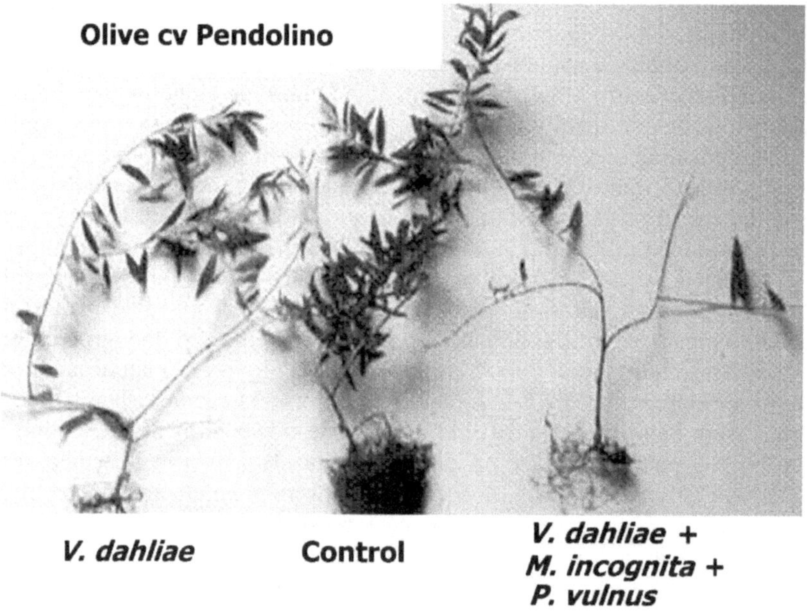

Figure 6. Effect of V. dahliae *alone or in combination with* M. incognita *and* P. vulnus *on olive cuttings cv. Pendolino.*

Table 10. Effect of nematode and fungus inoculation on growth of olive cuttings.

Pathogen	Increase from initial values (%)							
	Stem diameter		Main shoot diameter		Main shoot length		Node numbers	
	Leccino	Pendolino	Leccino	Pendolino	Leccino	Pendolino	Leccino	Pendolino
Control	101 a A	112 a A	133 a A	184 a A	579 a A	816 a A	340 a A	493 a A
M. incognita	82 ab AB	86 ab AB	68 cd BC	130 b AB	367 b B	366 b B	198 bc BC	257 b B
P. vulnus	101 a A	93 ab AB	111 ab AB	127 b AB	365 b B	411 b B	282 ab AB	266 b B
V. dahliae	62 bc AB	76 b AB	60 d BC	99 b B	258 b BC	379 b B	279 ab AB	257 b B
M. incognita + *P. vulnus*	78 abc AB	79 b AB	101 abc ABC	92 b B	366 b B	281 b B	282 ab AB	241 b B
M. incognita + *V. dahliae*	68 abc AB	76 b AB	76 bcd BC	128 b AB	231 bc BC	242 b B	217 b ABC	149 b B
P. vulnus + *V. dahliae*	66 abc AB	62 b B	84 bcd ABC	93 b B	271 b BC	373 b B	218 b ABC	249 b B
M. incognita + *P. vulnus* + *V. dahliae*	46 c B	65 b B	56 d C	80 b B	100 c C	265 b B **	92 c C	161 b B

Data flanked in columns by the same letters are not statistically different according to Duncan's Multiple Range Test (small letters: P=0.05; capital letters: P=0.01).
** Significant for P=0.01, according to Student's *t* test.

Also, root gall index was less severe on cv. Pendolino (Table 12) than on cv. Leccino (Table 11) and on this last cultivar it was slightly suppressed when both nematode species were present (Table 11). Reproduction of *M. incognita* was in fact, much suppressed by the concomitant presence of *P. vulnus* on the cv. Leccino or *V. dahliae*, alone or ensemble, on cv. Pendolino (Table 12). On the other hand, reproduction of *P. vulnus* was reduced by the simultaneous presence of *M. incognita* on cv. Pendolino, and by the simultaneous presence of *M. incognita* or *V. dahliae* singly or concomitantly on the cv. Leccino (Table 11).

Longer duration of the experiment, in the order of three years or more, would have more dramatically shown the interaction among these pathogens. In general, it seems that both cvs. are equally susceptible to *V. dahliae* and nematodes, since the means comparison by Student's *t* test did not reveal significant differences. *Meloidogyne incognita* and *P. vulnus* certanly reproduce better on Leccino than on Pendolino and the former seems to be inhibited by the simultaneous presence of the latter on both cultivars and of *V. dahliae* on Pendolino. *Pratylenchus vulnus* is also inhibited on both cvs. by the simultaneous presence of *M. incognita*, but it is disturbed by *V. dahlaie* only on cv. Leccino.

Table 11. Reproduction of Meloidogyne incognita *and* Pratylenchus vulnus *on cv. Leccino.*

Pathogen	Root fresh weight (g)	Gall index (0-5)	*M. incognita*		*P. vulnus*	
			n/g root	$r = Pf/Pi$	n/g root	$r = Pf/Pi$
Control	13.7 ab AB [(1)]	0.0 a A	-	-	-	-
M. incognita	13.8 ab AB	4.5 bc B	2912 ab AB	13.0 a A	-	-
P. vulnus	16.3 a A	-			59 a AB	10.9 a A
V. dahliae	7.3 cd CD	-			-	
M. incognita + *P. vulnus*	10.4 bc BC	3.8 b B	409 b B	1.1 c B	72 a A	8.2 b AB
M. incognita + *V. dahliae*	4.6 d D	4.8 c B	4984 a A	10.8 ab AB		-
P. vulnus + *V. dahliae*	7.4 cd CD	-			20 b B	4.7 c C
M. incognita + *P. vulnus* + *V. dahliae*	5.4 d D	4.0 bc B	1818 b AB	4.0 bc AB	77 a A	6.0 bc BC

(1) Data flanked in columns by the same letters are not statistically different according to Duncan's Multiple Range Test (small letters: P=0.05; capital letters: P=0.01).

Table 12. Reproduction of Meloidogyne incognita *and* Pratylenchus vulnus *on cv. Pendolino*

Pathogen	Root fresh weight (g)	Gall index (0-5)	*M. incognita*		*P. vulnus*	
			n/g root	$r = Pf/Pi$	n/g root	$r = Pf/Pi$
Control	14.7[(1)] a A	0.0 a A	-	-	-	-
M. incognita	11.2 abc AB	3.3 b B	602 a A	2.7 a A	-	-
P. vulnus	13.8 ab A	-	-	-	47 ab AB	5.2 a A
V. dahliae	6.9 cd BC	-	-	-	-	
M. incognita + *P. vulnus*	9.7 bc ABC	2.8 b B	49 c B	0.2 b B	12 b B	2.5 b B
M. incognita + *V. dahliae*	5.1 d C	3.3 b B	386 ab AB	0.8 b B	-	-
P. vulnus + *V. dahliae*	4.4 d C	-	-	-	103 a A	4.5 a A
M. incognita + *P. vulnus* + *V. dahliae*	4.6 d C	3.4 b B	254 bc AB	0.2 b B	32 b AB	4.8 a AB

(1) Data flanked in any column by the same letters are not statistically different according to Duncan's Multiple Range Test (small letters: P=0.05; capital letters: P=0.01).

4. MANAGEMENT AND CONTROL

As for soil pathogens and weeds, control of plant-parasitic nematodes can be achieved by applying a wide range of available technologies. In any case, due to the increasing attention to environment safety and human health, alternative control strategies, environmentally sound and at the same time economically convenient, are required (Bridge, 1996). Nematodes management can be achieved in sustainable agricultural systems by the integration of different tactis that fall into the following strategies.

4.1. Prevention and Use of Nematodes-Free Propagation Material

One of the main dissemination routes of phytoparasitic nematodes over great distances throughout the world and within countries is given by the use of infected or contaminated propagation rootstocks. The majority of infections by root-knot (*Meloidogyne* spp.) and root-lesion nematodes (*Pratylenchus* spp.) on olive tree plantations originate from unsanitized propagative material, produced in uncertified nurseries (Lehman, 1994).

Major nematode pests of olive trees have been detected for many years in propagative plant material in leading olive-producing countries in the Mediterranean basin (Inserra & Vovlas, 1981; Castillo et al., 1999). The use of pathogen-free planting material and uninfested nematode soil during olive seedlings propagation is essential for reducing or minimizing the effects of single or concomitant infections by soilborne pathogens, particularly *V. dahliae*, and nematodes, during the early years of olive cultivation, as well as to prevent or avoid their spreading. The potential of these pests in olive production was recognized by Italian legislation through the DM 14 April, 1997 and DM 9 August, 2000, as also in Spain (BOJA, 1997; BOE, 1999), and in the European Union (UE) (OEPP/EPPO, 1993), as well as by certification schemes for olive seedlings and rootstocks (Table 13).

Table 13. Diseases and pests damaging olive of quarantine concern.

Disease	*Causal agent*
Verticillium wilt	*Verticillium dahliae*
Olive knot	*Pseudomonas savastanoi* pv. *savastanoi*
Viruses	SLRSV ArMV CLRV OLYaV
Root galls	*Meloidogyne incognita, M. javanica*
Root lesions	*Pratylenchus vulnus* *Xiphinema diversicaudatum*

Many methods can be adopted in nurseries or farms to avoid nematode attacks, as te use of planting material free of infestations, physical destruction of nematodes in roots, rotation of seedbed sites or their selection among areas without a previous history of cultivation, eradication of seedbeds weeds (potential hosts), burning of plant debris, introduction of biological control agents in seedbeds and use nematode-free soil for potting (Bridge, 1996).

4.2. Use of Resistant Cultivars or Rootstocks

This method can represent a desirable and effective alternative to chemicals for control of phythoparasitic nematodes. The evaluation of large number of plant cultivars for reaction to nematodes is facilitated by a screening method that is reliable, inexpensive and rapid, based on the use of in-vitro olive explants in screening trials (Fig. 7) (Sasanelli et al., 2000). Previous olive cultivars screenings for resistance to *Meloidogyne* spp. were conducted by growing rooted woody cuttings in nematode infected soil. This last technique, although effective, involved repeated tests with a single nematode population, many plant replications and require at least more than one year to complete a screening assay (Esmenjaud et al., 1994; Sasanelli et al., 1997).

Rooted explants of two olive cutivars Ascolana and Moraiolo, and a selection of the wild-olive DA 12I, showed interesting results on *M. incognita* reproduction (Table 14). Among the different substrates (agar, compost or liquid) used in in-vitro trials for screening resistance to root-knot nematodes, best results (Fig. 8) were given by agar (Table 15) (Sasanelli et al., 2002b). On the basis of screening trials among the most diffuse Italian cultivars for resistance to root-knot nematodes, cv. Coratina appeared resistant to *M. incognita* and *M. javanica* and cv. Leccino only to *M. javanica*.

Figure 7. In vitro *olive explants in screening trials for resistance to phytoparasitic nematodes.*

Figure 8. Different substrates used in in-vitro *screening trials for resistance to root-knot nematodes: agar (left), compost (centre) or liquid substrate (right).*

Table 14. Reproduction of M. incognita *on different* in vitro *explant olive roots.*

Parameters	*Inoculum*[a]	*Ascolana*	*Moraiolo*	*DA 12I (rootstock)*
Root weight (g)	E	1.2 a A	1.2 a A	1.1 a A
	EM	1.4 ab AB	1.8 b B	0.6 a A
Galls/ g root	E	1.6 b B	1.1 b B	11.4 a A
	EM	4.3 b B	2.5 b B	14.6 a A
Females/g root	E	26.5 ab A	32.8 b B	11.0 a A
	EM	36.0 ab A	31.4 b B	79.5 a A
Juveniles/g root	E	30.7 b B	20.8 b B	216.4 a A
	EM	116.4 bA	22.4 b A	373.0 a A
Eggs/g root	E	31.1 b B	9.1 b B	220.1 a A
	EM	64.0 b B	20.9 b B	454.0 a A
Total population/g	E	61.7 b B	29.9 b B	436.5 a A
	EM	180.4 b B	43.3 b B	827.0 a A

[a] Each value is average of ten replicates; E= eggs; EM= egg masses.
Data flanked by the same letters across row are not statistically different according to Duncan's Multiple Range Test (small letters: P=0.05; capital letters: P=0.01).

4.3. Grafting

Grafting is a technology applied in vegetable crops not only for nematodes control but also for managing other soil borne diseases (Colombo et al., 2003). This technique may represent an alternative to soil fumigation or chemicals when susceptible cultivars should be planted in nematode infested soils or in presence of virulent populations.

Susceptible cultivars can be grafted on resistant, moderately resistant or moderately susceptible rootstocks. Among resistant rootstocks, DA 12I is moderately resistant to *M. incognita*, *M. javanica* and resistant to *X. index* (Sasanelli et al., 1997; 1999).

Table 15. Effect of three rooting media on root growth and infestation parameters by a population of Meloidogyne incognita, *for* in vitro *explants of cvs Ascolana and Moraiolo.*

Parameters	Cultivar	Rooting media[a]			
		Agar	Compost	Liquid	Mean
Root weight (g)	Ascolana	0.93 aA	0.11 cC	0.37 bB	0.5
	Moraiolo	1.14 aA	0.09 cC	0.49 bB	0.6
Infected roots (%)	Ascolana	96	20	100	72
	Moraiolo	96	36	100	77
Galls/g root	Ascolana	3.8 aA	1.8 aA	12.3 bB	5.9
	Moraiolo	1.1 aA	0.8 aA	7.7 bB	3.2
Females	Ascolana	0.0 aA	0.0 aA	0.0 aA	0.0
	Moraiolo	0.0 aA	0.0 aA	4.0 aA	1.3
Juveniles/g root	Ascolana	10.0 aAB	0.0 aA	79 bB	29.6
	Moraiolo	10.0 aA	49 aA	204 bB	87.6
Eggs/g root	Ascolana	136 aA	106 aA	2117 bB	786.3
	Moraiolo	206 aA	231 aA	1729 bB	722.0
Total population /g root	Ascolana	147 aA	106 aA	2196 bB	816.3
	Moraiolo	216 aA	280 aA	1933 bB	806.3

[a] Each value is mean of twenty five replicates.
Data flanked by the same letters in rows are not statistically different according to Duncan's Multiple Range test (small letters: $P = 0.05$; capital letters: $P = 0.01$).

4.4. Biological Control

Natural biological control of nematodes is common in soil, since naturally occurring antagonists (nematophagous fungi, endoparasitic fungal parasites, AMF arbuscolar mycorrhizal fungi, the obligate parasite *Pasteuria penetrans* and predaceous organisms such as collembola, and other invertebrates) may provide effective control of phytoparasitic nematodes especially in traditional farming systems (Stirling, 1991).

The nematophagous fungi *Pochonia chlamydosporia* and *Paecilomyces lilacinus* are frequently isolated from soil infested with cyst or root-knot nematodes. They may infect eggs and juveniles of plant nematodes. However, soil application of nematophagous fungi often fails because of several factors, including competition for space, production of inhibitors or changes in the environment, as well as direct lysis of hyphae. If the inoculum of biocontrol agent is unable to compete in soil it will fail to establish in the soil environment, with failure of the biocontrol strategy (Monfort et al., 2006). Effective biological control against the root-knot nematode *M. javanica* was, however, observed for several organisms, including *Trichoderma harzianum* (Stirling, 1991; Sharon et al., 2001). This antagonist, and in particular strain T-22 and *Streptomyces griseoviridis* strain K-61, appeared suitable when plant-parasitic nematodes interact with other soil borne fungi (Percoco & Amenduni, 2001).

Biological control agents can be artificially introduced into the soil, and some microorganisms antagonistic to nematodes are commercially formulated. In the last years effective results against root-knot nematodes were shown by *Aphanocladium album*, isolate MX-95 (patent n. MI2006A 000503, University of Bari, Italy) (Fig. 9). Due to its strong chitynolitic activity it can be used also against pathogenic fungi causing wilt diseases or corky root, and therefore it appears suitable in presence of attacks by nematodes and pathogenic fungi (Sasanelli et al., 2006; 2008; Ciccarese et al., 2008). Effective control by *Aphanocladium album* was also demonstrated on wheat rusts (Biali et al., 1972; Yaniv et al., 1979).

Figure 9. Aphanocladium album *isolate MX 95(insert), and culture growing on potato dextrose agar in Petri dish.*

4.5. Soil Solarization

Soil or substrates solarization may be used instead of chemical fumigation in olive nursery operations in Mediterranean climatic areas (June-August) or in any region characterized by high summer temperatures (Fig. 10). Many experiments and applications showed effective results of this method against phytoparasitc nematodes, *V. dahliae* and many other pests (Katan & De Vay, 1991). The relationship between nematode populations and exposure times to a range of temperature was also investigated and modeled (Sasanelli & Greco, 2000).

The efficacy of solarization depends on the highest temperatures achievable, the length of the effective thermic exposure, and the depth at which highest temperatures penetrate in soil. In a field trial with moist soils naturally infested with the citrus nematode (*T. semipenetrans*), lesion nematode (*P. vulnus*), or ring nematode (*C. xenoplax*), (Stapleton et al., 1999) soil placed in black polyethylene planting sleeves or left in 30 cm high piles was subjected to solarization for one to four weeks. Treatments included 1) no solarization – untreated control, 2) daily exposure to open sun, 3) as in treatment 2 but also covered with a single layer of transparent polyethylene film; 4) as in 2 but also covered with two layers of transparent polyethylene sheets. Data showed that soil temperatures reached 50, 69, and 73°C in treatments 2, 3, and 4, respectively. The density of each test pathogen in soil or olive roots was reduced by 89–100% by the various solarization techniques. The potential of solarization to control *M. incognita* in soil piles used in olive nurseries in Southern Spain was also studied by Nico et al. (2002; 2003).

Figure 10. Example of soil solarization in seedbed.

Soil infested with free eggs and egg masses of *M. incognita* was buried in nylon bags 20 and 40 cm deep inside conical soil piles 80 cm high and with a base diameter of 1 m, solarized for 3 weeks in July and August. The effect of various periods of solarization was assessed by egg hatching bioassays in sterile water, and by infectivity on tomato plants. Maximum soil temperature at 20 cm depth in solarized piles was 47 °C in 1999 and 48 °C in 2000, compared with 33 °C and 32 °C in nonsolarized piles. Solarization reduced egg hatching more than 95%, when compared to controls, irrespective of type, burial depth and inoculum location in the pile. Bioassay on tomato plants the second year confirmed the reduction of infectivity of free eggs buried in solarized soil piles. Therefore, under the conditions in southern Spain or in Italy, solarization of 40 cm-high piles of soil for 3 weeks can be used for olive nursery production to control root-knot nematodes in potting soil (Nico et al., 2003).

As mentioned above, this technique can also be used to control Verticillium wilt in orchards. Olive trees with Verticillium wilt could be recovered after soil solarization of individual 10- to 15-yr-old trees. Rate of recovery in solarized soil significantly exceeded natural recovery of untreated control trees and was attributed to the lack of root re-infections. Microsclerotia of *V. dahliae* were eliminated in the soil around treated trees, whereas propagules of *Talaromyces flavus* (antagonist of the pathogenic fungus) not only survived solarization but also increased in treated soil, compared with untreated controls (Tjamos et al., 1991).

In a further assay, solarization treatments were applied to trees rows for either one or two consecutive years. Solarization significantly reduced pathogen populations in the top 20 cm of soil for at least 3 years in relation to untreated control plots. Only in orchards with medium or high initial inoculum densities disease severity was reduced. A second soil solarization treatment did not improve the effect of single solarization on Verticillium wilt control. In orchards with low inoculum densities, soil solarization did not result in significant differences in disease incidence and severity, but improved recovery of trees from the disease (Lopez-Escudero & Blanco-Lopez, 2001).

4.6. Biofumigation

Soil biofumigation, based on the incorporation in soil of biomasses of plants like *Brassicaceae* (*Brassica oleracea*, *B. nigra*, *Brassica juncea*, *Raphanus sativum*, *Crambe abyssinica*), is an ecological alternative to chemical fumigation against phytoparasitic nematodes and soil borne pathogens (Mojtahedi et al., 1991). Biofumigation is based on the release of glucosinolate-derived compounds able to develop a natural soil fumigation. Plant tissues with a high content of glucosinolates and incorporated in soil release these compounds when damaged. Glucosinolates then get in contact with the endogenous enzyme myrosinase, which catalyses their hydrolysis into various products, especially isothiocyanates, thiocyanates, nitriles or oxazolidine2-ethione, depending on the reaction conditions (Fahey et al., 2001; Laegdsmand et al., 2007). Isothiocyanates are analogous sulphonated of isocyanates with the general structure R-N=C=S, which reacts promptly with soil water. The carbamic acid spontaneously loses CO_2 yielding amines that react with the other isocyanates in a serial reaction, producing urea and ammonia products.

Several studies and trials showed that isothiocyanates are toxic to a range of pathogenic soil-borne organisms including fungi, bacteria and nematodes (Lazzeri et al., 1993; Brown & Morra, 1997; Smith & Kirkegaard, 2002). Several amendments based on cruciferous species showed a suppressive action on *M. incognita* on tomato (D'Addabbo et al., 2004). All tested amendments significantly reduced root gall index, number of nematode eggs and nematode reproduction rate (r = Pf / Pi) on tomato roots. Few data are, however, avilable on efficacy of biofumigation on olive trees in nematodes infested fields.

Many cruciferous genotypes seem to be suitable for biofumigation as green manures, but their suitability for the preparation of more easily available formulations, i.e. dry pellets is in progress and needs more investigation (Lazzeri et al., 2002).

4.7. Soil Amendments

The addition of organic materials to soil infested with phytoparasitic nematodes or other pests can effectively control many phytoparasitic nematodes, especially in developing regions, due to the materials low cost and availability (Rodriguez-Kabana, 1986; D'Addabbo, 1995; Akhtar & Mahmood, 1996; Akhtar & Malik, 1996). Soil amendments include a much broader category, normally consisting of various waste materials. These are mainly bioproducts and wastes from agricultural and other activities, and include oilseed cakes, plant crop residues, plant composts, green manure, agroindustrial wastes, ashes, animal manure, crustacean shells and other wastes (Fig. 11).

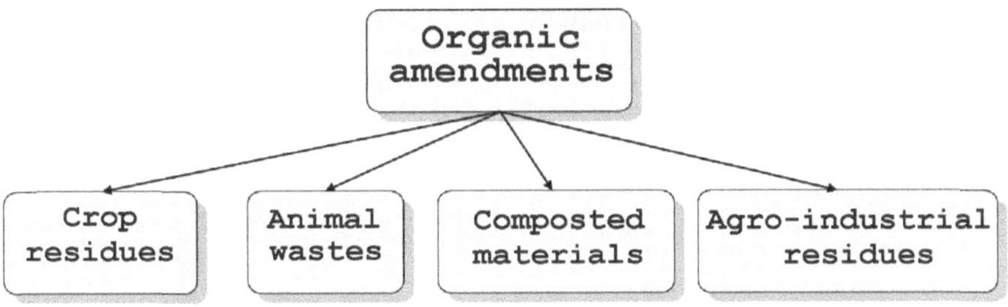

Figure 11. Subdivision of organic amendments according to their origin.

The effects of these materials on agroecosystem vary and involve changes in soil chemistry (soil fertility), physical properties (porosity, aggregates stability, water exchange) and microbiology, with positive effect on plant growth (Sequi et al., 1986; Ocio et al., 1991; Sasanelli et al., 2002c). Some materials directly release compounds like chitin, phenols, tannins, azadirachtin, ricinin, terpens, which are toxic to nematodes (Mian & Rodriguez-Kabana, 1982; Spiegel et al., 1987; Rich et al., 1989) or others deriving from soil decomposition processes, i.e. ammonia, nitrites or hydrogen sulphide (Rodriguez-Kabana, 1986). Amendments also provide a favourable substrate for sustenance and growth of soil microflora and microfauna which include parasites (fungi and bacteria) (Galper et al., 1990) or nematode predators (microarthropods), or may suppress soil nematode populations by toxic metabolites, such as antibiotics or enzymes (D'Addabbo, 1995; Sikora, 1992).

Many reports also describe the use of amendments derived from olive or oil industry against nematodes in glasshouse or field. Tests were carried out mainly in Italy, Spain and Greece due to the local large amount and availability of these materials, with interesting results (D'Addabbo & Sasanelli, 1996a; 1996b; 1997; D'Addabbo et al., 1997; 1999; 2000; 2003; Sasanelli et al., 2002b; 2003c; Manios, 2004; Nico et al., 2004; Piedra Buena et al., 2007).

A field assay was undertaken in Southern Italy to evaluate the effect of the incorporation of olive composted pomaces, both fresh and exhausted, and raw sewage (solid or liquid olive mill wastes - OMW) on *M. incognita* (Sasanelli et al., 2002c). Fresh and exhausted composted pomace was distributed on the soil surface at 10, 20 and 40 t/ha and then incorporated in soil at 25 – 30 cm depth. Composition and chemical characteristics of the composts are shown in Table 16. Raw sewage was added at 40 and 80 m^3/ha. Untreated soil and 300 kg/ha fenamiphos applied before transplanting were used as controls. All composts increased the yield of the test plant (tomato) compared to the untreated control, although less than the chemical treatment (Table 17). All treatments reduced the root infestation index and, with the exception of the lowest rates of raw sewage and exhausted pomace compost, the soil nematode population. Results from this experiment confirm that incorporation of OMW into the soil may induce a suppressive effect on nematode populations also increasing crop yields or growth. The practice of recycling olive wastes or material in olive orchards may hence help in the management of nematodes and soil borne diseases. It appears, however, of practical use only in areas where large amounts of amendments are available.

Table 16. Composition and chemical characteristics of composts.

Composition	Exhausted olive pomace compost	Fresh olive pomace compost
Fresh solid cake (%)	-	92.6
Farmyard manure (%)	-	7.4
Exhausted solid cake (%)	91.0	-
Poultry manure (%)	1.7	-
Wheat straw (%)	7.3	-
Chemical parameters:		
Humidity 105° C	40.8	43.8
pH in H_2O 1:10	6.8	6.5
EC estr. 1:10 25° C (dS/m)	0.9	0.9
Ammonia nitrogen g/kg	0.9	0.9
Total nitrogen g/kg	18.5	12.5
Total carbon (TOC) (%)	37.8	47.6
Extracted carbon (TEC) (%)	16.9	14.8
Humic carbon (HA+FA) (%)	11.5	12.9
Humic degree (DH) (%)	68.1	87.2
Humic rate (HR) (%)	30.4	26.9
Humic index (HI)	0.5	0.1

4.8. Nematicidal Plants

Natural compounds represent a potential alternative to chemicals for the control of plant pathogens and parasites (Grainge & Ahmed, 1988). Many nematode species are already reported to be killed or suppressed by several plants or plant products (Chitwood, 2002). Hills (1962) and Omidvar (1962) reported reduction of the potato cyst nematode *Globodera rostochiensis* in presence of *Tagetes* spp., although Omidvar (1961) also showed that root diffusates of *T. minuta*, *T. florida* or *T. signata*, at different concentrations, had no effect on hatching. Leaf and root extracts and root leachates, from *T. erecta* and *T. signata* showed inhibiting effects on an Italian population of *G. rostochiensis* (Sasanelli & Di Vito, 1991). The nematicidal effect of *T. erecta* leaf extracts was also investigated on *Heterodera schachtii* (Sasanelli & D'Addabbo, 1992). Extracts from *T. erecta* showed effective nematicidal activity also against *M. arenaria*, *M. hapla* and *M. javanica*, but not on *M. incognita* (Sasanelli & D'Addabbo, 1993a; Ploeg, 1999).

Table 17. Effects of different olive mill wastes amendments on Meloidogyne incognita *on tomato cv. Tondino di Zagaria.*

Treatment	Dose ($t \cdot ha^{-1}$)	Tomato yield ($t \cdot ha^{-1}$)		Root gall index (0 – 5)		Eggs and juveniles per cm^3 soil	
Raw sewage	4	32.3[1]	bc[2]	2.4	a	15.1	cd
	8	29.0	b	2.5	a	5.2	a
Exhausted olive pomace compost	10	36.8	d	2.1	a	13.7	bcd
	20	36.6	d	2.2	a	9.2	abc
	40	36.3	d	2.2	a	11.2	abc
Fresh olive pomace compost	10	32.1	bc	2.8	a	5.3	a
	20	33.7	cd	2.2	a	7.1	ab
	40	35.8	cd	2.6	a	9.3	abc
Fenamiphos	0,3	40.5	e	2.7	a	10.6	abc
Untreated control	-	25.4	a	4.5	b	19.8	d

(1) Each value mean of four replications.
(2) Data followed by the same letter in any columns are not significantly different according to Duncan's Multiple Range Test (P = 0.01).

Effective reduction of *Meloidogyne* species and *H. schachtii* hatching or nematode population in pot or field experiments has been demonstrated for *Cineraria maritima* and *Ruta graveolens* (Sasanelli & D'Addabbo, 1992; 1993a; 1993b; Sasanelli et al., 2003a). *R. graveolens* leaf extracts on *Meloidogyne* species has shown a nematicidal effect more efficient than fenamiphos (Sasanelli & D'Addabbo, 1993a; Sasanelli et al., 2007). Nematicidal activities of aqueous extracts from leaves of *R. graveolens* or from pods of *Capsicum annuum* have been shown against *Xiphinema index* (Sasanelli & Catalano, 1991; Sasanelli, 1992). Furanocoumarins, flavonoids, alkaloids, terpens and essential and volatile oils are the principal active compounds found in these species and especially in rue plants (Gray & Waterman, 1978; Kostova et al., 1999).

In many experiments extracts from rue leaves resulted more effective than root extracts. As concentration of the furanocoumarins (bergaptene, xanthotoxin, psoralen and isopimpinellin) is higher in leaves than in roots (Zobel & Brown, 1989), these components could be the main responsible for the nematicidal effect of rue or other nematicidal plants. Furanocoumarins have been identified as phytoalexins by Johnson et al. (1973), and have antibacterial, antifungal, antiviral and insecticide activities (Chaudhary et al., 1985; Zangerl, 1990). Their localization can vary among plant species, during plant life and by tissue types. Also, environmental conditions, i.e. seasonal temperature changes or plant pathogens attack can increase plant furanocoumarin content (Milesi et al., 2001). Seeds of neem plant *Azadirachta indica* or oil extracted from seeds contain large amount of active principles classified as limonoid triterpens, azadirachtin, with nematicidal activity (Akhtar, 2000; Javed et al., 2007). Compounds with nematicidal effect have been also found in *Quillaja saponaria* (D'Addabbo et al., 2005; 2008), *Chrysanthenum* spp. and *Calendula* spp. (Perez et al., 2003). Among plants with nematicidal activity only few have been marketed (azadirachtins, quillay, extracted maringold oil) (D'Addabbo et al., 2008). Although this method appears suitable for olive nematodes control, no data are available from experimental trials in olive nurseries or field conditions.

4.9. Soil Steam Sterilization

Steam sterilization relies on the direct application of steam to soil, in open field or greenhouse conditions. The method holds several benefits since steam may destroy insects, nematodes, weeds as well as reduce the incidence of fungi and bacteria. Sterilization is achieved by the high temperatures reached by steam under pressure. Introduction in the soil profile occurs by a negative pressure created by a fan directing the air flow out of soil, through buried perforated polypropylene pipes. This system requires a permanent installation of perforated pipes in soil, at least at 50-60 cm depths in order to be protected from ploughing.

To determine the exact treatment time, soil temperature must be monitored during the treatment along the soil profile. In alternative special steamining equipments are available, which are transported by tractors and diffuse steam directly into the soil. Some experiences showed that the addition of compounds yielding exothermic reactions (CaO or KOH) may improve the sterilization effect. The possibility of adding these materials at rates of 1000 – 4000 kg/ha, depends on the soil chemical characteristics (Triolo et al., 2004). In nurseries, although this can improve production costs, soil or potting mixtures may be sterilised in a self locking sterilizator for 6-7 hours.

4.10. Chemical Control

In integrated pest management (IPM), the use of nematicides can be appropriately combined with the other tactics previously described, aiming at applications of lower doses.

The nematicides can be fumigant or not fumigant. Fumigants must be applied at least 3 weeks before plantation, whereas the use of non fumigants is possible one day before or at transplantig. Actually, in spite of their short-term efficacy, fumigants show some drawbacks as the difficulty of application, the availabilty of specific equipments and phytotoxicity. Some non fumigant nematicides are systemic and have no phytotoxic effects, and therefore they can be applied in nurseries or with transplanted plants. It is worth to recall that the use of specific active ingredients with nematicidal activity depends on the legislation applied by each country and the corresponding available crop registrations.

4.10.1. Fumigants

These products may have a potential application in soil treatments in nurseries, due to their efficacy. 1,3-D (1,3-dichloropropene) is a liquid chemical applied into the soil with a fumigator, and must be applied 1 week/100 l of product/ha before transplanting or sowing. The nematicide 1,3-D (94% a.i.) is also in an emulsifiable formulation applied by irrigation (drip fumigation). Its use is also possible in protect conditions with doses between 150 a 250 l/ha.

Chloropicrin (trichloronitrometan: NO_2CCl_3), is a liquid chemical applicable, at doses between 350 e 500 l/ha (c.p.), by injection into the soil. It is a very effective fumigant with a wide spectre of action. To reduce immision in the environment the treatment must be completed by covering the soil with a PVC plastic film. A concentrated emulsifiable formulation in water (Italian registration in 2002) can be applicable by drip fumigation in protect conditions. Effective results in the control of fungi and plant parasitic nematodes can be obtained by mixtures with 1,3 D.

Dazomet (3,5 dimethyl 1,3,5-thiodiaziname-2-thione: $C_6H_{10}N_2S_2$) is a granular with fumigant action. It must be broadcasted on soil as a normal fertilizer and then incorporated in it. After the treatments it is necessary to wet the soil by a rain irrigation or sub-irrigation. Also for this fumigant it is convenient to cover the soil. Dose varies between 400 and 700 Kg /ha depending on the nature of the parasite or phatogen to control. This chemical also has a fungicidal activity. Its phytotoxic activity does not allow its use in presence of plants. Before transplanting or potting it is necessary to wait for 3–5 weeks, depending on temperature (never use below 10 °C), soil humidity and structure.

Metam-sodium (Na methilditiocarbammate: CH_3NHCS_2Na) is a chemical with fumigant activity. It can be applied by irrigation or sub-irrigation, more effective if used with a PVC plastic films, to avoid enviromental diffusion of the active ingredient, methylisothiocianate (MIT). It is convenient to apply the chemical with temperatures between 10 and 30 °C. Doses vary between 400 and 1,500 l/ha depending on the nature of the nematode or phatogen to control.

Ozone (O_3) treatments of soil or growth substrates is a further alternative to pesticides for control of phytoparasitic nematodes and pathogenic fungi, since this gas has an effective biocidal action (Francis, 1997; Ciccarese et al., 2007). A two year set of trials in open field were undertaken on tomato, carrot and strawberry infected by root-knot nematodes in California, using different dosages (Pryor, 1999).

In a plastic greenhouse infested by the root-knot nematode *M. incognita* on melon soil treatment with ozone (4 mg/m^3 soil) significantly reduced the nematode infestation on roots and its final population density in the soil, but no yield increase was recorded. A similar reduction of galls formation on the roots following ozone treatment was found in a trial on tomato infested by *M. incognita* (unpublished data).

International literature on the nematicidal use of ozone is rather scarce. In some trials soil treatments with 250 kg/ha ozone mixture (concentration = 4.35% w/w) reduced *M. javanica* and free nematofauna populations by 68% and 58%, respectively, compared to untreated soil (Qui et al., 2001). Significant nemtodes density decreases (-24 and -19%, respectively) were also observed at 50 kg/ha dosage (4.35% O_3).

Although no data are available on olive nematodes, this product may have a potential for soil treatment in nurseries. On other crops, subirrigation treatments with 280 kg/ha ozone mixture (4.35% O_3) significantly increased (14.5-79.5, according to the year) tomato yields in comparison to untreated soil. Treatments resulted also better than 1,3 D fumigation for nematode control, increasing carrot yields (by 46-92%) in comparison to untreated control. Satisfactory results were obtained on both crops also with a lower dose of ozone (56 kg/ha) and best results were recorded when ozone treatment was preceded by soil irrigation.

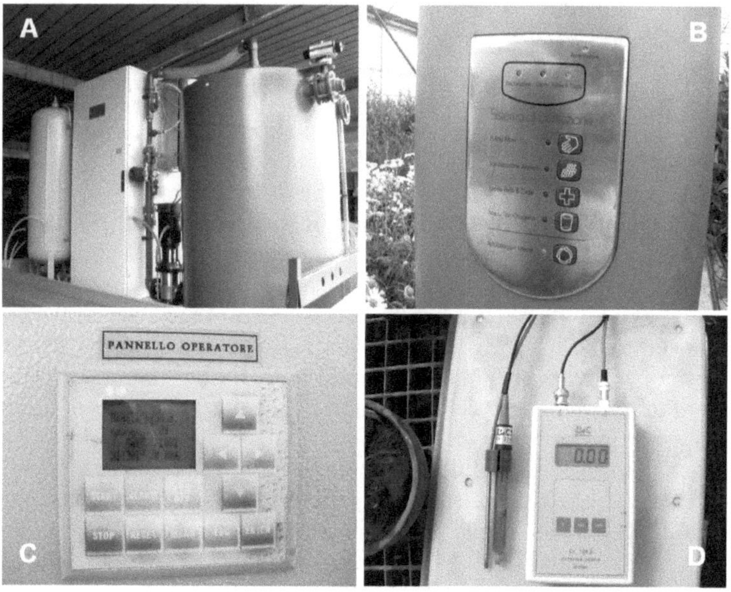

Figure 12. Ozone generators. Mobile ozone generator for filed applications (A); for smaller applications (B); panel control to adjust ozone concentration (C) and ozone meter to control ozone concentration in ozonated water (D).

On strawberry, in soil infested by root-knot nematodes and *Verticillium* sp., 450 kg/ha ozone treatment (4.35 % O_3) increased yield by 6.1-51.1 % compared to the untreated soil. Treatments with ozone resulted effective also for the disinfection of propagation material in nurseries. In a trial carried out in California, treatment with ozone of bulbs of *Lilium longiflorum* infested by the root lesion nematode *Pratylenchus penetrans* resulted in a significant increase of the bulb growth (Giraud et al., 2001). Moreover the production of ozone as gas, it is possible to directly produce ozonated water for soil irrigation or treatments. In both cases it is possible to adjust the ozone concentration in the mixuture air-ozone or in the ozonated water. In Fig. 12 some ozone generators for different application types are shown. All these trials showed that attention should be focused on the dosages and application time, in order to avoid phytotoxicity (especially in protected conditions).

4.10.2. Non Fumigant Nematicides

A first group of non fumigant nematicides are carbamic nematicides. Aldicarb is systemic and develops metabolites in soil with nematicidal and insecticidal properties. Due to its penetration in plant tissues and accumulation its use has been restricted in several countries and/or is allowed on a limited number of crops. It has low toxicity and therefore it is suitable at transplanting or in new plantations. Its high solubilty in water allows the product to penetrare in the soil by irrigation (http://www.fao.org/ag/AGP/AGPP/Pesticid/JMPR/Download/94_eva/aldicarb.pdf).

Carbofuran (2,3-dihydro-2,2dimethylbenzofuran-7-methylcarbammate), is a carbamic nematicide, not phytotoxic and systemic. For its high water solubility it can be easily lost, especially in soils with low organic matter contents. Soil absorbing is reversible and therefore with an exact combination of doses and watering volumes it is possible to obtain a uniform vertical distribution in soil, improving nematodes control at different depth (Greco et al., 1979).

Oxamyl is a liquid formulation with a limited persistance, soluble in irrigation water. A soil pH close to 4.5 is optimal for adsorption by the root system and availability. Application by fertirrigation reduces costs and operators risks (http://www.fao.org/ag/AGP/AGPP/Pesticid/Specs/docs/Pdf/new/Oxamyl08.pdf).

Among phosphorganics, cadusafos is a nematicide-insecticide moderately persistent in soil, not phytotoxic and with low soil mobility, effective against *Meloidogyne* spp. (Sasanelli et al., 1996; Sasanelli & D'Addabbo, 1999) (http://www.efsa.europa.eu/cs/BlobServer/PRAPER_Conclusion/praper_cadusafos_addendum_final1.pdf).

Fenamiphos is a phosphorus ester with systemic activity, available as granular or liquid. The product is not phytotoxic. Doses range between 10 and 15 kg/ha of a.i. The liquid formulation can be applied by means of water irrigation (http://www.fao.org/docrep/W8141E/w8141e0p.htm).

Finally, fosthiazate is an organothiophosphate nematicide recently included in Annex I of the Directive 91/414/EEC under the clause that it should be used with special care in soils vulnerable to leaching. Effective control of *Meloidogyne*, *Globodera pallida* and *G. rostochiensis* have been obtained by the use of this nematicide (Woods et al., 1999; Chabrier & Hubervic, 2000a; 200b) (http://www.epa.gov/opprd001/factsheets/fosthiazate.pdf).

5. CONCLUSIONS

The most important problem caused by nematodes on olive crops is registered in nurseries for the production of nematode-free olive propagative materials or trees. The routine application of techniques aiming at preventing the introduction and spread of nematodes is very important in the management of nurseries, and several control methods are also available to manage infestations by root-knot or other important nematode groups. Before planting young olive trees in the field a soil nematode analysis is always helpful, in order to avoid problems especially during the early years of tree growth. In case of severe nematode infestations or synergic effects with other pests, soil disinfection before planting is convenient, applying physical,or integrated treatments, according the local availability of resistant/tolerant plants, of organic amendments and/or chemicals, as allowed by the country legislation.

REFERENCES

Abrantes, I. M. De O., Vovlas, N., & Santos, M. S. N. De A. (1987). Morphological studies on six Tylenchid species associated with olive in Portugal. *Ciência Biologica, Ecologia e Systematica*, 7, 1-9.

Abrantes, I. M. De O. (1980). Nematode problems of olive trees. *XVth International Nematology Symposium of the European Society of Nematologists*, Bari, Italy, 27-28 (Abstract).

Abrantes, I. M. De O., & Santos, M. S. N. De A. (1991). *Meloidogyne lusitanica* n. sp. (Nematoda: Meloidogynidae), a root-knot nematode parasitizing olive tree. *Journal of Nematology*, 23, 210-224.

Abrantes, I. M. De O., Vovlas, N., & Santos, M. S. N. De A. (1992). Host-parasite relationships of *Meloidogyne javanica* and *M. lusitanica* with *Olea europaea*. *Nematologica*, 38, 320-327.

Abu-Gharbieh, W. I., Maccouk, K. M., & Saghir, A. R. (1978). Response of different tomato cultivars to the root-knot nematode, tomato yellow leaf curl virus, and *Orobanche* in Jordan. *Plant Disease Reporter*, 62, 263-266.

Akhtar, M., & Mahmood, I. (1996). Control of plant parasitic nematodes with organic and inorganic amendments in agricultural soil. *Applied Soil Ecology*, 4, 243-247.

Akhtar, M., & Malik, A. (1996). Roles of organic soil amendments and soil organisms in the biological control of plant parasitic nematodes: a review. *Bioresource Technology*, 74, 35-47.

Akhtar, M. (2000). Nematicidal potantial of the neem tree *Azadirachta indica* (A Juss). *Integrated Pest Management Reviews*, 5, 57-66.

Almeida, M. T. M., Waele, D. De, Santos, M. S. N. De A., & Sturhan, D. (1989). Species of *Trichodorus* from Portugal. *Revue de Nematologie*, 12, 219-233.

Arias, M. (1975). New information on the genus *Xiphinema* and its distribution in Spainish soils. *Anales Edafologia y Agrobiologia*, 34-198 (in Spanish).

Baines, R. C. (1951). Citrus-root nematode: an olive pest pathologically and morphologically similar to that on orange roots infests and reproduces on olive roots. *California Agriculture*, 5, 11.

Biali, M., Dinoor, A., Eshed, N., & Kenneth, R. (1972). *Aphanocladium album*, a fungus inducing teliospore production in rusts. *Annals Applied Bilogy*, 72, 37-42.

BOE (1999). Modificacion del reglamento tecnico de control y certification de plantas de vivero de frutales. Boletin Oficial Estado N° 276, 40077-40079.

BOJA (1997). Reglamento Especifico de Produccion Integrada del Olivar. Boletin Junta Andalucia N° 100,10543-10555.

Bridge, J. (1978). Plant Nematology in Jordan. Report of Scientific Liaison Office. Overseas Development Ministry. UK: London.

Bridge, J. (1996). Nematode management in sustainable and subsistence agriculture. *Annual Review of Phytopatholog*, 34, 201-225.

Brown, P. D., & Morra, R. (1997). Control of soil-borne plant pests using glucosinalates-containing plants. *Advances in Agronomy*, 61, 167-231.

Buhrer, E. M., Cooper, C., & Steiner, G. (1933). A list of plants attacked by the root-knot nematode (*Heterodera marioni*). *Plant Disease Reporter*, 17, 64-96.

Castillo, P., Vovlas, N., Nico, A. & Jimenez-Diaz, R. M. (1999). Infection of olive trees by *Heterodera mediterranea* in orchards in southern Spain. *Plant Disease*, 83, 710-713.

Castillo P., Vovlas, N., Subbotin, S., & Troccoli, A. (2003). A new root-knot nematode, *Meloidogyne baetica* n. sp. (Nematoda: Heteroderidae), parasitizing Wild Olive in southern Spain. *Phytopathology*, 93, 1093-1102.

Castillo P., Vovlas, N., & Troccoli, A. (2003). The reniform nematode, Rotylenchulus macrosoma, infecting olive in southern Spain. *Nematology*, 5, 23-29.

Chabrier, C., & Hubervic, J. (2000a). Evaluation of fosthiazate (Nemathorin 10G) for the control of nematodes on bananas in Martinique. *Nematropica*, 30, 117-118.

Chabrier, C., & Hubervic, J. (2000b). The effect of granular nematicide incorporation depth and potato planting depth on potatoes grown in land infested with the potato cyst nematodes *Globodera rostochiensis* and *G. pallida*. *Annals of Applied Biology*, 136, 27-33.

Chaudhary, S. K., Ceska, O., Warrington, P. J., & Ashwood-Shmith, M. J. (1985). Increased furanocoumarin content of celery during storage. *Journal Agricultural Food Chemistry*, 33, 1153-1157.

Chitwood, D. J. (2002). Phytochemical based strategies for nematode control. *Annual Review of Phytopathology*, 40, 221-249.

Ciccarese, F. (1998). Verticillium wilt of olive: new acquirements and possibility of control. *Olivo e Olio*, 6, 40-45 (in Italian).

Ciccarese, F., Sasanelli, N., Ciccarese, A., Ziadi, T., Ambrico, A., & Papajova, I. (2007). Control of *Pyrenochaeta lycopersici* on tomato by ozone disinfestation. IOA AINIA Conference & Exbition, October 29-31, Sustainable Agri-Food Industry, Use of Ozone & Related Oxidants. Valencia, Spain, 4.2, 1-6.

Ciccarese, F., Sasanelli, N., Gallo, M., Papajova, I., & Renco, M. (2008). Biological control of Fusarium-wilt and the root-knot nematode *Meloidogyne incognita* on *Cucumis melo* subsp. *Melo* conv. Adzhur (Pang.) Grebensch. Proceedings: Biotechnology 2008, Cezch Budejovice. Cezch Republic: 33-35.

Colbran, R. C. (1955). A preliminary survey of plant nematodes in Queensland. *Journal of the Australian Institute of Agricultural Science*, 15, 167-169.

Colbran, R. C. (1964). Studies of plant and soil nematodes. 7. Qeensland records of the order Tylenchida and the genera *Trichodorus* and *Xiphinema*. *Qeensland Journal of Agricultural and Animal Science*, 21, 77-123.

Colombo, A., Serges, T., Assenza, M., Donzella, G., Minuto, A., & Garibaldi, A. (2003). Use of erbaceous graft to control soil pathogens and parasites attacks on tomato and eggplant: actual situation and prospective. *Informatore Fitopatologico*, 53 (2), 13-19.

Condit, I. T., & Horne, W. T. (1938). Nematode infestation on olive roots. *Phytopathology*, 28, 756-757.

D'Addabbo, T. (1995). The nematicidal effect of organic amendments: a review of the literature 1982-1994. *Nematologia Mediterranea*, 23, 299-305.

D'Addabbo, T., & Sasanelli N. (1996a). Effect of olive pomace soil amendment on *Meloidogyne incognita*. *Nematologia Mediterranea*, 24, 91-94.

D'Addabbo, T., & Sasanelli N. (1996b). The effect of olive pomace soil amendment on *Heterodera carotae*. *Nematologia Mediterranea*, 24, 205-208.

D'Addabbo, T., & Sasanelli N. (1997). Suppression of *Meloidogyne incognita* by combinations of olive pomace or wheat straw with urea. *Nematologia Mediterranea*, 25, 159-164.

D'Addabbo, T., Fontanazza, G., Lamberti, F., Sasanelli, N., & Patumi, M. (1997). The suppressive effect of soil amendments with olive residues on *Meloidogyne incognita*. *Nematologia Mediterranea*, 25, 195-198.

D'Addabbo, T., Sasanelli, N., & Coiro, M. I. (1999). Suppression of *Xiphinema index* by olive and grape pomace. *Nematologia Mediterranea*, 27, 257-260.

D'Addabbo, T., Sasanelli, N., Lamberti, F., & Carella, A. (2000). Control of root-knot nematodes by olive and grape pomace soil amendments. *Proceedings of the Fifth International Symposium on Chemical and Non-Chemical Soil and Substrate Disinfestation*. Turin, Italy: 53-57.

D'Addabbo, T., Sasanelli, N., Lamberti, F., Greco, P., & Carella, A. (2003). Olive pomace and chicken manure amendments for control of *Meloidogyne incognita* over two crop cycles. *Nematropica*, 33, 1-7.

D'Addabbo, T., De Mastro, G., Sasanelli, N., Di Stefano, A., & Omidbaigi, R. (2004). Suppressive action of different crocuferous crops on the root-knot nematode *Meloidogyne incognita*. *Agroindustria*, 3, 379-380.

D'Addabbo, T., Curto, G., Greco, P., Di Silvestro, D., Coiro, M. I., Lamberti, F., et al. (2005). Preliminary trials to control root-knot-nematodes by *Quillaja saponaria* Molina. *Nematologia Mediterranea*, 33 (Suppl.), 29-34 (in Italian).

D'Addabbo, T., Greco, P., & Radicci, V. (2008). Effectiveness of plant commercial formulations for the control of root-knot nematodes. *Atti Giornate Fitopatologiche 2008*, 1, 317-322 (in Italian).

Dasgupta, D. R., Raski, D. J., & Sher, S. A. (1968). A revision of the genus *Rotylenchulus* Linford & Oliveira, 1940 (Nematoda: Tylenchidae). *Proceedings of the Helminthological Society of Washington*, 35, 169-192.

D'Errico, F. P., Lamberti, F., & Fiume, F. (1977). Discovery of *Dolichodorus heterocephalus* Cobb in Southern Italy. *Nematologia Mediterranea*, 5, 99-101 (in Italian).

Diab, K. A., & El-Eraki, S. (1968). Plant parasitic nematodes associated with olive decline in the United Arab Republic. *Plant Disease Reporter*, 52, 150-154.

Edongali, E. A. (1989). Plant-parasitic nematodes associated with olive trees in Libya. *International Nematology Network Newsletter*, 6, 36-37.
Esmenjaud, D., Minot, J. C., Voisin, R., Pinochet, J., & Salesses, G. (1994). Inter and intraspecific variability in plum, peach and peach-almond rootstock using 22 root-knot nematode populations. *Journal of the American Society for Horticultural Science*, 119, 94-100.
Fahey, J. W., Zalcman, A. T., & Talalay, P. (2001). The chemical diversity and distribution of glucosinolates and isothiocyanates among plants. *Phytochemistry*, 56, 5-51.
Francis, A. W. (1997). Ozone. In: McGraw-Hill Encyclopedia of Science and Tecnology 8th Ed. McGraw-Hill, N.Y., USA: 12, 683-686.
Franc, L. J., & Wheller, T. A. (1993). Interaction of plant parasitic nematodes with wilt-inducing fungi. In: Nematode Interactions (Ed. M.W. Khan). Chapman & Hall, London, UK: 79-103.
Fiume, F. (1978). The genera of nematodes living in the rhizosphere of olive in Calabria. *Informatore Fitopatologico*, 28, 11-14 (in Italian).
Gallo, D. P., & Jimenez, R. M. (1976). The phytoparasitic nematofauna associated with olive in Azapa valley. *IDESIA* no. 4, 105-109 (in Spanish).
Galper, S., Cohn E., Spiegel, Y., & Chet, I. (1990). A collagenolytic fungus, *Cunninghamella elegans*, for biological control of plant parasitic nematodes. *Journal of Nematology*, 23, 269-274.
Giraud, D. D., Westerdahl, B., Riddle, L., Anderson, C., & Pryor, A. (2001). Hot water and ozone treatments of Easter lily for the management of lesion nematode, *Pratylenchus penetrans*. *Phytopathology*, 91 (6)S, 134 (Abstract).
Grainge, M., & Ahmed, S. (1988). Handbook of Plants with Pest-Control Properties. J. Wiley & Sons, New York: 238-248.
Graniti, A. (1955). A dieback of olive in Sicily associated with two nematode species. *Olearia*, 9, 114-120.
Gray, A. I., & Waterman, P. G. (1978). Coumarins in the *Rutaceae*. *Phytochemestry*, 17, 845-864.
Greco, N., Di Vito M., & Basile, M. (1979). Actual knowledge on the control of plant parasitic nematodes. *Atti Giornate Nematologiche 1979*, 223-263.
Hashim, Z. (1979). A preliminary report on the plant parasitic nematodes in Jordan. *Nematologia Mediterranea*, 7, 177-186.
Hashim, Z. (1983). Description of *Trophotylenchulus saltensis* n. sp. with a comment on the status of *Trophotylenchulus* Raski, 1957 and a proposal for *Ivotylenchulus* n. gen. (Nematoda: Tylenchida). *Revue de Nématologie*, 6, 179-186.
Hashim, Z. (1984a). Re-diagnosis and key to species *Neolobocriconema* Metha & Raski, 1971 (Nematoda: Tylenchida), with description of *N. olearum* n. Sp. From Jordan. *Systematic Parasitology*, 6, 69-73.
Hashim, Z. (1984b). Description of *Tylenchorhynchus tenuis* n. sp. and observation on *Rotylenchus cypriensis* Antoniou, 1980 (Nematoda: Tylenchida) from Jordan. *Systematic Parasitology*, 6, 33-38.
Hills, L. D. (1962). The 1961 *Tagetes* experiment. Henry Doubleday Research Association, Braintree, Essex, England: 10 pp.
Hirschmann, H., Paschalaki-Kourzi, N., & Triantaphyllou, A. C. (1966). A survey of plant-parasitic nematodes in Greece. *Annales de l'Institut Phytopathologique Benaki*, 5, 144-156.
Inserra, R. N., Vovlas, N., Lamberti, F., & Bleve, T. (1976). Plant parasitic nematodes with declining olive trees in Italy. *Poljoprivedna Znanstvena Smorta, Agriculturae, Conspectus Scientificus*, 39, 419-424.
Inserra, R. N., & Vovlas, N. (1977). Parasitic habits of *Gracilacus peratica* on olive feeder roots. *Nematologia Mediterranea*, 5, 345-348.
Inserra, R. N., & Vovlas, N. (1978). *Tylenchulus semipenetrans* on olive trees in Northern Italy. In: *Third International Congress of Plant Pathology* (abstract of papers), Munchen, DE: 151.
Inserra, R. N., Vovlas, N., & Golden, A. M. (1979a). *Helichotylenchus oleae* n. sp. and *H. neopaxili* n. sp. (Hoplolaimidae) two new spiral nematodes parasitic on olive trees in Italy. *Journal of Nematology*, 11, 56-62.
Inserra, R. N., Zepp A. & Vovlas N. (1979b). The *Pratylenchus* spp. of Southern Italy. *Nematologia Mediterranea*, 7, 137-162 (in Italian).
Inserra, R. N., Vovlas, N., & O'Bannon, J. H. (1980). A classification of *Tylenchulus semipenetrans* biotypes. *Journal of Nematology*, 12, 97-102.
Inserra, R. N., & Vovlas, N. (1981). Data on the geographical distribution of nematode parasites of olive in Italy. *Informatore Fitopatologico*, 31, 117-179 (in Italian).

Inserra, R. N., Vovlas, N., Fontanazza, G., & La Casta, G. (1981). Performance of some olive cultivars infested with four nematode species. *Rivista di Ortoflorofrutticoltura Italiana*, 65, 143-148.

Javed, N., Gowen S. R., Inam-ul-Haq, M., Abdullah, K., & Shahina, F. (2007). Systemic and persistent effect of Neem (*Azadirachta indica*) formulations against root-knot nematodes, *Meloidogyne javanica* and their storage life. *Crop Protection*, 26, 911-916.

Jimenez, R. M. (1982). Phytoparasitic nematodes and olive growing. *In Primeras Jornadas Olivicolas Nazionales*. Arica, Chile.

Johnson, C., Brannon, D. R., & Kuc, J. (1973). Xanthotoxin: a phytoalexin of *Pastinaca sativa* root. *Phytochemistry*, 12, 2961-2962.

Katan, J., & De Vay, J. E. (1991). *Soil Solarization* CRC Press, Boca Raton, Florida.

Koliopanos, C. N., & Vovlas, N. (1977). Records of some plant parasitic nematodes in Greece with morphometrical descriptions. *Nematologia Mediterranea*, 5, 207-215.

Kostova, I., Ivanova, A., Mikhova, B., & Klaiber, I. (1999). Alkaloids and Coumarins from *Ruta graveolens*. *Monatshefte für Chemie*, 130, 703-707.

Laegdsmand, M., Gimsing, A. L., Strobel, B.W., Sorensen, J. C., Jacobsen, O. H., & Hansen, H. C. B. (2007). Leaching of isothiocyanates trough intact soil following simulated biofumigation. *Plant & Soil*, 291, 81-92.

Lamberti, F., & Lownsbery, B. F. (1968). Olive varieties differ in reaction to the roo-knot nematode *Meloidogyne javanica*. *Phytopathologia Mediterranea*, 7, 91-106.

Lamberti, F. (1969a). Presence in Italy of a decline of olive caused by the nematode *Pratylenchus vulnus*. *Phytopatologia Mediterranea*, 8, 232-234 (in Italian).

Lamberti, F. (1969b). The olive as host for *Xiphinema americanum* Cobb. *Phytopathologia Mediterranea*, 8, 230.

Lamberti, F., & Baines, R. C. (1969). Effect of *Pratylenchus vulnus* on the growth of Ascolana and Manzanillo olive trees in a glasshouse. *Plant Disease Reporter*, 53, 557-558.

Lamberti, F., & Baines, R. C. (1970). Infectivity of three biotypes of the citrus nematode (*Tylenchulus semipenetrans*). *Plant Disease Reporter*, 54, 717-718.

Lamberti, F., & Di Vito, M. (1972). Sanitation of the root-knot nematode infected olive stocks. In *Proceedings of the 3rd Congress of the Mediterranean Phytopathology Union*, Oeiras, Portugal: 401-411.

Lamberti, F., Bleve-Zacheo, T., & Martelli, G. P. (1975a). A case of intersexuality in *Xiphinema ingens* Luc & Dalmasso (Nematoda: Longidoridae. *Nematologia Mediterranea*, 3, 181-183. (in Italian).

Lamberti, F., Greco, N., & Zauchi, H. (1975b). A nematological survey of date palms and other major crops in Algeria. *FAO Plant Protection Bulletin*, 23, 156-160.

Lamberti, F., Vovlas, N., & Torre, A. (1976). Infectivity and pathogenicity of three Italian populations of *Tylenchulus semipenetrans* on *Citrus* and other hosts. *Meeting SOIF "Rootstocks of fruit tress"*, Pisa, Italy: 259-265 (in Italian).

Lamberti, F., Roca, F., Agostinelli, A., & Bleve-Zacheo, T. (1986). *Xiphinema barense* n. sp. (Nematoda: Dorylaimida) from Italy. *Nematologia Mediterranea*, 14, 101-106.

Lamberti, F., Roca, F. & Agostinelli A. (1989). *Xiphinema macroacanthum* (Nematoda: Dorylaimida) a new species from Southern Italy closely resembling *X. ingens* Luc & Dalmasso. *Nematologia Mediterranea*, 17, 115-119.

Lamberti, F., & Vovlas, N. (1993). Plant parasitic nematodes associated with olive. *Bulletin OEPP/EPPO Bulletin*, 23, 481-488.

Lamberti, F., Vouyoukalou, E. & Agostinelli, A. (1996). Longidorids (Nematoda: Dorylaimoidea) occurring in the rhizosphere of olive trees in Western Crete, Greece. *Nematologia Mediterranea*, 24, 79-85.

Lamberti, F., D'Addabbo, T., Sasanelli, N., & Carella, A. (2001a). Control of *Pratylenchus vulnus* in stone fruit nurseries. *Mededelingen van de Faculteit Landbouwwetenschappen / Rijksuniversiteit Gent*, 66/2b, 629-632.

Lamberti, F., Ciccarese, F., Sasanelli, N., Ambrico, A., D'Addabbo, T., & Schiavon, D. (2001b). Relationships between plant parasitic nematodes and *Verticillium dahliae* on olive. *Nematologia Mediterranea*, 29, 3-9.

Lazzeri, L., Tacconi, R., & Palmieri, S. (1993). *In vitro* activity of some glucosinolates and their reaction-products toward a population of the nematode *Heterodera schachtii*. *Journal of Agricultural and Food Chemistry*, 41, 825-829.

Lazzeri, L., Leoni, O., & Manici, L. M. (2002). Biocidal plant dried pellets for soil biofumigation. *Proceedings of International Congress & Trade Show Products*. The Floriade, NL.

Lehman, P. E. (1994). Dissemination of phytoparasitic nematodes. Nematology Circular N° 208. Florida Department of Agriculture and Consumer Services, Gainesville, FL, USA.

Lopez-Escudero, F. J., & Blanco-Lopez, M. A. (2001). Effect of a Single or Double Soil Solarization to Control Verticillium Wilt in Established Olive Orchards in Spain. *Plant Disease*, 85, 489-496.

Lownsbery, B. F., & Serr, E. F. (1963). Fruit and nut tree rootstocks as hosts for the root lesion nematode *Pratylenchus vulnus*. *Proceedings of the Society for Hoticultural Science*, 82, 250-254.

Macara, A. M. (1971). A importância agrícola dos nemátodos *Meloidogyne* spp. no espaço portugues. *Boletim Agronómico Nitratos de Portugal Agran*, 9, 3-15.

Manios, T. (2004). The composting potential of different organic solid wastes: experience from the island Crete. *Environment International*, 29, 1079-1089.

Martelli, G. P., Cohn, E., & Dalmasso, A. (1966). A redescription of *Xiphinema italiane* Meyl, 1953 and its relationship to *Xiphinema arenarium* Luc & Dalmasso, 1963 and *Xiphinema conurum* Siddiqi, 1964. *Nematologica*, 12, 183-194.

McLeod, R., Reay, F., & Smyth, J. (1994). Plant nematodes of Australia listed by plant and by genus. *NSW Agriculture and RIRDC*: 201.

Metha, U. K., & Raski, D. J. (1971). Revision of the genus *Criconema* Hofmanner & Menzel, 1914 and other related genera (Criconematidae: Nematoda). *Indian Journal of Nematology*, 1, 145-198.

Mian, I. H., & Rodriguez-Kabana, R. (1982). Organic amendments with high tannin and phenolic contents for control of *Meloidogye arenaria* in infested soil. *Nematropica*, 12, 221-234.

Milesi S., Massot B., Gontier F., Bourgaud F. & Guckert A. (2001). *Ruta graveolens* L.: a promising species for the production of furanocoumarins. *Plant Science*, 161, 189-199.

Minz, G. (1961). Additional hosts of the root-knot nematode, *Meloidogyne* spp. recorded in Israel during 1958-1959. *Israel Journal of Agricultural Research*, 11, 69-71.

Mojtahedi, H., Santo, G. S., Hang, A. N., & Wilson, J. H. (1991). Suppression of root-knot nematode populations with selected rapeseed cultivars as green manure. *Journal of Nematology*, 23, 170-174.

Monfort E., Lopez-Llorca, L. V., Jansson, H. B., & Salinas, J. (2006). *In vitro* soil receptivity assays to egg-parasitic nematophagous fungi. *Mycological Progress*, 5, 18-23.

Nico, A. I., Rapoport, H. F., Jiménez-Dìaz, R. M., & Castillo, P. (2002). Incidence and population density of plant-parasitic nematodes associated with olive planting stocks at nurseries in Southern Spain. *Plant Disease*, 86, 1075-1079.

Nico, A. I., Jimenez-Diaz, R. M., & Castillo, P. (2003). Solarization of soil in piles for the control of *Meloidogyne incognita* in olive nurseries in southern Spain. *Plant Pathology*, 52, 770-778.

Nico, A. I., Jiménez-Dìaz, R. M., & Castillo, P. (2004). Control of root-knot by composted agro-industrial wastes in potting mixtures. *Crop Protection*, 23, 581-587.

Nyczepir, A. P., & Halbrendt, J. M. (1993). Nematode pests of deciduous fruit and nut trees. In: Plant parasitic nematodes in temperate agriculture. (Eds. K. Evans, D. L.Trudgill and J.M. Webster). CAB International, University Press, Cambridge, UK: 381-425.

Ocio, J. A., Brookes, P. C., & Jenkinson, D. S. (1991). Field incorporation of straw and its effects on soil microbial biomass and soil inorganic N. *Soil Biology & Biochemistry*, 23, 171-176.

OEPP/EPPO (1993). Certification Schemes. N° 7. Nursery requirements - recommended requirements for establishments participating in certification of fruit or ornamental crops. *Bulletin OEPP/EPPO Bulletin*, 23, 513-516.

Omidvar, A. M. (1961). On the effects of root diffusates from *Tagetes* spp. on *Heterodera rostochiensis* Woll. *Nematologica*, 6, 123-129.

Omidvar, A. M. (1962). The nematicidal effects of *Tagetes* spp. on the final population of *Heterodera rostochiensis* Woll. *Nematologica*, 7, 62-64.

Peña-Santiago, R. (1990). Plant-parasitic nematodes associated with olive (*Olea europaea* L.) in the province of Jaen, Spain. *Revue de Nématologie*, 13, 113-115.

Percoco, A., & Amenduni, M. (2001). Biological control of corky root on tomato by composted olive pomace activated with biocontrol agents. *Atti Progetto POM, B-10*, 105 - 107

Perez, M. P., Navas-Cortes, J. A., Pascual-Villalobos, M. J., & Castillo, P. (2003). Nematicidal activity of essential oils and organic amendments from Asteraceae against root-knot nematodes. *Plant Pathology*, 52, 395-401.

Philis, J., & Siddiqi, M. R. (1976). A list of plant parasitic nematodes in Cyprus. *Nematologia Mediterranea*, 4, 171-174.

Piedra Buena, A., Garcia-Alvarez, A., Diez-Rojo, M. A., Ros, C., Fernandez, P., Lacasa, A., & Bello, A. (2007). Use of pepper crop residues for the control of root-knot nematodes. *Bioresource Technology*, 98, 2846-2851.

Ploeg, A. T. (1999). Greenhouse studies on the effect of marimgolds (*Tagetes* spp.) on four Meloidogyne species. *Journal of Nematology*, 31, 62-69.
Pryor, A. (1999). Results of 2 years of field trials using ozone gas as a soil treatment. *Annual International Research Conference on Methyl Bromyde Alternatives and Emission Reductions*. 1 – 4 November, San Diego, California (U.S.A.).
Qui, J. J., Westerdahl, B. B., Pryor, A., & Anderson, C. E. (2001). Reduction of root-knot nematode, *M. javanica*, in soil treated with ozone. *Phytopathology*, 91, (6)S, 141.
Rich J. R., Rahi G. S., Oppermann, C. H. & Davis, E. L. (1989). Influence of the castor bean (*Ricinus communis*) lectin (ricin) on motility of *Meloidogyne incognita*. *Nematropica*, 19, 99-103.
Ridolfi, M., Sasanelli, N., Patumi, M., D'Addabbo, T., Fontanazza, G., & Lamberti, F. (1998). Phenolic and peroxidase metabolism in olive trees attacked by root-knot nematodes (*Meloidogyne* spp.). *Italus Hortus*, 5(4), 22-26 (in Italian).
Ridolfi, M., Patumi, M., D'Addabbo, T., Sasanelli, N., & Lemos, R. J. (2001). Enzymatic response of olive varieties to parasitism by *Xiphinema index* (Nematoda: Longidoridae). *Russian Journal of Nematology*, 9, 25-32.
Roca, F., & Lamberti, F. (1988). *Xiphinema aequum* sp. n. (Nematoda: Dorylaimida) from Italy, with description of the male of *Longidorus eridanicus*. *Nematologia Mediterranea*, 16, 87-91.
Rodriguez-Kabana, R. (1986). Organic and inorganic amendments to soil as nematode suppressants. *Journal of Nematology*, 18, 129-135.
Romero, M., & Arias, M. (1969). Nematodes of Solanaceae in the Mediterranean zone of southern Spain. 1.Tylenchida. *Boletin de la Real Sociedad Española de Historia Natural (Biologia)*, 67, 121-142.
Saad, A. T., & Nienhaus, F. (1969). Plant diseases in Lebanon. *Zeitschrift für Pflanzenkrakheiten und Pflanzenschutz*, 76, 537-551.
Santiago, R. P., & Geraert, E. (1991). New data on *Aorolaimus perscitus* (Doucet, 1980) and *Gracilacus teres* Raski, 1976 (Nematoda: Tylenchida) associated with olive in the province of Jaén, Spain. *Nematologica*, 36, 408-416.
Santos, S. M. N. De A., & Abrantes, De O. I. M. (1980). Root nematodes in Portugal. In *Proceedings of the Second Research Planning Conference on Root-knot Nematodes*, *Meloidogyne* spp. Athens, Greece: 17-23.
Santos, S. M. N. De A. (1982). Studies on the root-knot nematodes, *Meloidogyne* spp. from olive trees in Portugal. *Nematologica*, 28, 169 (Abstract).
Saponari, M., Savino, V., & Stano, B. (2001). Phytopathological cards: Viruses and similar-virus agents. In: Olive protection. Interreg II – Italy-Albania. Introduction of technological innovations in the productive processes, 9, 31-38.
Sasanelli, N., & Di Vito, M. (1991). The effect of *Tagetes* spp. on the hatching of an Italian population of *Globodera rostochiensis*. *Nematologia Mediterranea*, 19, 135-137.
Sasanelli, N., & Catalano, L. (1991). *In vitro* nematicidal activity of *Capsicum annuum* on *Xiphinema index*. *Informatore fitopatologico*, 10, 55-56 (in Italian).
Sasanelli, N., (1992). Nematicidal activity of aqueous extracts from leaves of *Ruta graveolens* on *Xiphinema index*. *Nematologia Mediterranea*, 20: 53-55.
Sasanelli, N., & D'Addabbo, T. (1992). The effect of *Cineraria maritima*, *Ruta graveolens* and *Tagetes erecta* extracts on the hatching of *Heterodera schachtii*. *Nematologia Mediterranea*, 20, 49-51.
Sasanelli, N., & D'Addabbo, T. (1993a). Effect of *Cineraria maritima*, *Ruta graveolens* and *Tagetes erecta* leaf and root extracts on Italian populations of *Meloidogyne* species. *Nematologia Mediterranea*, 21, 21-25.
Sasanelli, N., & D'Addabbo, T. (1993b). Potential application of the leaves of *Ruta graveolens* for controlling *Meloidogyne javanica* on sunflower. *Russian Journal of Nematology*, 1, 117-120.
Sasanelli, N., D'Addabbo, T., Basile, M., & Carella, A. (1996). Effect of cadusafos on the reproduction of *Meloidogyne incognita*. *Afro Asian Journal of Nematology*, 6, 36-39.
Sasanelli, N., Fontanazza, G., Lamberti, F., D'Addabbo, T., Patumi, M., & Vergari, G. (1997). Reaction of olive cultivars to *Meloidogyne* species. *Nematologia Mediterranea*, 25, 183-190.
Sasanelli, N., & D'Addabbo, T. (1999). Modelling of the *in vitro* effect of cadusafos on *Meloidogyne incognita*. *Nematologia Mediterranea*, 27, 193-201.
Sasanelli, N., Coiro, M. I., D'Addabbo, T., Lemos, R. J., Ridolfi, M., & Lamberti, F. (1999). Reaction of an olive cultivar and an olive rootstock to *Xiphinema index*. *Nematologia Mediterranea*, 27, 253-256.
Sasanelli, N., & Greco, N. (2000). Formulation of a model to relate nematode populations with exposure times to a range of temperatures. *Proceedings of the fifth International Symposium on Chemical and*

Non-Chemical Soil and Substrate Disinfestation. (Eds. M. L. Gullino, J. Katan and A. Matta). *Acta Horticulturae*, 532, 131-135.

Sasanelli, N., D'Addabbo, T., Dell'Orco, P., & Mencuccini, M. (2000). The *in vitro* use of olive explants in screening trials for resistance to the root-knot nematode, *Meloidogyne incognita. Nematropica* 30, 101-106.

Sasanelli, N., & D'Addabbo, T. (2002). Reaction of olive to *Pratylenchus vulnus* infections in Italy. *Nematology*, 4, 259 (Abstract).

Sasanelli, N., D'Addabbo, T., & Lemos, R. M. (2002a). Influence of *Meloidogyne javanica* growth of olive cuttings in pots. *Nematropica*, 32, 59-63.

Sasanelli, N., D'Addabbo, T., Dell'Orco, P., & Mencuccini, M. (2002b). The effect of different rooting media in *in vitro* screening trials for resistance to *Meloidogyne incognita*. Proc. 4th IS on Olive Growing. (Eds. C. Vitagliano & G. P. Martelli). *Acta Horticulurae*, 586, 845-848.

Sasanelli, N., D'Addabbo, T., Convertini, G., & Ferri, D. (2002c). Soil Phytoparasitic Nematodes Suppression and Changes of Chemical Properties Determined by Waste Residues from Olive Oil Extraction. *Proceedings of 12th ISCO Conference*, May 26-31, 2002 Beijing China. Vol. III: 588-592.

Sasanelli N., D'Addabbo, T., & Greco, P. (2003a). Nematicidal activity of *Eruca sativa* and *Ruta graveolens* on the root-knot nematode *Meloidogyne incognita.* Proceedings *XXXV Congress SIA*; N° 23; Naples, 16-18 september, 361-362 (in Italian).

Sasanelli, N., Greco, P., D'Addabbo, T., Coiro, M. I. &. Lamberti, F. (2003b). The use of olive mill wastes for the control of root-knot nematodes. 55th International Symposium on Crop Protection. May 6, Ghent Belgium. *Communications in agricultural and applied biological sciences, Ghent University*, 68(4a), 135-138 (Abstract).

Sasanelli, N., Ciccarese, F., Ambrico, A., Longo, O., Schiavone, D., & Ziadi, T. (2006). Control of root-knot nematode *Meloidogyne incognita* by *Aphanocladium album*, a new promising biocontrol agent. *Proceedings of the 12th Congress of the Mediterranean Phytopathological Union*, 11-15 June. Rhodes, Greece: 540-542.

Sasanelli, N., Attila, A., D'Addabbo, T., & Takacs, T. (2007). Nematicidal properties of leaf extracts of *Ruta graveolens* inoculated with arbuscular mycorrhizal fungi. *Russian Journal of Nematology*, 15, 65-73.

Sasanelli, N., Ciccarese, F., & Papajova, I. (2008). *Aphanocladium album* by via sub-irrigation in the control of *Pyrenochaeta lycopersici* and *Meloidogyne incognita* on tomato in a plastic-house. *Helminthologia*, 45, 137-142.

Scognamiglio, A., Talamè, M., & Giandomenico, N. (1968). Data on nematodes living in the rhizosphere of olive (1st paper). *Bollettino del Laboratorio di Entomologia Agraria Filippo Silvestri*, 26, 205-226.

Scognamiglio, A., Talamè, M., & D'Errico, F.P. (1971). Data on nematodes living in the rizosphere of olive (2nd paper). *Bollettino del Laboratorio di Entomologia Agraria Filippo Silvestri*, 29, 43-59.

Seinhorst, J. W. (1965). The relationship between nematode density and damage to plants. *Nematologica*, 11, 137-154.

Seinhorst, J. W. (1979). Nematodes and growth of plants: formulation of the nematode-plant system. In: *Root-knot nematodes* (Meloidogyne species) *Systematics, Biology and Control* (Eds F. Lamberti, & C. E. Taylor). Academic Press, London: 231-256.

Sequi, P., De Nobili, M., Leita, L., & Cercignani, G. (1986). A new index of humification. *Agrochimica*, 30, 175-179.

Serr, E. F., & Day, L. H. (1949). Lesion nematode to California fruit and nut trees, and comparative tolerance of various species of Juglandaceae. *Proceedings of the American Society for Horticultural Science*, 53, 134-140.

Sethi, C. L., Ganz, M. S., Kauslial, K. K., Srivastava, A. N., & Khan, E. (1988). Occurence of root-knot nematodes of fruit plants in association with *Agrobacterium tumefaciens*. *International Nematology Network Newsletter*, 5, 12-13.

Sharon, E., Bar-Eyal, M., Chet, I., Herrera-Estrella, A., Kleifeld, O., & Spiegel, Y. (2001). Biological control of the root-knot nematode *Meloidogyne javanica* by *Trichoderma harzianum. Phytopathology*, 91, 687-693.

Sher, S. A. (1966). Revision of the Hoplolaiminae (Nematoda). VI. *Helicotylenchus* Steiner, 1945. *Nematologica*, 12, 1-56.

Siddiqi, M. R. (1976). New plant nematode genera *Plesiodorus* (Dolichodorinae), *Meiodorus* (Meiodorinae subfam. N.), *Amplimerlinius* (Merliniinae) and *Gracilacea* (Tylodoridae grad. N.). *Nematologica*, 22, 390-416.

Sikora, R. A. (1992). Management of the antagonistic potential in agricultural ecosystems for the biological control of plant parasitic nematodes. *Annual Review of Phytopathology*, 30, 245-270.

Smith, B. J., & Kirkegaard, J. A. (2002). *In vitro* inhibition of soil microorganisms by 2-phenylethyl isothiocyanante. *Plant Pathology*, 51, 585-593.

Spiegel, Y., Chet, I., & Cohn, E. (1987). Use of chitin for controlling plant parasitic nematodes. II. Mode of action. *Plant Soil*, 98, 337-345.

Stapleton, J .J., Ferguson, L., McKenry, M. V., Dougherty, D. S., & Stapleton, S. C. (1999). Using solarization to disinfest soil for olive nursery production. *Acta Horticulturae*, 474, 589-594.

Stirling, G. R. (1991). Biological Control of Plant Parasitic Nematodes. Progress, Problems and Prospects. CAB International. Wallingford, UK: 282 pp.

Tarjan, A. C. (1953). Geographical distribution of some *Meloidogyne* species in Israel. *Plant Disease Reporter*, 37, 315-316.

Tarjan, A. C. (1964). Plant parasitic nematodes in thr United Arab Republic. *FAO Plant Protection Bulletin*, 12, 49-56.

Tjamos, E. C., Biris, D. A., & Paplomatas, E. J. (1991). Recovery of olive trees with verticillium wilt after individual application of soil solarization in established olive orchards. *Plant Disease*, 75, 557-562.

Tobar-Jimenez, A. (1964). *Ditylenchus virtudesae* n. sp. (Nematoda: Tylenchida), an inhabitant of Granada soils. *Revista Iberica de Parasitologia*, 24, 51-56.

Triolo, E., Materazzi, A., & Luvisi, A. (2004). Exothermic reactions and steam for the manegement of soil-borne pathogens: five years of research. *Advances Horticultural Science*, 18 (2), 89-94.

Tzortzakakis, E. A., Peneva, V., Terzakis, M., Neilson, R., & Brown, D. J. F. (2001). *Longidorus cretensis* n. sp. (Nematoda: Longidoridae) from a vineyard infected with a foliar "yellow mosaic" on Crete, greece. *Systematic Parasitology*, 48, 131-139.

Vlachopoulos, E. (1991). Nematode species in nurseries of Greece. *Annales de l'Institut Phytopathologique Benaki*, 16, 115-122.

Vovlas, N., & Lamberti, F. (1974). New hosts of *Rotylenchulus macrodoratus* in the Mediterranean region. *Nematologia Mediterranea*, 2, 177-179 (in Italian).

Vovlas, N., & Inserra, R. N. (1976). Histopathology of roots infested with *Rotylenchulus macrodoratus*. *Nematologia Mediterranea*, 4, 223-230 (in Italian).

Vovlas, N., & Inserra, R. N. (1981). Parasitic habits of *Ogma rhombosquamatum* and description of the male. *Journal of Nematology*, 13, 87-90.

Vovlas, N., & Inserra, R. N. (1981). Notes on *Helicotylenchus dihystera* on olive in Sicily. *Informatore Fitopatologico*, 31, 23-25 (in Italian).

Vovlas, N. (1982). *Macroposthonia sicula* n.sp. (Nematoda: Criconematidae), a parasite of olive trees in Sicily. *Journal of Nematology*, 14, 95-99.

Vovlas, N., & Inserra, R. N. (1983). Biology of *Heterodera mediterranea*. *Journal of Nematology*, 15, 571-576.

Waele, D. De, Mancini, G., Roca, F., & Lamberti, F. (1982). *Trichodorus taylori* n. sp. (Nematoda: Dorylaimidae) from Italy. *Nematologia Mediterranea*, 10, 27-37.

Woods, S. R., Haydock, P. P. J., & Edmunds, C. (1999). Mode of action of fosthiazate used for the control of potato cyst nematode *Globodera pallida*. *Annals of applied biology*, 135, 409-415.

Yaniv, Z., Kenneth, R. G., & Miura, J. (1979). Teliospore formation in *Puccinia graminis* f. sp. *tritici* grown in axenic culture, induced by the fungus *Aphanocladium album*. *Physiological Plant Pathology*, 14, 153-156.

Yong, B. J., & Zhong, X. W. (1980). The identification of root-knot nematodes in *Olea europaea*. *Scientia Silvae Sinicae*, 16, 264-265.

Zangerl, A. (1990). Furanocoumarin induction in wild parsnip: evidence for an induced defence against herbivores. *Ecology*, 71, 1926-1932.

Zobel, A. M., & Brown, S. A. (1989). Histological localization of furanocoumarins in *Ruta graveolens*. *Canadian Journal of Botany*, 67, 915-921.

INDEX

A
AA genome, 5
AAA genome, 5
AAA-group triploids, 33
AAB genome, 5
AAB-group triploids, 33
AB genome, 5
ABB genome, 5
aberrans-group, 260
Abiotic constraints, 23, 29, 30
Abiotic factors, 14
Abundance, 9, 23, 26, 31, 52, 56
Acanthocereus pentagonus, 107
Accessions, 37
Acerola, 63, 64
Acetic acid, 98
Acrocarpus fraxinifolius, 109
Africa, 3, 5, 8–10, 12, 13, 15, 17, 20, 22–29, 34–36, 39, 41, 43–45, 47, 48, 51, 52, 55, 57–60, 160, 167, 168, 218, 238, 249, 251, 275
Agar, 294, 295, 298
Aglenchus agricola, 277
Agrobacterium tumefaciens, 41, 186, 193
Agroecosystem, 105, 112
Agro-forest system, 105
Agronomic management, 243
Agronomic practices, 64
AILV, 217, 218
Albania, 218
Albizia, 122
Aldicarb, 129, 201, 210
Algeria, 218
Allele, 198, 199
Allotrichodorus, 218
Almería, 227
Almond, 177, 220, 250
Alphonse Lavalle, 73
Alternate cropping, 3, 28
Altitude, 105, 108, 114
Amanohashidate, 270
Amarillo, 66, 107
America, 8, 9, 11–13, 15–17, 20, 28–36, 39, 43, 45, 49, 51, 53, 55, 121, 122
American grape species, 204
Americani, 39
AMF, 43
Amines, 300
Aminoacid sequence, 224
Ammonia, 98, 300, 301
Ampelidaceae, 204
Ampelopsis, 204
Ampelopsis aconitifolia, 204
Amphimixis, 196
Amplimerlinius amplus, 277
Amplimerlinius macrurus, 277
Amplimerlinius paraglobigerus, 277
AMV, 219, 221
Anacardiaceae, 244
Anacardium occidentale, 66
Ananas comosus, 79, 81, 107
Anatolian region, 247
Anchorage, 22, 23
Andaman Island, 19
Andong, 258
Anhui province, 258
Animal manure, 21, 301
Annona, 64, 106
Annona muricata, 109
Annual crops, 18, 125
Anoxia, 90
Anoxic soil, 94
Antagonist, 234, 237
Antagonistic organisms, 130
Anthurium, 20
Antibiotics, 301
Antibodies, 222, 223, 224
Antigens, 261
Antracnosis, 111
Ants, 111

Anus, 138
Aorolaimus holdeman, 68, 79
Aorolaimus levicaudatus, 74
Aphanocladium album, 298, 308, 314, 315
Aphelenchoides, 111, 277
Aphelenchus, 65, 74, 76, 79, 80
Aphelenchus avenae, 277
Aphids, 111
Apple, 220
Apricot, 177, 178, 186, 220, 250
Apulia, 219
Aquatic fungi, 234
Arabis mosaic nepovirus, 200, 216, 288
Arabis mosaic virus 210, 211
Araceae, 35
Aragua State, 64–74, 76, 79, 81
Arecanut, 18–20, 35
Arganil, 259
Argentina, 70, 178, 218, 275
Aristophanes, 275
Arizona, 139, 170
Armenia, 218, 275
ArMV, 200, 288, 294
Aroma, 105
Arthrobotrys oligospora, 188
Arthrobotrys, 42, 188
Artichoke italian latent nepovirus, 217
Artocarpus altilis, 66
Ascolana, 283, 294, 296, 297, 311
Asia, 3–6, 8–10, 16–19, 20–23, 28, 30, 31, 34–36, 39, 45, 48–50, 52, 54–56, 58–60, 119, 121, 122, 132, 134, 218, 243, 251, 253–255, 258, 259, 262, 275
Asiatic pyroid citrus nematode, 158
Asparagus officinalis, 233, 236
Asphyxia, 87, 90
Atta cephalotes, 111
Atta fervens, 111
Atta mexicana, 111
Australasia, 149
Australia, 5, 9, 10, 12, 15, 17, 18, 21, 22, 28, 35, 36, 39, 48, 58, 157, 159, 160, 168, 171, 188, 191, 192, 193, 207, 211, 218, 275, 279, 280, 312
Australimusa section, 4, 18
Austria, 218
Auxines, 92
Avocado, 20, 63, 64, 65, 106
Azadirachta indica, 130, 304, 308, 311
Azadirachtin, 301, 304
Azerbaijan, 218, 244

B

Bacillus mycoides, 236
Bacillus thuringiensis, 42
Backyard culture, 204
Backyard production system, 18
Bacteria, 234–236, 239, 241, 276, 300, 301, 304
Bacterial Canker Complex, 180
Bacterial Canker, 180
Bacterial endospores, 234
Bacterium, 141
Badnavirus, 40
Baermann funnel, 146
Bahia grass, 182, 192
Bahia, 124, 131, 133, 134
Balearic Islands, 218
Baluchistan, 246, 248, 252
Bamboo, 66
Banana, 3, 4, 6–61, 63–66, 149, 161
Banana breeding scheme, 36
Banana cultivation, 7
Banana hybrids, 3, 36
Banana nematodes, 3, 15, 17, 18, 34–36, 42
Banana production, 7, 13, 17, 21, 23, 25–27, 31, 43, 48, 51
Banana propagation, 38
Banana roots, 9, 13, 28, 31
Banana streak disease, 40
Banane Cochon, 38
Bangka, 9, 35
Bangladesh, 19, 54
Bare-fallow, 16
Barinas state, 65
Basamid, 128, 129, 131

BCC, 180
Bean, 18
Beauveria, 268
Beauveria bassiana, 107, 109
Belgium, 254
Belize, 9, 16, 55, 56
Bell pepper, 113
Belonolaimus longicaudatus, 157, 158, 160, 161, 166, 170
Bergaptene, 303
Bermuda grass, 182
Betel vine, 20
BHC, 266
Big Ebanga, 38
Biochemical factors, 11
Biocontrol, 42, 43, 127, 130, 131, 235, 239, 298, 314
Biocontrol agent, 42, 43, 298, 314
Biocontrol strategy, 298
Biodiversity, 8
Biofumigation, 177, 183, 193, 215, 232, 300, 311
Biological control, 3, 7, 34, 41, 101, 130, 188, 215, 237, 243, 248, 250
Biological control agent, 21, 98, 188, 191, 215, 294
Biological control measures, 267
Biological management, 237
Biological suppression, 234
Biotypes, 68
Bixa orellana, 130
Black head disease, 8
Black head toppling disease, 13
Black leaf streak, 23, 31
Black pepper, 20
Black pine, 269
Black root rot, 111
Black Sea region, 220
Black Sigatoka, 17, 23, 30, 31, 53
Black weevils, 27, 56
Blackcurrant, 220
Bluggoe, 5, 30
Bodles Altafort, 36, 54
Bolivia, 122, 123, 125, 131
borealis-group, 260
Börner, 204, 208
Bourbon, 103, 106, 107, 108
Brassicaceae, 300
Brassica juncea, 233, 300
Brassica nigra, 300
Brassica oleracea, 300
Brazil, 5, 8, 9, 30, 32, 33, 36, 38, 48, 60, 66, 75, 80, 122–125, 130, 131, 133, 134, 154, 157, 161, 166, 170, 218, 254
Breadfruit, 63, 66
Breadfruit tree, 66
Breeding, 17, 37, 39, 40, 41, 50, 55, 56, 195, 197, 202, 204, 208–213
Breeding incompatibilities, 37
Breeding programs, 37, 40, 131, 142
Breeding strategies, 196
Brewing banana, 4, 7
Brewing-cooking banana, 5
Brooming, 79
Brown lesion, 73, 281
Bulbous crops, 217
Bulgaria, 215, 218, 219, 220, 221, 228, 240
Bunch, 4, 13, 14, 22, 26, 37, 39
Bunch grapes, 204, 209
Burrowing, 3, 7, 9–11, 13–15, 17–19, 21–23, 25, 26, 31, 33–35, 39, 45–47, 49–52, 54–56, 58–61
Burrowing nematode, 3, 7, 9, 15, 19, 22, 34, 35, 37, 136
Bursaphelenchus, 254, 260, 270–272
Bursaphelenchus antoniae, 259
Bursaphelenchus baujardi, 260
Bursaphelenchus cocophilus, 71
Bursaphelenchus conicaudatus, 260
Bursaphelenchus doui, 260
Bursaphelenchus fraudulentus, 260
Bursaphelenchus kolymensis, 260
Bursaphelenchus luxuriosae, 260
Bursaphelenchus mucronatus, 260, 261, 272, 274
Bursaphelenchus singaporensis, 260
Bursaphelenchus xylophilus, 253, 254, 256, 257, 260–262, 271–274
Butane, 98
Butanoic acid, 98

C

C1613, 227
Cabernet Sauvignon, 204, 205, 208, 229, 230, 231, 237
Cabernet Sauvignon-157, 229
Cacambou, 5
Cacao, 119–134
Cacao germplasm, 127
Cacao-growing countries, 120
Cacao production, 127
Cacao propagation, 121
Cacao seedling, 130
Cacao tissue culture, 121
Cadusafos, 307, 313
Calcareous soil, 141, 147
Calcium, 245
Calendula, 304
California, 136, 140, 143, 145, 156, 158, 159, 161, 165, 168, 170–172, 180, 186, 191, 192, 195, 197–200, 203–205, 208, 210–213, 243, 246, 249, 251, 252, 275, 283, 305, 306, 308, 313, 314
Calliandra, 122
Callose, 40
Calocarpum mammosum, 106, 109
Cameroon, 10, 23–28, 37, 39, 47, 55, 120
Canada, 253, 256, 262, 271
Canary Islands, 8, 10, 14, 25, 30, 38, 39, 55, 218
Canavalia ensiformis, 28
Cap Verde, 30
Capsicum annum, 109, 303, 313
Capsicum pubescens, 109
Carabobo, 65, 67, 68, 70, 72, 75, 77, 79, 81
CARBAP, 27, 29, 39, 44
Carbofuran, 128, 307
Carbohydrate, 140, 143, 151, 163, 164, 165
Carbohydrate transfer, 140
Carbon, 302
Cardamon, 35
Cardinal, 73, 74, 82
Caribbean, 5, 6, 8, 9, 11, 13, 15–17, 20, 30–33, 35, 36, 39, 40, 45, 51, 53
Carica papaya, 76, 106, 109, 130
Carrizo, 68
Carrot, 305, 306
Caryological characteristics, 204
Cash crops, 113
Cashew, 63, 64, 66
Cassava, 16, 33
Castor, 226
Catch crops, 232
Catenaria anguillulae, 234, 236
Catimor, 103, 108
Catuai, 112, 115
Catual, 103, 108, 110
Caturra, 103, 104, 106, 107, 108, 110, 112, 115
Cavendish, 3, 6–10, 11, 14, 18, 19, 21, 23, 25, 30, 38, 39, 41, 46
Cavendish banana, 7, 8, 18, 30
Cavendish plantations, 10
Cavendish subgroup, 11
Cedrella mexicana, 109
Cedrus, 256
Cell, 11, 12, 22, 32, 40, 89
Cell hyperplasia, 89
Cellina di Nardò, 282, 284, 285
Cellular wall, 94
Cell wall, 40
Central Africa, 5, 6, 122, 123
Central America, 6, 8, 9, 11, 13, 15, 16, 22, 23, 28–32, 49, 53, 54, 55, 57, 119, 218
Central cylinder, 74, 80
Cerambycid beetle, 253
Ceratocystis, 257
Cercospora coffeicola, 110, 111
Chamaedorea elegans, 109, 118
Chamaedorea tepejilote, 109
Chamomilla recutita, 233
Chemical control, 3, 16, 45, 185, 187, 226, 304
Chemical destruction, 17
Chemical formulation, 34

Chemical fumigants, 232
Chemical management, 144, 153
Chemical measure, 266
Chemical products, 93, 94, 95, 98
Chemical treatments, 85, 92
Chemicals, 15, 21, 33, 41, 42, 124, 128, 131, 267
Chemoreceptors, 88
Chemotaxis, 40
Chenopodium ambrosioides, 233
Cheravirus, 216
Cherries, 177, 186
Cherry raspberry leaf nepovirus, 216
Cherry rosette nepovirus, 217
Chestnut, 220, 221
Chiapas, 101–103, 105, 107, 108, 110, 111, 115–118
Chicomuselo, 115, 116, 118
Chile, 85, 87, 88, 94, 98, 99, 178, 218, 237, 275, 278, 311
China, 8, 17, 19, 21, 34, 52, 158, 178, 243, 244, 253, 254, 258, 261, 262, 268, 270, 271, 274, 278, 314
Chinantecos, 106
Chitin, 301, 314
Chitinases, 94
Chlorella, 228, 229, 230, 238
Chlorella vulgaris, 228, 229, 230, 238
Chloropicrin, 184, 305
Chlorosis, 35, 123, 126, 204, 205
Chlorotic plants, 157
Choles, 106
Cholesterol, 245
Chrysanthenum, 304
Cima di Bitonto, 282, 284, 285
Cineraria maritima, 303, 313
Citrange Carrizo, 67, 69, 70
Citrons, 135
Citrumelo Swingle, 67, 69, 70
Citrus, 9, 11, 12, 20, 32, 34, 35, 51, 58, 63, 64, 66–71, 74, 81, 89, 90, 97, 98, 107, 135–173, 220
Citrus amblicarpa, 70
Citrus aurantifolia, 159
Citrus aurantium, 70, 159
Citrus biotype, 68, 143
Citrus-growing countries, 34
Citrus latifolia, 109
Citrus limon, 159
Citrus nematode, 136
Citrus paradisi, 67, 142, 159
Citrus reshnii, 67, 69, 70, 72
Citrus reticulata, 70, 109, 142, 159
Citrus rhizosphere, 136
Citrus roots, 142, 158, 159, 165
Citrus sinensis, 67, 68, 109, 142, 159
Citrus volkameriana, 67, 69, 70
Citrus, 66, 68, 69, 135, 283
Clay, 64
Cleopatra mandarin, 155, 165
Climate, 8, 14, 101, 102, 114, 183, 263
Clones, 4, 22, 27, 30, 39
CLRV, 221
Coat protein gene, 206, 209, 210, 211, 212, 213
Cobb, 8, 9, 18, 32, 34, 46–48, 50–52, 54, 55, 57–61
Cobb technique, 221
Coccus, 110
Coccus hesperidum, 111
Coccus viridis, 111
Cocoa, 18, 23, 35, 119, 131, 132, 133, 134
Cocoa genotypes, 121
Cocoa market, 119
Cocoa production, 120, 132
Cocoa Swollen Shoot Virus, 124
Coconut, 18–20, 35, 63, 64, 71–73
Coconut trees, 72
Cocos nucifera, 71
Coffee, 9, 18–20, 23, 30, 32–34, 52, 101, 110
Coffea arabica, 102–105, 113–118
Coffea canephora, 103, 106, 113–115, 117
Coffee leaf borer, 111
Coffee plantations, 103, 104, 108, 111, 117
Coffee plants, 106, 108, 109, 110, 111, 113, 115, 116

Coffee production, 101
Coffee seedlings, 108
Colima, 102
Colletotrichum coffeanum, 111
Colombia, 5, 8, 30, 31, 33, 36, 188
Commercial banana plantations, 3
Commercial plantains, 33
Commercial polyculture, 101
Compost, 294, 295, 302, 303
Conformation, 39
Congo, 122, 123
Conifer forests, 253, 254
Coniferous forests, 254, 271
Control, 3, 7, 13, 15–17, 21, 27–29, 33, 34, 38, 41–49, 53–58, 60, 253, 254, 258, 263–271
Control measures, 265, 266, 270
Control strategies, 101, 225
Cooking bananas, 5, 6, 30
Coratina, 284, 285
Corchosis, 112, 117, 118
Córdoba, 101, 105
Corm, 4, 12, 16, 26, 27, 31, 48
Corn, 4, 18, 33, 143, 158, 221, 226
Cortex, 11, 12, 32, 68, 80, 123, 124, 137, 138, 139, 151, 152
Cortical cells, 160, 199
Cortical parenchyma, 11, 32, 75
Cortical tissue, 155
Corticium koleroga, 107, 111
Corticium salmonicolor, 111
Coslenchus cancellatus, 277
Coslenchus costatus, 277
Coslenchus lateralis, 277
Cosmos bipinnatus, 233
Costa Rica, 8, 12, 14–17, 31, 38, 39, 46, 49, 51, 55
Côte d'Ivoire, 120, 122, 124, 125
Crambe abyssinica, 300
Crete, 10, 25, 60, 218, 241
Criconema, 247, 277
Criconema demani, 68, 79
Criconema mutabile, 247
Criconemella, 65, 111, 115
Criconemella goodeyi, 125
Criconemoides, 74, 76, 178, 192, 193, 277
Criconemoides curvata, 180
Criconemoides informis, 277
Criconemoides onoense, 68, 74, 125
Criconemoides ornatum, 68, 79, 80
Criconemoides siculum, 279
Criconemoides sphaerocephala, 68, 74
Criconemoides xenoplax, 178, 180, 182, 183, 186–190, 195, 207, 277
Criolla Negra, 74
Criollo amarillo, 79
CRLV, 216
Croatia, 218
Crop, 3, 7, 9, 12, 14, 16–18, 20, 26, 28, 29, 34, 45, 52, 54, 57, 58, 63, 64, 66, 67, 71, 73, 76–78, 80, 88, 140, 143, 144, 163, 181, 177, 182, 301, 312
Crop loss, 140, 143, 144, 163, 181
Crop production, 64
Crop repetition, 88
Crop residues, 301, 312
Crop rotation, 177, 182
Crop systems, 7, 16
Cropping system, 4, 8, 21, 44, 45, 59, 101, 105
Crotalaria, 158, 233
Cruciferous species, 300
Crustacean shells, 301
CRV, 217
Cryptobiosis, 12
CSSV, 124
Cuba, 30, 31, 32, 35, 39, 42, 49, 58
Cultivar, 4, 5, 8, 14, 18, 19, 21, 23, 26, 30, 33, 36–38, 40, 41, 45, 48–51, 53–55, 57–60, 77, 79, 82, 127, 177
Cultural practices, 3, 16, 34, 44, 45, 136, 153, 154, 171
Cutting, 121, 253, 269
Cyanogenic compounds, 183
Cylindrocladium, 112
Cynodon dactylon, 182
Cyprus, 14, 218, 244

Cyst nematode, 21, 276
Cystatin, 41, 46, 59
Cystein proteinase, 41
Cytoplasm structure, 89
Cytoplasm, 89
Czech Republic, 218, 239

D

1,3-D, 184, 226, 305
DA12I, 282, 284–287, 296
Dactylellina dactyloides, 188
Dactylellina ellipsospora, 188
Daegu, 258
Dagger nematode, 160, 177, 180, 195
Daily variation, 63
Damage, 3, 7, 9, 10, 11, 13, 14, 19, 21, 22, 26, 28, 29, 32, 34, 35, 37, 38, 43, 45–47, 53, 54, 57, 58, 63, 64, 66, 69, 73, 76, 80, 122, 123, 128, 129, 135, 136, 138, 141, 143, 144, 147, 150, 151, 153, 155–157, 159, 160, 162, 177–181, 184, 185, 187, 189, 192
Damage symptoms, 119, 123
Damaged roots, 160
Danube, 220
Dastarcus longulus, 267, 268
Date palm, 20, 63, 73
Dazomet, 184, 185, 227, 305
DBCP, 15, 51, 54, 128, 129, 148, 160, 161, 169–172, 201
D-D, 15, 227
DDT, 266
Decay, 22
Defense response, 94
Defoliation, 283
Deformations, 10, 12
Delaware, 226
Density, 135, 140, 141, 143, 144, 145, 146, 152, 160, 166, 167
Dessert, 3–5, 7–10, 12–15, 17, 18, 21, 23, 25, 28, 30–34, 36, 40, 44
Dessert bananas, 3–6, 33, 44
Detection, 19, 20, 215, 222–225, 238, 239, 241, 242
Development, 12, 18, 22, 27, 29, 34, 37, 38, 43–45, 50, 53, 55, 58
Diagnostic character, 260
Diagnostic procedures, 222, 225
1,2-dibromo-3-chloropropane, 15
Dibromochloropropane, 148
Dichloropropane, 15, 227
1,3-dichloropropene, 148, 178, 184, 191, 201, 226, 238, 305
Dichloropropene, 15, 171, 227
Dieback, 123, 126
Dieback conditions, 123
Diet, 4, 244
Digestion, 41
2,3-dihydro-2,2dimethylbenzofuran-7-methylcarbammate, 307
3,5 dimethyl 1,3,5-thiodiaziname-2-thione, 305
Diospyros, 68
Diploid, 4, 37, 38, 52, 36– 40
Discocriconemella limitanea, 125
Disease, 7, 22, 23, 29, 30, 36, 39, 40, 43, 46, 47, 58, 60, 66, 71, 72, 73, 79, 82, 87, 90, 91, 94, 119, 123, 124, 130, 131, 136, 142, 149, 150, 151, 160, 164, 165, 171
Disease complexes, 119, 124
Disease-free suckers, 66
Disease resistance, 36
Dispersal, 4, 7
Dissemination, 9, 12, 25, 27, 30, 31, 34, 35, 45, 54
Distribution, 3, 4, 8, 9, 10, 23, 25, 26, 28, 30, 34, 36, 44–47, 52, 54, 63, 64, 76, 215, 217–219, 237, 238, 240
Ditylenchus, 76, 79, 80, 277
Ditylenchus acutus, 79
Ditylenchus anchiliposomus, 277
Ditylenchus virtudeasae, 277
DNA, 223, 224, 239, 241
DNA fragments, 261
DNA hybridization probes, 261
DNA sequence, 261
DNA sequencing, 224

Dog Ridge, 198, 199, 202, 203, 207, 227
Dolichodorus heterocephalus, 277, 309
Dolichodorus minor, 124, 125
Dominant genes, 199
Dominican Republic, 30, 32, 33, 57
Donghae, 258
Dorylaimida, 215, 238, 240, 241
Dosage, 34, 46
Dose, 92, 93
Douglas fir, 256
Drainages, 64
Drip fumigation, 305
Drip irrigation, 188
Drought, 101, 102, 103, 105
Drupe, 177
Durability, 199
Durian trees, 18
Dwarf Cavendish, 5, 39
Dwarfism, 39

E

EAHB, 26, 39
East Africa, 5, 6, 10, 22, 23, 25
East African Highland bananas, 26
Eastern Asia, 17
Economic importance, 135, 143, 149, 153, 154, 157
Economy, 18
Ectoparasitic nematodes, 136, 215
Ecuador, 8, 30, 39, 63
Ecuadorus, 218
EDB, 15, 267
Egg, 11, 123, 257, 265
Egg hatching, 187, 299
Egg laying, 11
Egg mass, 12, 68, 69, 78, 123, 137, 138, 141, 159, 179
Egg stage, 92
Egypt, 8, 10, 25, 138, 170, 244, 251
eidmanni-group, 260
El Jarillo, 73
Electron microscopy, 222, 241
ELISA, 201, 203, 210, 222, 239
Elytra, 257
Emamectin benzoate, 267
Embrapa, 39
Emergence, 11
Endogenous, 11, 50
Endophytic fungi, 42, 43, 49, 55
Ensete, 10, 54, 58
Entomoparasitic fungi, 268
Entomopathogenic nematodes, 268
Environmental conditions, 7, 9, 21, 36
Environmental factors, 13
Environmental quality, 16
Enzyme, 94, 301
Epidermis, 12, 199
EPPO, 216, 222
Equator, 119
Equipment, 146, 147, 148, 153
Eradication, 253, 259, 269
Eriobotrya japonica, 109
Erythrina, 122
Esfahan, 246, 251
Etanoic acid, 98
Ethane, 98
Ethiopia, 10, 54
Ethoprop, 128
Ethoprophos, 129
Ethylene dibromide, 15
EU, 253, 254, 255, 259, 262, 263, 270
Eumusa, 4, 58
Eumusa section, 4
Europe, 6, 8, 36, 180, 187, 216, 217, 218, 241, 254, 259, 260, 263, 270, 272, 273, 275
European Union, 253, 254, 263, 271
Euvitis, 199, 203, 204, 208, 209
Exclusion, 146
Excretory pore, 138
Exogenous, 11
Export, 3, 4, 6–8, 12, 14, 15, 17, 18, 21–23, 30, 31, 33, 34, 44, 45, 66

F

F1 hybrids, 195, 204, 205
Faba bean, 20
Falcon State, 72, 76

Fallow, 16, 28, 182, 226
Fallow periods, 29
False Horn, 5, 23, 30, 33
Fanleaf degeneration, 205, 211, 213
Far East, 253
Farmers, 18, 22, 27, 28, 29, 36, 39, 41, 44
Farms, 131
Fars province, 246
Federal Republic of Yugoslavia, 218
Feeder roots, 68, 69
Feeding, 216, 220, 222, 226, 228
Feeding site, 86, 89, 137, 196, 179
Fenamiphos, 128, 129, 227, 234, 303, 307
Fenitrothion, 266
Fensulfothion, 128, 129
Fenthion, 266
Fermentation, 98
Fertiliser, 3, 28, 29, 57, 107–109, 305
Fertilization, 137
Fertilizing, 18, 86
FHIA, 36, 37, 38, 39, 44, 46, 56
FHIA-01, 37, 39
Fibrous roots, 136–138, 140, 144, 145, 149, 150, 151, 157, 158, 163
Ficus carica, 73
Field experiments, 33, 37, 39, 56
Field survey, 265
Fig, 63, 64, 68, 69, 70, 72, 73, 75–77, 286
Figue Pomme, 5, 30
Figue sucrée, 5, 30
Fiji, 9, 18, 20, 34, 47, 52
Filenchus, 247
Filenchus filiformis, 277
Filenchus sandneri, 277
Filenchus thornei, 277
First World War, 119
Flavonols, 40, 41
Flavour, 105
Flood, 16
Florida, 9, 11, 12, 20, 32, 34, 35, 47–49, 52, 53, 60, 135, 139, 141, 142, 144, 146–151, 153–157, 160, 161, 163–173, 178, 191
Flowering, 13
Flowers, 113, 177
Folate, 245
FONCOPAL, 73
Food, 4, 35, 43, 45
Food legume, 20
Forest disease, 263, 264
Formulations, 300, 309, 311
Fortunella, 135
Fosthiazate, 307, 308, 315
Fosthietan, 129
Fougamou, 5
France, 178, 187, 191, 195, 197, 203, 204, 205, 206, 208–213, 218, 219, 240
Frantoio, 285
Freedom, 198, 199, 200, 203, 206, 207
French Guiana, 31, 32, 48
French Horn, 23, 30, 33
French plantain, 5
French West Indies, 9, 10, 17, 57
Fruit, 13, 22, 45, 63, 64, 66, 73, 76, 77, 80, 82
Fruit characteristics, 36
Fruit tree crops, 64
Fruit trees, 87, 91, 92
FS17, 282, 284–287
Fujian, 19
Fumigant, 15, 16, 148, 153, 161, 177, 184, 185, 187, 304
Fumigant nematicides, 177
Fumigant residues, 153
Fumigation, 91, 92, 148, 158, 160, 167, 169, 170, 177, 178, 184, 190, 192
Fumigator, 305
Fungal diseases, 101, 106
Fungi, 148, 152, 168, 276, 297, 298, 300, 301, 304, 305, 310, 314
Fungicides, 108
Fungus, 107
Fungus inoculations, 289
Furanocoumarins, 303, 312, 315
Fusarium, 8, 43, 46, 47, 50, 52, 54, 58, 60

Fusarium oxysporum f. sp. coffeae, 112
Fusarium oxysporum f. sp. cubense, 8
Fusarium oxysporum, 11, 43, 75, 83, 112, 142, 152
Fusarium solani, 142, 152, 164, 167, 172

G

Gall, 12, 38, 66, 74–77, 79, 158, 159, 179, 196, 221, 281, 294, 296, 297, 305
Gall index, 74
Galling, 137
Gambia, 23
Gamma-aminobutyric acid, 267
Gangneung, 258
Garnica, 103, 104, 106, 107, 108, 110, 114–116, 118
Gas exchange, 64
Gelatinous matrix, 179
Genetic diversity, 245, 248, 251
Genetic resistance, 66
Genome, 5
Geocenamus, 247
Geocenamus rugosus, 247
Geocenamus brevidens, 247
Geographic origin, 39
Germany, 178, 188, 218, 223
Germplasm, 36, 38, 41, 46, 47, 49, 57, 59
GFLV, 195, 200–206, 209, 212, 215, 216, 219, 222, 223, 225, 227
GFLV B particles, 223
GFLV particles, 222
Ghana, 24–26, 45, 46, 120, 122–125
Giant Cavendish, 5
Giant cell, 74, 89, 123, 159, 196, 200
Ginger, 18–20, 34, 35
Gliocladium, 257
Gliricidia, 122
Globodera pallida, 41, 59
Globodera rostochiensis, 302, 303, 307, 308, 313
Glomus, 185
Glomus mosseae, 130, 131
Glucosinolate, 232, 300
Glyphosate, 201
Golden apple, 228
Goldfinger, 37
Gonytrichum, 112
Gracilacus, 275–277, 281, 310, 313
Gracilacus aculenta, 68
Gracilacus peratica, 276, 277, 285, 310
Gracilacus teres, 277, 285
Grafted plant, 113, 114, 115
Grafting, 121, 269, 296
Grande Naine, 14, 39
Grape, 68, 74, 200, 208, 209, 210, 211, 212, 213
Grapefruit, 67, 135, 160, 162, 164, 170, 171
Grapevine, 63, 73, 82, 195, 200–202, 206–213, 215, 216, 218, 219, 221, 223, 227–230, 233, 238, 241, 242, 286
Grapevine fanleaf nepovirus, 216
Grapevine Fanleaf Virus, 195, 208, 210
Grass, 221
Greece, 178, 217, 218, 219, 239–241, 243, 244, 275, 277, 278, 279, 280, 286, 301, 310, 311, 313–315
Greenhouse, 10, 14, 26, 39, 43, 59
Greenhouse studies, 226
Green-red, 5
Grenada, 10, 32
Gros Michel, 5, 8, 9, 11, 14, 26, 30, 31, 33, 36, 38, 40, 52, 53
Guam, 20
Guangdong province, 258
Guárico, 74
Guatemala, 102
Guava, 63, 64, 74, 75, 80
Guerrero, 102, 105, 110
Guinea, 17–20, 37, 47
Guineo, 33
Gulf of Mexico, 101
Gulf slope, 105
Gumi, 258
Gyeongbuk, 258

H

Hainan, 19
Halogenated hydrocarbons, 184
Handling, 15, 21
Hardiness, 39
Hardwood cuttings, 206
Harmony, 198, 199, 203, 207, 227
Harvest, 13, 29, 44, 63, 64
Harvesting, 102
Haryana, 218
Hatching, 11, 40, 43, 138, 146, 151
Hawaii, 9, 20, 35, 57
Heat shock protein genes, 261
Heat treatment, 12, 15
Heat-treated seeds, 16
Hedera, 220
Helicotylenchus, 64, 65, 66, 73, 74, 76, 77, 79–81, 111, 112, 115, 178, 275, 277, 278, 281, 314, 315
Helichotylenchus cavenessi, 125
Helicotylenchus crenacauda, 68
Helicotylenchus digonicus, 247, 248, 277
Helicotylenchus dihystera, 19, 68, 64, 66, 74, 124, 125, 131, 277, 281
Helicotylenchus erythrinae, 68, 125, 277, 281
Helicotylenchus multicinctus, 3, 7, 9, 14, 19, 21, 23, 24, 26, 31–33, 36, 38, 45–47, 51, 53, 54, 56, 60, 65, 66, 68
Helicotylenchus neopaxilli, 277
Helicotylenchus oleae, 277, 281
Helicotylenchus pseudorobustus, 277
Helicotylenchus tunisiensis, 278
Helicotylenchus vulgaris, 278
Hemicriconemoides, 64, 68, 76, 80, 81, 82
Hemicriconemoides cocophilus, 125
Hemicriconemoides communis, 68
Hemicriconemoides mangiferae, 64
Hemicriconemoides strictathecatus, 64, 66, 68, 76, 80
Hemicycliophora, 111, 159, 168, 172, 278
Hemicycliophora arenaria, 159
Hemicycliophora loofi, 125
Hemicycliophora nudata, 159
Hemicycliophora oostenbrinki, 125
Hemicycliophora paradox, 125
Hemileia vastatrix, 107, 111
Herbaceous plants, 138
Herbicides, 108
Heterodera, 125, 132, 247, 248, 250, 251, 252, 275, 276, 278, 281, 308, 309, 311, 312, 313, 315
Heterodera marioni, 247, 248
Heterodera mediterranea, 247, 276, 278, 288, 308, 315
Heterodera oryzicola, 10, 19, 21, 47
Heterodera schachtii, 41, 303
Hevea brasiliensis, 109
Hibiscus sabdariffa, 109
Highgate, 5
Highland bananas, 10, 23, 25, 26, 28, 29, 33, 39
Hirsutella rhossiliensis, 188, 191, 234, 238
Histochemical studies, 40
Histopathology, 283, 315
Home garden bananas, 25
Home gardens, 21
Homologation procedures, 43
Honduras, 10, 12, 15, 16, 17, 31, 32, 36, 39, 46, 54, 55
Honey bee, 268
Hoplolaimus, 178, 278
Hoplolaimus aorolaimoides, 278
Hoplolaimus galeatus, 125
Hoplolaimus pararobustus, 10, 24–26, 125
Hoplolaimus seinhorsti, 10, 64–66, 68, 74, 80
Hormones, 92
Horn plantain, 5
Horticultural characteristics, 39
Host, 7, 11, 12, 14, 27, 29, 33, 35, 38, 43, 45, 46, 50, 51, 52, 57, 58, 59
Household refuse, 21
hsp70, 261
HT-5203, 226
Hubei province, 258

Humic index, 302
Humid forest, 22
Humidity, 87, 90–92, 94–96, 120, 302
Hungary, 218, 237
hunti-group, 260
Hybrid, 4, 27, 36, 39, 40, 44, 45, 49, 52, 53, 59
Hybrid VR 039-16, 205
Hybrid VR 043-43, 205
Hybridization, 186
Hydrogen sulphide, 301
Hydrolysis, 300
Hyogo, 264
Hyperplasia, 123, 151, 157, 159
Hypersensitive resistance reactions, 199
Hypersensitive response, 159
Hypertrophy, 123
Hyphae, 298
Hyphomycetes, 234
Hypothenemus hampei, 107, 109, 110

I

Idiarthron subquadratum, 111
IITA, 27, 29, 39, 44, 47, 56, 57
India, 4, 5, 8, 9, 17, 19, 20–22, 34, 35, 36, 40, 52, 58, 120, 122, 124, 125, 132, 143, 154, 170, 218, 278
Indian subcontinent, 158
Indonesia, 5, 8, 9, 18–20, 34, 35, 36, 38, 50, 59, 120, 130, 132, 149
Indonesian peninsula, 22
Infection, 40, 48, 137, 141, 142, 148, 152, 157, 159, 161, 163, 164, 168
Infectivity, 299
Inga, 106, 109
Injectors, 15
INMECAFE, 108, 118
Inoculation, 14, 26, 43, 49, 202, 203, 204
Insecticide, 128, 253, 265, 266, 267
Insects pests, 64
Insect vector, 253, 254, 255, 257, 258, 262, 263, 274
Inspection, 73, 253, 270
Instituto Mexicano del Café, 103, 104, 108, 118
Integrated management, 119, 189
Integrated Nematode Management, 112
Integrated Pest Management, 127, 177
Intercropping, 215, 225, 227, 233
Intercrops, 28
International Plant Genetic Resources Institute, 8
Intracropping, 101
Introduction, 254, 255, 258, 259, 262, 263
Intundu, 5
Iodomethane, 184
IPGRI, 8, 57
IPM, 7, 27, 29, 33, 34, 57, 58, 112, 113, 127, 130, 178, 189, 190
IPM strategies, 27, 29, 33
IPPC, 34, 35
Iran, 218, 243, 244, 245, 246, 249–252, 275
Iraq, 218, 244, 275
Iron, 245
Irrigation, 69, 71, 72, 85, 87, 90, 91, 93–96, 98, 135, 147, 182–184, 187, 192, 305–307
Irrigation practices, 85
Irrigation systems, 145, 149
Irrigation water, 226, 241
Isocyanates, 300
Isopimpinellin, 303
Isothiocyanates, 300, 310, 311
Israel, 9, 14, 38, 138, 144, 160, 162, 165, 167, 168, 218, 278, 280, 312, 315
Italia, 73, 74, 82
Italy, 178, 187, 191, 215, 217–219, 227, 234, 241, 243, 244, 248, 254, 275, 277–280, 283, 286, 298, 299, 301, 308–311, 313–315
ITS-RFLP, 261, 271
Ivory Coast, 12, 16, 17, 23–26, 36, 39, 45, 49, 55, 56
Ivy, 220, 221

J
Jalisco, 102
Jamaica, 8, 13, 17, 20, 31, 33, 36, 38, 46, 51, 54
Japan, 154, 173, 253, 254, 258, 262, 264, 265, 266, 269, 270, 271, 272, 273
Japanese black pine, 258, 265, 269
Japanese pine forests, 269
Japanese red pine, 258, 269
Jarillazo, 79
Jasmonic acid, 94
Jiangsu Province, 258
Jordan, 275, 276, 277, 278, 279, 280, 288, 308, 310
Juglans, 109
Juvenile stages, 11

K
Kaleghochi, 246
Kashmir, 244
Kava, 20
Kenya, 24, 25, 50
Kerala district, 19
Kerman province, 245, 246, 251
Kernel, 244
Khorasan province, 245, 246
Kisubi, 5
Korea, 253, 258, 262, 263, 269, 272, 273, 274
Kuharski Carrizo citrange, 154
Kumquat, 135
Kunnan, 5
Kyoto University, 253, 269
Kyoto, 253, 269, 270, 271, 273
Kyrgyzstan, 244
Kyushu island, 264

L
Laboratory, 14, 39, 42
Lady's finger, 5
Lagenidium, 234
Lakatan, 5
Lake Maracaibo, 65
Laknao, 5
Land, 7, 15, 23, 28, 29
Land, clearing, 23
Lannate, 129
Lara State, 64, 73, 75, 77, 79, 81
La Réunion, 10, 60
Latin America, 5, 11, 15, 30, 33, 34, 119
Leaves, 14, 33
Lebanon, 10, 57, 278, 313
Leccino, 282, 284, 285, 288, 289, 291, 292, 295
Legislation, 34
Leguminous, 66
Lemon, 67, 135
Lesion, 283, 294
Lesion nematode, 3, 7, 9, 10, 12, 13, 17, 25, 26, 34, 37, 45, 54, 59, 136, 195, 206
Leucaena, 122
Leucoptera coffeella, 107, 111
Levamisol hydrochloride, 267
Libya, 278, 309
Lignification, 40
Lignin, 41, 94
Lilium longiflorum, 306
Lime, 135
Linoleic acid, 244
Litchi chinensis, 109
Longevity, 13, 17, 29, 45, 77
Longidoridae, 215, 217, 238, 239, 240, 241, 242
Longidorid nematodes, 216, 238, 242
Longidorus, 124, 125, 200, 209, 217, 218, 220–222, 232, 234, 238–242
Longidorus africanus, 247, 278
Longidorus apulus, 217, 218
Longidorus artemisiae, 217, 241
Longidorus arthensis, 217
Longidorus attenuatus, 200, 223
Longidorus balticus, 217
Longidorus carpathicus, 217
Longidorus closelongatus, 278
Longidorus cretensis, 217, 278
Longidorus cylindricapitatus, 217
Longidorus dalmassoi, 217
Longidorus danuvii, 217
Longidorus distinctus, 231

Longidorus elongatus, 200, 217, 220, 221, 223, 227, 240, 241
Longidorus fagi, 217
Longidorus fasciatus, 217, 219
Longidorus helveticus, 223
Longidorus juglandicola, 217
Longidorus macrosoma, 217, 221, 223
Longidorus macrosoma, 278
Longidorus piceicola, 217
Longidorus pisi, 220, 221
Longidorus profundorum, 223
Longidorus siddiqii, 278
Longidorus sturhani, 217, 223
Louisiana, 35
Lousã, 259
Lowlands, 8, 23
Lycopersicum esculentum, 109

M

Maça, 5
Macadamia, 107
Macadamia integrifolia, 109
Macadamia tetraphylla, 109
Macrophomina phaseolina, 75, 83
Madagascar, 10, 22, 46, 60, 243, 244
Magnesium, 245
Maize, 66
Malawi, 122, 123, 125
Malaysia, 5, 8, 10, 19, 20, 35, 37, 45, 49, 56, 60, 130, 131, 132, 133, 134
Malpighia glabra, 64
Malta, 218
Management, 3, 4, 7, 8, 11, 12, 15, 16, 17, 21, 22, 26–29, 33, 34, 40, 43–46, 48, 49, 55, 56–61, 63, 64, 69, 73, 74, 79, 85, 101, 106–109, 111–113, 115, 118, 127, 128, 130, 135, 137, 143, 145, 147, 149, 153, 160, 166, 167, 171, 177, 178, 181, 182, 189, 190, 192, 193, 215, 222, 225, 226, 233, 235, 237, 239, 248, 250, 293, 302, 304, 307, 308, 310
Management practices, 7, 18
Management strategies, 3, 28
Management tactics, 45
Mandarin, 67, 135
Manganese, 245
Mangifera indica, 75, 107, 109
Mango, 63, 64, 75
Manure, 301, 302, 309, 312
Maracay, 63, 68, 81, 82, 83
Maragogipe, 103, 107, 108
Marantaceae, 35
Marcus Tullius Cicero, 275
Margarita Island, 73
Maricongo, 33
Marigold, 226
Maringold oil, 304
Maritime pine, 255, 259, 273
Maritza, 220
Market, 4, 8, 18, 30
Martinique, 3, 12, 16, 30, 31, 39, 42, 47, 48, 52, 56
Mastic tree, 244
Matavia, 5
Maturation time, 13
Mauritius,243
Mazatecos, 106
Mbire, 5
Mediterranean Basin, 200
Mediterranean biotype, 68, 143
Meiotic parthenogenesis, 196, 201
Meloidogyne, 3, 7, 10, 12, 19, 21, 23–26, 31, 33, 36, 38, 42, 45–48, 51–57, 60, 64, 65, 68, 73–77, 79–83, 98, 99, 107, 110, 111, 114, 115, 117, 122, 123, 125, 131–134, 141, 158, 161, 162, 165, 166, 168–170, 172, 173, 178, 179, 182, 185, 186, 190–193, 195, 196, 198, 199, 207–211, 213, 214, 275, 276, 278, 279, 281–283, 292–294, 303, 307–309, 311–315
Melidogyne arenaria, 19, 32, 42, 65, 75, 122, 123, 130, 158, 178, 186, 196, 197, 198, 199, 214, 278, 303
Meloidogyne acrita, 122, 199, 278
Meloidogyne arabicida, 112
Meloidogyne baetica, 278, 282
Meloidogyne exigua, 68, 122, 123, 125
Meloidogyne floridensis, 178

Meloidogyne fujianensis, 158
Meloidogyne graminicola, 19
Meloidogyne hapla, 178, 196, 197, 207, 278, 303
Meloidogyne hispanica, 178
Meloidogyne incognita, 14, 21, 32, 38, 41, 64–66, 68, 74, 75, 76, 77, 79, 80–83, 112, 115, 122, 123, 125, 126, 158, 178, 186, 187, 189, 196, 197, 198, 199, 214, 246, 247, 249, 278, 281, 282, 284, 288–294, 296, 297, 299, 300, 301, 303, 305
Meloidogyne javanica, 19, 32, 65, 74, 76, 77, 79, 122, 123, 125, 130, 158, 161, 178, 186, 189, 192, 196–198, 214, 246–250, 252, 278, 281, 282, 284, 294–296, 298, 303, 305, 313
Meloidogyne lusitanica, 279, 282, 308
Meloidogyne mayaguensis, 64, 81
Meloidogyne morocciensis, 178
Meloidogyne oteifae, 158
Meloidogyne thamesi, 122, 125
Melon, 113
Mérida, 65, 77
Meristem, 16, 38, 44
Merlinius, 247
Merlinius brevidens, 279
Mesulfenfos, 267
Metabolism, 89
Metam potassium, 184
Metam sodium, 148, 184, 305
Metanoic acid, 98
Methane, 98
Methomyl, 227
Methyl bromide, 112, 128, 132, 148, 153, 178, 184, 185, 193, 201, 212, 226, 237, 267
Methyl iodide, 185
Methyl isothiocyanate, 184, 185
Methyl phenphos, 267
Mexican Institute of Coffee, 104
Mexican Republic, 102
Mexico, 8, 101, 102, 103, 104, 105, 106, 108, 109, 111, 112, 114, 115, 116, 117, 243, 256, 275
Michoacán, 101, 102
Microarthropods, 301
Microbial antagonist, 3
Microbial biodegradation, 44
Microbial degradation, 148
Microcitrus hybrid, 155
Microfauna, 301
Micronutrient, 14, 33, 67, 68
Microorganism, 42, 43, 130, 265
Microsclerotia, 300
Microsprinkler, 135
Middle East, 6, 143, 160, 219, 243, 244
Migration, 11
Migratory endoparasites, 7, 42, 136
Milam lemon, 152, 154
Milbemectine, 267
Minirhizotron, 85, 95, 97
Miranda, 67, 70, 73, 79
MIT, 184, 185
Mites, 64
Mitotic parthenogenesis, 196
Mixed-cropping production system, 18
Mixes, 106
Mixtecos, 106
Modelling, 234
Modelling studies, 39
Moisture, 135, 141, 144, 145, 151, 161, 172
Mokpo, 258
Moldova, 218
Molecular beacon, 223, 224
Molecular detection tools, 215
Molecular detection, 222
Molecular probe, 224
Monacrosporium ellipsosporum, 188
Monagas, 67, 68, 70, 76, 83
Monitoring programme, 15
Monochamus alternatus, 253, 257, 265, 266, 271, 272

Monochamus beetles, 257, 266, 267, 269
Monochamus galloprovincialis, 255, 259, 273
Monocultivation, 108, 110
Monoculture, 8, 18
Mononchids, 128
Monotrichodorus, 218
Monotrichodorus monohystera, 64, 66, 68
Montreal Protocol, 112
Moraiolo, 296, 297
Morantel tartarate, 267
Morelos, 103
Morocco, 10, 51, 178, 193
Morphological identification, 260
Morphological variations, 11
Moult, 216
Mountain System, 105
Mtp 3146-1-87, 205, 206
Mucuna pruriens, 28
Muhlenbergia schreberi, 189
Mujuba, 5
Mulberry, 220
Mulched plants, 29
Mulching, 3, 28, 48, 53
Multiplication rate of nematodes, 39
Mundo Novo, 103, 104, 106, 107, 108
Musa, 3, 4, 7, 8, 11, 14, 17–19, 21, 22, 23, 26, 30–33, 36–41, 44–60, 107, 121
Musa AA, 30, 36
Musa AAA, 11, 23, 30, 32, 65, 81
Musa AAB, 19, 30, 32, 33, 46, 65, 65, 82
Musa ABB, 23, 30, 32
Musa acuminata, 4, 109
Musa balbisiana, 4, 40
Musa genome AA, 18
Musa genome AAA, 18
Musa genome AAB, 18
Musa genome AB, 18
Musa genome ABB, 18, 31
Musa group AAA, 24
Musa group AAB, 24
Musa group ABB, 24
Musa group EAHB, 24
Musa group mixed, 24
Musa varieties, 4
Muscadinia, 195, 197, 202, 204, 205, 206, 208, 209, 213
Muscadinia rotundifolia, 195, 197, 198, 204–206, 210, 227
Muscle activity blockers, 267
Mycena citricolor, 106, 107, 111
Mycoleptodiscus terrestris, 124
Mycorrhizae, 42
Mycorrhizal infection, 130
Mycosphaerella fijiensis, 23
Myrobalan plum, 187
Myrosinase, 232
Mysore, 5

N

Nahuas, 106
Na methilditiocarbammate, 305
Natural hybridization, 4
Nayarit, 102, 105, 108, 110, 118
NCS, 267
Necrosis, 11, 13, 15, 23, 26, 35, 37, 43, 52, 79, 137, 141, 160, 281, 288
Nemadectin, 267
Nemaguard, 186
Nematicidal activity, 303, 304, 313
Nematicidal plants, 215, 232, 235, 239, 303
Nematicide, 3, 14–17, 21, 26, 28, 33, 34, 42, 44–46, 48, 51, 54, 59, 66, 75, 79, 87, 91–95, 101, 119, 124, 127, 128, 129, 142–145, 148, 149, 153, 154,158, 162, 168, 170, 171, 173, 178, 185, 187–189, 192, 226, 227, 253, 267, 305, 307, 308
Nematicide treatments, 14, 26
Nematode, 3, 4, 7–19, 21–61, 63–67, 69–71, 73–80, 85–93, 95, 96, 98, 99, 101, 110, 111, 114, 119, 122–124, 126–173, 177–202, 206–209, 212, 213, 215, 216, 218, 219, 221–228, 230, 231, 233–243,

246–255, 257, 258, 261, 263, 266, 267, 269–277, 281–283, 285, 286, 288, 289, 292–315
Nematode analysis, 95
Nematode attack, 86, 179
Nematode control method, 12
Nematode control, 124
Nematode countings, 15
Nematode counts, 16
Nematode density, 143, 144, 145
Nematode management strategies, 85
Nematode management, 143, 147
Nematode populations, 69, 74, 178, 181, 182, 183, 188, 189, 190
Nematode reproduction, 88
Nematode species, 7, 12, 13, 19, 23
Nematode survival, 7
Nematode transmitting viruses, 86
Nematode vectors, 215–219, 225, 237, 239, 240, 241
Nematology, 7, 47
Nematophagous fungi, 297, 312
Nemfix, 234
Neodiplogaster tropica, 125
Neolobocriconema olearum, 279
Neopsilenchus magnidens, 279
Nepovirus, 216, 180, 191
Nested-PCR, 269
New Caledonia, 35
New Delhi, 158
New South Wales, 160
New Zealand, 218
Ney poovan, 5
Niacin, 245
Nigeria, 23–27, 29, 36, 39, 47–49, 57, 58, 119, 120, 122–125, 128–133
Nitrile, 300
Nitrite, 301
Nitrogen, 302
Non-fumigant nematicide, 15, 185, 304, 307
Non-quarantine pests, 225
North Africa, 6, 243
North America, 6, 8, 143, 155, 160, 180, 216, 218, 253, 254, 255, 256, 258, 260, 275
North Carolina, 197, 211, 213
North Yemen, 10
Nothocriconema princeps, 279
Nshakara, 5
Nuccellus, 121
Nuclear genome, 40
Nucleotide polymorphisms, 223, 224
Nueva Esparta, 68, 73
Nurse cells, 89, 137, 138, 139
Nursery, 29, 33, 36, 43, 68, 76, 77, 119, 131, 136, 146, 147, 151, 153, 155, 156, 160, 167, 170, 288, 293, 294, 299, 304, 305, 306, 307, 311, 312, 315
Nutrient deficiencies, 64
Nutrients, 136, 139, 140, 144, 154
Nyoya, 5

O

Oaxaca, 102, 105, 107, 108, 110, 117
Oceania, 3, 8, 17, 18, 21, 22, 45, 218
OcI deltaD86, 41
Ocimum gratissimum, 130
Oesophagus, 216
Ogma, 275, 279, 281, 315
Ogma civellae, 279
Ogma rhombosquamatum, 279, 288, 315
Ohady, 246
Oilseed cakes, 301
Ojo de gallo, 106
Olea europaea L. subsp. *europaea*, 275
Olea europaea, 68
Oleic acid, 244
Oligonychus coffeae, 110, 111
Olive, 68, 136, 143, 250, 252, 275, 281, 283, 291, 294, 296, 300, 308, 309, 311–314
Olive mill wastes, 303
Olive pomace, 302, 303, 309, 312

Olive roots, 275, 276, 281, 285, 288, 299, 308, 309
Oman, 138, 154, 167
Orange, 64, 66, 135, 170
Organic amendments, 101, 119, 124, 128, 129, 130, 131, 134, 215, 228, 237
Organic coffee, 103, 104, 105, 108, 115
Organic management, 234
Organic matter, 98, 227, 230
Organic molecules, 98
Organic mulches, 29
Oriental States, 66, 68
Ornamental plants, 177
Ornamental species, 107
Ornamentals, 9, 10, 35, 46
Orography, 63
Osmotic pressure, 140
Otomis, 106
Ovipositing beetles, 257
Oxalis rosea, 233
Oxamyl, 129, 227, 307
Oxazolidine2-ethione, 300
Oxygen, 63, 87, 94
Ozonated water, 306
Ozone, 113, 305, 306, 309, 310
Ozone generator, 306

P
Pacamar, 103
Pacific, 4, 5, 20, 22, 34, 45, 47–50, 52, 54–56, 58–60, 101, 102, 105, 106, 110
Pacific Slope, 102
Paecilomyces lilacinus, 42, 58, 141, 170, 188, 297
Pakistan, 218, 243, 244, 248, 251
PAL, 287
Pamir, 275
Panama, 20
Panama disease, 8, 11, 54
Pangola grass, 16, 33
Pantothenic acid, 245
Papaya, 63, 64, 76–78, 80, 81
Papua New Guinea, 35
Paraguanera, 76, 77, 78
Paralongidorus, 124
Paralongidorus iberis, 218, 239
Paralongidorus litoralis, 247, 248, 252
Paralongidorus maximus, 220, 221, 223
Paralongidorus monegrensis, 218, 239
Paraphelenchus pseudoparietinus, 279
Parasitic fungi, 42
Parasitism, 3, 37
Parasitoid, 267
Paratrichodorus, 124, 160, 171, 216, 217, 218, 240, 242
Paratrichodorus christiei, 125
Paratrichodorus lobatus, 160
Paratrichodorus minor, 68, 160, 233, 279
Paratrichodorus pachydermus, 217, 235
Paratrichodorus porosus, 160
Paratrichodorus teres, 279
Paratrophurus loofi, 279
Paratylenchus, 68, 74, 76, 77, 79, 80, 178, 246, 247, 279
Paratylenchus arculatus, 125, 279
Paratylenchus baldacii, 279
Paratylenchus elachistus, 68
Paratylenchus hamatus, 246, 247
Paratylenchus minutus, 68
Paratylenchus nawadus, 79
Paratylenchus projectus, 247
Paratylenchus vandenbrandei, 279
Parenchyma, 11
Parenchyma cells, 12, 123
Paring, 12, 15, 26, 27
Parthenocissus, 204
Paspalum notatum, 182
Passiflora, 77, 82, 83
Passiflora edulis f. sp. flavicarpa, 77, 78, 83
Passiflora edulis, 77
Passiflora quadrangularis, 78
Passionfruit, 63, 77

Pasteuria, 141, 164, 170, 171, 234, 236, 238
Pasteuria penetrans, 42, 188, 243, 249, 251
Pathogen, 77, 86, 87, 90, 94, 181
Pathogenic fungi, 298, 305
Pathogenicity, 7, 16, 26, 33, 49, 51, 54, 57, 142, 149, 155, 160, 162, 167, 170, 172, 253, 266
PCR, 200, 201, 210, 213, 214, 222, 223, 224, 238, 239, 242, 261, 262, 271, 273
PCR-based diagnostics, 261
PCR amplification, 261
Pea early-browning tobravirus, 217
Peach, 63, 78, 79, 177, 178, 186, 192, 199, 209, 213, 216, 220, 221, 227, 234
Peach nematode, 79
Peach rosette mosaic nepovirus, 200, 216
Peach Tree Short Life, 180
PEBV, 217
Pelipita, 5, 30
Pellets, 300, 311
Peltamigratus, 111
Peltamigratus holdemani, 125
Peltamigratus macbethi, 79
Pendolino, 288–292
Pepper, 9, 19, 35, 199, 210, 220, 221
Pepper ringspot tobravirus, 217
PepRSV, 217
Perennial crop, 124, 125, 188
Perennials, 4, 18, 140, 143, 199
Permanent farming system, 22
Permanent plantain crop, 28
Perolera, 79
Persea americana, 65, 109
Persimmon, 20, 68
Peru, 218, 275
Pest avoidance, 190
Pest, 3, 4, 7, 10, 13, 19, 21–23, 25–27, 29, 30, 34–36, 39, 43, 45, 47, 48, 50, 51, 54, 55, 58, 60, 190, 225
Pesteh, 243
Pest management, 7, 48
Pest risk analyses, 225
Pesticide, 7, 29, 34, 36, 39, 127, 128
Pesticide application, 27, 29
Peyan, 5
Phaseolus vulgaris, 109
Phenol, 40, 286, 287, 301
Phenological stages, 69
Phenology, 71
Phenotypic differences, 39
Phenylalanine ammonia lyase, 140
Phenylpropanoid compounds, 40
Phenyphenalenone, 40
Phialophora, 112
Philippines, 5, 8, 9, 10, 14, 16, 18, 19, 21, 35, 37, 38, 42, 45, 48, 50, 52, 54–56, 58, 59, 60
Phloem, 151
Phoenix dactylifera, 73
Phoma costarricensis , 111
Phosphorganics, 307
Phosphorus, 230, 245
Phylloxera, 196, 197, 204, 205, 208, 210
Physalis ixocarpa, 109
Phytoalexins, 40
Phytoparasitic nematodes, 86, 90, 92, 94, 95
Phytophthora, 137, 142, 147, 163, 164, 165, 234, 246
Phytophthora cinnamomi, 65, 79
Phytophthora citrophthora, 142, 163
Phytophthora megakarya, 124
Phytophthora nicotianae, 137, 142, 147, 164, 165
Phytophthora nicotianae, 142, 147
Phytosanitary legislation, 217
Phytosanitary risk, 255
Phytotoxic activity, 305
Phytotoxic effects, 234
Phytotoxicity, 185, 15, 184, 304, 306
Picea, 256
Pimienta dioica, 107, 109
Pine, 220
Pineapple, 11, 16, 35, 41, 48, 63, 64, 79

Pine stand, 268, 269
Pine tissues, 267
Pine trees, 258, 259, 260, 265, 266, 267, 268, 269, 271, 273, 274
Pine wilt disease, 253, 263, 269, 270, 271, 272, 273
Pine wilt nematode, 253
piniperdae-group, 260
Pink blight, 111
Pinus, 253, 255, 256, 258, 268, 269, 273
Pinus densiflora, 253, 258, 269
Pinus luchuensis, 258
Pinus pinaster, 253, 255, 258
Pinus thunbergii, 253, 258, 265, 269
Pisang awak, 26, 57
Pisang berengan, 5
Pisang jari buaya, 36–38, 40
Pisang kelat, 5
Pisang lilin, 5
Pisang mas, 5
Pisang masak hijan, 5
Pisang nangka, 5
Pisang raja, 5
Pistachio, 243–252
Pistacia atlantica, 244, 246, 248, 249
Pistacia cabulica, 244
Pistacia chinensis, 244
Pistacia falcata, 244
Pistacia integerrima, 244, 246, 249
Pistacia khinjuk, 244, 245, 249
Pistacia kurdica, 244
Pistacia lentiscus, 244, 248
Pistacia mutica, 244, 245, 249
Pistacia palaestina, 244, 249
Pistacia terebinthus, 244, 246, 248
Pistacia vera, 243–252
Pit observation, 95
Pithecolobium, 122
Plagiohammus maculosos, 111, 107
Plagiohammus spinipensis, 111
Planococcus citri, 110, 111
Plant, 63, 65, 66, 67, 69, 71, 72, 73, 74, 76, 77, 78, 79
Plantain hybrids, 30
Plantain, 10, 25, 26, 30–33, 38, 47, 49, 50, 51, 53, 54, 55, 56, 63, 64, 65, 66, 121
Plantation longevity, 11, 23, 26, 32, 48
Plantation, 3, 8, 9, 10, 13, 15, 16, 17, 21, 25, 46, 54, 60, 73, 75, 76
Plant defence mechanisms, 3
Plant defence, 3, 40, 41, 42
Plant growth, 13, 14, 21, 33, 50, 52, 77
Plant Health Directive 2000/29/EC, 216, 225
Plant health measures, 3, 34
Plant husbandry, 127
Planting, 3, 7, 9, 12, 13, 15, 17, 21, 22, 25–27, 29, 31, 33, 35, 45, 48, 51, 59, 60
Planting material, 3, 9, 12, 15, 21, 22, 25–27, 29, 31, 33, 35, 48, 59, 60, 127, 293, 294
Plant-parasitic nematodes, 13, 36, 40–43, 45, 52, 53, 57–60
Plant physiology, 14
Plant products, 302
Plant Protection Convention, 34
Plant resistance, 85, 88, 94
Plant size, 13
Plant tissues, 257
Plant toppling, 66
Plant varieties, resistant, 7
Plant vigor 27, 85, 95
Plant viruses, 124, 133
Pliny the Elder, 275
Plot, 32, 44
Plum, 177, 186, 192, 199, 209–211
Plum hybrid, 187
Pluma Hidalgo, 106, 107
Pochonia chlamydosporia, 188, 243, 249, 250, 252, 297
POD, 287
Pogonomyrmex barbatus, 111
Poland, 218, 238
Pollen, 37
Polyclonal antibodies, 261
Polycultivation system, 105

Polycultivation, 106, 108, 109, 111, 114
Polyculture, 101
Polyculture system, 106, 109
Polyethylene films, 250
Polynesia, 36
Polyphenol oxidase, 40
Pome, 30
Pomelos, 135
Poncirus, 135, 147, 161, 165, 166, 168, 171
Poncirus biotype, 68
Poncirus trifoliata, 67–70, 142, 143, 147, 152, 155, 159,
Poovan, 5
Popoulou, 5
Population density, 13, 108, 110, 114, 144, 146, 177, 179, 180, 181, 182, 183, 187, 189
Population development, 138, 152
Population dynamics, 14, 39, 44, 59
Population growth, 136, 138, 139, 140, 141, 143, 151, 155, 157
Population increase, 88
Portugal, 218, 241, 253–256, 259, 262, 263, 270, 272, 273, 277–280, 308, 311–313
Portuguesa, 65, 75, 76
Potassium, 245
Potato, 217, 227, 235, 238, 239, 240
Poteau, 5
Poyo, 12, 39, 40, 47, 49, 53, 55, 56, 57
PPO, 40
Prata, 5, 30
Pratylenchoides, 247
Pratylenchoides ritteri, 279
Pratylenchus, 3, 7, 9, 12, 25, 31, 33, 37, 42, 46, 48, 50, 51, 53–55, 58, 60, 65, 73, 74, 76, 79, 80–82, 89, 98, 110–112, 114, 115, 178, 179, 182, 192, 193, 246–249, 252, 275, 279, 281, 293, 294, 306, 310, 311, 312, 314
Pratylenchus thornei, 247
Pratylenchus brachyurus, 68, 74, 79, 80, 125, 124, 154, 155, 161, 165, 169–171, 179
Pratylenchus coffeae, 9, 12, 14, 19, 21, 23–26, 31, 32, 34, 36, 37, 39, 40, 42, 44, 65, 66, 124, 125, 146, 153–157, 163, 165, 166, 169, 170, 173, 279
Pratylenchus convallariae, 179
Pratylenchus crenatus, 179, 279
Pratylenchus fallax, 279
Pratylenchus goodeyi, 7, 9, 10, 14, 23–26, 28, 34, 36, 37, 39
Pratylenchus hamatus, 247
Pratylenchus hexincisus, 179
Pratylenchus jaehni, 154, 166
Pratylenchus microdorus, 279
Pratylenchus neglectus, 179, 246–248, 279
Pratylenchus penetrans, 73, 178, 179, 186, 207, 233, 247–249, 275, 279, 306, 310
Pratylenchus pratensis, 179
Pratylenchus thornei, 179, 279
Pratylenchus vulnus, 178, 179, 186, 190, 195, 206, 207, 212, 246, 247, 252, 275, 279, 283, 285, 288, 289, 290–293, 299
Pratylenchus zeae, 68, 179
Precociousness, 87
Predators, 301
Pre-planting measures, 177, 181, 182
Prevention, 85, 177, 178, 181, 182, 190
Primary root, 12, 13
PRMV, 216
Proanthocyanidins, 40
Probe, 223, 224
Production, 135, 140, 145, 154, 161, 166, 168, 169, 170, 171
Production cycle, 86
Productivity, 39, 40, 45, 129, 131
Progeny, 4
Propagation, 52
Propagation material, 68, 219, 225
Propane, 98

Prophylactic measures, 36
Prophylactic methods, 34
Protein, 41
Prunes, 177, 178
Pruning, 108
Prunus, 177, 178, 179, 180, 181, 186, 187, 190, 191, 192, 193
Prunus armeniaca, 177
Prunus avium, 177, 186
Prunus cerasifera, 177, 186
Prunus cerasus, 177
Prunus davidiana, 188
Prunus domestica, 177, 186
Prunus dulcis, 177
Prunus insititia, 186
Prunus mahaleb, 185, 186
Prunus munsoniana, 186
Prunus persica, 78, 107, 177, 186
Prunus salicina, 177, 186
Pseudococcus cryptus, 110
Pseudomonas, 42, 180, 188, 193, 236
Pseudomonas 'BG33R', 188
Pseudomonas syringae, 180
Pseudotsuga, 256
Psidium fiedrichsthalianum, 75
Psidium guajava, 74, 75, 83, 107
Psilenchus, 279
Psoralen, 303
PTSL, 180, 182, 186, 187, 188
Puebla, 102, 103, 106–108, 110, 117, 118
Puerto Rico, 31, 33, 56, 57, 157
Purple passion fruit, 77
PVC plastic films, 305
PWD, 253, 254, 258, 263, 264, 265, 266, 267, 268, 269, 270
PWN, 253, 254, 255, 256, 257, 258, 259, 261, 262, 263, 264, 266, 267, 268, 269, 270
Pythium, 234, 235

Q

Quantitative PCR, 223
Quarantine, 11, 30, 34, 35, 45, 114, 177, 182, 181, 190, 217, 221, 224, 225, 226, 227, 253, 256, 259
Quiescence, 12
Quillaja saponaria, 304, 309
Quillay, 304

R

Radiation, 108
Radical tissue, 88, 89
Radopholus citri, 149
Radopholus similis, 3, 7–21, 23–26, 28, 30–61, 65, 66, 125, 146, 149, 150, 152–155, 161, 163–169, 171, 172
Radopholus, 3, 7, 9, 25, 31, 34, 45, 46, 47, 48, 49, 50, 51, 52, 53, 54, 55, 56, 57, 58, 59, 60, 61, 111, 279
Rafsanjan, 246, 251
Rainfall, 102, 120
Ramsey, 198, 199, 202, 203, 207
RAPD, 261, 262, 272, 274
RAPD markers, 142
Raphanus sativum, 300
Raspberry, 216, 219, 220, 221, 227, 240, 241
Raspberry ringspot nepovirus, 217
Ratoons, 13, 29
rDNA ITS regions, 261
rDNA sequences, 157
Real-time PCR, 261
Rectum, 138
Red, 5
Red pine, 269
Red ring disease, 71, 72, 73
Red Spanish, 79, 80
Replanting, 16, 28, 44
Reproduction potential, 203
Reproductive capabilities, 11
Republic of Georgia, 218
Republic of South Africa, 38
Research institute, 8, 44
Research station, 36
Resin exudation, 265
Resistance, 3, 14, 27, 29, 33, 36–41, 43, 45–50, 53–55, 57–60, 75, 79, 81, 85, 88, 89, 93, 94, 101, 113, 141, 142, 143, 148, 152, 154, 155, 163–168, 172, 177, 178, 185–187,

191–199, 202–214, 225, 227, 253, 258, 263, 267, 268, 270, 294, 295, 314
Resistance-breaking biotypes, 198
Resistance-breaking populations, 198, 199, 207
Resistance durability, 199
Resistance genes, 148, 154, 199
Resistance mechanisms, 199, 200
Resistance spectra, 199
Resistance trait, 142
Resistant cultivars, 294
Resistant grapes, 196
Resistant grapevine, 195
Resistant hybrids, 142
Resistant planting materials, 119, 127
Resistant rootstocks, 147, 152, 153, 185, 186, 190, 296
Resistant varieties, 34
Retention, 216, 222, 241
Reverse transcription, 224
Rhabditis, 128
Rhizobacteria, 42, 43, 52, 56
Rhizoctonia solani, 235
Rhizome, 4, 8, 11, 12, 15, 21, 27, 32, 54–56
Rhizoplane, 139
Rhizosphere, 135, 138, 139, 140, 141, 142, 151, 152, 154, 159, 160, 163, 164, 178, 181, 190, 192, 286
Riboflavin, 245
Ribosomal DNA, 223, 241
Ribosomal probes, 261
Rice, 4, 19, 41, 56
Ricinin, 301
Ridge Pineapple, 152, 154
Ring nematode, 180, 195, 207
Rizosphere, 96
RKN, 195, 196, 197, 198, 199, 200, 202, 204, 205, 207
RKN resistant rootstocks, 207
RNA-1 molecule, 223
RNA-2 molecules, 223
Robusta, 5
Robusta coffee, 107
Rojo, 66, 107
Romania, 218
Root, 4, 9, 10–16, 18, 22, 25, 26, 28, 32, 34, 41, 42, 44, 46, 48, 49, 50, 52, 54, 58, 60, 61, 65, 85–87, 93, 123, 137, 139, 140, 144, 146, 147, 149, 150, 151, 154, 157, 160, 179, 180, 183, 286–288, 291, 300
Root abundance, 160
Root cell compounds, 41
Root cell structure, 86
Root cells, 179, 180
Root disease, 65
Root gall index, 288, 291, 300
Root growth, 87, 93, 140, 147, 149, 151, 157
Root-knot nematode, 3, 10, 12, 14, 19, 23, 32, 38, 48, 54, 58, 59, 73, 101, 119, 122, 123, 124, 127, 130, 133, 134, 136, 195, 196, 201, 204, 208, 210, 211, 213
Root lesion nematodes, 179
Root mass density, 137, 144, 146
Root pathogens, 42
Root phenols, 286
Root sampling, 14
Rootstock, 67, 69, 70, 74, 75, 78, 79, 85, 89, 90, 98, 127, 141–144, 147, 152, 154, 155, 157–159, 162–164, 166–172, 195, 196, 197, 199, 202–213, 243, 246–249, 252, 275, 276, 281, 283–286, 293, 294, 296, 310, 312, 313
Root surface, 123
Root system, 85, 139, 150, 154, 179, 180, 183
Root weight, 14
Rose, 220
Rossellinia bunodes, 111
Rossetti, 73
Rotation, 10, 11, 16, 17, 21, 28, 29, 33, 34, 44, 45, 57, 66, 294
Rotylenchulus macrodoratus, 247, 248, 252, 276, 280, 286, 315
Rotylenchulus macrosomus, 247, 280

Rotylenchulus reniformis, 10, 19, 32, 41,48, 59, 65, 66, 68, 74, 76–83, 124, 125, 280, 286
Rotylenchulus, 65, 73, 74, 76–79, 81–83, 275, 276, 280, 281, 308, 309, 315
Rotylenchus caudaphasmidius, 68
Rotylenchus cypriensis, 280
Rotylenchus microstriatus, 125
Rotylenchus robustus, 280
Rotylenchus, 247, 280
RRSV, 217, 221, 223, 227
RT-PCR, 223
Rubber, 18
Rue plant, 303
140 Ruggeri, 205
Russia, 178, 218, 241
Rust spot, 111
Ruta graveolens, 234, 303, 311–315
Ruta graveolens, 303
Rutaceae, 135, 152
Rwanda, 24

S
Saba, 5
Safet velchi, 5
Saissetia, 110
Saissetia coffea, 111
Saissetia oleae, 111
Salicylic acid, 94
Salinity, 137, 140, 144, 147, 164, 167, 245, 246
Salt Creek, 198, 202, 203
Sampling, 85, 95, 96
Sampling method, 14
Sanitation, 136, 146, 147, 153, 266
Sanliurfa province, 247
Sao Paulo State, 154, 161
São Tomé, 122
Sapodilla, 63, 80, 82
SAR, 41, 92, 94
Sarakhs, 245
Satellite DNA, 261, 262, 273
Saturna, 235
Sclerodermus, 268
Scorpions, 224
Screening, 14, 36, 38, 49, 53, 54, 55, 294, 295, 314
Scutellonema brachyurum, 68, 125
Scutylenchus quettensis, 247, 248
Seasons, 63
Secondary metabolites, 43, 60
Secondary pathogens, 151
Sedentary endoparasites, 136
Sedentary nematodes, 41, 250
Seed, 12, 22, 37, 44, 89, 90, 119, 121, 127
Seedbeds, 121, 294
Seedling, 76, 119, 121–124, 126, 127, 128, 129, 130, 131, 142, 152, 155, 156, 160, 164, 165, 168–172, 185, 202, 269
Seed plants, 15
Seed treatment, 15
Selection, 269
Selection program, 249
Selenaspidus articulatus, 111
Semnan, 246
Senescence, 179
Sesamun indicum, 233
Setúbal Peninsula, 255, 256
Severinia buxifolia, 142, 159
28S gene, 261
Shade coffee, 101
Shade trees, 121, 122
Shandong Province, 258
Shoots, 123
Sibling species, 11
Sicily, 248
Silk, 5
Silk Fig, 30
Silk worm, 268
Sinan, 258
Sinapis alba, 233
Sistan, 246
Slovakia, 218, 240
Slovenia, 218, 241
Slow decline of citrus, 136, 137
Slow decline symptoms, 136
SLRSV, 200, 219, 221, 227, 216, 288, 294
SLRV, 240

SO4, 204, 205, 207
Sodium, 136, 140
Sodium azide, 184
Soil, 3, 8–10, 12, 14–18, 23, 27–32, 35, 41–44, 46, 48, 50, 51, 53, 55, 58, 60, 87, 88, 90, 91, 92, 93, 94, 95, 96, 97, 98
Soil amendment, 130, 133, 301
Soil-borne organisms, 300
Soil-borne pathogens, 183, 184, 246
Soil-borne viruses, 122
Soil chemical characteristics, 304
Soil conditions, 139
Soil conservation, 107
Soil eutrophication, 265
Soil fertility, 28, 32
Soil fumigant, 128, 178, 190
Soil fumigation, 159, 160, 184, 194
Soil fungi, 77
Soil horizon, 154, 157
Soil microbial population, 250
Soil microflora, 301
Soil microspaces, 95
Soil moisture, 135, 139
Soil organisms, 141, 152
Soil pH, 144
Soil preparation, 183, 184
Soil samples, 74
Soil solarization, 177, 178, 183, 243, 248, 252, 300
Soil submersion, 87
Soil temperature, 299, 304
Soil water potential, 157
Solarization, 183, 193, 298, 299, 300, 315
Solid cake, 302
Solomon Islands, 18, 19, 35
Sorghum, 33, 183
Source of resistance, 37
South Africa, 24, 25, 51, 143, 145, 147, 148, 154, 155, 160, 162, 167, 168, 188, 218,
South America, 6, 8, 119, 120, 218, 275
South Asia, 20
South Carolina, 180, 192, 193
Southeast Asia, 8, 10, 17, 18, 35
Southern Asia, 20
Sowing, 63
Soybean severe stunt virus, 226
Soybean severe stunt, 226
Spain, 138, 141, 146, 165, 168, 170, 171, 172, 177, 178, 187, 191, 193, 207, 212, 218, 239, 240, 254, 275, 277–280, 285, 294, 299, 301, 308, 309, 312, 313
Species-specific primers, 261
Spicule structure, 260
Spiral nematode, 3, 9, 12, 19, 33
Spondias purpurea, 107
Spraing disease, 217, 227
Spraying, 253, 265, 266, 267
Spreading decline symptoms, 150
Spreading decline, 149, 150, 152, 153, 154, 163, 164, 171
Spring treatment, 96
Sri Lanka, 19, 20, 34, 35, 50
SSS, 226
SSS severity, 226
SSSV, 226
State of Mexico, 103
Steam, 183
Steam applications, 177
Steam sterilization, 304
Stele, 11
Stellar structures, 288
Stenotrophomonas maltophilia, 236
Sterculiaceae, 119
Stone fruit orchards, 180, 182, 184, 188, 190, 192
Strawberry latent ringspot sadwavirus, 200, 216, 288
Strawberry, 113, 227, 305, 306
Stress, 68, 74, 79
Struma, 220
Stunting, 10, 12, 13, 35, 64, 123, 124
Stylet, 86, 88, 216, 276, 285
Suberin, 94
Sub-group Bluggoe, 5
Sub-group Cavendish, 5
Sub-group Chuoi Xiem, 5
Sub-group Gros-Michel, 5

Sub-group Ibota, 5
Sub-group Laknao, 5
Sub-group Matooke, 5
Sub-group Mbire, 5
Sub-group Mutika Lujugira, 5
Sub-group Mysore, 5
Sub-group Ney poovan, 5
Sub-group Pelipita, 5
Sub-group Peyan, 5
Sub-group Pisang awak, 5
Sub-group Pisang kelat, 5
Sub-group Pisang nangka, 5
Sub-group Pisang raja, 5
Sub-group Plantain, 5
Sub-group Pome, 5
Sub-group Popoulou, 5
Sub-group Red, 5
Sub-group Saba, 5
Sub-group Silk, 5
Sub-group sucrier, 5
Sub-Saharan Africa, 22, 23
Substrates, 294, 295, 298, 305
Suckers, 21, 27, 66, 108
Sucre State, 71
Sudan, 160, 173
Sudan grass, 16
Sudden death, 127
Sugarcane, 11, 16, 20, 33, 34, 47, 48
Sukari, 5
Summer, 63
Sun exposure, 27
Sunlight system, 101, 109, 111, 114, 115
Suppression, 129, 132, 134
Suppressiveness, 3, 42
Suppressive soils, 190
Surface roots, 152, 154
Surinam, 9, 16, 31, 53
Survival strategy, 12
Susceptibility, 7, 36, 37, 38, 39, 49, 253, 263, 270
Susceptible rootstock, 69
Sustainable nematode management, 17
Swam taro, 20
Swaziland, 24
Sweet dessert bananas, 5
Sweet orange, 154, 168, 172
Sweet potato, 18, 20, 28, 158
Swiettenia, 109
Swingle citrumelo, 142, 147, 166
Switzerland, 218
Symptom, 13, 67, 68, 74, 76, 79, 80, 178–180, 193, 219, 221, 222
Syncytia, 288
Synthetic banana hybrids, 39
Syria, 243, 244, 251
Systemic acquired resistance, 41, 92
Systemic herbicides, 201
Systemic pesticide, 156

T

Táchira, 79
Tagetes erecta, 302
Tagetes florida, 302
Tagetes minuta, 302
Tagetes signata, 302
Tagetes, 233, 302, 310, 312, 313
Tahiti lime, 69
Taiwan, 10, 38, 39, 52, 154, 158, 172
Tajikistan, 218
Talaromyces flavus, 300
Tamarind, 20, 63, 80
Tamarindus indica, 80
Tamil, 40
Tamil Nadu Agricuture University, 39
Tangelo, 67
Tannins, 301
Tanzania, 24, 26, 47, 57, 60
TaqMan, 223, 238
Taro, 18, 20, 34
TBRV, 200, 217, 223, 227
Tea, 9, 20, 34, 50
Technical production systems, 88
Tehran, 243, 246, 252
Temperature, 63, 78, 120
Tephrosia vogelli, 28
Terpens, 301, 303
Test pit, 85, 87, 95, 96
Test pit inspection, 85
Tetraploid AAAB-group banana, 33

Tetraploid banana, 36
Tetraploid cultivars, 37, 38
Texas, 138, 146, 162, 171
Texture, 141, 144, 145, 151, 172
Thailand, 5, 19, 35, 55, 59
Theobroma cacao, 107, 119
Theobroma, 119, 131, 132, 133, 134
Thermophilic pathogens, 266
Thiamin, 245
Thiruvanthapuram, 5
Thymus vulgaris, 234, 236
Tissue culture, 10, 16, 21, 30, 34, 36, 38, 39, 42, 43, 49
Tithonia diversifolia, 29
Tobacco, 33, 113
Tobacco ringspot nepovirus, 200, 216
Tobravirus, 216, 217
Togo, 120, 131
Tojolobales, 106
Tolerance limit, 75, 76
Tolerance, 3, 36, 38, 39, 40, 43, 57, 58, 202, 203, 205, 207, 209, 212, 227
Tolerant varieties, 34
Tomato, 113, 199, 213, 214, 299, 300, 302, 303, 305, 306, 308, 309, 312
Tomato black ring nepovirus, 200, 217
Tomato ringspot nepovirus, 200, 216
TomRSV, 216, 223
Tondino di Zagaria, 303
Toppling, 13, 22, 26, 52
ToRSV, 216, 217, 219
Totonacas, 106
Toxicity, 128, 266, 267
Toxic compounds, 92
Toxic metabolites, 301
Toxoptera aurantii, 110, 111
Trade, 66
Traders, 22, 30, 34
Transformation, transgenic, 41
Transgenic plants, 206
Transgenic resistance, 3
Transmission, 215, 216, 217, 218, 219, 222, 227, 233, 234, 236, 238, 240, 241, 257, 272
Transmission assays, 215
Transmission control, 227
Transmission process, 201
Transpiration, 157
Trap, 73
Trapping fungi, 268
Tree biomass, 130
Tree canopy, 136, 138, 145, 149
Tree growth, 307
Trichloronitrometan, 305
Trichoderma harzianum, 298, 314
Trichoderma, 43, 112, 130
Trichodoridae, 215, 222, 239
Trichodorus, 74, 124, 160, 162, 216, 217, 218, 238, 242, 247
Trichodorus aequalis, 280
Trichodorus giennensis, 280
Trichodorus monohystera, 125
Trichodorus primitivus, 217
Trichodorus primitivus, 235
Trichodorus primitivus, 280
Trichodorus taylori, 280
Trifoliate orange, 135
Trinidad, 9, 32, 49, 51, 53–55, 71
Triploid, 4, 8, 37
Triplonchida, 215, 239
Tristeza virus, 67, 142
Triticum aestivum, 182
Trogossita japonica, 267, 268
Trophotylenchulus saltensis, 276, 280, 288, 310
Trophurus, 247
Trophurus imperialis, 125
Tropical fruit crops, 63
Tropics, 8, 9, 25, 29
Troyer, 67, 68, 69, 70
TRSV, 216, 217, 223
True grape, 204
Trujillo State, 65, 79, 81
Trunk, 73, 144, 149, 150, 167
TRV, 217, 218, 227, 235
Tsotziles, 106
Thiocyanates, 300
Tuber, 25, 35
Tucupita, 73, 74
Tunisia, 218, 243, 244

Turkestan, 275
Turkey, 178, 218, 243, 244, 247, 248, 250, 252
Turkmenistan, 218
Turmeric, 20
Tuxtla Chico, 101, 115
Tylenchorhynchus, 64, 66, 68, 73, 76, 79, 80, 111, 115, 247
Tylenchorhynchus aduncus, 280
Tylenchorhynchus annulatus, 64, 66, 68
Tylenchorhynchus capitatus, 68
Tylenchorhynchus clarus, 280
Tylenchorhynchus contractus, 74
Tylenchorhynchus dubius, 280
Tylenchorhynchus goffarti, 280
Tylenchorhynchus huesingi, 280
Tylenchorhynchus mamillatus, 280
Tylenchorhynchus martini, 125
Tylenchorhynchus striatus, 280
Tylenchorhynchus tenuis, 280
Tylenchulus, 275, 276, 280, 281, 310, 311
Tylenchulus graminis, 143, 166
Tylenchulus palustris, 143, 166
Tylenchulus semipenetrans, 67–72, 74, 81, 82, 83, 89, 97, 98, 135, 136–148, 152, 155, 160–173, 195, 207, 276, 280, 283, 310, 311
Tylenchulus similis, 9, 18, 48
Tylenchus, 65, 74, 76, 79, 111, 247
Tylenchus arcuatus, 280
Tylenchus coffeae, 9, 125, 132
Tylenchus multicinctus, 9
Tylenchus musicola, 10
Type B phylloxera, 205
Typica, 103, 104, 107, 108
Tzelzales, 106

U
Uganda, 22, 24, 28, 29, 36, 39, 46, 49, 52, 57, 58
Ukraine, 218
United Nations, 112
United States, 8, 135, 154, 157, 178, 180, 183, 193
Upper Amazon region, 119
Urea, 300, 309
USA, 195, 197, 207, 210, 211, 218, 226, 240, 243, 244, 245, 249, 250, 251, 253, 256, 258, 262, 275, 278, 279, 280, 283, 310, 311
Uttar Pradesh, 218
Uzbekistan, 218, 243, 244

V
Valera Amarilla, 79
Valera Roja, 79, 80
Valles Altos, 70
Vanilla planifolia, 109, 117
Vascular discoloration, 289, 290
Vascular parenchyma, 75
Vascular tissues, 32, 155
Vection, 201
Vector, 253, 262, 265, 266, 267, 268, 269, 270, 271, 272
Vector endemism, 226
Vegetative activity, 69, 70, 71
Vegetative growth, 26
Vegetative phase, 13
Venezuela, 31, 52, 59, 60, 63–69, 71–83, 122, 134, 157
Veracruz, 101–103, 105, 107, 108, 109, 110, 112, 114–118
Vermiform stages, 93
Vernonia amygdalina, 130
Verticillium dahliae, 275, 276, 288–294, 298, 300, 311
Verticillium wilt, 246, 294, 300, 309
Verticillium, 112, 306
Vicia villosa, 233
Vietnam, 19, 21, 38, 47, 54
Vigna unguiculata, 233
Vigor, 27, 38, 39, 178, 180, 189
Villa Nueva, 74
Violeta, 73
Viral DNA, 40
Virulence, 253, 270
Virulent populations, 199, 200
Virus, 180, 181, 215, 216, 217, 218, 219, 222, 223, 225, 226, 227, 237, 238, 239, 240, 241, 242

Virus dispersal, 219, 225
Virus-free certified plants, 225
Virus-infected plants, 180
Virus persistance, 226
Virus transmission, 203
Virus vector nematodes, 200, 221, 228, 237
Virus vector species, 195
Vitamin B6, 245
Vitamin C, 245
Vitis, 68, 73, 82, 195, 197, 198, 202, 203, 204, 205, 206, 208, 209, 210, 211, 212, 213
Vitis aestivalis, 197
Vitis arizonica, 202, 203, 204
Vitis berlandieri, 197, 200, 205, 212
Vitis candicans, 198, 202, 203
Vitis champinii, 197, 198, 199, 207
Vitis cinerea, 197, 204
Vitis doaniana, 197
Vitis germplasm, 206
Vitis longii, 197, 198
Vitis mustangensis, 198, 199
Vitis riparia 'Gloire', 203
Vitis riparia 'Grand glabre', 203
Vitis riparia 'Messner n. 9', 203
Vitis riparia, 197, 198, 200, 202, 203, 204, 207
Vitis rotundifolia, 204
Vitis rubra, 202
Vitis rufotomentosa, 197, 202, 203
Vitis rupestris 'du Lot', 204
Vitis rupestris, 197, 198, 202, 205, 207, 210
Vitis slavinii, 202
Vitis solonis, 197, 202, 203
Vitis species, 197, 198, 202
Vitis vinifera, 68, 73, 82, 197, 198, 203, 204, 205, 212
Volatile chemicals, 184
Vulva, 138

W

Wastes, 301, 312, 314
Water contamination, 44
Water deficiency, 246
Water deficit, 150, 151
Watering, 86, 87
Watermelon, 113
Weed control, 232, 233
Weed fallow, 16
Weed management, 154
Weeding, 18
Weeds, 9, 11, 35, 56, 64, 101, 106, 107, 110, 108, 114, 182, 183, 189, 293, 294, 304
West Africa, 5, 6, 119, 120, 122, 132
West African beans, 120
West African cocoa, 120
West Bengal, 218
Wheat, 182, 192, 226
Wild plums hybrids, 186
William, 39, 54
Wilting, 123, 126, 136
Wind, 13
Windward Islands, 15, 16, 49
Winter treatment, 96
Wood, 105, 106
Woody cuttings, 294

X

Xanthomonas campestris pv. *pruni*, 180
Xanthosema, 109
Xanthotoxin, 303
Xiphinema aequum, 280, 313
Xiphinema, 64, 65, 73, 74, 76, 77, 79, 82, 111, 124, 160, 162, 167, 173, 178, 180, 191, 195, 200, 203, 206, 208–221, 225, 226, 229, 230, 233–242, 246, 248, 249, 275, 276, 280, 281, 294, 303, 308, 309, 311, 312, 313
Xiphinema americanum group, 221, 225
Xiphinema americanum sensu stricto, 216, 217, 219, 222, 226, 234, 236
Xiphinema americanum sensu lato, 74, 76, 216, 217, 218, 221, 222, 233, 237, 239

Xiphinema americanum, 180, 200, 207, 216, 217, 219, 221, 223, 225, 226, 233, 234, 236, 239, 240, 246, 247
Xiphinema attorodorum, 125
Xiphinema barense, 280
Xiphinema basiri, 180
Xiphinema brasiliense, 64, 66, 68, 80
Xiphinema brevicollum, 68, 125, 160, 180, 236, 237
Xiphinema bricolense, 216
Xiphinema bricolensis, 200, 216
Xiphinema californicum, 180, 217, 280
Xiphinema diversicaudatum, 180, 200, 216, 219, 220, 223, 227, 231, 232, 234–236, 238–242, 288
Xiphinema ebriense, 125
Xiphinema elongatum, 125, 276, 280, 286
Xiphinema ifacolum, 125
Xiphinema index, 160, 195, 196, 198, 200–216, 218–220, 223–231, 233, 234, 236–239, 241, 242, 247, 276, 280, 286, 287, 296, 303, 309, 313
Xiphinema index-resistant hybrid, 198
Xiphinema ingens, 280, 311
Xiphinema insigne, 125
Xiphinema italiae, 200, 213, 219, 220, 223, 238, 242, 280
Xiphinema krugi, 68
Xiphinema longicaudatus, 125
Xiphinema macroacanthum, 280
Xiphinema nigeriense, 125, 247
Xiphinema pachtaicum, 280
Xiphinema pacificum, 200, 210
Xiphinema peruvianum, 68
Xiphinema pirinense, 218
Xiphinema rivesi, 180, 191, 216, 217, 218, 234, 236
Xiphinema sahelense, 280
Xiphinema setariae, 125
Xiphinema silvesi, 218
Xiphinema simillimum, 68, 80
Xiphinema turcicum, 280
Xiphinema vuittenezi, 180, 223, 239, 242, 247, 280
Xiphinema vulgare, 68, 160
xylophilus–group, 260, 261

Y
Yam, 20, 25
Yangambi Km5, 5, 37, 50
Yaracuy, 65, 67, 70, 72, 75–77, 81
Yazd, 246
Yellow Creole, 79
Yellow disease of pepper, 9
Yellow passionfruit, 77
Yeongam, 258
Yield, 11, 13–15, 21, 23, 26, 28, 29, 33, 42, 45, 46, 48, 50–52, 54, 56, 57, 58, 64, 121, 122, 127, 128, 129, 131, 134,136, 137, 143–148, 153, 154, 155, 161, 163, 167, 170, 171, 173, 216, 248, 250
Yield loss, 26, 145, 146
Yusti, 282, 284, 285

Z
Zanjan, 246
Zanzibar, 24, 25
Zea mays, 109
Zhejiang Province, 258
Zimbabwe, 34
Zinc, 245
Zingiberaceae, 35
Zoosporangia, 235
Zoospores, 235, 236
Zoosporic fungi, 234, 239
Zoques, 106
Zulia State, 64–68, 70, 73–75, 77, 80–82
Zygotic embryo, 121
Zygotylenchus guevarai, 247, 280
Zygotylenchus taomasinae, 10

Kolibri 360 GmbH
Ferdinand-Jühlke-Straße 7
99095 Erfurt, Deutschland
produktsicherheit@kolibri360.de